MASTER THE

AP* CALCULUS AB & BC

TESTS

W. MICHAEL KELLEY
MARK WILDING,
CONTRIBUTING AUTHOR

TEACHER-TESTED STRATEGIES AND

TECHNIQUES FOR SCORING HIGH

2002

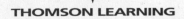

ARCO

THOMSON LEARNING™

Australia • Canada • Mexico • Singapore • Spain • United Kingdom • United States

ARCO
THOMSON LEARNING

An ARCO Book

ARCO is a registered trademark of Thomson Learning, Inc., and is used herein under license by Peterson's.

About Peterson's

Founded in 1966, Peterson's, a division of Thomson Learning, is the nation's largest and most respected provider of lifelong learning online resources, software, reference guides, and books. The Education Supersite at petersons.com—the Web's most heavily traveled education resource—has searchable databases and interactive tools for contacting U.S.-accredited institutions and programs. CollegeQuest (CollegeQuest.com) offers a complete solution for every step of the college decision-making process. GradAdvantage (GradAdvantage.org), developed with Educational Testing Service, is the only electronic admissions service capable of sending official graduate test score reports with a candidate's online application. Peterson's serves more than 55 million education consumers annually.

Thomson Learning is among the world's leading providers of lifelong learning, serving the needs of individuals, learning institutions, and corporations with products and services for both traditional classrooms and for online learning. For more information about the products and services offered by Thomson Learning, please visit www.thomsonlearning.com. Headquartered in Stamford, Connecticut, with offices worldwide, Thomson Learning is part of The Thomson Corporation (www.thomson.com), a leading e-information and solutions company in the business, professional, and education marketplaces. The Corporation's common shares are listed on the Toronto and London stock exchanges.

For more information, contact Peterson's, 2000 Lenox Drive, Lawrenceville, NJ 08648;800-338-3282; or find us on the World Wide Web at: www.petersons.com/about

ISBN 0-7689-0737-3

Printed in the United States of America

10 9 8 7 6 5 4 3 2 1   02 01 00

Author's Dedication

So many people have inspired and supported me that I balk to think of how lucky and blessed my life has been. The wisdom, love, and support of these people have shaped and made me, and without them, I am nothing. Special thanks to:

- My beautiful bride Lisa, who teaches me every day that being married to your best friend is just the greatest thing in the world.

- Mom, who taught me to rise above the circumstance.

- Dad, who taught me always to tip the barber.

- My brother Dave ("The Dawg"), who taught me that life is pretty funny but never as funny as a joke that references the human butt.

- My best friends Rob Halstead, Chris Sarampote, and Matt Halnon, who taught me what true friendship means (and why housecleaning is for the weak).

- My principals, George Miller and Tommy Tucker, who have given me every opportunity and then seven more.

- My students, who make my job and my life happier than they could ever imagine.

- My English teachers Ron Gibson, Jack Keosseian, Mary Douglas, and Daniel Brown, who showed me how much fun writing can be.

- Sue Strickland, who inspired in me a love of teaching and who taught my methods class over dinner at her house.

- Mark Wilding, for not saying, "Are you crazy?"

- The Finley boys: James, for his great questions, written specifically to challenge, deceive, nauseate, and "ensmarten" students, and Tim, for his computer prowess and ability to produce crazy three-dimensional diagrams in the blink of an eye.

- The Big Mathematician in the Sky, who (mercifully) makes straight the paths of us mathematicians here on earth.

- Finally, wrestling legend Koko B. Ware, because Chris thought that would be pretty funny.

CONTENTS

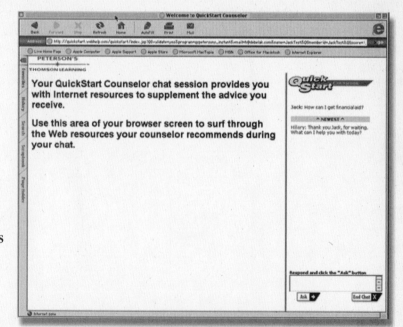

PART

1

PREVIEW

How to Use This Book

AP Calculus Formulas

Digging In

INTRODUCTION: HOW TO USE THIS BOOK

THE CHESS DILEMMA

I like chess. It's a fascinating game that, although easy to grasp, is nonetheless very difficult. Once you learn how to move the pieces around the board, you are still light years away from actually being any good (chess geniuses excepted). I have known how to move chess pieces since I was about nine, but I have still yet to win a single chess game against a living person. I am just terrible at chess, even though I have always wanted to be good. In my holy grail–like quest to outsmart someone at this dastardly game, I have sought wisdom from books. I can still remember my last trip to the bookstore, standing with my mouth agape before the shelf allotted to chess instruction books. I was hoping to find something to fortify (resuscitate) my chess game, only to find books with titles such as *200 Pawn and Bishop Endings, 50 of the Greatest Chess Games Ever, and 35 Tactics to Capture the Queen's Rook in Six Moves or Less.* None of these books were of any use to me at all! None of them even claimed to try to teach me what I needed to know.

I wanted a book to teach me how to play. I knew that a horsey moves in an 'L' shape, but I didn't know where he (or she) should go. Often, friends gave me such advice as "control the middle of the board." I still am not sure what that means or how to go about doing it. In my frustration, I bought children's chess games for my computer, hoping to benefit from the tutorials that, surely, I had a better shot at understanding. Once I had completed the tutorials, I was confident, ready to attack the toy men and finally take my first step toward Chess Grand Master—or whatever they call those smart guys. I was destroyed by the smiling toy chess pieces—beaten in under ten moves, on the easiest level. I don't think I blinked for fifteen minutes. I have seen that look on my students' faces before. It's a look that says, "I have no idea what I'm doing wrong, and I think I'm more likely to sprout wings than to ever understand." Perhaps you, too, have felt that expression creep across your face, shadowed by an oily, sick feeling in the pit of your stomach. Perhaps you have a knack for math and are not inspired to fear and cold sweat by calculus but are just looking for practice or to tie up a few loose ends before the AP Calculus test. Either way, this book will help you master calculus and prepare you for the infamous test day.

UNDERSTANDING VERSUS MASTERING: CALCULUS REFORM

Calculus is a subject, like chess, that requires more than a simple understanding of its component parts. Before you can truly master calculus, you need to understand how each of its basic tenets work, what they mean, and how they interrelate. Such was not always the case, however. In fact, you may have even heard the urban legend about the AP Calculus student who simply took the derivative of each equation and set it equal to zero and still got a three rating on the AP test! Such rumors are not true, but they underscore the fundamental change gripping the mathematics world—a change called reform. In fact, for many years, calculus was treated as merely "advanced arithmetic," rather than a complex and even beautiful logical system. Calculus is so much more than formulas and memorizing, and when you see this, you begin to actually understand the rationale behind the mathematics involved. Having these connections makes understanding flourish and, before very long, banishes math phobias to the dark places, inhabited heretofore only by your car keys when you were in a hurry.

WHAT MAKES THIS BOOK DIFFERENT

This book is unlike most math textbooks published today. It not only presents formulas and practice problems, but it also actually helps you understand through hands-on activities and detailed explanations.

It also puts difficult formulas into everyday English, to shed some light on their meanings. In relation to my chess metaphor, my book intends not only to tell you that bishops move diagonally, but also to offer you advice on where to move them, how soon, and in what circumstances they are most effective. (Although I cannot actually give you the corresponding chess advice, since my deep understanding of bishops ends with the knowledge that they wear pointy hats.)

Traditional math test-preparatory books state formulas, prove them, and present exercises to practice using them. There is rarely any explanation of how to use the formulas, what they mean, and how to remember them. In short, traditional books do not offer to teach, when that is what you truly need. I have compiled in this text all of the strategies, insights, and advice that I have amassed as an instructor. Most of all, I have used common sense. For example, I do not think that proving a theorem always helps students understand that theorem. In fact, I think that the proof can sometimes cloud the matter at hand! Thus, not every theorem in the book is accompanied by a proof; instead, I have included it only when I think it would be beneficial. Also, there are numerous activities in this book to complete that will teach you, as you progress, basic definitions, properties, and theorems. In fact, you may be able to devise them yourself! I not only want you to succeed—I want you to *understand*.

In order to promote understanding, it is my belief that a book that intends to teach must offer answers as well as questions. How useful are 100 practice problems if only numerical answers and maybe a token explanation are given? Some calculus review books even confuse calculus teachers with their lack of explanation (although I bet none of your teachers would ever admit it!). *Every single question in this book has a good explanation and includes* all *of the important steps*; this is just one of the characteristics that makes this book unique. No longer will you have to spend fifteen minutes deciding how the author got from one step to the next in his or her computations. I know I am not the only one who, gnashing his teeth, has exclaimed, "Where did the 4 go in the third step!? How can they just drop the 4? Did the 4 just step out for a bite to eat and will be back later? Am I so stupid that it's obvious to everyone but me?"

You should also find that the problems' difficulties increase as you progress. When topics are first introduced, they are usually of easy or medium difficulty. However, the problems at the end of each section are harder and will help you bind your understanding to previous topics in the book. Finally, the problems at the end of the chapter are the most challenging of all, requiring you to piece together all the important topics and involve appropriate technology along the way.

NOTE

It is rare for math textbooks to show many steps in their calculations. Authors often use the word *clearly* to justify omitted mystery steps, as in "this complicated expression *clearly* simplifies to 3." It is a common joke among math majors that the use of the word clearly means that the textbook's author couldn't figure out how to prove it either.

Herein lies the entire premise and motivation of the book: *You can prepare for the AP Calculus test by understanding what's involved rather than simply practicing skills out of context.* This is not to say that you won't have to memorize any formulas or that the concepts themselves will be presented any less rigorously than in a textbook. This is not, as some might claim, "soft mathematics." Instead, it is mathematics presented in a way that ensures and promotes understanding.

THE GAME PIECES

The following are included to help you in your study of calculus:

- Hands-on activities or guided practice to introduce and teach all of the major elements of calculus, including calculus reform topics

- "Target Practice" examples with detailed solutions to cement your understanding of the topics (each target practice problem is accompanied by an icon denoting its difficulty—see below—so that you can constantly monitor your progress)

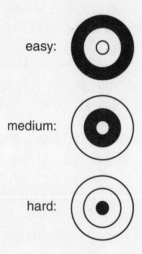

easy:

medium:

hard:

- "Common Errors" denoted during practice problems and notes so that you can avoid them

- Exercises at the end of each section to practice the skills you just learned

- A technology section in each chapter with step-by-step directions for the TI-83 series calculator to ensure that you are using it correctly and in accordance with College Board guidelines. (The TI-83, TI-83 Plus, and TI-83 Plus Silver Edition calculators constitute the vast majority of calculators used on the AP test, so that's why we chose them.)

- Additional problems at the end of each chapter to review your skills and challenge you

- James' Diabolical Dilemma Problems at the end of each chapter are written by a former AP Calculus student of mine, James Finley; as a former 5-er on the AP test, he has created these problems to push your understanding to the very limits—consider these problems concentrated, with all the pulp left in for flavor.

NOTE

These provide interesting calculus facts, connections, and notations to improve your understanding of the material. Don't overlook these little gems —they are an important part of the text, highlighting important information and putting things into the proper context.

TIP

These are the authors' personal pointers to you, so that you can stay on track and in focus as you study the material. Typically, they contain pieces of information to help you with the many tasks you'll be required to perform on the AP Calculus Test.

CAUTION

Steer clear of these common errors of judgement in mathematics. Think of horror movies: If the doorknob of the room you are about to enter is smeared with blood, rethink your plan. Don't go into that room!

CONCLUSION

This book can be utilized for numerous reasons and toward many ends. It is best used as a study guide to supplement your calculus textbook . As a teaching tool, it can help you to learn calculus or fill in gaps in your understanding. As a resource, it can provide numerous practice problems with full solutions. However you choose to use the book, it is my hope that it can help unlock some of the mysteries of calculus for you, although it's almost certainly going to do nothing for your chess game.

Mike Kelley

AP Calculus Formulas

These pages contain just about all the formulas you need to know by heart before you can take the AP Test. It does *not* contain all the theorems and techniques you need to know. An asterisk (*) indicates a Calculus BC-only formula.

LIMITS

$$\lim_{x \to \infty} \frac{c}{x^n} = 0$$

$$\lim_{x \to 0} \frac{\sin x}{x} = 1$$

$$\lim_{x \to 0} \frac{\cos x - 1}{x} = 0$$

$$\lim_{x \to \infty} \left(1 + \frac{1}{x}\right)^x = e$$

DERIVATIVES

Power Rule: $\dfrac{d}{dx}\left(x^n\right) = nx^{n-1}$

Product Rule:
$$\frac{d}{dx}\big(f(x) \cdot g(x)\big) = f(x)g'(x) + g(x)f'(x)$$

Quotient Rule:
$$\frac{d}{dx}\left(\frac{f(x)}{g(x)}\right) = \frac{g(x)f'(x) - f(x)g'(x)}{\big(g(x)\big)^2}$$

Chain Rule: $\dfrac{d}{dx}\big(f(g(x))\big) = f'(g(x)) \cdot g'(x)$

$$\frac{d}{dx}(\sin x) = \cos x$$

$$\frac{d}{dx}(\cos x) = -\sin x$$

$$\frac{d}{dx}(\tan x) = \sec^2 x$$

$$\frac{d}{dx}(\cot x) = -\csc^2 x$$

$$\frac{d}{dx}(\sec x) = \sec x \, \tan x$$

$$\frac{d}{dx}(\csc x) = -\csc x \, \cot x$$

$$\frac{d}{dx}(\ln x) = \frac{1}{x}$$

$$\frac{d}{dx}\left(e^x\right) = e^x$$

$$\frac{d}{dx}\left(\log_a x\right) = \frac{1}{(\ln a)x}$$

$$\frac{d}{dx}\left(a^x\right) = (\ln a)a^x$$

$$\frac{d}{dx}(\arcsin x) = \frac{1}{\sqrt{1 - x^2}}$$

$$\frac{d}{dx}(\arccos x) = -\frac{1}{\sqrt{1 - x^2}}$$

$$\frac{d}{dx}(\arctan x) = \frac{1}{1 + x^2}$$

$$\frac{d}{dx}(\text{arc}\cot x) = -\frac{1}{1 + x^2}$$

$$\frac{d}{dx}(\text{arc}\sec x) = \frac{1}{|x|\sqrt{x^2 - 1}}$$

$$\frac{d}{dx}(\text{arc}\csc x) = -\frac{1}{|x|\sqrt{x^2 - 1}}$$

$$\left(f^{-1}\right)' x = \frac{1}{f'\left(f^{-1}(x)\right)}$$

*Parametric derivatives: $\dfrac{dy}{dx} = \dfrac{\frac{dy}{dt}}{\frac{dx}{dt}}$;

$$\frac{d^2 y}{dx^2} = \frac{\frac{d}{dt}\left(\frac{dy}{dx}\right)}{\frac{dx}{dt}}$$

DERIVATIVE APPLICATIONS

Mean Value Theorem: $f'(c) = \dfrac{f(b) - f(a)}{b - a}$

$s(t)$ is the position function; $s'(t) = v(t)$, the velocity function; $v'(t) = a(t)$, the acceleration function

Projectile position equation:

$s(t) = -\frac{1}{2}gt^2 + v_0 t + h_0$; g = 9.8 m/s² or 32 ft/s²

INTEGRATION

Power Rule for Integrals: $\int x^n \, dx = \frac{x^{n+1}}{n+1} + C$

$\int \cos x \, dx = \sin x + C$

$\int \sin x \, dx = -\cos x + C$

$\int \tan x \, dx = -\ln|\cos x| + C$

$\int \cot x \, dx = \ln|\sin x| + C$

$\int \sec x \, dx = \ln|\sec x + \tan x| + C$

$\int \csc x \, dx = -\ln|\csc x + \cot x| + C$

$\int \frac{du}{\sqrt{a^2 - u^2}} = \arcsin\frac{u}{a} + C$

$\int \frac{du}{a^2 + u^2} = \frac{1}{a}\arctan\frac{u}{a} + C$

$\int \frac{du}{u\sqrt{u^2 - a^2}} = \frac{1}{a}\text{arcsec}\frac{|u|}{a} + C$

Trapezoidal Rule: $\frac{b-a}{2n}(f(a) + 2f(x_1) + 2f(x_2) + ...$
$+ 2f(x_{n-1}) + f(b))$

Fundamental Theorem (Part 1):

$\int_a^b f(x)\,dx = F(b) - F(a)$, if F is the antiderivative of $f(x)$

Fundamental Theorem (Part 2):

$\frac{d}{dx}\left(\int_a^x f(t)\,dt\right) = f(x)$

Average Value: $f(c) = \frac{1}{b-a}\int_a^b f(x)\,dx$

*Integration by Parts: $\int u\,dv = uv - \int v\,du$

INTEGRATION APPLICATIONS

Disk Method: $\pi\int_a^b (r(x))^2\,dx$

Washer Method: $\pi\int_a^b \left[(R(x))^2 - (r(x))^2\right]dx$

Shell Method: $2\pi\int_a^b d(x)h(x)\,dx$

*Arc Length:

$\int_a^b \sqrt{1 + (f'(x))^2}$; $\int_a^b \sqrt{\left(\frac{dx}{dt}\right)^2 + \left(\frac{dy}{dt}\right)^2}$

*Polar Area: $\frac{1}{2}\int_a^b (r(\theta))^2\,d\theta$

DIFFERENTIAL EQUATIONS

Exponential growth/decay: $\frac{dy}{dt} = ky$; $y = Ne^{kt}$

*Logistic growth:

$\frac{dy}{dt} = ky(L-y)$; $y = \frac{L}{1 + ce^{-Lkt}}$

* $\Delta y = \Delta x \cdot m$

*SEQUENCES AND SERIES

Sum of a geometric series: $\frac{a}{1-r}$

Ratio Test: $\lim\limits_{n \to \infty} \frac{a_{n+1}}{a_n}$

Limit Comparison Test: $\lim\limits_{n \to \infty} \frac{a_n}{b_n}$

Taylor series for $f(x)$ centered at $x = c$:

$f(x) = f(c) + f'(c)(x-c) + \frac{f''(c)(x-c)^2}{2!} + $
$\frac{f'''(c)(x-c)^3}{3!} + ... + \frac{f^{(n)}(c)(x-c)^n}{n!} + ...$

$\sin x = x - \frac{x^3}{3!} + \frac{x^5}{5!} + ...$

$\cos x = 1 - \frac{x^2}{2!} + \frac{x^4}{4!} + ...$

$e^x = 1 + x + \frac{x^2}{2!} + \frac{x^3}{3!} + ...$

$\frac{1}{1-x} = 1 + x + x^2 + x^3 + x^4 + ...$

Digging In

Your goal and vision in any Advanced Placement class should be to take the AP test, pass it with a sufficiently high score, jump up and down like a lunatic when you receive your score, and attain credit for the class in the college or university of your choice. All AP tests are graded on a scale from 1 to 5, with 5 being the highest possible grade. Most colleges will accept a score of 3 or above and assign credit to you for the corresponding course. Some, however, require higher scores, so it's important to know the policies of the schools to which you are applying or have been accepted. An AP course is a little different from a college course. In a college course, you need only pass the class to receive credit. In an AP course, you must score high enough on the corresponding AP test, which is administered worldwide in the month of May. So, it's essential to know that all-important AP test inside and out.

FREQUENTLY ASKED QUESTIONS ABOUT THE A.P. CALCULUS TESTS

Below are common questions that students pose about the AP Calculus tests. For now, this test is your foe, the only thing standing in your way to glorious (and inexpensive) college credit. Spend some time understanding the enemy's battle plans so that you are prepared once you go to war.

GENERAL QUESTIONS

What topics are included on the test?

The list of topics changes a little bit every couple of years. The College Board Web site (www.collegeboard.org/ap/calculus) always has the current course description. As your academic year draws to a close, use it as a checklist to make sure you understand everything.

What's the difference between Calculus AB and BC?

The Calculus BC curriculum contains significantly more material than the AB curriculum. Completing Calculus BC is equivalent to completing college Calculus I and Calculus II courses, whereas AB covers all of college Calculus I and about half of Calculus II. The AB and BC curricula cover the same material with the same amount of rigor; BC simply covers additional topics. However, if you take the BC test, you will get both an AB and BC score (the AB score excludes all BC questions from the test).

CAUTION
The World Wide Web is constantly in a state of flux. Web sites come and go, and their addresses change all the time. All of these links were active at publication time.

Of the topics on the course description, which actually appear the most on the AP test?

This will vary, of course, but at the conclusion of the 2000 AP Exam, I asked my students to list the topics they saw the most. This is the list of topics they generated (BC topics are denoted with an asterisk): relative extrema (maximums and minimums), the relationships between derivatives of a function, the difference quotient, basic integration, integral functions with variables as limits of integration, volumes of solids with known cross-sections, motion (position, velocity, and acceleration functions), differential equations, area between curves, power series*, elementary series* (e^x, cos x, sin x), Taylor polynomials*, radius of convergence*, and integration by parts*.

How is the test designed?

The test is split into two sections, each of which has a calculator-active and a calculator-inactive portion. Section I has 45 multiple-choice questions and lasts 105 minutes. Of that time, 55 minutes are spent on 28 non-calculator questions, and 50 minutes are dedicated to 17 calculator active questions. Section II has 6 free-response questions and lasts 90 minutes. Three of the free-response questions allow the use of a calculator, while 3 do not.

Should I guess on the multiple-choice questions?

You lose a fraction of a point for every multiple-choice question you answer incorrectly; this penalizes random guessing. If you can eliminate even one choice in a question, the odds are in your favor if you guess. If you cannot eliminate any choices, it is best to omit the question.

Should I have the unit circle memorized?

Oh, yes. The unit circle never dies—it lives to haunt your life.

Should my calculator be in degrees or radians mode?

Unless specifically instructed by the question, set your calculator for radians mode.

I have heard that the AP Calculus test is written by scientists living in the African rainforest, and that many tests are lost each year as couriers are attacked and infected by virulent monkeys. Is this true?

Definitely.

How many questions do I have to get right to get a 3?

There is no set answer to this, as the number varies every year based on student achievement. Unofficially, answering approximately 50 percent of the questions correctly usually results in a 3.

Why is the test so hard?

Many students are shocked that a 50 percent is "passing" on the AP test, but the exam is constructed to be a "super test" that tests not only your knowledge but also your ability to apply your knowledge under extreme pressure and in very difficult circumstances. No one expects you to get them all right. You're shooting for better than 50 percent, so don't panic.

How will I feel when the test is over?

Hopefully, you will still be able to function. Most students are completely exhausted and drained at the end of the ordeal. Students who are well prepared (like those who buy my book) experience less depression than others. In general, students have a vague positive feeling when they exit the test if they dedicated themselves to studying all year long. I have found that the way you feel when exiting the test is independent of how you will actually perform on the test. Feeling bad in no way implies that you will score badly.

Are there any Web sites on the Internet that could help me prepare for the AP test?

There are a few good sites on the Web that are free. Among them, one stands clearly above the rest. It offers a new AB and BC problem each week, timed to coordinate with a year-long curriculum to help you prepare for the test. Furthermore, each problem is solved in detail the following week. Every problem ever posted is listed in an archive, so it's a very valuable studying tool for practicing specific skills and reviewing for tests throughout the year. To top it off, the site is funny, and the author is extremely talented. The site? *Kelley's AP Calculus Web Page*, written and shamefully advertised here by yours truly. You can log on at www.geocities.com/Athens/Oracle/2613 or www.geocities.com/impactedcolon. Enjoy the same hickory-smoked flavor of this book on line for free each and every week. Another good problem of the week site is the *Alvirne Problems of the Week* (www.seresc.k12.nh.us/www/alvirne.html).

FREE-RESPONSE QUESTIONS

How are the free-response questions graded?

Each free-response question is worth up to nine points. Free response questions usually have multiple parts, typically two or three, and the available points are dispersed among them. Many points are awarded for knowing how to set up a problem; points are not only given for correct answers. It is best to show all of the setup and steps in your solution in an orderly fashion to get the maximum amount of credit you can. The College Board has examples of excellent, good, and poor free response answers given by actual test takers on their Web site (www.collegeboard.org/ap/calculus). In addition, they have the most recent free-response questions with their grading rubrics. You should try these problems, grade yourself according to the rubrics, and see how you stack up to the national averages.

TIP

The past free-response questions, course outline, and grade rubrics are all available free of charge on line. The College Board Web site is updated far more frequently than printed material. Check there for breaking news and policies.

TIP

Cross out any of your errant work instead of erasing it. Erasing takes more time, and time is money on the AP test.

NOTE

$-\ln 3 = \ln 3^{-1} = \ln \frac{1}{3}$ according to the log property that states $n\log_a x = \log_a x^n$.

Are free-response questions the same as essays?

No. Free-response questions are most similar to questions you have on a typical classroom quiz or test. They require you to solve a problem logically, with supporting work shown. There's no guessing possible like on the multiple-choice questions. There's really no need to write an essay—it just slows you down, and you need every last second on the free response—it's really quite meaty.

Should I show my work?

Yes, indeed. The AP graders (called *readers*) cannot assign you partial credit if you don't give them the opportunity. On the other hand, if you have no idea what the problem is asking, don't write a detailed explanation of what you *would* do, and don't write equations all over the paper. Pick a method and stick to it—the readers can definitely tell if you are trying to bluff your way through a problem you don't understand, so don't pull a Copperfield and try to work magic through smoke and mirrors. Also, keep in mind that any work erased or crossed off is *not* graded, even if it is completely right.

What if the problem has numerous parts and I can't get the first part?

You should do your best to answer the first part anyway. You may not get any points at all, but it is still worth it. If the second part requires the correct completion of the first, your incorrect answer will not be penalized again. If you complete the correct sequence of steps on an incorrect solution from a previous answer, you can still receive *full* credit for the subsequent parts. This means you are not doomed, so don't give up.

Is it true that a genetically engineered chicken once scored a 4 on the AB test, but the government covered it up to avoid scandal?

I am not allowed to comment on that for national security reasons. However, I can say that free-range poultry typically score better on free-response questions.

Should I simplify my answers to lowest terms?

Actually, no! The AP readers will accept an answer of $\frac{39}{3}$ as readily as an answer of 13. Some free-response questions can get a little messy, and you're not expected to make the answers pretty and presentable. However, you still need to be able to simplify for the multiple-choice questions. For example, if you reach a solution of $\ln \frac{1}{3}$ but that is not listed among the choices, you should be able to recognize that $-\ln 3$ has the same value, if you apply logarithmic properties.

How accurate should my answers be?

Unless specified otherwise, the answer must be correct to *at least* three decimal places. You may truncate (cut off) the decimal there or round the decimal there. For example, a solution of $x = 4.5376219$ may be recorded as 4.537 (truncated) or 4.538 (rounded). If you want to write the entire decimal, that is okay, too, but remember that time is money.

Should I include units in my answer?

If the problem indicates units, you need to include the appropriate units in your final answer. For example, if the problem involves the motion of a boat and phrases the question in terms of feet and minutes, velocity is in ft/min, and the acceleration will be in ft/min^2.

CALCULATOR QUESTIONS

When can I use the calculator to answer questions?

You may use the calculator only on calculator-active questions, but you probably figured that out, Mr. or Ms. Smarty Pants. Occasionally, you may use a calculator to completely answer a question and show no work at all. You can do this only in the following circumstances: graphing a function, calculating a numerical derivative, calculating a definite integral, or finding an x-intercept. In fact, your calculator is *expected* to have these capabilities, and you are expected to know how to use them. Therefore, in these four cases, you need only show the setup of the problem and jump right to the solution. For example, you may write

$$\int_5^2 (x^2 e^x)dx = 2508.246$$

without actually integrating by hand at all or showing any work. In all other circumstances, you must show supporting work for your solutions.

How should I write an answer if I used my calculator?

As in the above example, $\int_5^2 (x^2 e^x)dx = 2508.246$ is all you should write; the readers understood and expected you to use your calculator. Never write "from the calculator" as a justification to an answer. Also, never write calculator language in your answer. For example, a free-response answer of fnInt($x \char`\^ 2*e \char`\^ (x),x,2,5$) = 2508.246 cannot get a point for the correct setup, though it may get points for the correct answer.

What calculators can I use on the AP test?

The most current list of calculators can be found on the College Board Web site. Most favored among the calculators are the Texas Instruments 83 and 83+ (and probably the TI 89 before too long). It's a matter of preference. Some people live and die by HP calculators and will jump down your throat in the blink of an eye if you suggest that the TI calculators are better. Calculators like the TI-92 cannot be used because they have QWERTY keyboards. Make sure to check the Web site to see if your calculator is acceptable.

CAUTION
Never, never, never round a number in a problem unless you are giving the answer. If you get a value of 3.5812399862 midway through a problem, use the *entire* decimal as you complete the problem. Rounding or truncating during calculations almost always results in inaccurate final answers.

TIP
Because you can use your calculator to find x-intercepts, you can also use it to solve any equation without explaining how. See the Technology section in Chapter Two for a more detailed explanation.

NOTE

Only in the four listed circumstances can you use the calculator to reach an answer. For instance, most calculators can find the maximum or minimum value of a function based on the graph, but you cannot use a calculator as your justification on a problem such as this.

I recently made a calculator out of tinfoil, cat food, and toenail clippings. Are you telling me I can't use it on the AP test?

Sorry, but you can't. By the way, I shudder to think about how the toenail clippings were put to use.

Can I have programs stored in my calculator's memory?

Yes. Programs are not cleared from the calculator's memory before the test begins. Many of my students have stored various programs, but I don't think a single student has ever used a program on the test. The test writers are very careful to construct the calculator portions of the test so that no calculator has an advantage over another. It's really not worth your time to load up your calculator.

If I can store programs in the calculator memory, can't I store formulas and notes? Why do I need to memorize formulas?

Technically, you can enter formulas in the calculator as programs, but the test writers also know you can do this, so it is highly unlikely that such a practice could ever be useful to you. Remember that more than half of the test is now calculator inactive! Don't become so calculator dependent that you can't do basic things without it.

NOTE

A QWERTY keyboard, for those not in the know, has keys in the order of those on a computer keyboard.

PART 2

PREVIEW

Calculus Prerequisites

This chapter is meant to help you review some of the mathematics that lead up to calculus. Of course, all mathematics (and your entire life, no doubt) up until this point has simply been a build-up to calculus, but these are the most important topics. Since the focus of this book must be the actual content of the AP test, this chapter is meant only to be a review and not an in-depth course of study. If you find yourself weak in any of these areas, make sure to review them and strengthen your understanding before you undertake calculus itself. Ideally, then, you should plod through this chapter early enough to address any of your weaknesses before it's too late (read with a scary voice).

FUNCTIONS AND RELATIONS

Calculus is rife with functions. It is unlikely that you can find a single page in your textbook that isn't bursting with them, so it's important that you understand what they are. A function is a special type of *relation*, much as, in geometry, a square is a special type of rectangle. So, in that case, what is a relation? Let's begin with a simple relation called r. We will define r as follows: $r(x) = 3x + 2$. (This is read "r of x equals 3 times x plus 2.") Our relation will accept some sort of input (x) and give us something in return. In the case of r, the relation will return a number that is two more than three times as large as your input. For example, if you were to apply the rule called r to the number 10, the relation would return the number 32. Mathematically, this is written $r(10) = 3(10) + 2 = 30 + 2 = 32$. Thus, $r(10) = 32$. We say that the relation r has solution point $(10,32)$, as an input of 10 has resulted in an output of 32.

 Example 1: If $g(x) = x^2 - 2x + 9$, evaluate $g(4)$ and $g(-3)$.

Solution: Simply substitute 4 and –3 in for x, one at a time, to get the solutions:

$$g(4) = (4)^2 - 2(4) + 9 = 16 - 8 + 9 = 17$$
$$g(-3) = (-3)^2 - 2(-3) + 9 = 9 + 6 + 9 = 24.$$

We call g and r relations because of the way they relate numbers together. Clearly, r related the input 10 to the output 32 in the same way that g related 4 to 17 and -3 to 24. It is conventional to express these relationships as ordered pairs, so we can say that g created the relationships (4,17) and (-3,24).

There are numerous ways to express relations. They don't always have to be written as equations, though most of the time they are. Sometimes, relations are defined simply as the sets of ordered pairs that create them. Here, we have defined the relation k two ways that mean the same thing:

$$k: \{(-3,9),(2,6),(5,-1),(7,12),(7,14)\}.$$

x	−3	2	5	7	7
$k(x)$	9	6	−1	12	14

You can also express a relation as a graph of various ordered pairs that create it, in the form (x,y). Below we have graphed the function we defined earlier, $r(x) = 3x + 2$.

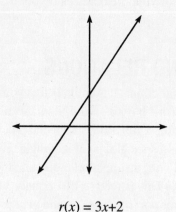

$$r(x) = 3x+2$$

Sometimes, the rule for a relation changes depending on the input of the relation. These are called *piecewise-defined relations* or *multi-ruled relations*.

 Example 2: Graph the piecewise-defined relation

$$h(x) = \begin{cases} x^2, & x \le -1 \\ 2, & -1 < x \le 2 \\ 2x - 4, & x > 3 \end{cases}$$

and evaluate $h(-2)$, $h(-1)$, $h(0)$, $h(2)$, $h(2.5)$, and $h(6)$.

Solution: The graph begins as the parabola x^2, but once $x = -1$, the new rule takes over, and the graph becomes the horizontal line $y = 2$. This line stops, in turn, at $x = 2$. The function is undefined between $x = 2$ and $x = 3$, but for all $x > 3$, the line $2x - 4$ gives the correct outputs.

Note: If you didn't recognize that $y = x^2$ is a parabola, that's OK (for now). Later in the chapter, we discuss how to recognize these graphs.

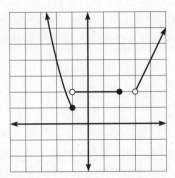

$h(-2) = (-2)^2 = 4$, since –2 falls within the definition of the first of the three rules.

$h(-1) = (-1)^2 = 1$, for the same reason.

$h(0) = 2$, as 0 is between –1 and 2, the defined region for the second rule.

$h(2) = 2$, as it just falls within the definition of the second rule.

$h(2.5)$ is not defined for this function—no inputs between 2 and 3 are allowed.

$h(6) = 2(6) – 4 = 8$, since $6 > 3$, the defining restriction for the last rule.

So, a relation is, in essence, some type of rule that relates a set, or collection, of inputs to a set of outputs. The set of inputs for a relation is called the *domain*, whereas the set of outputs is called the *range*. Often, it helps to look at the graph of the relation to determine its domain and range, as the *x* values covered by a graph represent its domain, and the *y* values covered by a graph represent its range. Alternatively, you can think of the domain as the numbers covered by the "width" of the graph, and the range as the numbers covered by the "height" of the graph.

 Example 3: Find the domain and range for the functions *k* and *h*, as already defined above.

Solution: Finding domain and range for *k* is quite easy. The domain is $\{-3,2,5,7\}$ and the range is $\{-1,6,9,12,14\}$. The order of the numbers does not matter in these sets. The domain of *h* comes right from the function—written in interval notation, the domain is $(-\infty,2] \cup (3, \infty)$. This is true because any number up to or including 2 is an input for the relation, as is any number greater than 3. The range is easily determined from the graph we created already. Look at the height or vertical span of the graph. Notice that it never dips below a height of 1, but above 1, every single number is covered. Even though there are holes in the graph at (–1,2) and (3,2), the height of 2 is covered by numerous other x values. Thus, the range of the graph is [1, ∞).

Now that you know quite a bit about relations, it's time to introduce functions. *Functions are simply relations such that every input has a unique output.* In other words, every element of the domain must result in only one output.

 Example 4: Which of the relations, *r, g,* and *k,* as defined in the preceding examples, are functions?

Solution: Take *r* as an example. When you evaluated $r(10)$, the result was 32. Is there any chance that something other than 32 could result? No. In a similar fashion, any number, when substituted for *r,* will result in only one output. Thus, *r* is a function. Similarly, *g* is a function. However, *k* is not a function, since $k(7) = 12$ and 14. Because the domain element 7 results in two range elements, 12 and 14, *k* is not a function. Consider what's happening in function *k* graphically:

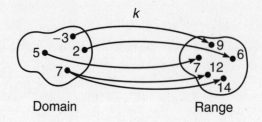

Graphically, the road (or map) from 7 in the domain to the range forks, whereas the inputs –3, 2, and 5 have only one path to follow. Because of this fork in the road from 7, *k* is not a function.

Often, it is unfruitful to ponder whether or not a relation is a function merely by remembering the definition. A shortcut to determining whether or not a given relation is a function is the *vertical line test.* To use this test, imagine vertical lines passing through the graph of the relation. If any vertical line you could possibly draw intersects the graph in more than one place, the relation is not a function. If any vertical line resembles former U.S. President James K. Polk, seek professional help.

Example 5: Explain why the relation described by the equation $x^2 + y^2 = 9$ is not a function.

Solution: First, you need to remember that $x^2 + y^2 = 9$ is the equation of a circle centered at the origin with radius 3. Look at the graph of the relation:

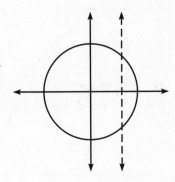

Notice the line $x = 2$ (one of many vertical lines you could imagine intersecting the graph) intersects the circle in two places. By substituting 2 for x in the equation you can determine that the two points of intersection are $\left(2, \sqrt{5}\right)$ and $\left(2, -\sqrt{5}\right)$. Since $x = 2$ has two distinct outputs, this is not a function.

Even if you weren't sure what a function was by definition, you undoubtedly know many functions already. You should know, at the minimum, 15 specific functions and their properties by heart. Nine of these are listed below with any important characteristics you should either memorize or be able to determine from the graph. The remaining 6 are presented later in this chapter. These functions and their defining characteristics need to become really familiar to you.

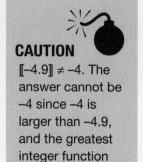

Important Functions to Memorize

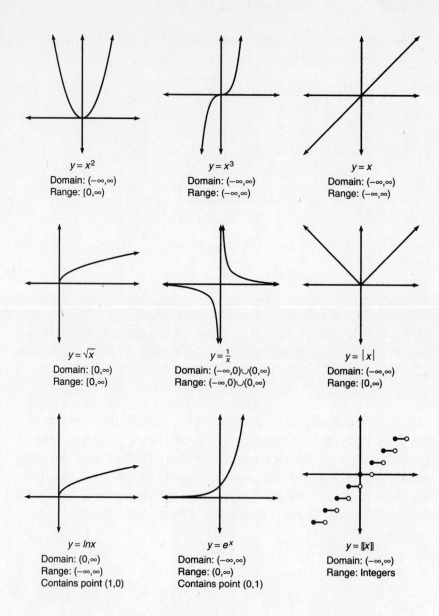

$y = x^2$
Domain: $(-\infty, \infty)$
Range: $[0, \infty)$

$y = x^3$
Domain: $(-\infty, \infty)$
Range: $(-\infty, \infty)$

$y = x$
Domain: $(-\infty, \infty)$
Range: $(-\infty, \infty)$

$y = \sqrt{x}$
Domain: $[0, \infty)$
Range: $[0, \infty)$

$y = \frac{1}{x}$
Domain: $(-\infty, 0) \cup (0, \infty)$
Range: $(-\infty, 0) \cup (0, \infty)$

$y = |x|$
Domain: $(-\infty, \infty)$
Range: $[0, \infty)$

$y = \ln x$
Domain: $(0, \infty)$
Range: $(-\infty, \infty)$
Contains point $(1,0)$

$y = e^x$
Domain: $(-\infty, \infty)$
Range: $(0, \infty)$
Contains point $(0,1)$

$y = [[x]]$
Domain: $(-\infty, \infty)$
Range: Integers

Among these functions, the strangest might be $y = [[x]]$, called the *greatest integer function*. This function takes any real number input and returns the largest integer that is less than or equal to it. For example, $[[5.3]] = 5$ since among the integers that are less than or equal to 5.3 $(5,4,3,2,1,0,-1,-2,\ldots)$, the largest is 5. Likewise, $[[4]] = 4$. However, $[[-3.6]] = -4$, since among the integers *less* than -3.6 $(-4,-5,-6,-7,\ldots)$, -4 is the largest.

PROBLEM SET

You may use a graphing calculator on problems 3 through 6.

1. Which of the following relations are functions?

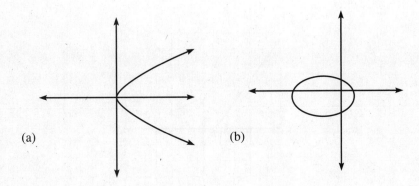

(a)   (b)

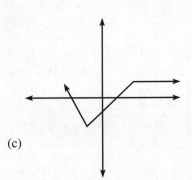

(c)

(d) $y = \pm(2x) + 3$

(e) $y = \begin{cases} \frac{1}{x}, & x < 0 \\ \sqrt{x}, & x \geq 0 \end{cases}$

(f) $y = \begin{cases} x^2, & x \leq 1 \\ \ln x, & x \geq 1 \end{cases}$

2. If $f(x) = x^2 - 25$, $g(x) = x^2 + 9x + 20$, and $h(x) = \dfrac{f(x)}{g(x)}$, what is the domain of h?

3. If $f(x) = |x| + 1$ and $g(x) = \dfrac{1}{3x^2 + 4}$, find $f(4) - g(3) + (fg)(0)$.

4. Write the function, m, whose graph is given below. Also, find the domain and range of m.

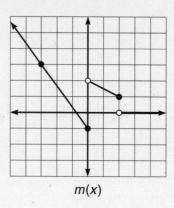

$m(x)$

5. If $p(x)$ is defined by the graph below, evaluate $[\![m(-3)]\!]$, $[\![m(0)]\!]$, and $[\![m(4)]\!]$.

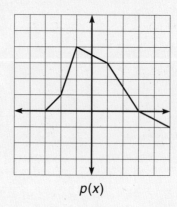

$p(x)$

6. Graph some function $s(x)$ such that

-   $s(1) = 4$
-   $s(-2) = s(2) = s(5) = 0$
-   s is increasing on $(-6,1) \cup (4,6)$
-   s has domain $[-6,6]$ and range $(-2,4]$

SOLUTIONS

1. Only c and e are functions. Notice that a, b, and f (when graphed) fail the vertical line test. In d, every input except zero will result in two outputs.

2. According to the problem, $h(x) = \dfrac{x^2 - 25}{x^2 + 9x + 20}$. When the denominator is zero, the function is undefined (since it is illegal in this arm of the Milky Way to divide by zero). Factoring the denominator gives $(x + 4)(x + 5)$, which means

that function is undefined when $x = -4$ or $x = -5$. Thus, the domain of h is all real numbers except -4 and -5.

3. The function $(fg)(x)$ is created by the product of f and g. Since $f(x) \cdot g(x) = \frac{|x|+1}{3x^2+4}$, $(fg)(0) = \frac{1}{4}$. Thus, $f(4) - g(3) + (fg)(0) = 5 - \frac{1}{31} + \frac{1}{4} = \frac{647}{124}$ or, approximately 5.218.

4. By examining the slopes and y-intercepts of these lines, it is not too hard to get the piecewise-defined function.

$$m(x) = \begin{cases} -\frac{4}{3}x - 1, & x \le 0 \\ -\frac{1}{2}x + 2, & 0 < x \le 2 \\ 0, & x > 2 \end{cases}$$

The domain of $m(x)$ is $(-\infty,\infty)$, and the range is $(-1,\infty)$.

5. From the graph, you can see that $m(-3) = 0$, $m(0) = 3.5$, and $m(4) = -.5$, so the greatest integer function values are not too difficult: $[[m(-3)]] = 0$, $[[m(0)]] = 3$, and $[[m(4)]] = -1$.

6. There are a number of solutions to the problem. One is given here:

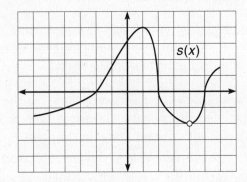

FUNCTION PROPERTIES

Once a relation is classified as a function, the fun is only just beginning. Functions exhibiting other specific properties allow us to deepen our ever-widening puddle of understanding. For example, some functions can be classified as *even*. An even function is so mathematically robust that a negative input does not affect the function at all. Consider the function $f(x) = x^2 - 3$, and compare $f(4)$ to $f(-4)$. Upon substituting, it is clear that $f(4) = f(-4) = 13$. Opposite elements of the domain result in the same output. Thus, $f(x)$ exhibits the property that allows us to classify it as *even*. Mathematically, we write: If $f(-x) = f(x)$, then $f(x)$ is even, which means that evaluating the function at a negative input, $-x$, results in the original function, $f(x)$.

Example 6: Prove that $m(x) = 5x^4 - 2x^2 + 7$ is even, whereas $b(x) = x^2 + 8x$ is not even.

Solution: In order to test m, you substitute $-x$ into the function to get

$$m(-x) = 5(-x)^4 - 2(-x)^2 + 7 = 5x^4 - 2x^2 + 7 = m(x).$$

Because $m(-x) = m(x)$, the function is even. Similarly, test b:

$$b(-x) = (-x)^2 + 8(-x) = x^2 - 8x \neq b(x).$$

Thus, b is not even.

Be careful! It is just plain wrong to assume that because b is not even, it must be an odd function. Though that assumption is true with integers, it is *not* true with functions. In fact, for a function to be odd, it must satisfy a completely different property: If $g(-x) = -g(x)$, then $g(x)$ is said to be *odd*. Whereas the terms in an even function stay exactly the same for an input of $-x$, each of the terms of an odd function will become its opposite.

 Example 7: Show that the function $d(x) = x^5 - 3x$ is odd.

Solution: Notice that $d(-x) = (-x)^5 - 3(-x) = -x^5 + 3x = -d(x)$. Therefore, d is odd. All of the terms must change sign (as they did in this problem) in order for the function to be odd.

CAUTION
If a function does not fulfill the requirements to be even or odd, the function is classified as "neither even nor odd."

Another important property of a function is the symmetry, if any, that is evident in its graph. A graph is called symmetric if it is exactly the same (or mirrored) on both sides of some arbitrary line or point. For example, the graph of $y = x^{2/3}$, below, is described as y-symmetric because the graph is identical on either side of the y-axis. All y-symmetric graphs have this defining property: if a point (x, y) lies on the graph, so must the point $(-x, y)$.

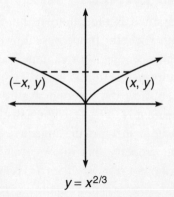

$y = x^{2/3}$

CAUTION
Objects in the margin are larger than they appear.

Notice that the sign of the input to the function does not matter—both x and $-x$ result in the same y. Does that sound familiar? It should! That's the definition of an even function. Therefore, *all even functions are y-symmetric and vice versa.*

The other type of symmetry that is very common to calculus is *origin-symmetry*. If a graph (or function) is origin-symmetric, then all points (x,y) on the graph must have a corresponding $(-x,-y)$, as displayed in the graph below, the graph of $y = x^3 - x$.

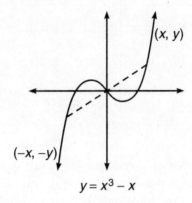

$y = x^3 - x$

Using the same reasoning as above, *all odd functions are origin-symmetric.*

 Example 8: Determine what symmetry, if any, is evident in the function $h: \{(3,5),(4,-1),(0,0),(-3,-5),(-4,1)\}$.

Solution: The function h is origin-symmetric, since every (x,y) is paired with a corresponding $(-x,-y)$.

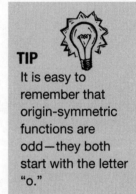

TIP

It is easy to remember that origin-symmetric functions are odd—they both start with the letter "o."

PROBLEM SET

You may use a graphing calculator on problems 7 through 9.

In problems 1 through 6, determine if the given function is even, odd, or neither.

1. $g(x) = x^4 - 3x^2 + 1$

2. $p(x) = 2x^3 + \sqrt[3]{x}$

3. $m(x) = \dfrac{-3x^3\left(1 + 4x^2\right)}{2x^4 - 5x^2}$

4. $b(x) = x^7 + 5x^3 - 17$

5. $v:\{(-1,4),(2,6),(1,4)\}$

6.

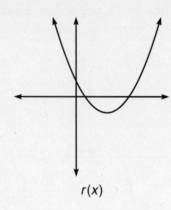

$r(x)$

7. The equation $x = y^2$ has an x-symmetric graph. In x-symmetric graphs, if the point (x,y) is contained, then so is $(x,-y)$. Why aren't x-symmetric functions used as often as y- and origin-symmetric functions in calculus?

8. Complete the below graph of f if ...

    (a) f is odd

    (b) f is even

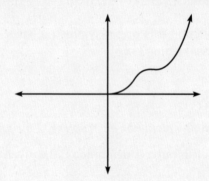

9. If $j(x)$ and $k(x)$ are odd functions and $h(x) = \dfrac{j(x)}{k(x)}$, what kind of symmetry characterizes h?

SOLUTIONS

1. g is even: $g(-x) = (-x)^4 - 3(-x)^2 + 1 = x^4 - 3x^2 + 1 = g(x)$.

2. p is odd: $p(-x) = 2(-x)^3 + \sqrt[3]{-x} = -2x^3 - \sqrt[3]{x}$. Remember that $\sqrt[3]{-1} = -1$.

3. m is odd. No signs will change in $m(-x)$, except $-3x^3$ will become $3x^3$. Thus, the entire fraction (all one term) changes from negative to positive, and $m(-x)=-m(x)$.

4. b is neither even nor odd: $b(-x)=(-x)^7 + 5(-x)^3 - 17 = -x^7 - 5x^3 - 17 \neq b(x)$.

Because the –17 does not change signs like all the other terms, *b* is not odd; *b* is clearly not even, either.

5. *v* is neither even nor odd. You might think the function is even since the pair of points (–1,4) and (1,4) are present. However, (–2,6) would also have to be present to make the function even.

6. *r* is neither even nor odd since there is no *y*- or origin-symmetry.

7. The vast majority of *x*-symmetric graphs are not functions, as they fail the vertical line test. Note that the single input *x* results in two outputs, *y* and –*y*, which is forbidden in the happy land of functions.

8.

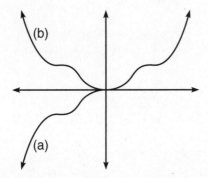

9. To test *h(x)*, you proceed as usual: $h(-x) = \dfrac{j(-x)}{k(-x)}$. Since *j* and *k* are odd, you know that $h(-x) = \dfrac{-j(x)}{-k(x)} = \dfrac{j(x)}{k(x)} = h(x)$. Therefore, *h* is even, as the negative signs will cancel out, making *h(x)* *y*-symmetric by definition.

INVERSE FUNCTIONS

Everything has its eventual undoing—for every high there is a low, and for every up, there is a down. The specific element that actually plays the role of spoiler depends on the situation, of course; Superman has kryptonite and the Wicked Witch of the West has the bucket of water. Functions are no exception to this rule—every function, *f(x)*, has its inverse function, $f^{-1}(x)$—a rule that completely "undoes" the function, effectively destroying it, leaving behind only a smoldering, wide-eyed variable. This is a little too much imagery for most mathematicians, who choose, instead, to write $f(f^{-1}(x)) = f^{-1}(f(x)) = x$. Translated, this means that any function of *x*, when plugged into its inverse function (or vice versa), leaves behind only *x*.

Any functions that satisfy this condition are inverse functions. Some inverse functions are pretty obvious. For example, the inverse of *g(x)* = *x* + 3 is a piece of cake: $g^{-1}(x) = x - 3$. (The opposite of adding 3 to a number is subtracting 3). However, it's not always so obvious to determine if functions are inverses of each other.

NOTE
Plugging one function into another is called *composing* the function with another. It it usually written as *f(g(x))*, where the function *g* is being plugged into *f*. The notation *(f ∘ g)(x)* is also used; it means the same thing.

Example 9: Prove that $h(x) = 2x^2 - 3, x \geq 0$ and $j(x) = \sqrt{\dfrac{x+3}{2}}$ are inverse functions.

Solution: If the functions fit the above rule, they are inverse functions, but if the glove don't fit, we must acquit.

$$h(j(x)) = j(h(x)) = x$$

$$h\left(\sqrt{\frac{x+3}{2}}\right) = j(2x^2 - 3) = x$$

$$2\left(\sqrt{\frac{x+3}{2}}\right)^2 - 3 = \sqrt{\frac{\left(2x^2 - 3\right) + 3}{2}} = x$$

$$2 \cdot \frac{x+3}{2} - 3 = \sqrt{\frac{2x^2}{2}} = x$$

$$x + 3 - 3 = \sqrt{x^2} = x$$

$$x = x = x$$

Therefore, h and j are inverses.

You may be wondering how functions and their inverses actually "undo" each other, as we have said. The answer is that, in essence, inverse functions reverse the domain and range of the "host" function, making the inputs the outputs and the outputs the inputs. Consider the simple function $r:\{(1,5),(2,-3),\left(7,\frac{1}{2}\right),(9,2\pi)\}$. Clearly, the domain is $\{1,2,7,9\}$ and the range is $\{5,-3,\frac{1}{2},2\pi\}$. It is quite easy to construct r^{-1}—simply switch the domain and range: $r^{-1}:\{(5,1),(-3,2),\left(\frac{1}{2},7\right),(2\pi,9)\}$. Now, the defining property of inverse functions becomes more obvious. Consider $r^{-1}(r(x))$ for the ordered pair $(1,5)$. Because $r(1) = 5$, we write $r^{-1}(r(1)) = r^{-1}(5) = 1$. You end up with a 1, just as you started, and the functions have, in essence, canceled one another out. The same property is at work when you commute from home to school in the morning and then back home in the afternoon. The rides to and from school are inversely related, and you end up back where you started.

Because the purpose of inverse functions is to reverse the domain and range of a function—changing the order of the function's (x,y) pair— the graphs of inverse functions have a special property as well. The graphs of inverse functions are reflections of one another across the line $y = x$. Below, the functions h and j from Example 9 are graphed. Notice how they are symmetric to each other across the $y = x$ line.

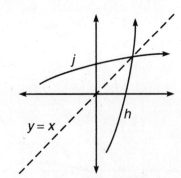

h and *j* are symmetric about *y* = *x*.

It's worth discussing why you had to limit the domain of *h* in order to even qualify *h* for an inverse function. The graph of $h(x) = 2x^2 - 3$, with no restriction, has two roots. Easily found, the two roots are $\left(\sqrt{\dfrac{3}{2}}, 0\right)$ and $\left(-\sqrt{\dfrac{3}{2}}, 0\right)$. Because inverse functions reverse ordered pairs, h^{-1} must include the points $\left(0, \sqrt{\dfrac{3}{2}}\right)$ and $\left(0, -\sqrt{\dfrac{3}{2}}\right)$. If this were allowed, h^{-1} would have two outputs for the domain element 0 and, therefore, could not be a function.

The visual test used to ensure that a function has an inverse is called the *horizontal line test*, and it works much like the vertical line test. If any horizontal line drawn through a function graph intersects the graph at more than one place, the function cannot have an inverse. (In the function *h* above, because the parabola has two roots, the horizontal line *y* = 0 intersects the parabola in two places—one of many such lines that ensures *h* fails the test.) If a function passes the horizontal line test, it is said to be *one-to-one*. In other words, for every input there is one output and vice versa.

It still remains to actually find an inverse function, but the process is rather simple and is directly based on the properties dicussed above.

 Example 10: Find the inverse function of $f(x) = \sqrt[4]{2x - 7}$.

Solution: Because inverse functions, in effect, switch the *x* and *y* variables, do so in the function, and solve the resulting equation for *y*. This is the inverse function.

$$y = \sqrt[4]{2x - 7}$$

$$x = \sqrt[4]{2y - 7}$$

$$x^4 = 2y - 7$$

$$\frac{x^4 + 7}{2} = y = f^{-1}(x)$$

NOTE

By restricting the graph of *h* to $x \geq 0$, you are removing the left half of the parabola. The resulting graph only increases from left to right, as the portion that decreased was included on $x < 0$. Because the restricted graph moves only in one direction, it is termed *monotonic*. Many monotonic graphs have inverses because of their unidirectional nature.

However, you must restrict $f^{-1}(x)$ to $x \geq 0$ to make the graph monotonic and ensure the existence of an inverse.

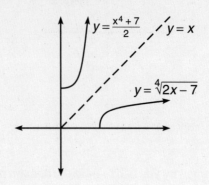

PROBLEM SET

You may use a graphing calculator on all of these problems.

1. If $f(x) = x^2 + 3x + 1$ and $g(x) = \sqrt{x-2}$, find

    (a) $(f \circ g)(x)$

    (b) $g(g(x))$

    (c) $g^{-1}(f(x))$

    (d) $g(f(4))$

Use the chart below for problems 2 and 3.

x	$r(x)$	$s(x)$
−3	1	2
−2	0	4
−1	2	6
0	5	−1
1	3	1
2	−1	−3
3	−3	4

2. If $r(x)$ and $s(x)$ are functions, as defined above, evaluate

    (a) $r(s(2))$

    (b) $s(r^{-1}(0))$

    (c) $r^{-1}(r^{-1}(s(1)))$

3. Why does $s^{-1}(x)$ not exist?

4. Find the inverse functions of each (if possible):

    (a) $p(x) = 2x^3 - 1$

    (b) $y = [[x]]$

5. If $h(x) = x^5 + 3x - 2$, find $h^{-1}(4)$.

6. Using the definition of one-to-one functions, explain why function m, as defined in the function map below, has no inverse.

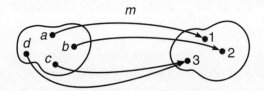

Solutions

1. (a) $(f \circ g)(x) = f(g(x)) = f(\sqrt{x-2}) = (\sqrt{x-2})^2 + 3(\sqrt{x-2}) + 1$
   $= x + 3(\sqrt{x-2}) - 1$.

   (b) $g(g(x)) = g(\sqrt{x-2}) = \sqrt{\sqrt{x-2} - 2}$.

   (c) First, find $g^{-1}(x)$. Using the same method as Example 10, $g^{-1}(x) = x^2 + 2$. Now, $g^{-1}(x^2 + 3x + 1) = (x^2 + 3x + 1)^2 + 2 = x^4 + 6x^3 + 11x^2 + 6x + 3$.

   (d) Clearly, $f(4) = 29$ (by substitution), so $g(f(4)) = g(29) = \sqrt{27}$.

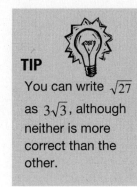

TIP

You can write $\sqrt{27}$ as $3\sqrt{3}$, although neither is more correct than the other.

2. This problem is simplified by finding $r^{-1}(x)$ (by reversing the ordered pair of r); $r^{-1}:\{(1,-3),(0,-2),(2,-1),(5,0),(3,1),(-1,2),(-3,3)\}$.

   (a) $r(s(2)) = r(-3) = 1$.

   (b) $s(r^{-1}(0)) = s(-2) = 4$.

   (c) $r^{-1}(r^{-1}(s(1))) = r^{-1}(r^{-1}(1)) = r^{-1}(-3) = 3$.

3. Function s has no inverse because it is not one-to-one. Notice that both $x = -2$ and $x = 3$ result in the same output of 4. In a one-to-one function, every output must have only one input that maps to it.

4. (a) Reverse the x and y to get $x = 2y^3 - 1$, and solve for y; $p^{-1}(x) = \sqrt[3]{\dfrac{x+1}{2}}$.

   (b) There is no inverse function for $y = [[x]]$ since the greatest integer function fails the horizontal line test. Because the graph is a collection of horizontal line segments, a horizontal line overlapping one of these segments will intersect an infinite amount of times.

5. This problem is difficult because you cannot find $h^{-1}(x)$—if you try to use the method of Example 10, you are unable to solve for y. Thus, a different approach is necessary. Note that if 4 is in the domain of h^{-1} (as evidenced by the fact that we are able to plug it into h^{-1}), then it must be in the *range* of h. Thus, h^{-1} contains some point $(4,a)$ and h contains $(a,4)$, where a is a real number. Substitute the point $(a,4)$ into h to get $a^5 + 3a - 2 = 4$, and use your graphing calculator to solve the equation. Therefore, $a = 1.193$ and h^{-1} contains the point $(4,1.193)$. Hence, $h^{-1}(4) = 1.193$. See the technology section at the end of this chapter to review solving equations with the graphing calculator.

6. Notice that $m(c) = m(d) = 3$. Therefore, two inputs result in the same output, m is not one-to-one, and only one-to-one functions can have inverses. One-to-one functions map one domain element to one range element; visually, this means that only one road can lead away from each input and only one road can lead into each output.

TRANSFORMING FUNCTIONS: HANDS-ON ACTIVITY 2.1

The process of graphing functions becomes much easier once you can see how the presence of additional numbers affects a graph. You can move, reflect, and stretch a graph simply by tinkering with the constants in a function, as you will learn in this activity. You will find Hands-on Activities like this throughout the book. Sometimes, the best way to learn something isn't by reading but by doing. In such cases, we have foregone notes and substituted these laboratory activities. In order to complete this activity, you will need to be able to graph functions on a calculator.

TIP

In the Hands-on Activities in this book, exact graphs are not important. It is more important to get the general shape of the graph correct and pay more attention to the concepts behind the actvities instead.

1. Draw the graphs of $y = x^2$, $y = x^2 - 1$, and $y = x^2 + 2$ on the grids below. The first you should know by memory (as one of the nine important functions described earlier in the chapter). You should graph the others with your graphing calculator.

2. How do the numbers -1 and 2 seem to affect the graph of x^2?

_____

_____

3. Complete this conclusion: When graphing the function $f(x) + a$, the a value causes the graph of $f(x)$ to

_____

_____

4. Using the same process as you did for Question 1 above, graph the functions $y = x^3$, $y = (x - 3)^3$, and $y = (x + 1)^3$.

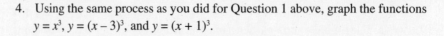
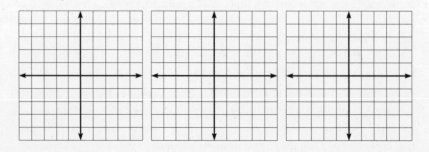

5. How do the numbers −3 and 1 seem to affect the graph of x^3?

    _____

    _____

6. Complete this conclusion: When graphing the function $f(x + a)$, the a value
    causes the graph of $f(x)$ to

    _____

    _____

7. Try to draw the graphs of $y = (x + 2)^2 − 3$ and $y = (x − 1)^3 + 1$ below, based on
    the conclusions you made in Questions 3 and 6. Check your result with your
    graphing calculator.

    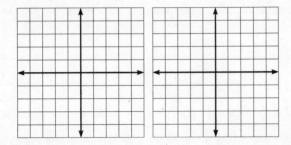

8. Graph the functions $y = |x + 1|$ and $y = −|x + 1|$ below. You should be able
    to graph the first, but use the graphing calculator for the second.

    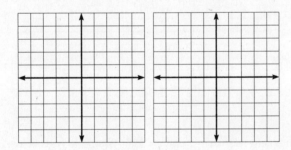

9. Complete this conclusion: Multiplying $f(x)$ by a negative, $−f(x)$, causes the
    graph to

    _____

    _____

TIP

You may need to
consult your
owner's manual if
you're not sure
how to graph
absolute values on
your calculator.
On a TI-83, the
command is *abs*(
and is found on
the Math→Number
menu.

10. Graph the functions $y = \sqrt{x}$ and $y = \sqrt{-x}$ below, again using your graphing calculator for the second graph.

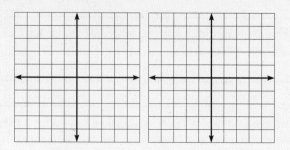

11. Complete this conclusion: Substituting $-x$ into the function $f(x)$ causes its graph to

_____

_____

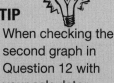

TIP
When checking the second graph in Question 12 with your calculator, make sure you enter the function as $\dfrac{-1}{x-1}$. If $x - 1$ is not in parentheses, the calculator interprets the equation as $\dfrac{1}{x} - 1$.

12. Try to draw the graphs of $y = -|x + 2| + 3$ and $y = -\dfrac{1}{x-1} - 2$ based on all of your conclusions so far. Once again, check your results with your graphing calculator.

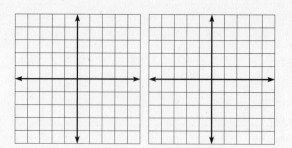

13. Graph the functions $y = x^2 - 2$ and $y = |x^2 - 2|$ on the grids below, again using your calculator only for the second graph, if possible.

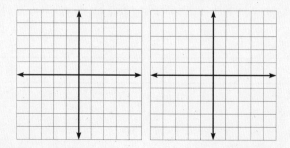

14. Complete this conclusion: The graphs of $f(x)$ and $|f(x)|$ differ in that $|f(x)|$

_____

_____

15. Graph the functions $y = \sqrt{x}$ and $y = \sqrt{|x|}$, using your calculator for the second only.

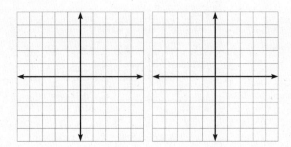

16. Complete this conclusion: The graph of $f(|x|)$ fundamentally changes the graph of $f(x)$ by

_____

_____

SELECTED SOLUTIONS TO HANDS-ON ACTIVITY 2.1

2. The graph is moved down one and up two, respectively.

3. …move up if a is positive, and move down if a is negative.

5. The graph is moved right 3 and left 1, respectively.

6. …move left if a is positive, and move right if a is negative (although this might be the opposite of what you expected).

9. Flip upside down, or (more mathematically) reflect across the x-axis.

11. Reflect across the y-axis.

14. …has no negative range. Visually, any negative portion of the graph is "flipped" so it lies above the x-axis.

16. …replacing all the values of $x < 0$ with a mirror image of $x > 0$. The graphs $f(x)$ and $f(|x|)$ are always y-symmetric.

PROBLEM SET

Do not use a graphing calculator on these problems.

1. Graph the following:

(a) $y = (x - \pi)^2 + 2$

(b) $y = \dfrac{1}{x+1} - 1$

(c) $y = |x-1| + 3$

(d) $y = -(|x|)^3$

(e) $y = \left|\sqrt{-(x-3)} - 2\right|$

2. Explain mathematically why the graphs of $y = (-x)^3$ and $y = -(x^3)$ are identical.

3. Given $f(x)$ as defined in the below graph, graph the indicated translations:

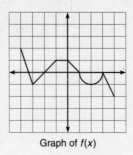

Graph of $f(x)$

(a) $f(x) - 2$

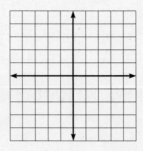

(b) $f(x + 1)$

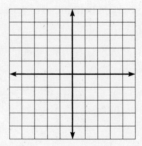

(c) $-f(x)$

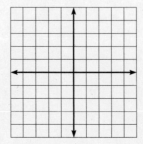

(d) $f(-x)$

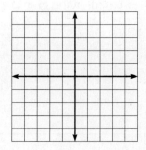

(e) $|f(x)|$

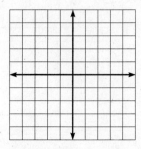

(f) $f(|x|)$

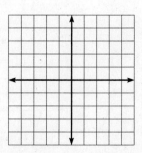

(g) $g^{-1}(x)$, if $g(x) = f(x)$ when $-3 \le x \le -1$

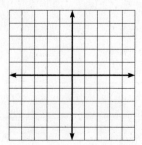

4. Why does $y = \sqrt{-x}$ have a graph if you cannot find a real square root of a negative number? (answer based on the graph)

SOLUTIONS

1.

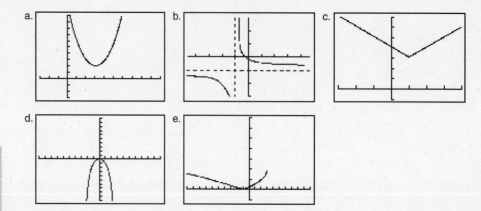

a.   b.   c.

d.   e.

2. The graphs are identical because $y = x^3$ is odd; therefore $(-x)^3 = -x^3$.

3.

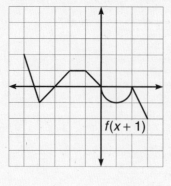

$f(x + 1)$

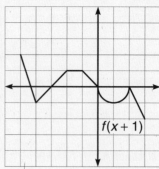

$f(x + 1)$

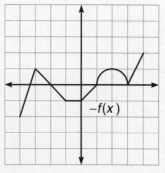

$-f(x)$

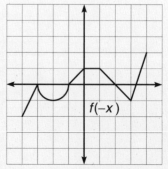

$f(-x)$

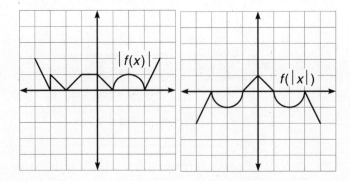

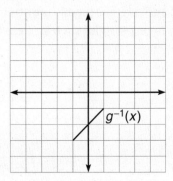

Translate each of the graphs' defining points, as indicated in the graph. To find the inverse function in 3(g), simply reverse the order of the coordinate pair of $g(x)$; note that g is made by the points $(-3,-1)$, $(-2,0)$, $(-1,1)$ and the line segments connecting them.

4. You graphed this function in Activity 2.1, Question 10. Notice that the domain of the resulting function is $(-\infty,0]$. A negative input will eliminate the problem of a negative radicand, as the product of two negative numbers is positive.

TRIGONOMETRY

It is extremely important to make sure you are very proficient in trigonometry, as a great deal of calculus uses trigonometric functions and identites. We have whittled the content down to only the most important topics. Familiarize yourself with all these basics. Learn them. Love them. Take them out to dinner, and pick up the tab without grimacing.

The study of triangles and angles is fundamental to trigonometry. When angles are drawn in a coordinate plane, their vertices sit atop the origin, and the angles begin on the positive x-axis. A positive angle proceeds counterclockwise from the axis, whereas a negative angle winds clockwise. Both angles in the below figure measure $\frac{\pi}{3}$ radians (60°), although they terminate in different quadrants because of their signs.

NOTE
Radicand is the fancy-pants term for "stuff beneath the radical sign."

NOTE
In order to convert from radians to degrees, multiply the angle by $\frac{180}{\pi}$.
To convert from degrees to radians, multiply the angle by $\frac{\pi}{180}$.

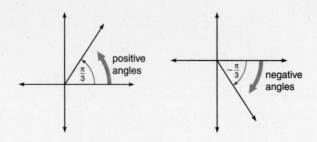

On the other hand, some angles, although unequal, look the same. In the figure below, the angles A = $\frac{5\pi}{4}$ and B = $\frac{13\pi}{4}$ terminate in the exact same spot, but B has traveled an extra time around the origin, completing one full rotation before coming to rest. Angles such as these are called *coterminal angles* (since they terminate at the same ray).

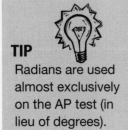

TIP
Radians are used almost exclusively on the AP test (in lieu of degrees).

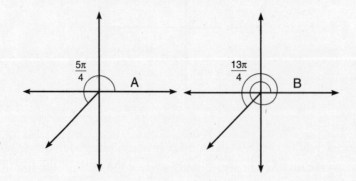

Another major topic from yonder days of precalculus is the unit circle—the dreaded circle with center at the origin and radius one —that defines the values of many common sines and cosines. It is essential to memorize the unit circle in its entirety.

NOTE
In order to find coterminal angles, add or subtract 2π (or 360°) to or from the given angle. This adds a full "loop" to the angle but will not affect where the angle ends.

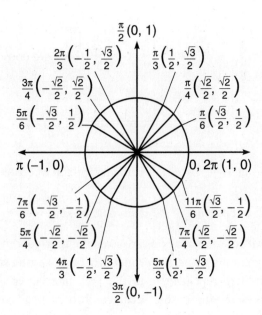

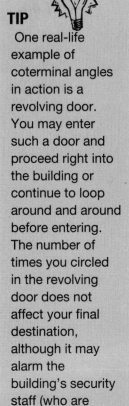

The points next to each of the angles are the cosine and sine of that angle, respectively. The remaining trigonometric functions are directly based on sine and cosine:

$$\tan x = \frac{\sin x}{\cos x}$$

$$\cot x = \frac{\cos x}{\sin x}$$

$$\sec x = \frac{1}{\cos x}$$

$$\csc x = \frac{1}{\sin x}$$

 Example 11: Evaluate all six trigonometric functions if θ

$$= \frac{11\pi}{6}.$$

Solution: From the unit circle, $\cos \frac{11\pi}{6} = \frac{\sqrt{3}}{2}$ and $\sin \frac{11\pi}{6} = \frac{-1}{2}$. From there, we use the definitions of the other functions:

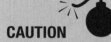

CAUTION

Some students record unit circle values in their calculator's memory instead of memorizing them. Remember that approximately 50 percent of the AP exam is taken without a calculator, and those precious notes will be inaccessible. Make sure to memorize the unit circle!

TIP

You are not required to rationalize fractions on the AP test (like when calculating secant and tangent in Example 11), and it's generally a good idea not to waste time doing it. However, a non-rationalized fraction may not always be listed among multiple-choice options, whereas a rationalized answer could be. Make sure you can express your answer either way.

$$\tan \frac{11\pi}{6} = \frac{\sin \frac{11\pi}{6}}{\cos \frac{11\pi}{6}} = \frac{-\frac{1}{2}}{\frac{\sqrt{3}}{2}} = -\frac{1}{\sqrt{3}} \text{ or } -\frac{\sqrt{3}}{3} \text{ (both are correct)}$$

$\cot \frac{11\pi}{6} = \frac{-3}{\sqrt{3}}$, as cotangent is the reciprocal of tangent

$\sec \frac{11\pi}{6} = \frac{2}{\sqrt{3}}$ or $\frac{2\sqrt{3}}{3}$, as secant is the reciprocal of cosine

$\csc \frac{11\pi}{6} = -2$, as cosecant is the reciprocal of sine

The reason that coterminal angles are so useful in trigonometry is that all trigonometric functions are *periodic functions*, because after some period of time, the graphs will repeat themselves. That interval of time is called (surprise!) the *period*. Sine, cosine, secant, and cosecant all have a period of 2π, whereas tangent and cotangent have a period of π. Thus, tan

$\left(\frac{\pi}{4}\right) = \tan\left(\frac{5\pi}{4}\right)$, since there is an interval of $\frac{4\pi}{4}$ (π) between the inputs, and tangent has begun to repeat itself.

Perhaps the most important aspects of trigonometry you will use are the trigonometric identities. These are used to rewrite expressions and equations that are unsolvable in their current form. In other cases, through the substitution of an identity, an expression becomes much simpler. You'll need to memorize these formulas, too. Below are listed the most important trigonometric identites:

Pythagorean identities: $\cos^2 x + \sin^2 x = 1$ (Mamma Theorem)

$1 + \tan^2 x = \sec^2 x$ (Pappa Theorem)

$1 + \cot^2 x = \csc^2 x$ (Baby Theorem)

Even and odd identities: $\sin(-x) = -\sin x$

$\csc(-x) = -\csc x$

$\tan(-x) = -\tan x$

$\cot(-x) = -\cot x$

$\cos(-x) = \cos x$

$\sec(-x) = \sec x$

Double-angle formulas: $\sin 2x = 2\sin x\cos x$

$$\cos 2x = \cos^2 x - \sin^2 x$$
$$= 2\cos^2 x - 1$$
$$= 1 - 2\sin^2 x$$

$\tan 2x = \frac{2\tan x}{1-\tan^2 x}$

Power-reducing formulas:

$$\cos^2 x = \frac{1 + \cos 2x}{2}$$

$$\sin^2 x = \frac{1 - \cos 2x}{2}$$

$$\tan^2 x = \frac{\sin^2 x}{\cos^2 x} = \frac{1 - \cos 2x}{1 + \cos 2x}$$

Sum and difference formulas:

$$\sin (x \pm y) = \sin x \cos y \pm \cos x \sin y$$

$$\cos (x \pm y) = \cos x \cos y \mp \sin x \sin y$$

$$\tan (x \pm y) = \frac{\tan x \pm \tan y}{1 \mp \tan x \tan y}$$

 Example 12: Simplify the expression $\frac{\sec^2 x}{\tan x} - \tan x$.

Solution: The first order of business is common denominators so we can add the terms—multiply the tan x term by $\frac{\tan x}{\tan x}$ to do so. You then use a result of the Big Pappa Theorem to simplify.

$$\frac{\sec^2 x}{\tan x} - \frac{\tan^2 x}{\tan x}$$

$$\frac{\sec^2 x - \tan^2 x}{\tan x}$$

$$\frac{1}{\tan x}$$

$$\cot x$$

 Example 13: Rewrite the expression $\sqrt{1 - \cos x}$ in terms of sine.

Solution: If you multiply the second power-reducing formula by 2, the result is $1 - \cos 2x = 2\sin^2 x$. Therefore, $1 - \cos x = 2\sin^2 \frac{x}{2}$. This is still true, as the cosine angle is still twice as large as the sine angle. Therefore, we can substitute $2\sin^2 \frac{x}{2}$ for $1 - \cos x$:

NOTE

Because the Pappa Theorem says $1 + \tan^2 x = \sec^2 x$, subtract $\tan^2 x$ from both sides to get $1 = \sec^2 x - \tan^2 x$. This substitution is made in the second step.

TIP

It helps to read "arc" as "where is...". For example, the function arcsin 1 is asking "Where is sine equal to 1?". The answer is $\frac{\pi}{2}$.

TIP

We will refer to the restricted trig ranges as "bubbles" for convenience and, well yes, fun. For example, $\arctan 1 \neq \frac{5\pi}{4}$ because $\frac{5\pi}{4}$ is not in the arctan bubble (i.e., the angle does not fall in the intervals given by The Restricted Ranges diagram).

$$\sqrt{2 \sin^2 \frac{x}{2}}$$

$$\sqrt{2} \sin \frac{x}{2}$$

Another major important trigonometric topic is inverse functions. Inverse trigonometric functions can be written one of two ways. For example, the inverse of cosine can be written $\cos^{-1} x$ or arccos x. Because the first format looks like $(\cos x)^{-1}$ (which equals sec x), the latter format (with the *arc-* prefix) is preferred by many. Both mean the same thing. The trickiest part of inverse trig functions is knowing what answer to give.

If asked to evaluate arccos 0, you might answer that cosine is equal to 0 when $x = \frac{\pi}{2}$ or $\frac{3\pi}{2}$. It is true that $\cos \frac{\pi}{2} = \cos \frac{3\pi}{2} = 0$. However, this means that the function $y = \arccos x$ has two outputs when $x = 0$, so arccos x is not a function! This is remedied by restricting the ranges of the inverse trig functions, as shown below.

The Restricted Ranges (The Bubbles)

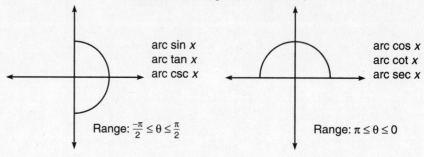

Any answer for arcsin x, arctan x, or arccsc x will have to fall within the interval $[-\frac{\pi}{2}, \frac{\pi}{2}]$, whereas outputs for arccos x, arccot x, and arcsec x must fall in the interval $[0, \pi]$. For our previous example of arccos 0, the correct answer is $\frac{\pi}{2}$, as $\frac{3\pi}{2}$ does not fall within the correct interval (or "bubble"— see margin note).

 Example 14: Evaluate the following expressions without the use of a calculator:

(a) $\arcsin \dfrac{\sqrt{3}}{2}$

(b) $\operatorname{arcsec} -2$

(c) $\arctan -\dfrac{\sqrt{3}}{3}$

Solution: (a) The sine function has the value $\frac{\sqrt{3}}{2}$ when $x = \frac{\pi}{3}$ and $\frac{5\pi}{3}$; only $\frac{\pi}{3}$ falls in the arccos bubble.

(b) If an angle has a secant value of –2, it must have a cosine value of $-\frac{1}{2}$, as the functions are reciprocals. Cosine takes this value when $x = \frac{2\pi}{3}$ and $\frac{4\pi}{3}$. Only the first falls in the arccos bubble, so the answer is $\frac{2\pi}{3}$.

(c) Notice that $-\frac{\sqrt{3}}{3}$ is the same as $-\frac{1}{\sqrt{3}}$ (the latter is just not rationalized). This is the same as $-\frac{\frac{1}{2}}{\frac{\sqrt{3}}{2}}$. Tangent is negative in the second and fourth quadrants, but only the fourth quadrant is in the arctan bubble. Thus, only angle $\frac{11\pi}{6}$ has the appropriate values (recall that tangent is equal to sine divided by cosine). However, to graph $\frac{11\pi}{6}$, you have to pass outside the arcsin bubble, since $\frac{11\pi}{6} > \frac{\pi}{2}$, the largest value allowed for arctan, and passing outside the bubble is not allowed. Therefore, the answer is the coterminal angle $-\frac{\pi}{6}$, which ends in the same spot.

Finally, as promised earlier in the chapter, here are the final six functions you need to know by heart. They are—big shocker—the trigonometric functions:

The Trigonometric Functions

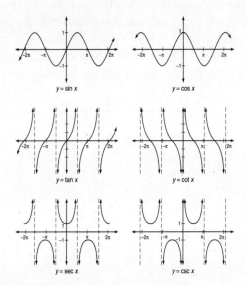

$y = \sin x$ $y = \cos x$

$y = \tan x$ $y = \cot x$

$y = \sec x$ $y = \csc x$

TIP

When giving solutions to inverse trigonometric functions, always remember the bubble. The bubble is a happy place. Outside the bubble, the people are not nice, and the dogs bite. Also, it's hard to find stylish shoes that are still comfortable outside the bubble.

PROBLEM SET

Do not use a graphing calculator on these problems.

1. If $\sin \beta = -\frac{2}{5}$ and $\pi < \beta < \frac{3\pi}{2}$, find the values of the other five trigonometric functions for β.

2. Use coterminal angles to evaluate each of the following:

    (a) $\sin(11\pi)$

    (b) $\cos \dfrac{21\pi}{4}$

    (c) $\tan \dfrac{13\pi}{3}$

3. Evaluate $\cos(\arcsin(-2x))$.

4. Simplify: (a) $\tan^4 x + 2\tan^2 x + 1$

    (b) $\dfrac{\cos x}{\sec x} + \dfrac{\sin x}{\csc x}$

5. Verify that $\csc^2 x = \dfrac{1}{\cos^2 x - \cos 2x}$.

6. Solve this equation algebraically: $\cos^2 x - \cos 2x + \sin x = -\sin^2 x$, and give answers on the interval $[0, 2\pi)$.

7. Graph $y = -\dfrac{3}{2}\sin(2x) + 1$.

SOLUTIONS

1. The given interval for β makes it clear that the angle is in the third quadrant (sine is also negative in the fourth quadrant, so this information is necessary). This allows you to draw a reference triangle for β, knowing that sine is the ratio of the opposite side to the hypotenuse in a right triangle. Be careful to make the legs of the triangle negative, as both x and y are negative in the third quadrant. Using the Pythagorean Theorem, the remaining (adjacent) side measures $\sqrt{21}$. From the diagram, you can easily find the other 5 trigonometric ratios.

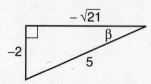

$\cos \beta = -\dfrac{\sqrt{21}}{5}$, $\tan \beta = \dfrac{2}{\sqrt{21}}$, $\sec \beta = -\dfrac{5}{\sqrt{21}}$, $\csc \beta = -\dfrac{5}{2}$, $\cot \beta = \dfrac{\sqrt{21}}{2}$.

2. (a) $\sin(11\pi) = \sin(\pi) = 0$

    (b) $\cos\left(\dfrac{21\pi}{4}\right) = \cos\left(\dfrac{5\pi}{4}\right) = \dfrac{-\sqrt{2}}{2}$

(c) $\quad \tan\left(\dfrac{13\pi}{3}\right)=\tan\left(\dfrac{4\pi}{3}\right)=\dfrac{\sin\dfrac{4\pi}{3}}{\cos\dfrac{4\pi}{3}}=\dfrac{\dfrac{\sqrt{3}}{2}}{\dfrac{1}{2}}=\sqrt{3}$

3. This problem is similar to Number 1, but the additional restriction we had there is replaced by the bubble. You know that sine is negative, but since this is the arcsin funcion, only the fourth quadrant is in the bubble. Thus, your reference triangle is drawn as if in the fourth quadrant.

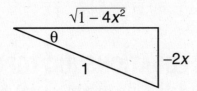

From the diagram, $\cos\theta = \dfrac{\text{adjacent}}{\text{hypotenuse}} = \sqrt{1-4x^2}$

4. (a) The quadratic expression can be factored to get $(\tan^2 x + 1)(\tan^2 x + 1)$, or $(\tan^2 x + 1)^2$. By the Pappa Theorem, this equals $(\sec^2 x)^2 = \sec^4 x$.

(b) Substituting the values of $\sec x$ and $\csc x$, you get $\dfrac{\cos x}{\dfrac{1}{\cos x}}+\dfrac{\sin x}{\dfrac{1}{\sin x}}=\cos^2 x$

+ $\sin^2 x$, which equals 1 by the Mamma Theorem.

5. $\quad \dfrac{1}{\sin^2 x}=\dfrac{1}{\cos^2 x-\cos 2x}$

$$\cos^2 x - \cos 2x = \sin^2 x$$
$$\cos^2 x - \sin^2 x - \cos 2x = 0$$
$$\cos^2 x - \sin^2 x = \cos 2x$$

6. Note that this question specifies the interval $[0,2\pi)$. Thus, we do not ignore answers outside of the arcsin bubbles in the final step but give all the answers on the unit circle. Note also that we made the problem easier by substituting 1 for $\cos^2 x + \sin^2 x$ in the second step:

$$\cos^2 x + \sin^2 x - \cos 2x + \sin x = 0$$
$$1 - (1 - 2\sin^2 x) + \sin x = 0$$
$$2\sin^2 x + \sin x = 0$$
$$\sin x(2\sin x + 1) = 0$$
$$x = 0, \pi, \frac{7\pi}{6}, \frac{11\pi}{6}.$$

7. The coefficient of the x affects the graph by stretching or shrinking the period. The number—here 2—explains how many full graphs of sine will fit where one used to. Since the period of sine is 2π, now two full graphs will occupy the period instead of one. Had the coefficient been 3, three graphs would squeeze into the same space. The $\frac{3}{2}$ gives the amplitude of the sine wave. The rest of the translations work the same as they did earlier in the chapter.

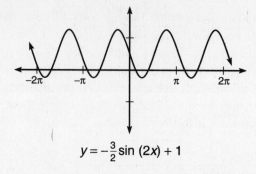

$$y = -\frac{3}{2}\sin(2x) + 1$$

PARAMETRIC EQUATIONS (BC TOPIC ONLY)

There are numerous ways to express relations, and the calculus AB test focuses almost exclusively on rectangular representation. However, the BC test includes three others: parametric, polar, and vector. Luckily, there is a lot of overlap among the three, so come on back in off of that ledge, BCer—you have so much to live for!

Parametric equations use a *parameter* (a third variable) in their definition. For example, the parametric representation of a circle centered at the origin with radius one is given by: $x = \cos t$, $y = \sin t$. You get the ordered pair to draw the graph by substituting successive values of t into the expressions for x and y (the arrows give you a sense of direction in the graph but are not actually part of the graph).

NOTE

In this example, you should use t values on the interval $[0, 2\pi)$, since x and y involve trigonometric functions. The graph you get as a result is the unit circle! If you think about it, this makes a lot of sense.

t	$x = \cos t$	$y = \sin t$
0	1	0
$\frac{\pi}{4}$	$\frac{\sqrt{2}}{2}$	$\frac{\sqrt{2}}{2}$
$\frac{\pi}{2}$	0	1
$\frac{3\pi}{4}$	$-\frac{\sqrt{2}}{2}$	$\frac{\sqrt{2}}{2}$
π	-1	0
$\frac{5\pi}{4}$	$-\frac{\sqrt{2}}{2}$	$-\frac{\sqrt{2}}{2}$
$\frac{3\pi}{4}$	0	-1
$\frac{7\pi}{4}$	$\frac{\sqrt{2}}{2}$	$-\frac{\sqrt{2}}{2}$
2π	1	0

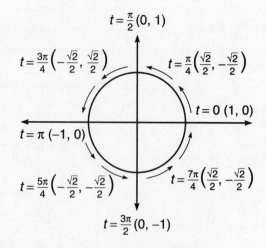

The graph tells you a lot more than the equivalent rectangular equation of $x^2 + y^2 = 1$. With this set of parameters, you can easily tell that the graph begins and ends at the point $(1,0)$ and proceeds in a counter-clockwise path. You can even get some notion of the speed traveled in the path. These are the major endearing qualities of parametric equations.

Your TI calculator can easily draw parametric equations. Press the Mode button and select Parametric mode. The "Y=" screen will now show "$x_{1T} =$" and "$y_{1T} =$". Type your parametric equations there. The "x,t,θ" button will now display a t, since you are in parametric mode. In order to adjust the values through which t will cycle, press Window, and set the maximum and minimum values of t.

 Example 15: Graph the parametric equation $x = 1 + 2t$, $y = 2 - t^2$ and convert it to rectangular form.

Solution: Using a table of values or the calculator, you get the graph of a parabola. If you got only half of the parabola, make sure you are substituting in negative values of t. A good range for t on the calculator is $-10 \leq t \leq 10$ if there aren't any trigonometric functions, although sometimes 10 is unnecessarily high. Better safe than sorry, though.

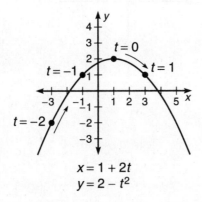

$$x = 1 + 2t$$
$$y = 2 - t^2$$

In order to convert to rectangular form, solve either x or y for t and plug into the other. The x expression seems easier, so it's better to start by solving it for t:

$$x - 1 = 2t$$

$$\frac{x-1}{2} = t$$

$$y = 2 - \left(\frac{x-1}{2}\right)^2$$

$$y = 2 - \frac{(x-1)^2}{4}$$

$$y = -\frac{1}{4}(x-1)^2 + 2$$

NOTE
If you graph the parametric equations $x = \cos(2t)$, $y = \sin(2t)$, the graph is the same as $x = \cos t$ and $y = \sin t$, but the circle is actually drawn twice! In fact, the circle is fully completed when $t = \pi$. There can be numerous ways to express a graph in parametric form, differing in direction, path, and the speed of the graph.

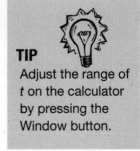

TIP
Adjust the range of t on the calculator by pressing the Window button.

This is a parabola with vertex (1,2) opening downward, as is verified in our graph.

 Example 16: Graph the parametric equation $x = \cos^2 t$, $y = \sin t$, and find its corresponding equation in rectangular form.

Solution: Again, graph using a table of values or a calculator, but make sure to reset your t values to $[0,2\pi)$. In order to convert to rectangular form, utilize the Mamma Theorem and substitute.

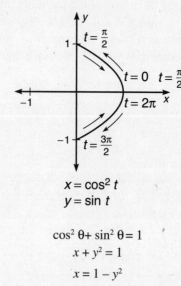

$$\cos^2 \theta + \sin^2 \theta = 1$$
$$x + y^2 = 1$$
$$x = 1 - y^2$$

Notice that the parametric graph does not include the entire parabola. Therefore, it is important that you restrict the rectangular equation so that its graph matches exactly. The final answer is $x = 1 - y^2$, $x \geq 0$.

PROBLEM SET

Do not use a graphing calculator on these problems.

1. Draw the graphs of the following parametric equations and rewrite each in rectangular form:

    (a) $x = 3\cos \theta$, $y = 2\sin \theta$

    (b) $x = t - 1$, $y = 2 - \dfrac{2}{t}$

2. Given the graphs below, draw the graph of the parametric equations $x = a(t)$, $y = b(t)$.

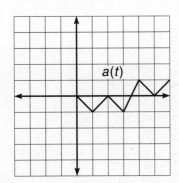
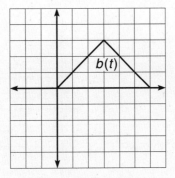

3. Find the parametric equations whose graph is an ellipse centered at the origin with horizontal major axis of length 8 and minor axis of length 4.

4. Based on your work on the above problems, name another benefit of parametric equations versus rectangular functions.

5. Create as many parametric representations of $y = ax + b$ as you can.

SOLUTIONS

1. (a) The graph is easy to find via chart or graphing calculator. In this instance, it is just as easy to find by converting the equation to rectangular form. Because the Mamma Theorem states that $\cos^2\theta + \sin^2\theta = 1$, you should solve the parametric equations for $\cos\theta$ and $\sin\theta$. If you do, the result is $\frac{x}{3} = \cos\theta$ and $\frac{y}{2} = \sin\theta$. By substitution into the Mamma Theorem, $\frac{x^2}{9} + \frac{y^2}{4} = 1$, which is standard form for an ellipse.

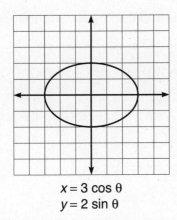

$$x = 3\cos\theta$$
$$y = 2\sin\theta$$

(b) Again, the challenge is the conversion to rectangular form. In this problem, it is easy to solve the x equation for t: $t = x + 1$. Substitute this into the y equation to get $y = 2 - \frac{2}{x+1}$. Simplify this to get the rectangular equation $y = \frac{2x}{x+1}$.

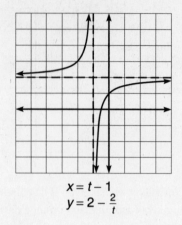

$$x = t - 1$$
$$y = 2 - \frac{2}{t}$$

2. Make an ordered pair to begin graphing. For example, consider $t = 2$. At this value of t, $a(2) = 0$ and $b(2) = 2$. Therefore, the graph will contain the point $(0,2)$. Use a similar process for the other values of t between 0 and 6.

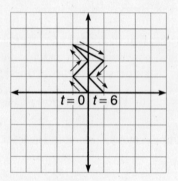

3. Your work in 1(a) makes this much easier. The ellipse described has equation $\frac{x^2}{16} + \frac{y^2}{4} = 1$. Because the Mamma Theorem also has two terms being summed to equal 1, you can do the following:

$$\cos^2 t = \frac{x^2}{16} \qquad \sin^2 t = \frac{y^4}{4} \text{ (because of Mamma)}$$

$$\cos t = \frac{x}{4} \qquad \sin t = \frac{y}{2}$$

$$x = 4\cos t \qquad y = 2\sin t$$

4. Parametric equations can easily describe non-functions, as best evidenced in Solution 2 above. That crazy-looking thing certainly does not pass the vertical line test.

5. The simplest representation of any function in parametric form is accomplished by setting $x = t$ and to make that substitution in the y equation: $y = at + b$. However, there are many ways to express that function. Notice that $x = t - b$ and $y = at - ab + b$ also results in the same graph. Simply pick anything for the x equation and adjust y accordingly.

POLAR EQUATIONS (BC TOPIC ONLY)

Polar equations are a very handy way to express complex graphs very simply. However, the polar system utilizes a completely different coordinate axis and graphing system. Once you warm up to polar coordinates, it's not too hard to bear them.

All polar points are given in the form (r, θ), where θ is an angle in standard position and r is a distance along the terminal ray of that angle. Therefore, in order to graph a polar coordinate, first draw the angle specified by θ, and then count r units along the terminal ray.

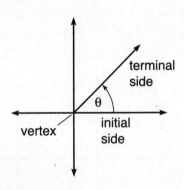

 Example 17: Graph the polar coordinates $A = (\frac{\pi}{2}, 3)$, $B = (-\frac{3\pi}{4}, 1)$, and $C = (\frac{5\pi}{6}, -2)$.

Solution:

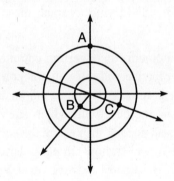

Point A is graphed on the terminal ray of $\frac{\pi}{2}$ (or 90°), the positive y-axis. Count 3 units along this ray from the origin to place A. B is one unit away from the origin at $\pi = -\frac{3\pi}{4}$. C is just a little trickier. After drawing the angle $\pi = \frac{5\pi}{6}$, you count two units *backward* from the origin, as r is given as –2. Therefore, C ends up in the fourth quadrant rather than the second, as you might have hypothesized.

NOTE
Remember that an angle is said to be in *standard position* if its vertex lies on the origin and its initial (or beginning) side coincides with the positive *x*-axis. For example, the angle $\theta = \frac{\pi}{4}$ is in standard position at left:

NOTE
The origin is also referred to as the *pole* when graphing—hence the name of polar equations.

TIP

Choose polar mode by pressing the "Mode" button on your calculator. The "Y=" screen now functions as as the "r=" screen, and pressing the x,t,θ button will display θ.

Graphing polar equations consists of nothing more than plotting a series of polar points (as you just did) and connecting the resulting dots. Just as was the case for parametric graphs, a table of values and the calculator are your main graphing tools.

 Example 18: Graph $r = 4\sin 2\theta$ without weeping.

Solution: To solve this, we will allow θ to take on values from 0 to 2π and find the corresponding r values. The table below gives the major angles of the first quadrant, and the result, r, when those angles, θ, are plugged into $r = 4\sin 2\theta$. If you continue this process for the other three quadrants, you will obtain a similar shape.

θ	$r = 4\sin\theta$	point on graph, (r, θ)
0	0	$(0, 0)$
$\frac{\pi}{6}$	$\frac{4\sqrt{3}}{2} \approx 3.464$	$\left(3.464, \frac{\pi}{6}\right)$
$\frac{\pi}{4}$	4	$\left(4, \frac{\pi}{4}\right)$
$\frac{\pi}{3}$	$\frac{4\sqrt{3}}{2} \approx 3.464$	$\left(3.464, \frac{\pi}{6}\right)$
$\frac{\pi}{2}$	0	$\left(0, \frac{\pi}{2}\right)$

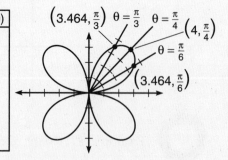

NOTE

The graph of $r = 4\sin 2\theta$ is called a *rose curve*, since it sort of looks like a flower. More common polar graphs are given in the exercises.

It is important that you know how polar graphs are created, but it is expected that you will typically graph them on your calculator. The AP test is more focused on applying different skills to polar and parametric equations, rather than focusing on simply graphing them. So, don't spend too much time honing your skills and trying to graph these quickly—make use of your technology.

Although you practiced converting parametric equations to rectangular form, no such practice follows suit for polar equations. As you saw in Example 18, the graphs can be quite complicated, and the major strength of polar graphing is the ease with which such complicated graphs can be formed. There's no need to strong-arm the equations into rectangular form. It is interesting, however, how simply polar coordinates can be converted into rectangular coordinates and vice versa. Well, at least it is interesting to me. A little.

 Example 19: Convert the polar coordinate $(-2, \frac{5\pi}{4})$ into a rectangular coordinate.

Solution: Any polar coordinate (r,θ) can be transformed into the corresponding rectangular coordinate (x,y) with the formulas $x = r\cos\theta$ and $y = r\sin\theta$. Therefore, $(\frac{5\pi}{4}, -2)$ becomes the rectangular coordinate $\left(-2 \cdot -\frac{\sqrt{2}}{2}, -2 \cdot -\frac{\sqrt{2}}{2}\right) = \left(\sqrt{2}, \sqrt{2}\right)$. The diagram below gives a visual proof of this—the right triangle created by the polar coordinate and the axes is a 45-45-90 isosceles right triangle with hypotenuse 2. Basic geometry verifies that the legs have length $\sqrt{2}$.

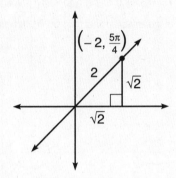

Example 20: What polar coordinate corresponds to the rectangular coordinate $(-3,-3)$?

Solution: Notice that $\tan\theta = \dfrac{\text{opposite}}{\text{adjacent}} = \dfrac{-3}{-3} = 1$. Tangent has a value of 1 at $\theta = \frac{\pi}{4}$ and $\frac{5\pi}{4}$, but the correct answer is $\frac{5\pi}{4}$ since you are working in the third quadrant. All that remains, then, is to find r, which is very easy to do thanks to good old Pythagoreas (or Pith-a-GORE-us, as Regis Philbin recently pronounced it on nationwide television). By the Pythagorean Theorem, $(-3)^2 + (-3)^2 = r^2$, so $r = \sqrt{18} = 3\sqrt{2}$. Thus, the correct polar coordinate is $(3\sqrt{2}, \frac{5\pi}{4})$.

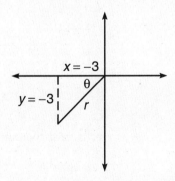

PROBLEM SET

You may use a graphing calculator on problems 4 and 5.

1. Graph the following polar equations without a table of values:

    (a) $r = 2.5$

    (b) $\theta = \dfrac{3\pi}{4}$

    (c) $r = \csc \theta$

2. Predict the graph of $r = \theta$, $0 \le \theta \le 2\pi$, and justify your prediction.

3. Below are examples of common polar curves. Match the graphs to the correct equations below.

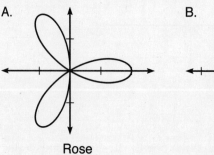

A.

Rose

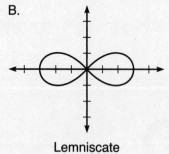

B.

Lemniscate

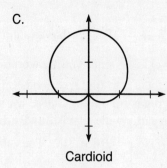

C.

Cardioid

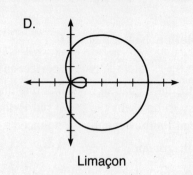

D.

Limaçon

$r = 1 + \sin \theta$

$r = 2 + 3\cos \theta$

$r^2 = 9\cos 2\theta$

$r = 2\cos 3\theta$

4. Graph $r = 2\sin \theta$.

5. Convert:

    (a) $(r, \theta) = (-2, \dfrac{11\pi}{6})$ to rectangular coordinates

    (b) $(x, y) = (-\sqrt{3}, 1)$ to polar coordinates.

SOLUTIONS

1. (a)

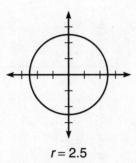

$$r = 2.5$$

The graph will have a radius of 2.5, regardless of the angle θ. Does this sound familiar? Yep…it's a circle with radius 2.5.

(b)

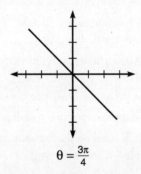

$$\theta = \frac{3\pi}{4}$$

In this case, only the angle comes into play. Along that angle, any radius is fair game. The result is a line that corresponds to the angle indicated. Notice that the line stretches into the fourth quadrant (because the negative radii are possible).

(c)

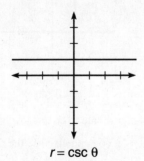

$$r = \csc \theta$$

If you rewrite $r = \csc \theta$ as $r = \frac{1}{\sin \theta}$ and cross-multiply, the result is $r \sin \theta = 1$. However, according to Example 19, we know that $y = r \sin \theta$. Therefore, $r \sin \theta = 1$ becomes the equation y = 1, a horizontal line.

2. Clearly, you begin the graph at the pole, as $r = \theta = 0$. As θ increases (moving counterclockwise), your radius (being equal) increases also. Therefore, the further you rotate, the longer the radius, forming a spiral.

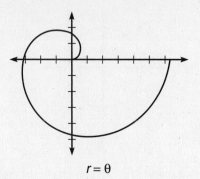

$r = \theta$

3. These problems are simply a study in testing points. Once you choose a significant number of angles and plug them into each of the formulas, you can begin to see which graphs could represent those equations.

    (A) $r = 2\cos 3\theta$: A *rose* curve has "petals," but not necessarily only three petals, as in this particular graph.

    (B) $r^2 = 9\cos 2\theta$: A *lemniscate* is in the shape of a figure *eight* that you can *skate*.

    (C) $r = 1 + \sin\theta$: A *cardioid* curve looks a little bit like a heart, hence the name. To me, it looks more like a tush print in a recliner.

    (D) $r = 2 + 3\cos\theta$: A *limaçon* may or may not have the puckered loop evident in this graph—without the loop, a *limaçon* looks like a less rounded cardioid.

4. A calculator or a table of values results in the graph below. Make sure you can do these by either method.

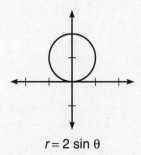

$r = 2\sin\theta$

5. (a) Use the polar coordinate to create a right triangle, from which it is clear that $x = -\sqrt{3}$ and $y = 1$. You can also use the formulas from Example 19. For instance $x = r\cos\theta = -2 \cdot \cos\dfrac{11\pi}{6} = -2 \cdot \dfrac{\sqrt{3}}{2} = -\sqrt{3}$. The y-coordinate will work just as easily.

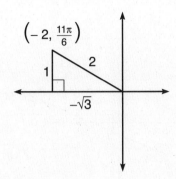

(b)   Once again, a graph is very helpful; from the diagram below,

$\tan\theta = \dfrac{1}{-\sqrt{3}} = -\dfrac{\frac{1}{2}}{\frac{\sqrt{3}}{2}}$. Therefore, $\theta = \dfrac{5\pi}{6}$. Use the Pythagorean Theorem to

find r: $1^2 + (-\sqrt{3})^2 = r^2$, so $r = \sqrt{4}$. One appropriate polar coordinate is $(2, \dfrac{5\pi}{6})$.

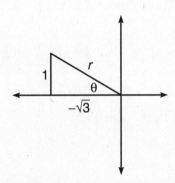

VECTORS AND VECTOR EQUATIONS (BC TOPIC ONLY)

Vector curves complete the triumvirate of BC graph representations. Although they have very peculiar and individual characteristics, they are very closely related (by marriage) to parametric equations, as you will see. A *vector* is, in essence, a line segment with direction. It is typically drawn as an arrow on the coordinate plane. The diagram below is the vector $\overrightarrow{AB}$, with initial point $A = (-3,-2)$ and terminal point $B = (4,1)$.

NOTE
Another correct polar point for 5(b) is $(-2, \dfrac{11\pi}{6})$—both coordinates represent the same point. While only one set of coordinates represents a rectangular point, an infinite number of coordinates can represent a polar point.

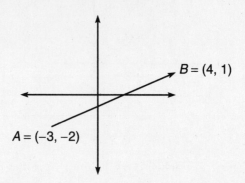

Although you can draw vectors anywhere in the coordinate plane, it's really handy when they begin at the origin—such a vector is said to be in *standard position*.

Example 21: Put vector $\overrightarrow{AB}$ (as defined above) in standard position.

Solution: To begin, calculate the slope of the line segment: $\frac{\Delta y}{\Delta x} = \frac{1-(-2)}{4-(-3)} = \frac{3}{7}$. Therefore, to get from point A to point B, you travel up 3 units and to the right 7 units.

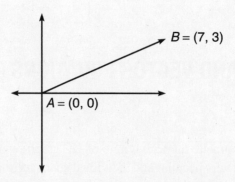

Because vector $\overrightarrow{AB}$ has the same length and direction wherever it is on the coordinate plane, moving it to standard position did not affect it at all. In fact, we can now write $\overrightarrow{AB}$ in *component form*: $\overrightarrow{AB} = <7,3>$. Graphing $<7,3>$ is almost equivalent to graphing the point $(7,3)$, except that $<7,3>$ will have a vector leading up to that point.

Thanks to Pythagoras (ya gotta love him), finding the length of a vector, denoted $\|\overrightarrow{AB}\|$, is very simple. For instance, $\|\overrightarrow{AB}\|$ from Example 21 is $\sqrt{58}$. To calculate the length, simply draw a right triangle with the vector as its hypotenuse and apply the Pythagorean Theorem.

NOTE

You really don't need a formula to put a vector in standard form—you can simply count the number of units you traveled along the graph.

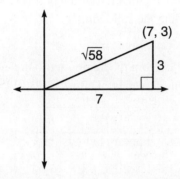

Perhaps the most common way a vector is represented is in *unit vector* form. A unit vector is defined as a vector whose length is 1. Clearly, **i** = <1,0> and **j** = <0,1> are unit vectors, and these vectors are the backbone of unit vector form.

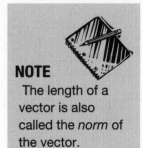

NOTE
The length of a vector is also called the *norm* of the vector.

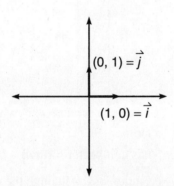

Let's put good old vector $\overrightarrow{AB}$ = <7,3> from previous examples in unit vector form. This vector is created by moving 7 units right and 3 units up from the origin. As demonstrated by the diagram below, this is the same as seven **i** vectors and three **j** vectors. Therefore, $\overrightarrow{AB}$ = 7**i** + 3**j** = 7**i** + 3**j**.

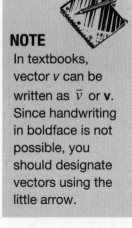

NOTE
In textbooks, vector *v* can be written as $\vec{v}$ or **v**. Since handwriting in boldface is not possible, you should designate vectors using the little arrow.

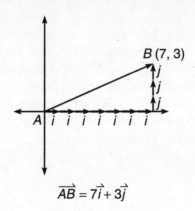

$\overrightarrow{AB} = 7\vec{i} + 3\vec{j}$

Example 22: If vector **v** has initial point (–2,6) and terminal point (1,–5), complete the following:

(a) Put **v** in component form.

In order to travel from the beginning to the end of the vector, you proceed right 3 units and down 11. Thus, **v** = <3,–11>.

(b) Find ||**v**||.

As pictured in the diagram below, ||**v**|| = $\sqrt{3^2 + (-11^2)} = \sqrt{130}$.

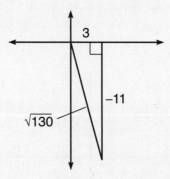

(c) Write **v** in unit vector form.

Any vector <*a,b*> has unit vector form *a***i** + *b***j** , so **v** = 3**i** – 11**j** .

Graphing vector curves is quite easy. Although the graphs are created by vectors, the graphs are not covered with arrows. Instead, you can graph vector equations *exactly the same* as parametric equations.

 Example 23: Graph the vector curve **r**(*t*) = (*t* + 1)**i** + *t*3**j**.

Solution: This vector function can be expressed parametrically. Remember that **i** refers to horizontal distance and **j** to vertical distance, exactly like *x* and *y*, respectively. Therefore, this vector function has exactly the same graph as the parametric equations $x = t + 1$, $y = t^3$, which you can graph using a calculator or table of values. Solving $x = t + 1$ for *t* and substituting *t* into the *y* equation gives you the rectangular form of this graph: $y = (x - 1)^3$.

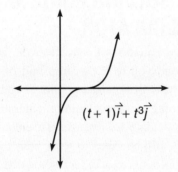

$(t+1)\vec{i}+t^3\vec{j}$

PROBLEM SET

You may use a graphing calculator on problem 4 only.

For numbers 1 through 3, $\mathbf{v} = <2,3>$, $\mathbf{w}$ has initial point $(3,-3)$ and terminal point $(-1,4)$, and $\mathbf{p} = 2\mathbf{v} - \mathbf{w}$.

1. Express $\mathbf{p}$ in component form.
2. Find $\|\mathbf{p}\|$.
3. What is the unit vector form of $\mathbf{p}$?
4. Given vectors $\mathbf{r}(t) = e^t\mathbf{i} + (2e^t + 1)\mathbf{j}$ and $\mathbf{s}(t) = t\mathbf{i} + (2t + 1)\mathbf{j}$, explain why $\mathbf{r}$ and $\mathbf{s}$ have the same rectangular form but different graphs.

SOLUTIONS

1. First, put $\mathbf{w}$ in component form. You can either count units from the start to the end of the vector or simply find the difference of the beginning and ending coordinates: $\mathbf{w} = <-1 - 3, 4 -(-3)> = <-4,7>$. Now, $\mathbf{p} = 2\mathbf{v} - \mathbf{w} = 2<2,3> - <-4,7> = <4,6> - <-4,7> = <8,-1>$.

2. If you like, you can draw a right triangle to justify your calculations, but it is not necessary. $\|\mathbf{p}\| = \sqrt{64+1} = \sqrt{65}$.

3. Once $\mathbf{p}$ is in component form, all of the hard work is done. In standard unit vector form, $\mathbf{p} = -8\mathbf{i} + 1\mathbf{j}$.

   You can express $\mathbf{r}(t)$ as the parametric equations $x = e^t$, $y = (2e^t + 1)$. By substitution, $y = 2x + 1$. You will get the very same parametric equations for $\mathbf{s}(t)$, but the graph of $\mathbf{s}(t)$ is the entire line $2x + 1$, whereas the graph of $\mathbf{r}(t)$ is $2x + 1, x > 0$. The reason for this is the domain of e^t. Think back to the major graphs to memorize for the chapter. Remember that e^t has a positive range and can't output negative numbers or 0. Therefore, after substituting $x = e^t$ into y to get $2x + 1$, that resulting linear function carries with it the restriction $x > 0$, limiting the graph accordingly.

CAUTION
Technically, the rectangular form of $\mathbf{r}(t)$ is $2x + 1, x > 0$, which does not equal $\mathbf{s}(t)$. Make sure to restrict the domain of a vector function's rectangular form, when appropriate.

TECHNOLOGY: SOLVING EQUATIONS WITH A GRAPHING CALCULATOR

You know how to solve equations; you've been doing it successfully since introductory algebra, and you're not even afraid of quadratic functions. "Bring 'em on," you say, with a menacing glint in your eye and a quadratic formula program humming in your calculator's memory. One problem: the College Board also knows that you have access to calculator technology, and some of the equations you'll be asked to solve aren't going to be the pretty little factorable ones you're used to. In fact, some equations on the calculator-active section will look downright ugly and frightening.

The College Board expects that you will have access to and proficiency with a graphing calculator that can do four major things, at the least, as described in Chapter 1. **You must be able to solve equations (even ugly ones) with your graphing calculator.** As always, instructions are included for the TI-83, currently the most common calculator used on the AP test. Consult your instruction manual if you have a calculator other than the TI-83.

 Example 24: Find the solutions to the equation $e^{2x} + \cos x = \sin x$ on $[-2\pi, 2\pi]$.

Solution: There is no easy way to solve this; it's just enough to shake your faith in the usefulness of the quadratic formula. However, there is hope! First, move things around so that the equation equals zero. If you subtract sin x from both sides of the equation, that goal is accomplished: $e^{2x} + \cos x - \sin x = 0$. Now, the solutions to the original equation will be the roots of this new equation. Put your calculator in radians mode ([Mode] →"Radian") and graph $Y = e^{2x} \cos x - \sin x$. If you press [Zoom] → "Ztrig", the window is nicely suited to trigonometric functions—the window is basically $[-2\pi, 2\pi]$ for the x-axis and $[-4,4]$ for the y-axis. If you did everything correctly, you should see this:

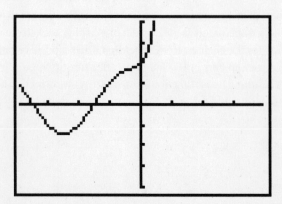

Finding those x-intercepts is our goal. It doesn't matter which you find first, but we'll start with the rightmost one. Clearly, the root falls between $x = -\pi$ and $-\frac{\pi}{2}$. To calculate the root, press [2nd]→[Trace] → "zero," as we are looking for zeros of the function. At the prompt "Left bound?" type a number to the left of this root; for example, $-\pi$. Similarly, for the prompt "Right bound?" you can type 0.

At the prompt "Guess?" you should take a stab at prognosticating the root. A good guess seems to be $-\frac{3\pi}{4}$. Once you press [Enter], the calculator does the rest of the work for you, and the root turns out to be $x = -2.362467$. The AP test only requests three-decimal place accuracy, so -2.362 is acceptable. Note, however, that 2.363 is not correct and is not accepted. You do not have to round answers—you can simply cut off (or truncate) decimals after the thousandths place.

Follow the same steps to get the second root. If you don't like typing in guesses for the boundaries, you can press the left and right arrrow buttons to move the little "X" turtle along the graph. For example, in the diagrams below, you move the little turtle to the left of the root at the prompt "Left bound?". Once you press Enter and get the "Right bound?" prompt, you move the turtle to the right of the root. You can even do this when asked for a guess—just move the turtle between the two boundaries and get close to the root.

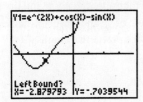 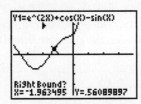 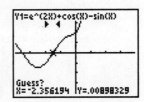

The second root of the equation is -5.497775, which can be written as -5.497 or -5.498 on the AP test.

PRACTICE PROBLEMS

You may use a graphing calculator on problems 9 and 10 only.

For problems 1 through 5, use the following graphs of *f* and *g*:

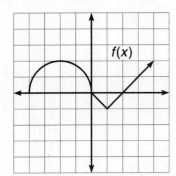 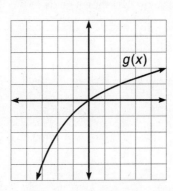

CAUTION
A bad choice for the bounds would have been [–3π,0], as this interval includes *both* roots. When you are trying to calculate a root, make sure only the root you are trying to find falls within the boundaries you assign.

1. Determine which of the two functions has an inverse and sketch it.
2. Graph $|f(x)|$ and $g(|x|)$.
3. If f is created with a semicircle and an absolute value graph, write the equation that represents $f(x)$.
4. Draw a function $h(x)$ such that $h(x) = g^{-1}(x)$ when $x \le 0$ and $h(x)$ is odd.
5. Evaluate $f(g(f(-2)))$.
6. Solve the equation $\cos 2x - \cos^2 x = 2\sin x$ and give solutions on the interval $[0, 2\pi)$.
7. If $m(x) = 2x^3 + 5x - 2$, find $m^{-1}(6)$.
8.* Graph $r = \cos\theta - \sin\theta\cos\theta$ and find the values of θ where the graph intersects the pole.
9.* Graph $x = e^t$, $y = t + 1$, and express the parametric equations in rectangular form.
10.* **James' Diabolical Challenge:** Write a set of parametric equations such that: at $t = 0$, $x = 0$ and $y = 0$ and at $t = 4$, $x = 4$ and $y = 6$. Then, find the rectangular inverse of your equations

SOLUTION

1   Only g has an inverse function; because f fails the horizontal line test, it is no one-to-one and thus has no inverse. To graph g^{-1}, either reflect g about the lin $y = x$ or choose some coordinates from the graph of g and reverse them —remember that if (a,b) is on the graph of g, then (b,a) is on the graph of g^{-1}.

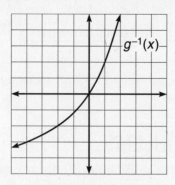

2. The graph of $|f(x)|$ will not extend below the x-axis, and $g(|x|)$ will be y-symmetric, as shown below:

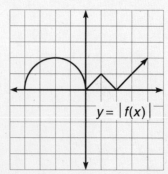 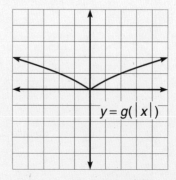

3. The circle has equation $(x + 2)^2 + y^2 = 4$, so solving for y and identifying only the upper half of the circle results in the equation $y = \sqrt{4 - (x + 2)^2}$. The absolute value graph is $y = |x - 1| - 1$, by graph translations. Therefore, we can write the multi-rule function.

$$y = \begin{cases} \sqrt{4 - (x+2)^2}, & x \le 0 \\ |x - 1| - 1, & x \ge 0 \end{cases}$$

4. The graph of h will look exactly like the graph of g^{-1} for $x \le 0$ (as the functions are equal there), but since h is odd, its graph will have to be origin-symmetric, which dictates the graph of h for $x > 0$.

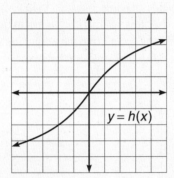

5. From the graphs, you see that $f(-2) = 2$, $g(2) = 1$, and $f(1) = -1$. Thus, $f(g(f(-2))) = -1$.

6. Using a double-angle formula for cos $2x$, rewrite the equation to get $\cos^2 x - \sin^2 x - \cos^2 x = 2\sin x$. This simplifies easily to $-\sin^2 x - 2\sin x = 0$. Now, factor the equation: $-\sin x(\sin x + 2) = 0$. The answer is $x = 0, \pi$ (remember that you cannot solve sin $x = -2$, since sine has a range of $0 \le y \le 1$).

7. First of all, if 6 is in the *domain* of m^{-1}, then it is in the *range* of m. Therefore, there is some number x such that $2x^3 + 5x - 2 = 6$. It is not easy to find that number, however, and you should resort to the graphing calculator to solve the equivalent equation $2x^3 + 5x - 8 = 0$ in the method described in the Technology section of this chapter. Doing so results in $x = 1.087$. Thus, $m(1.087) = 6$ and $m^{-1}(6) = 1.087$.

8. The graph, given below, will hit the pole whenever $r = 0$. Therefore, set the equation equal to 0 and factor to get cos $\theta(1 - \sin \theta) = 0$. This has solutions $\theta = \frac{\pi}{2}$ and $\frac{3\pi}{2}$.

NOTE
The inequality signs of the restrictions on Problem 3 could vary. For example, restrictions of $x < 0$ for the circle and $x \ge 0$ for the absolute value would have also been acceptable.

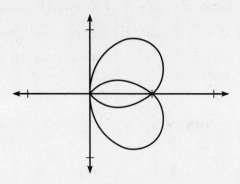

9. The graph is given below. In order to put in rectangular form, solve $x = e^t$ for t by taking ln of both sides. Doing so results in $\ln x = t$. (You could also solve the y equation for t, but this results in a *much* uglier rectangular form—a function of y.) Substituting $t = \ln x$ into the y equation gives $y = \ln x + 1$, $x > 1$. The restriction is caused by the range of e^t being positive, as denoted earlier in the chapter.

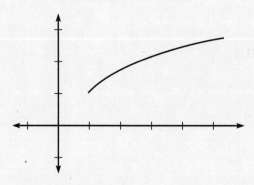

10. The simplest solution for this problem is $x = t$ and $y = \frac{3}{2}t$, as 6 is $\frac{3}{2}$ of 4. The rectangular form of this problem is $y = \frac{3}{2}x$, so the inverse is $y = \frac{2}{3}x$.

Limits and Continuity

The concepts of limits stymied mathematicians for a long, long time. In fact, the discovery of calculus hinged on these wily little creatures. Limits allow us to do otherwise illegal things like divide by zero. Since, technically, it is not ever acceptable to divide by zero, limits allow uptight math people to say that they are dividing by "basically" zero or "essentially" zero. Limits are like fortune tellers—they know where you are heading, even though you may not ever get there. Unlike fortune tellers, however, the advice of limits is always free, and limits never have bizarre names like "Madame Vinchense."

WHAT IS A LIMIT?: HANDS ON ACTIVITY 3.1

By completing this activity, you will discover what a limit is, when it exists, and when it doesn't exist. As in previous Hands-on Activities, spend quality time trying to answer the questions before you break down and look up the answers.

1. Let $f(x) = \dfrac{x^2 - 3x - 4}{x + 1}$. What is the domain of $f(x)$? Graph $f(x)$.

----

2. The table below gives x-values that are less than but increasingly closer and closer to -1. These values are said to be *approaching -1 from the left*. Use your calculator to fill in the missing values of $f(x)$ for each x.

x	-3	-2	-1.5	-1.25	-1.1	-1.001
$f(x)$	-7		-5.5			

3. The y-value (or height) you are approaching as you near the x value of -1 in the table above is called the *left-hand limit* of -1 and is written $\lim\limits_{x \to -1^-} f(x)$. What is the left-hand limit of $f(x)$?

----

4. The table below gives x-values that are greater than but increasingly closer to -1. These values are *approaching -1 from the right*. Use your calculator to fill in the missing values, as you did in Number 2.

x	1	0	$-.5$	$-.75$	$-.9$	$-.999$
$f(x)$		-4	-4.5			

TIP

Limits can help you understand the behavior of points not in the domain of a function, like the value you described in Number 1.

CAUTION

Just because the general limit may exist at one x value, that does not guarantee that it exists for all x in the function!

5. The y-value (or height) you are approaching as you near the x value of -1 in the table above is called the *right-hand limit* of -1 and is written $\lim_{x \to -1^+} f(x)$. What is the right-hand limit of $f(x)$?

_____

6. Graph $f(x)$ below, and draw the left- and right-hand limits as arrows on the graph.

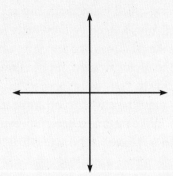

7. When the left- and right-hand limits as x approaches -1 both exist and are equal, the general limit at $x = -1$ exists and is written $\lim_{x \to -1} f(x)$. Does the general limit exist at $x = -1$? If so, what is it?

_____

8. Write a few sentences describing what a limit is and how it is found.

_____

_____

_____

9. Each of the following graphs has no limit at the indicated point. Use a graphing calculator and your knowledge of limits to determine why the limits do not exist.

(a) $\displaystyle\lim_{x \to 2} g(x)$ if $g(x) = \begin{cases} -\frac{1}{2}x + 3, & x < 2 \\ 3x - 1, & x \geq 2 \end{cases}$

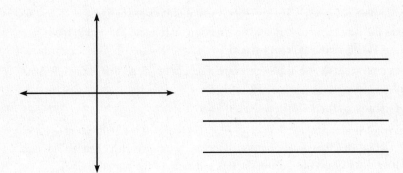

(b) $\displaystyle\lim_{x \to 0} \sin \frac{2\pi}{x}$

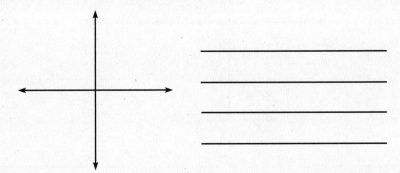

(c) $\displaystyle\lim_{x \to 0} \frac{1}{x^4}$

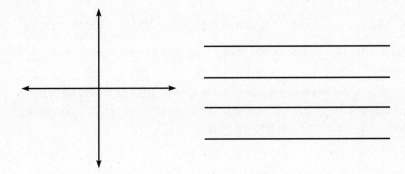

10. Complete this statement: A limit does not exist if...

A. _____

B. _____

C. _____

SELECTED SOLUTIONS TO HANDS-ON ACTIVITY 3.1

1. The domain is $(-\infty,-1)\bigcup(-1,\infty)$, or all real numbers excluding $x = -1$, as this makes the denominator zero and the fraction undefined. The graph looks like $y = x - 4$ with a hole at point $(-1,-5)$.

3. The graph seems to be heading toward a height of -5, so that is the left-hand limit.

5. The right-hand limit also appears to be -5.

6. The graph is identical to $y = x - 4$, except $f(x)$ is undefined at the point $(-1,-5)$. Even though the function is undefined there, the graph is still "headed" toward the height (or limit) of -5 from the left and right of the point.

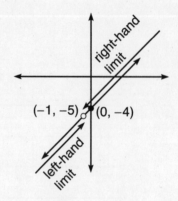

7. The general limit does exist at $x = -1$, and $\lim\limits_{x\to-1} f(x) = -5$.

8. A general limit exists on $f(x)$ at $x = c$ (c is a constant) if the left- and right-hand limits exist and are equal at $x = c$. Mathematically, $\lim\limits_{x\to c} f(x) = L$ if and only if $\lim\limits_{x\to c^-} f(x) = \lim\limits_{x\to c^+} f(x) = L$.

9. (a) $\lim\limits_{x\to2^-} g(x) = 2$ and $\lim\limits_{x\to2^+} g(x) = 5$. Because $\lim\limits_{x\to2^-} g(x) \neq \lim\limits_{x\to2^+} g(x)$, the general limt $\lim\limits_{x\to2} g(x)$ does not exist, according to your conclusion from Question 8.

(b) As you get closer to $x = 0$ from the left or the right, the function does not approach any one height—it oscillates infinitely between heights, as demonstrated in its graph on the following page.

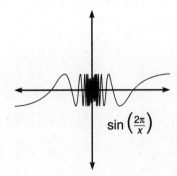

$\sin\left(\frac{2\pi}{x}\right)$

(c) As you approach $x = 0$ from the left or the right, the function grows infinitely large, never reaching any specified height. Functions that increase or decrease without bound, such as this one, have no general limit.

PROBLEM SET

For problems 1 through 6, determine if the following limits exist, based on the graph below of $p(x)$. If the limits do exist, state them.

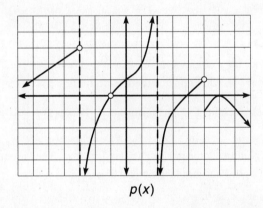

$p(x)$

1. $\lim\limits_{x\to 2^-} p(x)$

2. $\lim\limits_{x\to 3^+} p(x)$

3. $\lim\limits_{x\to 3} p(x)$

4. $\lim\limits_{x\to 5^-} p(x)$

5. $\lim\limits_{x\to 5^+} p(x)$

6. $\lim\limits_{x\to -1} p(x)$

For problems 7 through 9, evaluate (if possible) the given limits, based on the graphs below of $f(x)$ and $g(x)$.

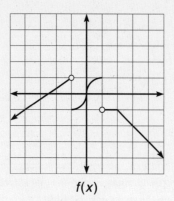

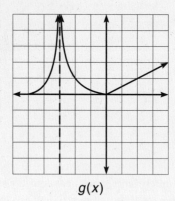

$f(x)$ $\qquad$ $g(x)$

7. $\quad \lim\limits_{x \to 2}(f(x) + g(x))$

8. $\quad \lim\limits_{x \to 0}(f(x) \cdot g(x))$

9. $\quad \lim\limits_{x \to -1} \dfrac{f(x)}{g(x)}$

SOLUTIONS

1. This limit does not exist, as p increases without bound as $x \to 2^+$. You can also say $\lim\limits_{x \to 2^-} p(x) = \infty$, which means there is no limit.

2. You approach a height of -1. In this instance, $\lim\limits_{x \to 3^+} p(x) = \lim\limits_{x \to 3^-} p(x) = \lim\limits_{x \to 3} p(x)$ $= -1$.

3. As stated in Number 2, the general limit exists at $x = 3$ and is equal to -1 (since the left- and right-hand limits are equal to -1).

4. $\lim\limits_{x \to 5^-} p(x) = 1$.

5. $\lim\limits_{x \to 5^+} p(x) = -1$. Notice that $\lim\limits_{x \to 5} p(x)$ cannot exist.

6. $\lim\limits_{x \to -1} p(x) = \lim\limits_{x \to -1} p(x) = 0$. Thus, the general limit, $\lim\limits_{x \to -1} p(x) = 0$. Even though the function is undefined at $x = -1$, p is still headed for a height of 0, and that's what's important.

7. $\lim\limits_{x \to 2}(f(x) + g(x)) = (-1 + 1) = 0$.

8. $\lim\limits_{x \to 0}(f(x) \cdot g(x)) = (0 \cdot 0) = 0$.

9. Although $\lim\limits_{x \to -1} g(x)$ appears to be approximately $\dfrac{1}{4}$, $\lim\limits_{x \to -1} f(x)$ does not exist. Therefore, $\lim\limits_{x \to -1} \dfrac{f(x)}{g(x)}$ cannot exist.

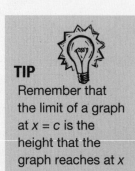

TIP
Remember that the limit of a graph at $x = c$ is the height that the graph reaches at $x = c$.

EVALUATING LIMITS ANALYTICALLY

At this point in your mathematics career, you possess certain polished skills. For example, if asked "How many is $4 + 5$?" you would not need nine apples in order to reach a solution. In the same way, you can evaluate most limits without actually looking at a graph of the given function. In fact, you don't need fruit of any kind to evaluate limits analytically. Instead, you need to learn the three major methods of finding limits: substitution, factoring, and the conjugate methods.

Remember that limits answer the question "Where is a function heading?" Luckily for math nerds all over the world (like me), a function usually reaches the destination for which it was heading. In such cases, you can use the *substitution method* for evaluating limits. It should be the first method you try in every limit problem you encounter.

NOTE

You can add, subtract, multiply, and divide limits without any hassle, as shown in Numbers 7, 8, and 9.

 Example 1: Evaluate the following limits:

(a) $\lim\limits_{x \to 2}(x^2 - 4x + 9)$

Substitute $x = 2$ into the function to get $2^2 - 4 \cdot 2 + 9 = 5$.

Thus, $\lim\limits_{x \to 2}(x^2 - 4x + 9) = 5$. That's all there is to it.

(b) $\lim\limits_{x \to \frac{\pi}{2}} \dfrac{x}{2} \cdot \cot 3x$

Substitution is again the way to go:

$$\frac{\frac{\pi}{2}}{2} \cdot \cot 3\frac{\pi}{2}$$

$$\frac{\pi}{4} \cdot \frac{\cos \frac{3\pi}{2}}{\sin \frac{3\pi}{2}}$$

$$\frac{\pi}{4} \cdot \frac{0}{-1} = 0$$

If all limits were possible via direct substitution, however, you would have done limits in basic algebra, and everyone would, in general, be happier people. Sometimes, substitution will not work, because the result is illegal. For example, consider the function with which you experimented in Hands-on Activity 3.1: $\lim\limits_{x \to -1} \dfrac{x^2 - 3x - 4}{x + 1}$. If you try direct substitution, the result is $\dfrac{(-1)^2 - 3(-1) - 4}{-1 + 1} = \dfrac{1 + 3 - 4}{-1 + 1} = \dfrac{0}{0}$. The technical term for a result of $\dfrac{0}{0}$ is *indeterminate*, which means that you cannot determine the

NOTE

The only appreciable difference between $\dfrac{x^2 - 3x - 4}{x + 1}$ and $x - 4$ is that $\dfrac{x^2 - 3x - 4}{x + 1}$ is undefined when $x = -1$. Though technically unequal, the functions share identical limit values everywhere.

limit using this method. (And you were just starting to feel confident …) Luckily, just in the knick of time, in rides the *factoring method* of evaluating limits on a shiny white steed.

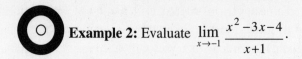

Example 2: Evaluate $\lim\limits_{x \to -1} \dfrac{x^2 - 3x - 4}{x + 1}$.

Solution: The factoring method entails factoring the expression and then simplifying it. This numerator factors easily, giving you $\lim\limits_{x \to -1} \dfrac{(x-4)(x+1)}{x+1}$. Next, cancel the common factors in the fraction to get $\lim\limits_{x \to -1} (x - 4)$. Therefore, $\lim\limits_{x \to -1} \dfrac{(x-4)(x+1)}{x+1} = \lim\limits_{x \to -1} (x - 4)$. Although substitution did not work before, it certainly will in the new expression, so $\lim\limits_{x \to -1} (x - 4) = -1 - 4 = -5$, the same result you reached in Activity 3.1. Huzzah!

The last limit evaluation method has a very specific niche in life—it attacks radical expressions in limits. This makes deciding to apply the *conjugate method* relatively easy. This technique is based on a simple complex number concept you have most likely already learned (although you won't be using it to *simplify* complex numbers in this application).

NOTE

The *conjugate* of the complex number *a – bi* is the complex number *a + bi*. To take the conjugate, change the sign that combined the terms to its opposite.

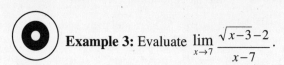

Example 3: Evaluate $\lim\limits_{x \to 7} \dfrac{\sqrt{x-3} - 2}{x - 7}$.

Solution: Substitution will result in the indeterminate answer $\frac{0}{0}$, and factoring isn't as fruitful as it was in Example 2. To evaluate this limit, multiply the fraction by the conjugate of the radical expression divided by itself: $\dfrac{\sqrt{x-3}+2}{\sqrt{x-3}+2}$.

$$\lim\limits_{x \to 7} \frac{\left(\sqrt{x-3} - 2\right) \bullet \left(\sqrt{x-3} + 2\right)}{(x-7) \bullet \left(\sqrt{x-3} + 2\right)}$$

$$\lim\limits_{x \to 7} \frac{(x-3) - 4}{(x-7) \bullet \left(\sqrt{x-3} + 2\right)}$$

$$\lim\limits_{x \to 7} \frac{x - 7}{(x-7) \bullet \left(\sqrt{x-3} + 2\right)}$$

It's best to leave the denominator alone initially, as you can now cancel the common $(x - 7)$ term to get

$$\lim_{x \to 7} \frac{1}{\sqrt{x-3}+2}$$

Now, substitution is possible, and the answer is

$$\frac{1}{\sqrt{7-3}+2} = \frac{1}{\sqrt{4}+2} = \frac{1}{4}$$

PROBLEM SET

Evaluate the following limits, if they exist. You may *not* use a calculator for numbers 1 through 6.

1. $\lim\limits_{x \to 4} \dfrac{2x^3 - 7x^2 - 4x}{x-4}$

2. $\lim\limits_{x \to 9} \dfrac{\sqrt{x}-3}{9-x}$

3. $\lim\limits_{x \to 1} \dfrac{x^2 - 2x - 5}{x+1}$

4. $\lim\limits_{x \to -2} \dfrac{x^3 + 8}{x+2}$

5. $\lim\limits_{x \to 4} f(x)$ if $f(x) = \begin{cases} \dfrac{x^2 - \sqrt{5}}{5 - x}, & x \neq 4 \\ 0, & x = 4 \end{cases}$

6. $\lim\limits_{x \to 0} \dfrac{|x|}{x}$

7. $\lim\limits_{x \to 0} x\sin\dfrac{1}{x}$

SOLUTIONS

1. Direct substitution results in the indeterminate form $\frac{0}{0}$ and cannot be used. The best option is factoring—doing so results in $\lim\limits_{x \to 4} \dfrac{x(2x+1)(x-4)}{x-4}$. Eliminating the common factor gives you $\lim\limits_{x \to 4} x(2x + 1)$. Substitution is now allowed, and the answer is 36.

2. Substitution (always your first method to attempt) fails. The presence of a radical alerts you to use the conjugate method:

$$\lim_{x \to 9} \frac{\left(\sqrt{x}-3\right)\cdot\left(\sqrt{x}+3\right)}{(9-x)\cdot\left(\sqrt{x}+3\right)}$$

TIP

While undertaking the factoring or conjugate methods, make sure to leave the "$\lim\limits_{x \to c}$" in front of each successive line—the limits of each step are equal, but the functions themselves are not always equal.

$$\lim_{x \to 9} \frac{x-9}{-(x-9)\left(\sqrt{x}+3\right)}$$

Notice that you have to factor a -1 out of $(9-x)$ in the denominator to be able to match the $(x-9)$ in the numerator.

$$\lim_{x \to 9} -\frac{1}{\sqrt{x}+3}$$

$$-\frac{1}{\sqrt{9}+3}$$

$$-\frac{1}{6}$$

3. Although you may want to apply factoring in this example because of the presence of polynomials, substitution works—don't forget to try substitution first:

$$\frac{1^2 - 2(1) - 5}{1+1} = \frac{-6}{2} = -3$$

4. Because substitution fails and there are no radicals, factoring is the method to use. In fact, the numerator is a sum of perfect cubes and factors easily.

$$\lim_{x \to -2} \frac{(x+2)\left(x^2 - 2x + 4\right)}{x+2}$$

$$\lim_{x \to -2} x^2 - 2x + 4 = (-2)^2 - 2(-2) + 4$$

$$4 + 4 + 4 = 12.$$

CAUTION
As demonstrated in Number 5, the limit of a function at $x = c$ is not always equal to the function value $f(c)$.

5. Don't be distracted by the fact that $f(4) = 0$. This does not mean that $\lim_{x \to 4} f(x)$ = 0. If you graph the function, you'll see that $f(x)$ is *not heading* for a height (limit) of zero when $x = 4$. The graph is headed for the height defined by the rule that is true for all $x \neq 4$. Substitution works for this function. So, the limit is $\frac{16 - \sqrt{5}}{5 - 4} \approx 13.764$.

6. None of your methods will apply in this problem, so you'll need to draw a graph (use a table of values):

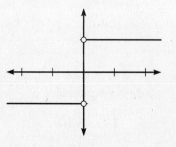

Once you do, you'll see that the left-hand limit at $x = 0$ is -1 and the right-hand limit at $x = 0$ is 1; thus, there is no general limit, and $\lim\limits_{x \to 0} \dfrac{|x|}{x}$ does not exist.

7. You currently have no good methods to evaluate this limit, so look at the graph as x approaches 0 from the left and the right:

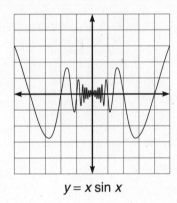

$y = x \sin x$

Visually, you can determine that the function approaches a limit (height) of zero as you near $x = 0$.

CONTINUITY

I have had some bad experiences at the movies. Perhaps you have also—right at the good part of the film when everything's getting exciting, the movie flickers off and the house lights come up. An usher informs the crowd that they are experiencing technical difficulties and that the movie should be up and running again in about 15 minutes. It ruins the whole experience, because the smooth, flowing, *continuous* stream of events in the film has been interrupted. In calculus, if a function experiences a break, then it, too, is said to be discontinuous. Any hole or jump in the graph of a function prevents the function from being classified as continuous. (Sometimes the sound in a theater prevents you from enjoying the film, but volume is not discussed until Chapter 9.)

Mathematically, a function $f(x)$ is said to be continuous at $x = a$ if *all three* of the following conditions are true:

(1) $\lim\limits_{x \to a} f(x)$ exists

(2) $f(a)$ exists

(3) $\lim\limits_{x \to a} f(x) = f(a)$.

In other words, a function must be headed toward some height at $x = a$, and when you reach $x = a$, the function must actually exist at the height you expected.

NOTE

Classifying functions as continuous or discontinuous is important, because the continuity of a function is often a prerequisite for important calculus theorems.

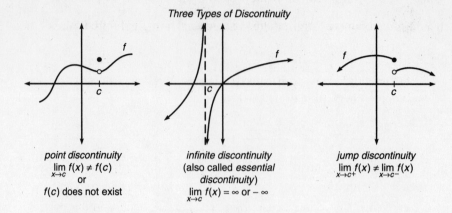

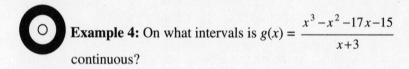

Example 4: On what intervals is $g(x) = \dfrac{x^3 - x^2 - 17x - 15}{x+3}$ continuous?

Solution: Visually, the graph is discontinuous at any hole or break, but a graphing calculator shows no obvious holes or breaks—it looks like a parabola. However, because $g(x)$ is a rational (fractional) function, it will be undefined whenever the denominator equals 0. Thus, $g(x)$ is discontinuous at $x = -3$, because $g(-3)$ does not exist, and that breaks the second requirement for continuity. There must be a hole (or *point discontinuity*) in the graph at $x = -3$, even though it wasn't obvious from the graph. So, the intervals of continuity for $g(x)$ are $(-\infty, -3) \cup (-3, \infty)$. Note that $x = -3$ is the only discontinuity of the function—g is continuous at all other values of x.

In Example 4, it is still true that $\lim\limits_{x \to -3} g(x)$ exists and is easily found (using the factoring method) to be 16, even though the function doesn't exist there. What if you rewrote the function as follows?

$$h(x) = \begin{cases} \dfrac{x^3 - x^2 - 17x - 15}{x+3}, & x \neq -3 \\ 16, & x = -3 \end{cases}$$

The new function, h, acts exactly as g did, except that $h(-3) = 16$. Redefining $x = 3$ "fixes" the discontinuity in $g(x)$, satisfies the final two continuity conditions that g did not, and makes h a continuous function. Whenever you are able to redefine a finite number of points like this and make a discontinuous function continuous, the function is said to have had *removable discontinuity*. If it is not possible to "fix" the discontinuity by redefining a finite number of points, the function is said to be *nonremovably discontinuous*. If it is possible to "fix" the discontinuity, but you don't feel like it, you are said to be a *lazy bonehead*.

Example 5: What type of discontinuity is exhibited by

$$r(x) = \frac{x+2}{x^2-4}?$$

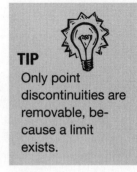
Solution: Factor the denominator to get $r(x) = \frac{x+2}{(x+2)(x-2)}$. Clearly, r is undefined for $x=-2$ and 2, and, hence, these are the discontinuities. Next, it is important to determine if $\lim\limits_{x \to 2} r(x)$ and $\lim\limits_{x \to -2} r(x)$ exist, because if a limit exists at a point of discontinuity, that discontinuity is removable. Remember: if no limit exists there, the discontinuity is nonremovable. Using the factoring method of evaluating limits,

$$\lim_{x \to -2} r(x) = \lim_{x \to -2} \frac{1}{x-2} = \frac{1}{-2-20} = -\frac{1}{4}.$$

However, $\lim\limits_{x \to 2} r(x)$ results in $\frac{1}{0}$, which does not exist. Therefore, there is a point discontinuity at $x=-2$ and an infinite discontinuity at $x=2$. The graph of the function verifies these conclusions.

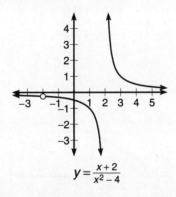

$$y = \frac{x+2}{x^2-4}$$

Substitution is a shortcut method for classifying types of discontinuity of a function that typically works very well. In the previous example, you substituted $x=2$ into r and got $\frac{1}{0}$. In the shortcut, a number divided by zero suggests that a vertical asymptote exists there, making it an infinite (or essential) discontinuity (the two terms are interchangeable). Substituting $x=-2$ results in $\frac{0}{0}$, which suggests point discontinuity.

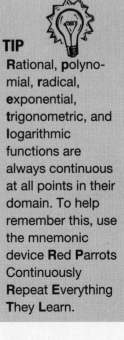

Example 6: Find the value of k that makes $f(x)$ continuous,

$$\text{given } f(x) = \begin{cases} \sqrt{2x+7}, \ x \leq 2 \\ x^2 - x + k, \ x > 2 \end{cases}$$

Solution: Because $f(x)$ is defined as a radical and a polynomial function, the two pieces of the graph will be continuous. The only possible discontinuity is at $x = 2$, where the graphs will have to meet. Otherwise, a jump discontinuity will exist. You know that $f(2) = \sqrt{11}$, according to the function. In order for a limit to exist at $x = 2$, the other rule in the function $x^2 - x + k$ must also reach a height of $\sqrt{11}$ when $x = 2$.

$$(2)^2 - 2 + k = \sqrt{11}$$

$$4 - 2 - \sqrt{11} = -k$$

$$-(2 - \sqrt{11}) = k$$

$$k \approx 1.3166$$

The graph below of $f(x)$ visually verifies our result—no holes, gaps, or jumps when $x = 2$:

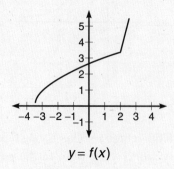

$$y = f(x)$$

PROBLEM SET

Do not use a graphing calculator for any of these problems.

For problems 1 through 3, explain why each function is discontinuous and determine if the discontinuity is removable or nonremovable.

1. $g(x) = \begin{cases} 2x - 3, \ x < 3 \\ -x + 5, \ x \geq 3 \end{cases}$

2. $b(x) = \dfrac{x(3x+1)}{3x^2 - 5x - 2}$

3. $h(x) = \dfrac{\sqrt{x^2 - 10x + 25}}{x - 5}$

4. Describe the continuity of the 15 functions you were to memorize in Chapter 2 without consulting any notes.

5. Draw the graph of a function, $f(x)$, that satisfies each of the following conditions. Then, describe the continuity of the function:

   - $\lim\limits_{x \to 2} f(x) = -1$

   - $\lim\limits_{x \to 0^+} f(x) = -\infty$

   - $\lim\limits_{x \to 0^-} f(x) = \infty$

   - $f(2) = 4$

   - $f(-1) = f(3) = 0$

   - f increases on its entire domain

6. Find the value of k that makes p continuous if

$$p(x) = \begin{cases} -|x - 2| + k, & x \le 4 \\ -x^2 + 11x - 23, & x > 4 \end{cases}$$

SOLUTIONS

1. g is made up of two polynomial (linear) segments, both of which will be continuous everywhere. However, the graph has a jump discontinuity at $x = 3$. Notice that $\lim\limits_{x \to 3^-} g(x) = 3$ (you get this by plugging 3 into the $x < 3$ rule). The right-hand limit of $g(x)$ at $x = 3$ is 2. Because the left- and right-hand limits are unequal, no general limit exists at $x = 3$, breaking the first condition of continuity. Furthermore, because no limit exists, the discontinuity is nonremovable.

2. Because b is rational, b will be continuous for all x in the domain. However, $x = -\frac{1}{3}$ and 2 are not in the domain. Using the factoring method of evaluating limits, you get that $\lim\limits_{x \to -\frac{1}{3}} \dfrac{x}{x-2} = \dfrac{1}{7}$, so $x = -\frac{1}{3}$ is a removable discontinuity. No limit exists at $x = 2$, an essential discontinuity.

3. The numerator of the fraction is a perfect square, so simplify to get

$$\mathrm{h}(x) = \dfrac{\sqrt{(x-5)^2}}{x-5}.$$

Remember that the square root function has a positive range, so the numerator must be positive:

$$\mathrm{h}(x) = \dfrac{|x-5|}{x-5}.$$

It helps to think about this graphically. After substituting some values of x into h, you get the following graph:

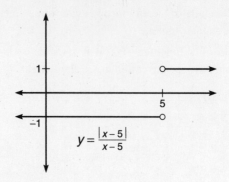

$$y = \frac{|x-5|}{x-5}$$

Thus, h has a jump discontinuity at $x = 5$.

4. $y = x$: continuous on $(-\infty, \infty)$

   $y = x^2$: continuous on $(-\infty, \infty)$

   $y = x^3$: continuous on $(-\infty, \infty)$

   $y = \sqrt{x}$: continuous on $[0, \infty)$

   $y = |x|$: continuous on $(-\infty, \infty)$

   $y = \dfrac{1}{x}$: continuous on $(-\infty, 0) \bigcup (0, \infty)$

   $y = [[x]]$: continuous for all real numbers x, if x is not an integer

   $y = e^x$: continuous on $(-\infty, \infty)$

   $y = \ln x$: continuous on $(0, \infty)$

   $y = \sin x$: continuous on $(-\infty, \infty)$

   $y = \cos x$: continuous on $(-\infty, \infty)$

   $y = \tan x$: continuous for all real numbers x, if $x \neq \dots, -\dfrac{3\pi}{2}$, $-\dfrac{\pi}{2}, \dfrac{\pi}{2}, \dfrac{3\pi}{2}, \dots$ (which can also be written $x \neq 2(n+1) \cdot \dfrac{\pi}{2}$, when n is an integer)

   $y = \cot x$: continuous for all real numbers x if $x \neq \dots, -\pi, 0, \pi, \dots$ (which can also be written $x \neq n\pi$, when n is an integer)

   $y = \sec x$: continuous for all real numbers x, $x \neq 2(n+1) \cdot \dfrac{\pi}{2}$, when n is an integer

   $y = \csc x$: continuous for all real numbers x, $x \neq n\pi$, when n is an integer

5. There is some variation in the possible answer graphs, but your graph should match relatively closely.

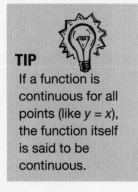

TIP

If a function is continuous for all points (like $y = x$), the function itself is said to be continuous.

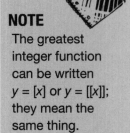

NOTE

The greatest integer function can be written $y = [x]$ or $y = [[x]]$; they mean the same thing.

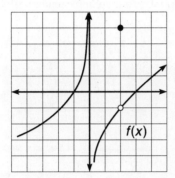

NOTE
The intervals of continuity for trigonometric functions are the intervals that make up their domains, as trigonometric functions are continuous on their entire domains.

6. Just like Number 1 in this problem set, the $x \leq 4$ rule evaluated at 4 represents $\lim\limits_{x \to 4^-}$, and the $x > 4$ rule evaluated at 4 represents $\lim\limits_{x \to 4^+}$. In order for p to be continuous, these limits must be equal.

$$\lim_{x \to 4^+} p(x) = -(4)^2 + 11(4) - 23 = 5$$

Therefore, $\lim\limits_{x \to 4^-} p(x) = 5$.

$$\lim_{x \to 4^-} p(x) = -|4 - 2| + k = 5$$

$$-2 + k = 5$$

$$k = 7$$

THE EXTREME VALUE THEOREM: HANDS-ON ACTIVITY 3.2

This and the next activity introduce you to two basic but important continuity theorems. Note that both of these are called *existence theorems*. They guarantee the existence of certain values but do not tell you where these values are—it's up to you to find them, and they're always in the last place you look (with your car keys, your wallet from two years ago, and the words to that Bon Jovi song you used to know by heart).

1. Given $f(x) = x^4 - 3x - 4$, justify that $f(x)$ is continuous on the x interval $[-1, 2]$.

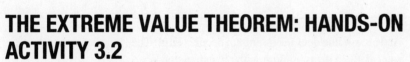

2. Draw the graph of $f(x)$. Use a graphing calculator if you wish.

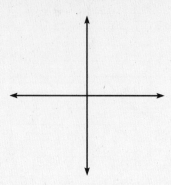

3. The maximum height reached by $f(x)$ on the interval $[-1,2]$ is called the *maximum of f* on the interval. At what value of x does $f(x)$ reach its maximum, and what is that maximum value?

_____

_____

NOTE
You are not allowed to use the maximum/minimum functions of your calculator on the AP exam. We will learn how to find them in different ways (which are acceptable on the test) later.

4. To calculate the minimum value of $f(x)$, use the $2^{nd} \rightarrow$ Trace$\rightarrow$minimum function on your calculator. Set bounds to the left and the right of the minimum and make a guess, as you did when finding x-intercepts in Chapter 2. What is the minimum value of $f(x)$ on $[-1,2]$?

_____

5. At what x-value does the minimum occur? (**Hint**: The value is displayed when you calculate the minimum with the calculator.)

_____

6. Graph $g(x) = \dfrac{1}{x}$ on the axes below. Can you find a maximum and a minimum for $g(x)$ on the x interval $(1,5)$?

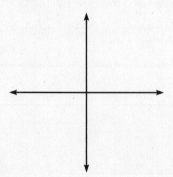

_____

7. Why do your results for *f* and *g* differ?

8. Graph $h(x) = \dfrac{x^3 - 4x^2 + 5x}{x}$ below. Why is there no maximum on [0,3]?

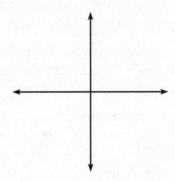

_____

_____

9. Complete the Extreme Value Theorem, based on your work above:

   Extreme Value Theorem: Any function, $f(x)$, will have a maximum and a minimum on the _____ interval _____ as long as $f(x)$ is _____.

10. Visually speaking, where can maximums and minimums occur on a graph?

_____

SELECTED SOLUTIONS TO HANDS-ON ACTIVITY 3.2

1. *f* is a polynomial whose domain is $(-\infty, \infty)$, and polynomials are continuous on their entire domain (remember the red **p**arrot?)

3. *f* reaches its highest point on [–1,2] when $x = 2$, and $f(2) = 6$. Thus, *f* has a maximum of 6.

4. *f* has a minimum of –6.044261. If you can't get this, use a left bound of 0, a right bound of 2, and a guess of 1.

5. The minimum occurs at $x = .9085621$.

6. You cannot find a maximum or a minimum. Although $g(1) = 1$ and $g(5) = \frac{1}{5}$ *would* be the maximum and minimum values, respectively, they are not included on the *open* interval (1,5). You cannot choose an *x* value whose function value is higher or lower than every other in the interval—try it!

7. You used a closed interval with *f* and an open interval with *g*.

8. *h* is removably discontinuous at $x = 0$, and because the maximum would have occurred there, the function will have no maximum.

9. Any function, $f(x)$, will have a maximum and minimum on the *closed* interval *[a,b]* as long as $f(x)$ is *continuous*. If the interval is not closed or the function is discontinuous, the guarantees of the Extreme Value Theorem do not apply.

CAUTION
In Number 2, a common mistake is to say that the maximum of the function is 2, but that is only the *x*-value at which the maximum occurs.

CAUTION

If the EVT conditions are not satisfied, a function can still have a maximum, a minimum, or both—their existence is just not guaranteed in that case.

10. The maximum and minimum (together called *extrema*, since they represent the *extreme* highest and lowest points on the graph) will occur at "humps" on the graph (like the minimum of *f*) or at the endpoints of the interval (like *f*'s maximum).

PROBLEM SET

For each of the following functions and intervals, determine if the Extreme Value Theorem applies, and find the maximum and minimum of the function, if possible. You may use your calculator.

1. $y = \tan x$, on $[0, \pi]$.

2. $y = \begin{cases} \sin x, & x \le -\dfrac{\pi}{2} \\ \csc x, & x > -\dfrac{\pi}{2} \end{cases}$, on $[-\pi, \dfrac{\pi}{4}]$.

3. $y = e^x - x^3$, on $[-2, 4]$.

4. $y = \ln(x - 1)$, on $[2, 5]$.

SOLUTIONS

1. The Extreme Value Theorem (EVT) does not apply because $\tan x$ is discontinuous on the given interval, specifically at $x = \dfrac{\pi}{2}$. No maximum or minimum values are possible on the closed interval, as the function both increases and decreases without bound at $x = \dfrac{\pi}{2}$.

2. The function is continuous on the interval. Both parts of the function are trigonometric and, therefore, continuous on their domain, and $[-\pi, -\dfrac{\pi}{4}]$ is included in the domain of each. The only possible discontinuity comes at $x = -\dfrac{\pi}{2}$, but both functions have the same value there, guaranteeing that the function is continuous. See the below graph—there is no break or jump. The EVT will apply. The maximum is 0 and occurs at $x = -\pi$, as the function decreases for the remainder of the interval. The minimum is $\dfrac{2}{\sqrt{2}} \approx -1.414$ and occurs at $x = -\dfrac{\pi}{4}$.

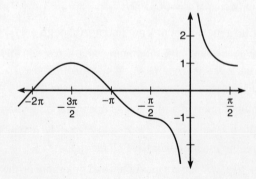

3. The EVT will not apply, but the function still has a maximum and minimum. The maximum occurs at the point (–2,8.135), and the minimum occurs at the point (3.733,–10.216). (The minimum is found using the minimum function on your graphing calculator.)

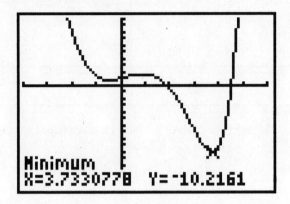

4. Although $\ln (x-1)$ is discontinuous at $x=1$, $\ln (x-1)$ *is* continuous on the given interval [2,5], so the EVT will apply. Because $\ln x$ is monotonic and increasing, the minimum (0) will occur at the beginning of the interval, and the maximum (1.386) will occur at the end of the interval.

THE INTERMEDIATE VALUE THEOREM: HANDS-ON ACTIVITY 3.3

Much like the Extreme Value Theorem guaranteed the existence of a maximum and minimum, the Intermediate Value Theorem guarantees values of a function but in a different fashion. Once again, continuity is a cornerstone of this theorem.

1. Consider a *continuous* function $f(x)$, which contains points (1,–4) and (5,3). You are not given an equation that defines $f(x)$—only these points.

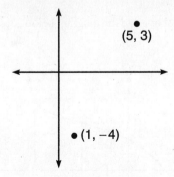

Draw one possible graph of $f(x)$ on the axes above.

2. Decide which of the following must occur between $x = 1$ and $x = 5$. Justify your answer.

    (A) maximum

    (B) minimum

    (C) root

    (D) y-intercept

3. Which of the following height(s) is the function guaranteed to reach, and why?

    (A) –5

    (B) –1

    (C) 2

    (D) 5

4. Draw three different graphs of $f(x)$ that are *discontinuous* and, therefore, do not fulfill the conclusion you drew in Number 2.

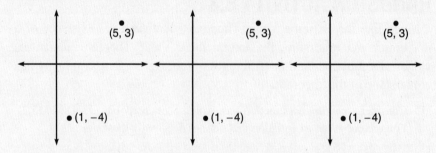

5. Based on your work above and the diagram below, complete the Intermediate Value Theorem below.

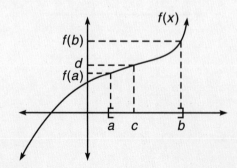

Intermediate Value Theorem: Given a function $f(x)$ that is _____ on the interval _____, if d is between _____ and _____, then there exists a c between _____ and _____ such that $f(c) =$ _____.

6. Rewrite the Intermediate Value Theorem in your own words to better illustrate its meaning.

_____

_____

_____

7. Give one real-life example of the Intermediate Value Theorem's guarantees.

_____

_____

SOLUTIONS TO HANDS-ON ACTIVITY 3.3

1. There are many possible answers, but here's one:

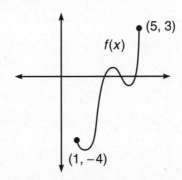

2. **The correct answer is (C),** root. If the function is continuous, it *must* cross the x-axis at some point in the interval. If the function begins with negative values at $x = 1$ but eventually has positive values at $x = 5$, the function *had* to equal zero (change from positive to negative) somewhere in [1,5].

3. Much like the function must reach a height of zero (as described in Number 2 above), the function must also reach all other heights between –4 and 3. Thus, the answer is both (B) and (C).

4. There are numerous ways to draw discontinuous functions that will not reach a height of zero. Below are three possible graphs. This highlights the importance of $f(x)$ being continuous.

CAUTION
The Intermediate Value Theorem guarantees that I will hit each *intermediate* height *at least once.* Note that in my solution for Number 1, *f* has 3 roots, although only 1 was guaranteed.

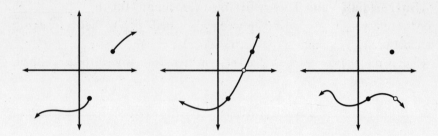

5. Given a function $f(x)$ that is *continuous* on the interval *[a,b]*, if *d* is between $f(a)$ and $f(b)$, then there exists a *c* between *a* and *b* such that $f(c) = d$.

6. If a continuous function begins at height $f(a)$ when $x = a$ and ends at height $f(b)$ when $x = b$, then the function will cover every single height between $f(a)$ and $f(b)$ at some x between a and b.

7. If I am two feet tall when I am 18 months old and 6 feet tall when I am 27 years old, then at some age between 18 months and 27 years, I was 4 feet tall. (In fact, since height according to age is continuous—you don't suddenly jump from 3 to 4 feet tall—I will cover all heights between 2 and 6 feet sometime in that time interval.)

PROBLEM SET

Do not use your calculator on problems 1 and 2.

1. On what interval *must* the function $g(x) = 2x^2 + 7x - 1$ intersect the line $y = 7$?

    (A) [–8,–6]

    (B) [–4,–1]

    (C) [0,2]

    (D) [6,9]

2. Given a function $h(x)$ continuous on [3,7] with $h(3) = 1$ and $h(7) = 9$, which of the following *must* be true?

    I. There exists a real number p such that $h(p) = 5$, $1 < p < 9$

    II. $h(5) = 5$

    III. h has a maximum and a minimum on [3,7]

    (A) I only

    (B) III only

    (C) I and III

    (D) I, II, and III

You may use your calculator for problem 3.

3. Given the continuous function $f(x) = \ln(-x) + \cos x$, prove that there exists a $c \in [-\pi, -\frac{\pi}{2}]$ such that $f(c) = .240$ and find c.

SOLUTIONS

1. This question is only the Intermediate Value Theorem (IVT) rephrased. Using the terminology of Number 5 from the hands-on exercise, 7 is the d value between $g(a)$ and $g(b)$. Each of the interval choices is a candidate for $[a,b]$. You should plug each pair into the function to see if the resulting pair $f(a)$ and $f(b)$ contains 7. Choice (C) results in $f(a) = -1, f(b) = 21$. Clearly, 7 falls between these numbers; this is not true for any of the other intervals.

2. It helps to draw the situation in order to visualize what's being asked.

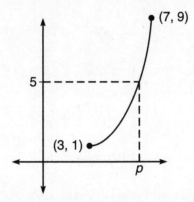

The correct answer is (C). Statement I is true. The IVT guarantees that p will exist on the interval $[3,7]$. So, if p exists on $[3,7]$, p will definitely exist on $[1,9]$, as $[3,7]$ is merely a subset of $[1,9]$. Statement II is not always true; in fact, it's not true in the representation of h in the diagram above. Statement III is true by the Extreme Value Theorem since h is continuous on a closed interval.

3. To apply the IVT (the prerequisites of the theorem are met as f is continuous on a closed interval), you note that $f(-\pi) \approx .1447$ and $f(-\frac{\pi}{2}) \approx .4516$. As .240 falls between these values, a $c \in [-\pi, -\frac{\pi}{2}]$ such that $f(c) = .240$ is guaranteed. To find c, set $f(x) = .240$ and solve with your calculator; $c = -2.0925$.

LIMITS INVOLVING INFINITY

You can learn a lot about a function from its asymptotes, so it's important that you can determine what kind of asymptotes shape a graph just by looking at a function. Remember that asymptotes are lines that a graph approaches but never reaches, as the graph stretches out forever. The two kinds of asymptotes with which you should concern yourself are vertical asymptotes and horizontal asymptotes; in the graph below of $g(x)$, $x = -2$ is a vertical asymptote and $y = 4$ is a horizontal asymptote.

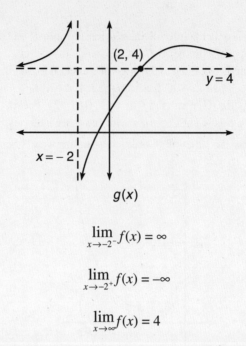

$$\lim_{x \to -2^-} f(x) = \infty$$

$$\lim_{x \to -2^+} f(x) = -\infty$$

$$\lim_{x \to \infty} f(x) = 4$$

Some students are confused by this diagram, since $g(x)$ actually intersects the horizontal asymptote. "I thought a graph can't hit an asymptote," they mutter, eyes filling with tears. A graph *can* intersect its asymptote, as long as it doesn't make a habit of it. Even though g intersects $y = 4$ at (2,4), g only gets closer and closer to $y = 4$ after that (for $x > 2$), and g won't intersect the line out there near infinity—it's the infinite behavior of the function that concerns us. It's the same with the criminal justice system—if you cross paths with the law a couple of times when you're very young, it's not that big a deal, but as you get much older, the police tend to frown upon you crossing them again.

Vertical asymptotes are discontinuities that force a function to increase or decrease without bound to avoid an x value. For example, consider the graph of $f(x) = \dfrac{1}{x-1}$.

$$y = \dfrac{1}{x-1}$$

In this case, the function has an infinite discontinuity at $x = 1$. As you approach $x = 1$ from the left, the function decreases without bound, and from the right, you increase without bound. In general, if $f(x)$ has a vertical asymptote at $x = c$, then $\lim_{x \to c} f(x) = \infty$ or $-\infty$. This is commonly called an *infinite limit*. This terminology is slightly confusing, because when f has an infinite limit at c, f has no limit at c!

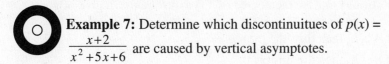

Example 7: Determine which discontinuitues of $p(x) = \dfrac{x+2}{x^2+5x+6}$ are caused by vertical asymptotes.

Solution: To begin, factor $p(x)$ to get $\dfrac{x+2}{(x+3)(x+2)}$. Because the denominator of a fraction cannot equal zero, p is discontinuous at $x = -2$ and -3. However, using the factoring method, $\lim_{x \to -2} p(x)$ exists and equals $\dfrac{1}{-2+3} = 1$. Therefore, the discontinuity at $x = -2$ is removable and not a vertical asymptote; however, $x = -3$ *is* a vertical asymptote. If you substitute $x = -3$ into the p, you will get $-\dfrac{1}{0}$. Remember, a constant divided by zero is the fingerprint of a vertical asymptote.

Horizontal asymptotes (or *limits at infinity)* are limiting heights that a graph approaches as x gets infinitely large or small. Consider the graph below of $s(x) = \dfrac{4x^2+1}{2x^2-8}$.

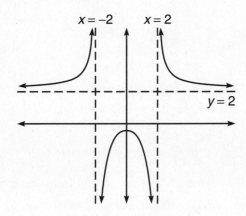

As x gets infinitely large (the extreme right side of the graph), the function is approaching a height of 2; in fact, the same is true as x gets infinitely negative (the left side of the graph). In this case, we write $\lim_{x \to \infty} s(x) = \lim_{x \to -\infty} s(x) = 2$. The AP Calculus test often features problems of the type $\lim_{x \to \infty} \dfrac{f(x)}{g(x)}$, and there is a handy trick to finding these limits at infinity of rational functions. We'll begin with a generic example to learn the technique.

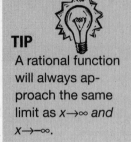

TIP
A rational function will always approach the same limit as $x \to \infty$ and $x \to -\infty$.

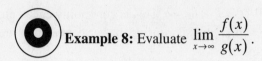

Example 8: Evaluate $\lim\limits_{x \to \infty} \dfrac{f(x)}{g(x)}$.

Solution: Let A be the degree of $f(x)$ and B be the degree of $g(x)$.

- If $A > B$, then then the limit is ∞.
- If $B > A$, then the limit is 0.
- If $A = B$, then the limit is the ratio of the leading coefficients of $f(x)$ and $g(x)$.

This technique only works for rational functions when you are finding the limit as x approaches infinity, and although it may sound tricky at first, the method is quite easy in practice.

Example 9: Evaluate the following limits.

(A) $\quad \lim\limits_{x \to \infty} \dfrac{x^4 - 3x^2 + 1}{x^2 + 5}$

Because the degree of the numerator is greater than the degree of the denominator (4 > 2), the limit is ∞. In other words, the function does *not* approach a limiting height and will reach higher and higher as x increases.

NOTE
Remember that the *degree* of a polynomial is its highest exponent, and the coefficient of the term with the highest exponent is called the *leading coefficient*.

(B) $\quad \lim\limits_{x \to \infty} \dfrac{\sqrt{x^6 - 5x^3 + 2x + 7}}{x^5 + 4x^4 - 11x^3}$

The degree of the numerator is 3 ($\sqrt{x^6} = x^3$), and the degree of the denominator is 5. Since the denominator's degree is higher, the limit is 0.

(C) $\quad \lim\limits_{x \to \infty} \dfrac{2x^3 - 7x^4 + 2}{5x^4 + 3x^2 + 1}$

The degrees of the numerator and denominator are both 4, so take the ratio of those terms' coefficients to get a limit of $-\frac{7}{5}$.

Example 10: Give the equations of the vertical and horizontal asymptotes of $k(x) = \dfrac{2x^2 - 5x - 12}{x^2 - 12x + 32}$.

Solution: The horizontal asymptotes are easy to find using the technique of the previous two examples. The degrees of the numerator and denominator are equal, so $\lim\limits_{x \to \infty} k(x) = \frac{2}{1} = 2$. Therefore, k has horizontal asymptote $y = 2$. In order to find any vertical asymptotes, begin by factoring k:

NOTE
Remember: If limit equals ∞, there technically is no limit, since ∞ is not a real number.

$$k(x) = \frac{(x-4)(2x+3)}{(x-4)(x-8)}$$

From this, you can see that k has discontinuities at $x = 4$ and 8. To determine which represents an asymptote, substitute them both into k. $k(4) = \frac{0}{0}$ and $k(8) = \frac{76}{0}$. Thus, $x = 8$ is an infinite discontinuity, and $x = 4$ is a point discontinuity, as verified by the graph.

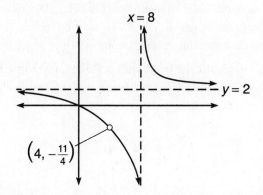

PROBLEM SET

1. Explain how horizontal and vertical asymptotes are related to infinity.

2. If m is an even function with vertical asymptote $x = 2$ and $\lim\limits_{x \to \infty} m(x) = 0$, draw a possible graph of $m(x)$.

3. Evaluate:

   (a) $\quad \lim\limits_{x \to \infty} \dfrac{1 - 4x^3 + 6x^4}{2x^5 - 3x^3 + 12}$

   (b) $\quad \lim\limits_{x \to \infty} \dfrac{5x^3 + 7x + 4}{\sqrt{3x + 8x^2 + 7x^6}}$

4. Given $f(x) = \dfrac{bx^2 + 14x + 6}{2\left(x^2 - a^2\right)}$, a and b are positive integers, f has horizontal asymptote $y = 2$, and f has vertical asymptote $x = 3$:

   (a) Find the correct values of a and b.

   (b) Find the point of removable discontinuity on f.

5. Draw a function $g(x)$ that satisfies all of the following properties:
   - Domain of g is $(-\infty, -2) \cup (-2, -\infty)$
   - g has a nonremovable discontinuity at $x = -2$
   - Range of g is $[-3, \infty)$
   - g has one root: $x = -4$

- $\displaystyle\lim_{x\to-2^-} g(x) = 3$

- $\displaystyle\lim_{x\to-\infty} g(x) = -1$

- $\displaystyle\lim_{x\to\infty} g(x) = 1$

SOLUTIONS

1. When a function approaches the vertical asymptote $x = c$, the function values either increase or decrease without bound (infinitely); for example: $\displaystyle\lim_{x\to c^+} f(x) = \infty$. A horizontal asymptote occurs when a function approaches a fixed height forever as x approaches infinity (for example: $\displaystyle\lim_{x\to\infty} f(x) = c$).

2. Because m is even, it must be y-symmetric, and, therefore, have a vertical asymptote at $x = -2$ as well. Furthermore, $\displaystyle\lim_{x\to-\infty} m(x)$ must also equal 0. Any solution must fit those characteristics; here is one:

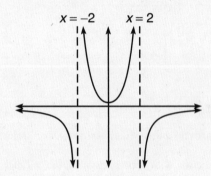

3. (a) Because the degree of the denominator is greater than the degree of the numerator, the limit at infinity is zero.

    (b) The numerator and denominator are both of degree 3 (since $\sqrt{x^6} = x^3$), so take the corresponding coefficients to find the limit of $\dfrac{5}{\sqrt{7}}$. Note that the radical remains around the 7.

4. (a) If f has vertical asymptote $x = 3$, we can find a. Remember the vertical asymptote fingerprint: a zero in the denominator but not in the numerator. The denominator equals zero when $3^2 - a^2 = 0$.

$$3^2 - a^2 = 0$$

$$a^2 = 9$$

$a = 3$ (since a has to be positive according to the problem)

Substitute a into the fraction to give you

$$\frac{bx^2 + 14x + 6}{2x^2 - 18}.$$

If f has horizontal asymptote $y = 2$, then $\displaystyle\lim_{x\to\infty} f(x) = 2$. The numerator and denominator have the same degree, so the limit is $\dfrac{b}{2} = 2$. Thus, $b = 4$.

(b)   Substituting both b and a gives you

$$\frac{4x^2+14x+6}{2x^2-18}$$

Factor completely to get

$$\frac{2(2x^2+7x+3)}{2(x^2-9)}=\frac{(2x+1)(x+3)}{(x+3)(x-3)}$$

The $(x+3)$ factor can be eliminated, meaning that $\lim_{x\to-3} f(x)=\frac{5}{6}$. Since $x=-3$ is a discontinuity at which a limit exists, f is removably discontinuous at the *point* $(-3,\frac{5}{6})$.

5. This problem is pretty involved, and all solutions will look similar to the graph below.

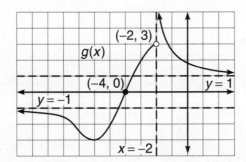

In order to get the range of $[-3,\infty)$, it's important that $\lim_{x\to-2^+} g(x)=\infty$. The graph must also reach down to and include the height of -3, although it need not happen at $(-6,-3)$ as on this graph.

SPECIAL LIMITS

You have a number of techniques available to you now to evaluate limits and to interpret the continuity that is dependent on those limits. Before you are completely proficient at limits, however, there are four limits you need to be able to recognize on sight. (BC students have still another topic to cover concerning limits—L'Hôpital's Rule—but that occurs in Chapter 5.) I call these "special" limits because we accept them without formal proof and because of the special way they make you feel all tingly inside.

Four Special Limits

- $\lim_{x\to\infty}\frac{c}{x^n}=0$, if c is a nonzero constant and n is a positive integer

Justification: In the simplest case ($c=1$, $n=1$), you are considering $\lim_{x\to\infty}\frac{1}{x}$.

You should know the graph of $\frac{1}{x}$ by heart, and its height clearly approaches zero as x approaches infinity. The same will happen with any c value. Remember that the denominator is approaching *infinity*, so it is getting very large, while the numerator is remaining fixed. The denominator will eventually get so large, in fact, that no matter how large the numerator is, the fraction has an extremely small value so close to zero that the difference is negligible. This limit rule works the same way for other eligible n values. Consider $\lim_{x \to \infty} -\frac{7}{x^3}$. The denominator is growing large more quickly than in the previous example, while 7 remains constant. Clearly, this limit is also equal to 0.

- $\lim_{x \to \infty} \left(1 + \frac{1}{x}\right)^x = e$

Justification: Using the first special limit rule, $\lim_{x \to \infty} \left(1 + \frac{1}{x}\right) = 1 + 0 = 1$. Technically, $\left(1 + \frac{1}{x}\right)$ is a number very close to, but not quite, one. That small difference becomes magnified when raised to the x power, and the result is e. To visually verify that the limit is accurate, use your graphing calculator to calculate $\left(1 + \frac{1}{x}\right)^x$ for a very large value of x. For example, if $x = 100,000,000$, $\left(1 + \frac{1}{x}\right)^x = 2.718281815$, which is approximately the value of e, accurate to seven decimal places.

- $\lim_{a \to 0} \frac{\sin a}{a} = 1$

Justification: This is easily proven by L'Hôpital's Rule, but that is outside the spectrum of Calculus AB, so ABers will have to satisfy themselves with the graph of $y = \frac{\sin x}{x}$ as proof.

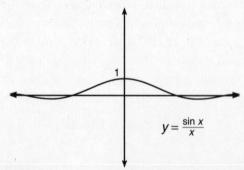

$$y = \frac{\sin x}{x}$$

Note that the formula above uses a instead of x, because the rule holds true for more than just $\frac{\sin x}{x}$. For example, you can set $a = 5x^3$ and the formula still works: $\lim_{x \to 0} \frac{\sin 5x^3}{5x^3} = 1$.

CAUTION
It is a mistake to assume that $1^\infty = 1$. In fact, 1^∞ is actually *indeterminate* in the same way that $\frac{0}{0}$ is. The second special limit rule essentially states that, in one case, $1^\infty = e$.

CAUTION
Special Limit Rules 3 and 4 work only if you are finding the limit as x approaches 0. Otherwise, plain old substitution usually works to evaluate the limits.

- $\lim\limits_{a \to 0} \dfrac{\cos a - 1}{a} = 0$

Justification: Again, L'Hôpital's Rule makes this easy, but the graph will suffice as proof; also, any value of a will make this true.

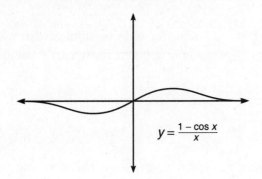

$$y = \frac{1 - \cos x}{x}$$

 Example 11: Evaluate each of the following limits.

(a) $\lim\limits_{x \to \infty} \left(3 - \dfrac{4}{x} + \dfrac{8}{x^2} \right)$

You can do each of these limits separately and add the results:

$$\lim_{x \to \infty} 3 = 3, \ \lim_{x \to \infty} \frac{4}{x} = 0, \ \lim_{x \to \infty} \frac{8}{x^2} = 0$$

$$\lim_{x \to \infty} 3 - \frac{4}{x} + \frac{8}{x^2} = 3 + 0 + 0 = 3$$

The last two limits are possible because of Special Limit Rule 1.

(b) $\lim\limits_{x \to \infty} \left(\dfrac{5}{x^3} + \dfrac{4}{x^{-2}} \right)$

You can rewrite this as

$$\lim_{x \to \infty} \frac{5}{x^3} + 4x^2$$

Although $\dfrac{5}{x^3}$ has a limit of 0 as x approaches infinity, $4x^2$ will grow infinitely large. The resulting sum will have no limiting value, so the limit is $0 + \infty = \infty$. (No limit exists.)

(c) $\lim\limits_{x \to 0} \dfrac{\sin 3x}{x}$

This is not quite the form of Special Limit Rule 3—the $3x$ and x have to match. However, if you multiply the fraction by $\dfrac{3}{3}$, you get

NOTE
You are allowed to multiply by $\frac{3}{3}$ in Example 11(c) because doing so is the same as multiplying by 1— you are not changing the value of the fraction.

$$\lim_{x \to 0} 3 \cdot \frac{\sin 3x}{3x}.$$

Now, Special Limit Rule 3 applies:

$$\lim_{x \to 0} 3 \cdot \frac{\sin 3x}{3x} = \lim_{x \to 0} 3 \cdot 1 = 3$$

Notice that $3 \cdot 1$ is not affected by the limit statement at all in the last step; the x may be approaching 0, but there are no x's left in the problem!

(d) $\lim_{x \to \infty} 2 \cdot \left(1 + \frac{1}{x}\right)^x + \cos \frac{\pi}{x}$

By Special Limit Rule 2, $\lim_{x \to \infty} 2 \cdot \left(1 + \frac{1}{x}\right)^x = 2 \cdot e$. Note that $\frac{\pi}{x}$ will follow Special Limit Rule 1, since π is a constant. Therefore you get

$$\lim_{x \to \infty} \cos \frac{\pi}{x} = \lim_{x \to \infty} \cos 0 = 1$$

Therefore, the solution is $2e + 1$.

Example 11 demonstrates that the presence of addition and subtraction does not affect the outcome of the limit. Each piece of the limit can be done separately and combined at the end.

PROBLEM SET

Do not use a graphing calculator on any of these problems.

1. $\lim_{x \to \infty} (4x^2 - x + 18)$

2. $\lim_{x \to 0} \frac{1 - \cos 7x}{x}$

3. $\lim_{x \to 0} \frac{3\cos x^2 - 3}{9x^2}$

4. $\lim_{x \to \infty} \frac{4x^{-2}}{x^3}$

SOLUTIONS

1. $\lim_{x \to \infty} (4x^2 - x + 18) = \infty$. Some texts propose very complicated means to prove this. However, the question is very easy if you consider the graph of $y = 4x^2 - x + 18$—it is a parabola facing upward. Therefore, $\lim_{x \to \infty} (4x^2 - x + 18) = \infty$. It is also correct to say that no limit exists because the function increases without bound.

2. This is almost of the form $\lim\limits_{x \to 0} \dfrac{\cos x - 1}{x}$. Multiply the numerator and denominator by 7, and factor out a negative to make it match that form:

$$\lim_{x \to 0} -\frac{\cos 7x - 1}{x} \cdot \frac{7}{7}$$

$$\lim_{x \to 0} (-7) \cdot \frac{\cos 7x - 1}{7x}$$

$$-7 \cdot 0 = 0$$

3. Again, some minor massaging is necessary to use Special Limit Rule 4:

$$\lim_{x \to 0} \frac{3\left(\cos x^2 - 1\right)}{9x^2}$$

$$\lim_{x \to 0} \frac{1}{3} \cdot \frac{\cos x^2 - 1}{x^2}$$

$$\frac{1}{3} \cdot 0 = 0$$

4. This fraction can be rewritten to give you $\lim\limits_{x \to 0} \dfrac{4}{x^5}$. According to Special Limit Rule 1, this is of form $\dfrac{c}{x^n}$, so the limit is 0.

TECHNOLOGY: EVALUATING LIMITS WITH A GRAPHING CALCULATOR

You can sum up the whole concept of limits in one statement: A limit is a height toward which a function is heading at a certain x value. With this in mind, a graphing calculator greatly simplifies the limit process. Many times, the function you are given is bizarre looking, and its graph is beyond the grasp of mere mortal men and women in the time allotted to answer an AP question. The majority of these limit questions appear on the non-calculator portion of the AP test, forcing you to use the substitution, factoring, and conjugate methods to reach an answer. However, evaluating limit problems will sometimes seep into the calculator-active section like a viscous, sticky goo. In these cases, limits are no match for you at all, as the calculator affords you numerous tools in your dual quests for a 5 on the AP test and peace in the universe.

CAUTION
Remember that your calculator can only be used to *approximate* limits. Using the calculator to approximate a limit may not be acceptable on a free-response question. This method is only your last resort!

Example 12: Evaluate $\lim\limits_{x \to 9} \dfrac{\sqrt{x} - 3}{9 - x}$ using your calculator.

Solution: You solved this problem in an earlier exercise using the conjugate method (which works just fine) and got $-\frac{1}{6}$. Graph it on your calculator—it looks almost like a straight line (but it's definitely not linear).

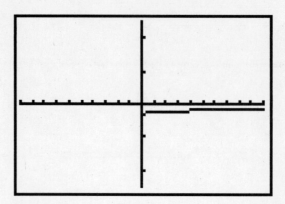

Use the [2nd]→[Trace]→"value" command on your calculator to find the value of the function at $x = 9$. The corresponding y value should come out blank! This makes sense, because 9 is not in the domain of the function. Therefore, substitution does not work. However, we can use the calculator to substitute a number *very close* to 9. This value is a good approximation of the limit for which you are looking. Again, use the [2nd]→[Trace]→"value" function of your calculator to evaluate the function at $x = 8.9999$

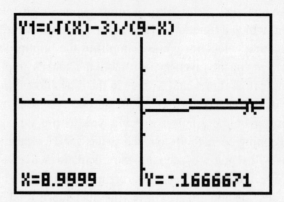

The calculator gives a limit of $x = -.1666671$. Using other x values even closer to 9 (e.g., 8.99999999999), the approximation looks more like $-.16666666667$, which is approximately $-\frac{1}{6}$, the exact value we received from using the conjugate method.

Example 13: Show that $\lim\limits_{x \to \infty} \dfrac{3x^2 - 7}{4 + 2x + 5x^2} = \dfrac{3}{5}$ using your calculator.

Solution: Using the rule for rational functions at infinity, the limit is clearly $\dfrac{3}{5}$ as the degrees of the numerator and denominator are the same. The graph certainly appears to approach that height as x approaches both ∞ and $-\infty$.

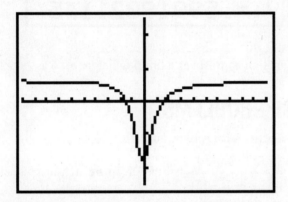

To verify this numerically, press [Trace] and repeatedly press the right arrow key. The function's height will slowly get closer to $\dfrac{3}{5} = .6$.

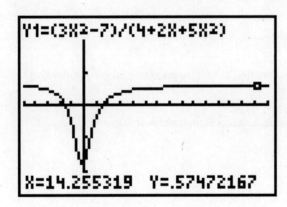

Another option open to you is the [2ⁿᵈ]→graph (or [Table]) function of the calculator, which lists function values quickly.

Clearly, the function approaches a limiting height of .6 as x approaches ∞.

PRACTICE PROBLEMS

TIP
You can change the "steps" taken by the table by pressing [2nd]→[Window] and adjusting the "ΔTbl" value.

1. If $\lim\limits_{x \to a} f(x)$ exists, what *must* be true?

2. Given $p(x) = \dfrac{\left(x^2 + 4x + b\right)}{\left(2x^2 + 7x - 15\right)}$, find the value of b that ensures p is nonremovably discontinuous only once on $(-\infty, \infty)$.

3. What conditions must be met if $f(x)$ is continuous at $x = c$?

4. $\lim\limits_{s \to -\infty} \dfrac{(5s+1)(s-4)}{\left(-6 + 3s - 7s^2 + 2s^3\right)}$.

5. Design two functions, $g(x)$ and $h(x)$, such that $g(x) + h(x) = 4x^2 + 3x + 1$ and $\lim\limits_{x \to \infty} \dfrac{g(x)}{h(x)} = 3$.

6. What three function behaviors prevent a limit from existing?

7. What value of α makes k continuous if $k(x) = \begin{cases} \dfrac{1 - \cos x}{x}, & x \neq 0 \\ \alpha, & x = 0 \end{cases}$

8. $\lim\limits_{x \to \infty} \dfrac{2x - \dfrac{4}{x^2}}{2 - 3x}$.

9. If $d(x)$ is defined by the graph below, answer the following questions:

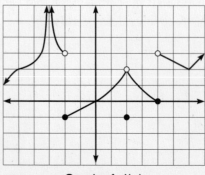

Graph of $d(x)$

(a) $\lim_{x \to 0} d(x)$

(b) $\lim_{x \to -1} d(x)$

(c) $\lim_{x \to -1} d(x)$

(d) $\lim_{x \to 2} d(x)$

(e) $\lim_{x \to 4^+} d(x)$

(f) List all x values where d is discontinuous.

(g) Which of your answers to part (f) represents removable discontinuities, and why?

(h) What value(s) of β make the following statement true?

$$\lim_{x \to \beta} d(x) = \infty$$

10. $\lim_{x \to 2} \dfrac{\sin x}{x}$

11. Give the equations of the horizontal and vertical asymptotes of $g(x) = \dfrac{3a^2 x^2 + 2abx - b^2}{6a^2 x^2 - abx - b^2}$, if a and b are real numbers.

12. $\lim_{x \to \infty} \left(\sqrt{4x^2 - 5x + 1} - 2x \right)$

13. The population, y, of the bacteria *Makeyoucoughus hurtyourthrouatus* is modeled by the equation $y = 50e^{.1013663t}$, where t is days and y is the number of colonies of bacteria. Use the Intermediate Value Theorem to verify that the bacteria will reach a population of 100 colonies on the time interval [4,7].

14. If $f(x)$ and $g(x)$ are defined by the graphs below, evaluate the following limits (if possible):

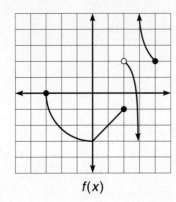

$f(x)$

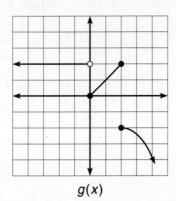

$g(x)$

(a) $\lim_{x \to 2} (f(x) - 3g(x))$

(b) $\lim_{x \to -3} \sqrt{\dfrac{f(x)}{g(x)}}$

(c) $\lim_{x \to 1} g(f(x))$

15. Find the values of c and d that make m(x)

continuous if $m(x) = \begin{cases} -\sqrt{4-(x+3)^2}, \, x \leq -1 \\ cx+d, \, -1, < x < 3 \\ \sqrt{x-3}+4, \, x \geq 3 \end{cases}$

16. **James' Diabolical Challenge Problem:**

Given $j(x) = \begin{cases} \dfrac{x^2-(4+A)x+4A}{x-4}, \, x \neq 4 \\ B, \, x = 4 \end{cases}$, $j(2) = 1$, and $j(x)$ is everywhere

continuous, find A and B.

SOLUTIONS

1. The three conditions for a limit to exist are: (1) $\lim\limits_{x \to a^+} f(x)$ exists, (2) $\lim\limits_{x \to a^-} f(x)$

exists, and (3) they are equal.

2. The denominator of p factors to $(2x-3)(x+5)$, so p will be discontinuous at

$x = \dfrac{3}{2}$ and -5. If the numerator contains one of these factors, our goal will be

achieved. It makes good sense to force $(x+5)$ to be a factor, rather than $(2x-3)$ because the leading coefficient of the numerator is 1. (In other words, the numerator will factor into $(x+A)(x+B)$ and $(x+5)$ is of this form.) To find b, we factor the numerator using $(x+5)$ as one of the factors:

$$(x+A)(x+5) = x^2 + 4x + b$$

$$x^2 + 5x + Ax + 5A = x^2 + 4x + b$$

Subtract x^2 from each side and factor to get

$$x(5+A) + 5A = 4x + b.$$

Thus, $5+A$ (the coefficient of x on the left side of the equation) must equal 4 (the coefficient of x on the right side of the equation).

$$5 + A = 4$$

$$A = -1$$

Therefore, $(x-1)$ is the remaining factor. Thus,

$$x^2 + 4x + b = (x+5)(x-1)$$

$$x^2 + 4x + b = x^2 + 4x - 5$$

$$b = -5.$$

Because $(x+5)$ is a factor of the numerator and denominator, $\lim\limits_{x \to -5} p(x)$

exists, and the discontinuity there is removable. The discontinuity at $x = \dfrac{3}{2}$

has no limit because $p\left(\dfrac{3}{2}\right) = \left(\dfrac{3.25}{0}\right)$, which indicates a vertical asymptote

(essential discontinuity).

3. If $f(x)$ is continuous at $x = c$, then (1) $\lim\limits_{x \to c} f(x)$ exists, (2) $f(c)$ exists, and (3) the two are equal.

4. Multiply the binomials in the numerator to get $\lim\limits_{s \to -\infty} \dfrac{5s^2 - 19s - 4}{-6 + 3s - 7s^2 + 2s^3}$.

   Because this is a rational function evaluated at infinity, you can use the shortcut method of examining their degrees. Because the degree of the denominator exceeds the degree of the numerator, the function's limit at infinity is 0. Don't be confused because $x \to -\infty$. Remember that a rational function approaches the same limits as $x \to \infty$ and $-\infty$.

5. Because $\dfrac{g(x)}{h(x)}$ is a rational function and $\lim\limits_{x \to \infty} = 3$, we know that the leading coefficients of $g(x)$ and $h(x)$ must be in the ratio 3:1. One possible $g(x)$ is $3x^2 + 2x + 1$. The matching $h(x)$ would have to be $x^2 + x$. Notice that the sum of the two functions is $4x^2 + 3x + 1$, as directed, and $\lim\limits_{x \to \infty} \dfrac{g(x)}{h(x)} = 3$. There are numerous possible answers, but they will work in essentially the same fashion.

6. If a function oscillates infinitely, increases or decreases without bound, or has right- and left-hand limits that are unequal, the function will not possess a limit there.

7. Special Limit Rule 4 gives us that $\lim\limits_{x \to 0} \dfrac{\cos x - 1}{x} = 0$. Thus, $\lim\limits_{x \to 0} k(x) = 0$, and the first condition of continuity is met. Notice that $k(0) = \alpha$, and (for k to be continuous) α must equal that limit as x approaches 0. Therefore, $\alpha = 0$.

8. Use Special Limit Rule 1 to simplify the fraction as follows:

$$\lim\limits_{x \to \infty} \frac{2x - 0}{2 - 3x}$$

This is a rational function being evaluated at infinity with equal degrees in the numerator and denominator, so the limit is equal to $\dfrac{2}{-3} = -\dfrac{2}{3}$.

9. (a) 0

   (b) −1: Remember the key is that you are approaching −1 *from* the right, not *to* the right.

   (c) 3

   (d) 2: even though f(2) = −1, the function leads up to a height of 2 when x = 2

   (e) 3

   (f) $x = -3, -2, 2, 4$

   (g) $x = 2$: If you redefine d such that $d(2) = 2$ (instead of −1), then the limit and the function value are equal to 2 and d is continuous there. No other discontinuities can be eliminated by redefining a finite number of points.

   (h) $\beta = -3, \infty$: d increases without bound as you approach 3 from the left and the right; the graph also increases without bound as x approaches ∞.

10. You don't have to use special limit rules here—substitution is possible. The answer is $\dfrac{\sin 2}{2}$, or .457.

11. Factor the fraction fully to get

$$\frac{(3ax-b)(ax+b)}{(3ax+b)(2ax-b)}.$$

The denominator of g will equal 0 when $x = -\dfrac{b}{3a}, \dfrac{b}{2a}$. However, the numerator will not equal zero simultaneously. Thus, vertical asymptotes are present, and the equations for the vertical asymptotes are $x = -\dfrac{b}{3a}$ and $x = \dfrac{b}{2a}$. To find the horizontal asymptotes, note that the numerator and denominator have the same degree, and find the limit at infinity.

$$\lim_{x \to \infty} g(x) = \frac{3a^2}{6a^2} = \frac{1}{2} \text{ if } a \neq 0.$$

From this, you know that the horizontal asymptote is $y = \dfrac{1}{2}$.

12. It's the revenge of the conjugate method! Multiply the numerator and denominator of the fraction by the conjugate of the expression.

$$\lim_{x \to \infty} \frac{\left(\sqrt{4x^2-5x+1}-2x\right) \cdot \left(\sqrt{4x^2-5x+1}+2x\right)}{\sqrt{4x^2-5x+1}+2x}$$

$$\lim_{x \to \infty} \frac{4x^2-5x+1-4x^2}{\sqrt{4x^2-5x+1}+2x}$$

$$\lim_{x \to \infty} \frac{-5x+1}{\sqrt{4x^2-5x+1}+2x}$$

Because the degrees of the numerator and denominator are both 1 (since $\sqrt{x^2} = x$), you find the limit by taking the coefficients of the terms of that degree and ignoring the rest of the problem (just as you've done in the past with rational limits at infinity).

$$\lim_{x \to \infty} \frac{-5x}{\sqrt{4x^2}+2x} = \frac{-5}{\sqrt{4}+2} = -\frac{5}{4}$$

13. The population at $t = 4$ is approximately 75, and the population at $t = 7$ is approximately 101.655. The Intermediate Value Theorem guarantees the existence of a $c \in [4,7]$ such that $f(c) = 100$.

14. (a)   No limit: $\lim_{x \to 2} f(x)$ does not exist, so $f(x) - 3g(x)$ cannot have a limit.

 (b)   0: $\lim_{x \to -3} f(x) = 0$ and $\lim_{x \to -3} g(x) = 2$. So, $\lim_{x \to -3} \sqrt{\dfrac{f(x)}{g(x)}} = \sqrt{\dfrac{0}{2}} = 0$.

 (c)   2: $\lim_{x \to 1} f(x) = -2$ and $g(-2) = 2$.

15.   The graph will begin as a semicircle of radius 2 centered at $(-3,0)$, will become a line (since $cx + d$ is linear) between the x values of -1 and 3, and will end as the graph of $\sqrt{x}$ shifted to the right 3 and up 4. To find the first point on the linear section, plug -1 into the first rule: $-\sqrt{4 - (-1 + 3)^2} = 0$. The line will begin at point $(-1,0)$. The line will end at $m(3) = \sqrt{3 - 3} + 4 = 4$. The focus of the problem, then, is to find the equation of the line that passes through $(-1,0)$ and $(3,4)$. Find the slope of the line and use point-slope form to get

$$y - 0 = \frac{4}{4}(x + 1)$$

$$y = x + 1.$$

Therefore, the correct c and d values are both 1. The solution is further justified by the graph of m.

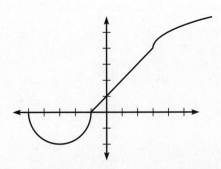

16.   To begin, factor the function and use the fact that $j(2) = 1$.

$$j(x) = \frac{(x - 4)(x - A)}{x - 4} = x - A$$

$$j(2) = 2 - A = 1$$

$$A = 1$$

Since the function is continuous, $\lim_{x \to 4} f(x)$ must be equal to $f(4)$. So, you get

$$f(4) = 4 - 1 = B \text{ (according to the given information)};$$

$$B = 3.$$

Differentiating

The study of limits and continuity only sets the stage for the first meaty topic of calculus: derivatives and differentiation. In fact, more than half of the questions on the AP test will involve derivatives in some way or another. Unlike limits, however, finding derivatives is an almost mechanical process full of rules and guidelines, which some students find a relief. Other students don't find any relief in any topic of calculus; these people have awful nightmares in which monsters corner them and force them to evaluate limits while poking them with spears.

DERIVATIVE AS A RATE OF CHANGE

The *derivative* of a function describes how fast and in what capacity a function is changing at any instant. You already know a bit about derivatives, though you may not know it. Consider the graph of $y = 2x - 1$. At every point on the line, the graph is changing at a rate of 2. This is such an important characteristic of a line that it has its own nifty term—slope. Because a slope describes the rate of change of a linear equation, the derivative of any linear equation is its slope. In the case of $y = 2x - 1$, we write $y' = 2$.

Not all derivatives are so easy, however. You'll need to know the definition of the derivative, also called the *difference quotient*. The derivative of a function $f(x)$ is defined as this limit:

$$\lim_{\Delta x \to 0} \frac{f(x + \Delta x) - f(x)}{\Delta x}$$

 Example 1: Use the difference quotient to verify that $\frac{dy}{dx} = 2$ if $y = 2x - 1$.

Solution: Set $f(x) = 2x - 1$, and apply the difference quotient. You need to substitute $(x + \Delta x)$ into the function, subtract $f(x)$, and then divide the whole thing by Δx.

$$\lim_{\Delta x \to 0} \frac{2(x + \Delta x) - 1 - (2x - 1)}{\Delta x}$$

NOTE

Differentiation is
the process of
taking derivatives.
If a function is
differentiable, then
it has derivatives.

NOTE

The notations y',
$\frac{dy}{dx}$, and Dx all
mean the derivative
of y with respect to
x. Don't concern
yourself too much
with what "with
respect to x"
actually means yet.

TIP

The difference
quotient has an
alterntaive form for
finding derivatives
at a specific value
$x = c$: $f'(c) =$
$\lim\limits_{x \to c} \dfrac{f(x)-f(c)}{x-c}$.

$$\lim_{\Delta x \to 0} \frac{2x+2\Delta x-1-2x+1}{\Delta x}$$

$$\lim_{\Delta x \to 0} \frac{2\Delta x}{\Delta x}$$

$$\lim_{\Delta x \to 0} 2 = 2$$

That wasn't so bad, now was it? Well, bad news—it gets worse. What about the graph $y = x^2 + 2$? That graph changes throughout its domain at different rates. In fact, when $x < 0$, the graph is decreasing, so its rate of change will have to be negative. However, when $x > 0$, the graph increases, meaning that its rate of change will need to be positive. Furthermore, can you discuss "slope" if the graph in question isn't a line? Yes. Derivatives allow us to discuss the slopes of *curves*, and we do so by examining tangent lines to those curves. The diagram below shows $y = x^2 + 2$ and three of its tangent lines.

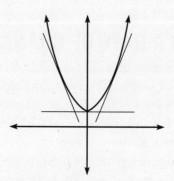

The graph of $y = x^2 + 1$
and some tangent lines.

Geometrically, the derivative of a curve at a point is the slope of its tangent line there. In the diagram above, it appears that the tangent line is horizontal when $x = 0$ and thus has slope zero. Is it true, then, that $y'(0) = 0$? You can use the alternate form of the difference quotient (given in the preceding margin note) to find out.

 Example 2: Prove that $y'(0) = 0$ if $y = x^2 + 1$ using the difference quotient.

Solution: Set $f(x) = x^2 + 1$. In this problem, you are finding $y'(0)$, so $c = 0$; therefore, $f(c) = f(0) = 1$:

$$\lim_{x \to 0} \frac{f(x)-f(0)}{x-0}$$

$$\lim_{x \to 0} \frac{x^2 + 1 - 1}{x}$$

$\lim_{x \to 0} x = 0$ (by substitution).

Sometimes, the AP test will ask you to find the average rate of change for a function. This is *not* the same thing as a derivative. The derivative is the *instantaneous* rate of change of a function and is represented by the slope of the tangent line. Average rate of change gives a rate over a period of time and is represented geometrically by the slope of a secant line. The next example demonstrates the difference.

Example 3: Given the function $f(x)$ defined by the graph below is continuous on $[a,b]$, put the following values in order from least to greatest: $f'(d)$, $f'(e)$, $f'(h)$, average rate of change of f on $[a,b]$.

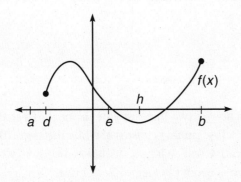

Solution: The function is not given, so you cannot use the difference quotient. Remember that derivatives are represented by slopes of tangent lines to the graph (drawn below), and average rate of change is given by the slope of the secant line (drawn below as a dotted line).

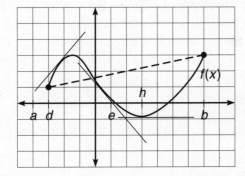

The tangent line at $x = e$ has the only negative slope, so it is ranked first. Of the remaining lines, you can tell which slope is greatest according to which line is the steepest. Thus, the final ranked order is $f'(e)$, $f'(h)$, average rate of change, $f'(d)$.

By now, you have an idea of what a derivative is (a rate of change), how it gets its value (the difference quotient), and what it looks like (the slope of a tangent line). As was the case with limits, there are times when a derivative does not exist. No derivative will exist on $f(x)$ at $x = c$ when any of the following three things happen:

- The graph of f has a sharp point (cusp) at $x = c$;
- f is discontinuous at $x = c$;
- The tangent line to f at $x = c$ is vertical.

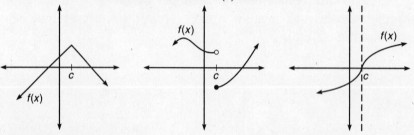

Conditions under which $f'(c)$ does not exist.

A sharp point exists at $x = c$.

In the first graph above, two linear segments meet at a sharp point at $x = c$. The slope of the left segment is positive, while the slope of the second is negative. The derivative will change suddenly and without warning ("Stand back, my derivative is about to change! Seek cover!") at $x = c$, and thus, it is said that no derivative exists. The same is true of the case of discontinuity at $x = c$ in the diagram. You will justify the nonexistence of a derivative at a vertical tangent line in the problem set.

The final fact you need to know before we get our feet wet finding actual derivatives without the big, bulky difference quotient is this important fact: If $f(x)$ is a differentiable function (in other words, f has derivatives everywhere), then f is a continuous function. (Remember from the discussion above that if a function is discontinuous, then it does not have a derivative.) However, the converse is *not* true: continuous functions do not necessarily have derivatives! Explore this further in the proceeding exercises.

PROBLEM SET

Do not use a graphing calculator on any of these problems.

1. Which function, among those you are to have memorized, is continuous but not differentiable on $(-\infty,\infty)$?

2. What is the average rate of change of $g(x) = \sin x$ on $\left[\dfrac{\pi}{2}, \dfrac{11\pi}{6} \right]$?

3. Knowing that the derivative is the slope of a tangent line to a graph, answer the following questions about $p(x)$ based on its graph.

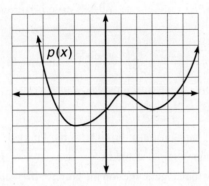

(a)   Where is $p'(x) = 0$?

(b)   Where is $p'(x) > 0$?

(c)   Where is $p'(x) < 0$?

(d)   Where is $p'(x)$ undefined?

4. Why does no derivative exist when the corresponding tangent line is vertical?

5. Set up, but do not evaluate, an expression that represents the derivative of $g(x)$ $= \csc x$ if $x = \frac{\pi}{4}$.

SOLUTIONS

1. $y = |x|$ is continuous but not differentiable at $x = 0$, because its graph makes a sharp point there.

2. The average rate of change is the slope of the line segment from $(\frac{\pi}{2}, 1)$ to $(\frac{11\pi}{6}, -\frac{1}{2})$, as shown in the diagram below.

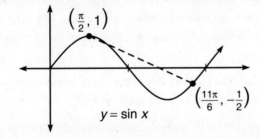

To find slope, calculate the change in y divided by the change in x:

$$\frac{y_2 - y_1}{x_2 - x_1} = \frac{-\frac{1}{2} - 1}{\frac{11\pi}{6} - \frac{\pi}{2}}$$

$$\frac{-\dfrac{3}{2}}{\dfrac{8\pi}{6}}$$

$$-\frac{9}{8\pi} \approx -.358.$$

3. (a) The derivative is zero when the tangent line is horizontal. This happens at $x = -2$, 1, and 3.

    (b) If you draw tangent lines throughout the interval, their slopes will be positive on the intervals $(-2,1)$ and $(3,\infty)$.

    (c) Again, drawing tangent lines on the graph shows tangents with negative slope on $(-\infty,-2)$ and $(1,3)$.

    (d) There are no cusps, discontinuities, or vertical tangent lines, so the derivative is defined everywhere.

4. The slope of a vertical line does not exist, and since the derivative is equal to the slope of a tangent line, the derivative cannot exist either.

5. Use either form of the difference quotient to get one of the following:

$$\lim_{\Delta x \to 0} \csc \frac{\left(\frac{\pi}{4}+\Delta x\right)-\csc\frac{\pi}{4}}{\Delta x}, \quad \lim_{x \to \frac{\pi}{4}} \frac{\csc x - \csc\frac{\pi}{2}}{x-\frac{\pi}{4}}, \text{ or } \lim_{x \to \frac{\pi}{4}} \frac{\csc x - 1}{x-\frac{\pi}{4}}.$$

TIP

Notice in 3(b) and 3(c) that $f'(x)$ is positive when $f(x)$ is increasing, and $f'(x)$ is negative when $f(x)$ is decreasing. More on this later—stay tuned.

THE POWER RULE

A few years ago, I attended a Faye Kellerman book signing with my father in California. The famous mystery writer politely asked what I did for a living as I handed her a copy of her most recent book, *Prayers for the Dead*, to sign for me. When she found out I was a calculus teacher, she confessed to taking many math classes in college, and that one of the most vivid things she remembered from calculus was how to take a derivative with the power rule. This sentiment is echoed by many of the people I meet—but only by those who will maintain eye contact with me when they find out I am a math teacher.

    The Power Rule is the most basic of derivative techniques, used when you encounter a variable raised to a constant power:

The Power Rule: If $y = x^c$, where c is a real number, then $y' = c \cdot x^{c-1}$.

TIP

Visually, you "pull the exponent down" in front of the variable and subtract one from the power.

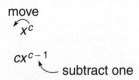

The Power Rule, so named because the variable is raised to a power, also works if a coefficient is present. In the case of $y = nx^c$, you still bring the original exponent c to the front and subtract one from the power, but now the c you brought to the front gets multiplied by the n that was already there. For example, if $y = 5x^4$, then by the Power Rule, $y' = 5 \cdot 4x^3 = 20x^3$.

That's the Power Rule in its entirety. Sometimes, the Power Rule will apply even though it is not obvious—some rewriting will be necessary first, as demonstrated below.

 Example 4: Use the product rule to find the derivative of each of the following:

(a) $y = x^8$

     This is a straightforward example: $y' = 8x^7$.

(b) $f(x) = 2x^3 + 9x^2$

This problem is a sum, so each term can be differentiated separately—the same would be true for a difference. Remember, when you "bring down" the powers, you need to multiply by the coefficients that are already present.

$$f'(x) = 3 \cdot 2x^2 + 2 \cdot 9x^1$$

$$f'(x) = 6x^2 + 18x$$

(c) $y = \dfrac{5}{x^2}$

     Rewrite this as $5x^{-2}$ to apply the Power Rule:

$$y' = 5(-2)x^{-3} = -10x^{-3}, \text{ or } -\frac{10}{x^3}.$$

(d) $y = -\dfrac{3p^2}{5p^6}$

This fraction can be simplified to $y = -\dfrac{3}{5}p^{-4}$, so $y' = -\dfrac{3}{5} \cdot (-4)p^{-5} = \dfrac{12}{5p^5}$.

(e) $y = 6$

You can rewrite this as $y = 6x^0$, since $x^0 = 1$ (as long as $x \neq 0$, but don't worry about that). If $y = 6x^0$, then $y' = 0 \cdot 6x^{-1} = 0$. This is an important fact: *The derivative of any constant term is zero.*

(f) $y = 7x + 5$

This is the same as $7x^1$, so $y' = 1 \cdot 7x^0 = 7 \cdot x^0 = 7 \cdot 1 = 7$. This is another important fact: *The derivative of a linear polynomial, $y = ax + b$, is a.*

Now that you have some practice under your belt, try some slightly trickier problems in Example 5. More rewriting is required here in order to apply the power rule.

 Example 5: Find $\dfrac{dy}{dx}$ for each of the following:

(a) $y = (x^2 - 1)(x + 5)$

You'll have to multiply these binomials together before you can apply the power rule. *You cannot simply find the derivative of each factor and then multiply those.* Remember, products and quotients have their own special rules, and we haven't gotten to those yet.

$$y = x^3 + 5x^2 - x - 5$$

$$\frac{dy}{dx} = 3x^2 + 10x - 1$$

(b) $y = (2x^3)^2$

Square the $(2x^3)$ term first to get $y = 4x^6$. Clearly, then, $\dfrac{dy}{dx} = 24x^5$.

(c) $y = \sqrt{x}\left(x^3 - \dfrac{3}{x^2}\right)$

Some rewriting and distributing is necessary to begin this problem:

$$y = x^{1/2}(x^3 - 3x^{-2})$$

$$y = x^{7/2} - 3x^{-3/2}.$$

Now the Power Rule applies:

$$y' = \frac{7}{2}x^{5/2} + \frac{9}{2}x^{-5/2}$$

$$y' = \frac{7}{2}x^{5/2} + \frac{9}{2x^{5/2}}$$

NOTE

The directions in Exercises 1 through 5 each ask you to find the derivative but in different ways. Be able to recognize different kinds of notation. More on the notation $\dfrac{d}{dx}$ and $\dfrac{dy}{dx}$ in the section entitled "A Word about Respecting Variables."

PROBLEM SET

Do not use a calculator on any of these problems.

1. Find $\dfrac{dy}{dx}$ if $y = 12x^\pi$.

2. If $f(x) = x^4 + 3x^2 - 1$, what is $f'(x)$?

3. Calculate $\dfrac{dy}{dx}\left(\dfrac{x^3}{4x^7}\right)$

4. Find $D_x\,[2x - 1]^3$.

5. Find y' if $y = \dfrac{\sqrt[3]{x} + 2x^5}{4\sqrt{x}}$.

6. What is the equation of the tangent line to $h(x) = 2x^3 - 3x + 5$ at the point $(1,4)$?

7. How many derivatives must you take of $g(x) = 4x^2 + 9x + 6$ until you get zero?

SOLUTIONS

1. π is just a real number, so you can still apply the power rule: $\dfrac{dy}{dx} = 12\pi x^{\pi-1}$.

2. Each of the terms can be differentiated separately, so $f'(x) = 4x^3 + 6x$. Don't forget that the derivative of any constant (in this case 1) will be 0.

3. Rewrite the expression to get $\dfrac{1}{4}\cdot\dfrac{1}{x^4}$ or $\dfrac{1}{4}x^{-4}$. The $\dfrac{1}{4}$ is now a constant and we can apply the power rule. The derivative will be $\dfrac{1}{4}\cdot(-4)x^{-5} = -\dfrac{1}{x^5}$.

4. A common mistake is to assume that the derivative is $3(2x - 1)^2$. This is *not* correct. Instead, expand the expression before applying the power rule:
$$y = (2x - 1)^3 = (2x - 1)\,(2x - 1)\,(2x - 1)$$
$$y = 8x^3 - 12x^2 + 6x - 1$$
$$y' = 24x^2 - 24x + 6.$$

5. This one is a little ugly to start with. Some rewriting and distributing will fix that:
$$y = \dfrac{1}{4}\cdot(x^{-1/2})(x^{1/3} + 2x^5)$$
$$y = \dfrac{1}{4}(x^{-1/6} + 2x^{9/2}).$$

Notice that the 4 in the denominator squirts out easily to become the coefficient $\dfrac{1}{4}$. This is not absolutely necessary, but it simplifies your calculations. As in past examples, leave the coefficient alone until the last step and then multiply it through:

CAUTION
Remember:
$x^{-1/2}\bullet x^{1/3} = x^{-1/6}$, because when exponential factors have the same base, you *add* the powers. Remember the rule from algebra: $x^a x^b = x^{a+b}$. In this case, multiplying the powers also gives you $x^{-1/6}$, but that's just a coincidence.

$$y' = \frac{1}{4}\left(-\frac{1}{6}x^{-7/6} + 9x^{7/2}\right)$$

$$y' = -\frac{1}{24 x^{7/6}} + \frac{9x^{7/2}}{4}$$

6. To create the equation of a line, you need a point and a slope; point-slope form will then easily follow. You are given the point (1,4), and the slope of a tangent line may be found by the derivative. Thus, the slope we need is given by $h'(1)$:

$$h'(x) = 6x^2 - 3$$

$$h'(1) = 6(1)^2 - 3 = 3.$$

Apply the point slope formula using $m = 3$ and $(x_1, y_1) = (1,4)$:

$$y - y_1 = m(x - x_1)$$

$$y - 4 = 3(x - 1)$$

$$y - 4 = 3x - 3$$

$$y = 3x + 1.$$

7. Each of the derivatives is pretty easy: $g'(x) = 8x + 9$, $g''(x) = 8$, $g'''(x) = 0$. The third derivative, then, will give you zero.

DERIVATIVES TO MEMORIZE

Before exploring other tools of differentiation, it is necessary to supplement your toolbox. If the Power Rule is the hammer of derivatives, it is pretty useless without a collection of different kinds of nails—one for every purpose. Calculus is full of derivatives, most of which cannot be nailed down by the Power Rule alone. In order to succeed in calculus and on the AP test, you'll need to be able to derive the functions in this section automatically, without even a second thought. In other words, memorize, memorize, memorize. Although memorizing may not be a glorious road to enlightenment, it has plenty of clean rest stops along the way with reasonably priced vending machines.

- Trigonometric Derivatives

Trigonometric functions and their derivatives are all over the AP test. By not memorizing these, you are crippling yourself and your chance to score well.

$$\frac{d}{dx}(\sin x) = \cos x$$

$$\frac{d}{dx}(\cos x) = -\sin x$$

$$\frac{d}{dx}(\tan x) = \sec^2 x$$

$$\frac{d}{dx}(\sec x) = \sec x \tan x$$

$$\frac{d}{dx}(\cot x) = -\csc^2 x$$

$$\frac{d}{dx}(\csc x) = -\csc x \cot x$$

Note the similarities between the tangent and cotangent functions as well as the cosecant and secant functions. It is also important to note that all the *co*-trigonometric derivatives are negative.

- Logarithmic Derivatives

In traditional courses, logarithmic and exponential derivatives are not introduced until the end of calculus. However, this is not the case any more. In fact, the functions are pervasive on the AP test, so it is common practice now to introduce them right away. The result is a lot more to memorize at first, but greater success and mastery later.

$$\frac{d}{dx}(\ln x) = \frac{1}{x}$$

$$\frac{d}{dx}(\log_a x) = \frac{1}{x \cdot (\ln a)}$$

You will see far more natural logs on the AP test, but logs with different bases occasionally make cameo appearances.

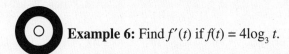 **Example 6:** Find $f'(t)$ if $f(t) = 4\log_3 t$.

Solution: Remember, when finding derivatives, you can ignore the coefficient and multiply through at the end. Therefore, according to the formula,

$$f'(t) = 4 \cdot \frac{1}{t(\ln a)}$$

$$f'(t) = \frac{4}{t(\ln 3)}$$

- Exponential Derivatives

It doesn't get much easier than exponential derivatives, especially e^x.

$$\frac{dy}{dx}(e^x) = e^x$$

$$\frac{dy}{dx}(a^x) = a^x \cdot \ln a$$

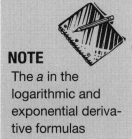

NOTE
The *a* in the logarithmic and exponential derivative formulas represents a constant.

Notice that the derivative of a^x involves multiplying by the natural log of the base used. This is similar to deriving $\log_a x$, in which you divide by the natural log of the base used. This makes sense, because exponential functions and logarithmic functions are inverse functions, just like multiplication and division are inverse operations.

Example 7: Find y' if $y = \csc x - \ln x + 5^x$ without consulting your derivative formulas.

Solution: If you are memorizing these formulas as you proceed, this problem is relatively simple: $y' = -\csc x \cot x - \dfrac{1}{x} + (\ln 5)5^x$.

• Inverse Trigonometric Function Derivatives

These functions revisit you later, during the integration section of the book. Knowing them now makes life so much easier down the road. Some textbooks use the notation $\sin^{-1}x$ to denote the inverse sine function, whereas many use $\arcsin x$. Both mean the same thing. I personally prefer $\arcsin x$ because $\sin^{-1}x$ looks hauntingly similar to $(\sin x)^{-1}$, or $\csc x$.

$$\frac{d}{dx}(\arcsin x) = \frac{1}{\sqrt{1-x^2}}$$

$$\frac{d}{dx}(\arccos x) = -\frac{1}{\sqrt{1-x^2}}$$

$$\frac{d}{dx}(\arctan x) = \frac{1}{1+x^2}$$

$$\frac{d}{dx}(\text{arccot } x) = -\frac{1}{1+x^2}$$

$$\frac{d}{dx}(\text{arcsec } x) = \frac{1}{|x|\sqrt{x^2-1}}$$

$$\frac{d}{dx}(\text{arccsc } x) = -\frac{1}{|x|\sqrt{x^2-1}}$$

Things start getting messy with arcsecant, but all these derivatives have things in common. Notice that the derivatives contain $(1+x^2)$, $\sqrt{1-x^2}$, and $\sqrt{x^2-1}$. Learning which denominator goes with which inverse function is the key to recognizing these later. Also note that the *co-* derivatives are again negative, as they were with ordinary trigonometric function derivatives.

PROBLEM SET

Do not use a calculator on any of these problems.

1.   Below is a list of all the functions whose derivatives were listed in this section. Complete the table without the use of notes.

Note: You may want to make a few photocopies of this problem and complete them every few weeks to keep the formulas fresh in your mind.

$\dfrac{d}{dx}(\cos x) = $ _____

$\dfrac{d}{dx}(e^x) = $ _____

$\dfrac{d}{dx}(\text{arccsc } x) = $ _____

$\dfrac{d}{dx}(\tan x) = $ _____

$\dfrac{d}{dx}(\arcsin x) = $ _____

$\dfrac{d}{dx}(\sec x) = $ _____

$\dfrac{d}{dx}(\arctan x) = $ _____

$\dfrac{d}{dx}(\log_a x) = $ _____

$\dfrac{d}{dx}(\cot x) = $ _____

$\dfrac{d}{dx}(\text{arcsec } x) = $ _____

$\dfrac{d}{dx}(\ln x) = $ _____

$\dfrac{d}{dx}(\sin x) = $ _____

$\dfrac{d}{dx}(\arccos x) = $ _____

$\dfrac{d}{dx}(\csc x) = $ _____

$\dfrac{d}{dx}(a^x) = $ _____

$\dfrac{d}{dx}(\text{arccot } x) = $ _____

2. Name each function whose derivative appears below:

    (a) $\sec x \tan x$

    (b) $-\dfrac{1}{x}$

    (c) $\dfrac{1}{x^2 + 1}$

    (d) e^x

    (e) $-\dfrac{1}{\sqrt{1 - x^2}}$

    (f) $\csc^2 x$

3. Find each of the following derivatives:

    (a) $\sec x + \operatorname{arcsec} x$

    (b) $2x^2 + \cos x - 8e^x + \operatorname{arccot} x$

    (c) $4^x + \log_4 x - \ln x$

4. Name a function, $g(x)$, other than e^x such that $g'(x) = g(x)$.

SOLUTIONS

1. Check answers with the formulas listed previously in this section.

2.   (a) $\sec x$

    (b) $-\ln x$: The -1 is a coefficient, so $\dfrac{d}{dx}(-\ln x) = -1 \cdot \dfrac{1}{x} = -\dfrac{1}{x}$.

    (c) $\arctan x$: By the commutative property of addition, $1 + x^2 = x^2 + 1$.

    (d) e^x

    (e) $\arccos x$

    (f) $-\cot x$: By the same reasoning as 2(b) above.

3.   (a) $\sec x \tan x + \dfrac{1}{|x|\sqrt{x^2 - 1}}$

    (b) $4x - \sin x - 8e^x - \dfrac{1}{1 + x^2}$

    (c) $(\ln 4)4^x + \dfrac{1}{(\ln 4)x} - \dfrac{1}{x}$

4. The easiest such function is $g(x) = 0$, since the derivative of any constant is 0.

THE CHAIN RULE

The Power Rule, although wonderful in some instances, falls short in others. For example, $\dfrac{d}{dx}(x^3)$ is solved easily using the Power Rule, and the

NOTE

It is no coincidence that chain rhymes with pain. You will use the chain rule often for the remainder of this course, so be sure to understand it, or the pain will be relentless.

answer is $3x^2$. However, if cos x is cubed instead of just x, the Power Rule fails! The derivative of $\cos^3 x$ is *not* $3(\cos x)^2$. Another rule is necessary, and it is called the *Chain* (rhymes with pain) *Rule*.

The Chain Rule applies to all composite functions (expressions in which one function is plugged into another function). For example, $(\cos x)^3$ is the cosine function plugged into the cubed function. Mathematically, you can write

$$f(x) = x^3; \ g(x) = \cos x$$

$$f(g(x)) = f(\cos x) = (\cos x)^3 .$$

To make things clearer, you can refer to cos x as the "inner function" and x^3 as the "outer function."

Example 8: Each of the following is a composition of functions. For each, identify which is the "inner" and which is the "outer" function.

(a) $\sqrt{\csc x}$

The inner function is csc x, and the outer function is $\sqrt{x}$. Note that the outer function always acts upon the inner function; in other words, the inner function is always "plugged into" the outer function.

(b) sec $3x$

Because $3x$ is plugged into secant, $3x$ is the inner function, and sec x is the outer.

(c) $\ln (x^2 + 4)$

The inner function is $x^2 + 4$, and the outer function is ln x. Again, you are finding the natural log of $x^2 + 4$, so $x^2 + 4$ is being plugged into ln x.

(d) $3^{\sin x}$

The inner function is sin x, and the outer function is 3^x. Here, the exponential function is being raised to the sin x power, so sin x is being plugged into 3^x.

With this terminology, the Chain Rule is much easier to translate and understand.

The Chain Rule: The derivative of $f(g(x))$, with respect to x, is $f'(g(x)) \cdot g'(x)$.

Translation: In order to find the derivative of a composite function, take the derivative of the outer function, leaving the inner function alone; then, multiply by the derivative of the inner function.

Now it is possible to find the derivative of $f(x) = \cos^3 x$. Remember, $\cos^3 x$ is the same as $(\cos x)^3$. The outer function is x^3, so we use the power rule to take the derivative, *leaving the inner function, cos x, alone*:

TIP
When applying the Chain Rule, use the mantra, "Take the derivative of the outside, leaving the inside alone, then multiply by the derivative of the inside." It becomes automatic if you say it enough.

$$f'(\text{x}) = 3(\cos x)^2$$

But, we already said that wasn't right! That's because we weren't finished. Now, multiply that by the derivative of the inner function:

$$f'(\text{x}) = 3(\cos x)^2 \cdot (-\sin x)$$

$$f'(\text{x}) = -3\cos^2 x \sin x.$$

 Example 9: Find $\dfrac{d}{dx}$ for each of the following:

CAUTION

Don't forget to leave the inner function alone when you begin the chain rule! For example,

$\dfrac{d}{dx}(\tan 4x) \neq$

$\sec^2(4)$. Instead, the derivative is $\sec^2(4x) \cdot 4$.

(a) $\sqrt{\csc x}$

This expression can be rewritten $(\csc x)^{1/2}$. You already know the inner and outer functions from Example 8. Thus, the derivative is as follows:

$$\frac{1}{2}(\csc x)^{-1/2} \cdot (-\csc x \cot x)$$

$$-\frac{\csc x \cot x}{\left(2\sqrt{\csc x}\right)}$$

$$-\frac{1}{2}\sqrt{\csc x}\cot x.$$

(b) $\sec 3x$

The derivative of $\sec x$ is $\sec x \tan x$, so the derivative of $\sec 3x$ is as follows:

$$\sec 3x \tan 3x \cdot (3),\ \text{or}$$

$$3\sec 3x \tan 3x,$$

since 3 is the derivative of the inner function 3x.

(c) $\ln (x^2 + 4)$

The derivative of $\ln x$ is $\dfrac{1}{x}$, so this derivative will be $\dfrac{1}{\left(x^2+4\right)}$ times the derivative of $(x^2 + 4)$:

$$\frac{1}{x^2+4} \cdot 2x = \frac{2x}{x^2+4}.$$

(d) $3^{\sin x}$

Remember, the derivative of 3^x is $3^x \cdot (\ln 3)$, so the derivative of $3^{\sin x}$ is as follows:

$$3^{\sin x} \cdot (\ln 3) \cdot (\cos x)$$

$$(\ln 3)\cos x\, 3^{\sin x}.$$

PROBLEM SET

Do not use calculators on any of these problems.

In questions 1 through 6, find $\dfrac{d}{dx}$.

1. $\ln(7x^3 + 2\sin x)$

2. $\sqrt[3]{x^3 + x - e^x}$

3. $\csc(\arcsin x^2)$

4. 4^{5e^x}

5. $\tan(\cos(3x + 4))$

6. $e^{6\ln(\operatorname{arcsec} e^x)}$

7. Given f is a continuous and differentiable function such that f and f' contain the values given by the below table:

x	0	1	2	3	4
$f(x)$	7	3	−1	2	5
$f'(x)$	$-\dfrac{1}{2}$	1	6	4	−3

   (a) Evaluate $m'(2)$ if $m(x) = (f(x))^3$.

   (b) Evaluate $g'(3)$ if $g(x) = \arctan(f(x))$.

   (c) Evaluate $k'(0)$ if $k(x) = f(e^x)$.

8. Let h be a continuous and differentiable function defined on $[0,2\pi]$. Some function values of h and h' are given by the chart below:

x	0	$\dfrac{\pi}{2}$	π	$\dfrac{3\pi}{2}$	2π
$h(x)$	3	$\dfrac{\pi}{6}$	$\dfrac{\pi}{4}$	0	−2
$h'(x)$	$-\dfrac{\pi}{3}$	$\dfrac{3\pi}{2}$	−1	$\dfrac{\pi}{2}$	1

If $p(x) = \sin^2(h(2x))$, evaluate $p'\left(\dfrac{\pi}{2}\right)$.

SOLUTIONS

1. $\dfrac{1}{7x^3 + 2\sin x} \cdot \left(21x^2 + 2\cos x\right) = \dfrac{21x^2 + 2\cos x}{7x^3 + 2\sin x}$.

2. Rewrite the expression as $(x^3 + x - e^x)^{1/3}$. The derivative will be

   $\dfrac{1}{3}(x^3 + x - e^x)^{-2/3} \cdot (3x^2 + 1 - e^x)$, or $\dfrac{3x^2 + 1 - e^x}{3\left(x^3 + x - e^x\right)^{2/3}}$.

NOTE
Problem 6 uses the logarithmic property $\log_a x^n = n \cdot \log_a x$.

3. $-\csc(\arcsin x^2)\cot(\arcsin x^2) \cdot \dfrac{1 \cdot 2x}{\sqrt{1 - \left(x^2\right)^2}} =$

$-\dfrac{2x\csc\left(\arcsin x^2\right)\cot\left(\arcsin x^2\right)}{\sqrt{1-x^4}}$. In this problem, the x^2 is plugged into the arcsin x function, which is then plugged into the csc x function. The Chain Rule is applied twice.

4. $4^{5e^x} \cdot (\ln 4) \cdot 5e^x$. When you derive 4^u, you get $4^u \cdot (\ln 4) \cdot u'$. The u is the inner function and is left alone at first. Once the outer function is differentiated, you mutliply by the derivative of u, u'.

5. The Chain Rule is applied twice in this problem. Begin with the outermost function (leaving the rest alone) and work your way inside:

$\sec^2(\cos(3x+4)) \cdot (-\sin(3x+4)) \cdot 3 = -3\sec^2(\cos(3x+4)) \cdot \sin(3x+4)$.

NOTE
It is not necessary to write $|e^x|$ in the arcsecant derivative formula because e^x has no negative range elements.

6. Using logarithmic properties, rewrite the expression as $e^{\ln\left(\arccsc e^x\right)^6}$. Because e^x and $\ln x$ are inverse functions, they cancel each other out, and you get $(\text{arcsec } e^x)^6$. The Chain Rule will be applied three times when you differentiate:

$$6(\text{arcsec } e^x)^5 \cdot \dfrac{1}{e^x\sqrt{\left(e^x\right)^2 - 1}} \cdot e^x$$

$$\dfrac{6e^x(\text{arcsec} e^x)}{e^x\sqrt{(e^x)^2 - 1}}$$

$$\dfrac{6e^x(\text{arcsec} e^x)}{\sqrt{(e^x)^2 - 1}}$$

7. (a) $m'(x) = 3(f(x))^2 \cdot f'(x)$, so $m'(2) = 3(f(2))^2 \cdot f'(2) = 3(-1)^2(6) = 18$.

(b) $g'(x) = \dfrac{1}{1 + \left(f(x)\right)^2} \cdot f'(x)$, so $g'(3) = \dfrac{1}{1 + f(3)^2} \cdot f'(3) = \dfrac{1}{1+4} \cdot 4$

$= \dfrac{4}{5}$.

(c) $k'(x) = f'(e^x) \cdot e^x$, so $k'(0) = f'(e^0) \cdot e^0 = f'(1) \cdot 1 = 1 \cdot 1 = 1$.

8. The Chain Rule will have to be applied three times in this example. To make it easier to visualize, we have underlined the "inner" functions to be left alone as we find the derivative. Notice how the $2x$ is left alone until the very end.

$$p'(x) = 2(\sin(h(2x))) \cdot \cos(h(2x)) \cdot h'(2x) \cdot 2$$

Because $2x = 2\left(\dfrac{\pi}{2}\right) = \pi$, you can write:

$$p'\left(\dfrac{\pi}{2}\right) = 2(\sin(h(\pi))) \cdot \cos(h(\pi)) \cdot h'(\pi) \cdot 2$$

$$p'\left(\dfrac{\pi}{2}\right) = 2\left(\sin\left(\dfrac{\pi}{4}\right)\right) \cdot \cos\left(\dfrac{\pi}{4}\right) \cdot (-1) \cdot 2$$

$$p'\left(\dfrac{\pi}{2}\right) = 2 \cdot \dfrac{\sqrt{2}}{2} \cdot \dfrac{\sqrt{2}}{2} \cdot -2$$

$$p'\left(\dfrac{\pi}{2}\right) = -2.$$

THE PRODUCT RULE

If asked to find the derivative of $f(x) = \sin x + \cos x$, you should have no trouble by now. You wouldn't need to furrow your brow and scratch your chin like a gorilla trying to determine how to file its federal tax return. You'd know that f is the sum of two functions, so the derivative is simply the sum of the individual funtions' derivatives: $f'(x) = \cos x - \sin x$.

Given the function $g(x) = \sin x \cos x$, you might be tempted to use the same strategy and give the derivative $g'(x) = (\cos x)(-\sin x)$, but *this is not correct*. Any time you want to find the derivative of a product of two non-constant functions, you must apply the Product Rule:

The Product Rule: If $h(x) = f(x)g(x)$, then $h'(x) = f(x)g'(x) + g(x)f'(x)$.

Translation: To find the derivative of a product, differentiate one of the factors and multiply by the other; then, reverse the process, and add the two results together.

For example, in order to differentiate $g(x) = \sin x \cos x$, you multiply $\sin x$ by the derivative of $\cos x$ and add to that $\cos x$ times the derivative of $\sin x$:

$$g'(x) = \sin x\,(-\sin x) + \cos x\,(\cos x)$$

$$g'(x) = \cos^2 x - \sin^2 x$$

$$g'(x) = \cos 2x \text{ (by a double angle formula)}.$$

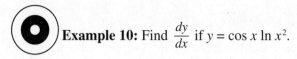 **Example 10:** Find $\dfrac{dy}{dx}$ if $y = \cos x \ln x^2$.

Solution: Because you are given a product of two functions, apply the Product Rule. I have denoted which derivatives to take below with a prime symbol ('). It's not wonderful notation, but it gets the point across:

CAUTION
Remember that the Product Rule is the *sum* (not the product) of $f(x)g'(x)$ and $g(x)f'(x)$.

NOTE
You can reverse $f(x)$ and $g(x)$ in the Product Rule by the commutative property of multiplication, and it still works. In other words, it doesn't matter which function is "first."

NOTE
Because of the commutative property, you can write $g(x) = \cos x \sin x$ and reverse the Product Rule. You will achieve the same result.

NOTE

To find the derivatives of ln x^2, arctan 2x, and e^{4x^2} in Examples 10 and 11, you need to apply the Chain Rule.

$$y' = \cos x \, (\ln x^2)' + \ln x^2 \, (\cos x)'$$

$$y' = \cos x \cdot \frac{1}{x^2} \cdot 2x + \ln x^2 \, (-\sin x)$$

$$y' = \frac{2\cos x}{x} - \sin x \ln x^2.$$

 Example 11: Find $\dfrac{dy}{dx}$ using the Product Rule if $y = \dfrac{\arctan 2x}{e^{-4x^2}}$.

Solution: You can rewrite the expression as the product

$$y = (\arctan 2x)(e^{4x^2})$$

and apply the Product Rule as follows:

$$y' = (\arctan 2x)(e^{4x^2})' + (e^{4x^2})(\arctan 2x)'$$

$$y' = (\arctan 2x)(e^{4x^2})(8x) + (e^{4x^2}) \cdot \frac{1}{1+4x^2} \cdot 2$$

$$y' = 8x e^{4x^2}(\arctan 2x) + \frac{2e^{4x^2}}{1+4x^2}$$

PROBLEM SET

Complete the following without a calculator.

1. Why don't you have to use the Product Rule to find the derivative of $y = 7x$?

2. Find $\dfrac{dy}{dx}$ if $y = x^2\sqrt{3x + \sin x}$.

3. Find $\dfrac{dp}{dx}$ if $p = \sec x \cdot \text{arcsec } x$.

4. Find $\dfrac{d}{dx}((x^2 + 3)^4 \cdot (2x^3 + 5x)^3)$.

5. Evaluate $y'(x)$ if $y = \sin x \cos x \tan x$.

6. Evaluate $h'(2)$ if $h(x) = 2^x \cdot \log_2 x$.

7. Given functions f and g such that $f(5) = 1, f'(5) = -2, g(5) = 4$, and $g'(5) = \dfrac{2}{3}$, if $k(x) = g(x)\sqrt[4]{f(x)}$, evaluate $k'(5)$.

SOLUTIONS

1. The Product Rule must be applied when deriving the product of non-constant functions. 7 is a constant. However, the Product Rule will work when finding $\dfrac{d}{dx}(7x)$—try it! Remember that the derivative of 7 is 0 since 7 is a constant.

2. $$\frac{dy}{dx} = x^2(\sqrt{3x+\sin x}\,)' + \sqrt{3x+\sin x}\,(x^2)'$$

    $$\frac{dy}{dx} = x^2 \cdot \frac{1}{2}(3x+\sin x)^{-1/2} \cdot (3+\cos x) + \sqrt{3x+\sin x} \cdot 2x$$

    $$\frac{dy}{dx} = \frac{x^2(3+\cos x)}{2\sqrt{3x+\sin x}} + 2x\sqrt{3x+\sin x}$$

3. These functions are *inverses*, not *reciprocals*, so their product is not 1.

    $$\frac{dp}{dx} = \frac{\sec x}{|x|\sqrt{x^2-1}} + \operatorname{arcsec} x \sec x \tan x$$

4. If you wish, you can expand each of the polynomials and multiply them together; only the Power Rule will be necessary, but you'll be multiplying all day long. Best to use the Power, Chain, and Product Rules all combined:

    $$(x^2+3)^4 \cdot 3(2x^3+5x)^2 \cdot (6x^2+5) + (2x^3+5x)^3 \cdot 4(x^2+3)^3 \cdot 2x.$$

    You can rewrite the expression to make it slightly more pretty, but that is not necessary.

5. You can generalize the Product Rule to any number of factors. The key is to take only one derivative at a time, leaving the other factors alone (see sidebar note).

    $$y' = \sin x \cos x\,(\sec^2 x) + \sin x\,(-\sin x)\tan x + (\cos x)\cos x \tan x$$

    $$y' = \sin x \cos x \sec^2 x - \sin^2 x \tan x + \cos^2 x \tan x$$

6. To find $h'(2)$, find $h'(x)$ and substitute 2 for x.

    $$h'(x) = 2^x \cdot \frac{1}{x\ln 2} + \log_2 x \cdot 2^x \ln 2$$

    $$h'(x) = 2^x \left(\frac{1}{x\ln 2} + \ln 2 \log_2 x \right)$$

7. The function can be rewritten as

    $$k(x) = g(x)(f(x))^{1/4}$$

    and now the Product Rule can be applied.

    $$k'(x) = g(x) \cdot \frac{1}{4}(f(x))^{-3/4} \cdot f'(x) + (f(x))^{1/4} \cdot g'(x)$$

    $$k'(5) = g(5) \cdot \frac{1}{4}(f(5))^{-3/4} \cdot f'(5) + (f(5))^{1/4} \cdot g'(5)$$

NOTE

The derivative of $h(x)$ in Number 5 is $f(x)g(x)h'(x) + f(x)g'(x)h(x) + f'(x)g(x)h(x)$. Only one derivative is taken at a time.

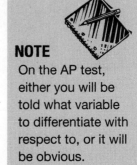

NOTE

On the AP test, either you will be told what variable to differentiate with respect to, or it will be obvious.

$$k'(5) = 4 \cdot \frac{1}{4} \cdot (1)^{-3/4} \cdot (-2) + (1)^{1/4} \cdot \frac{2}{3}$$

$$k'(5) = -2 + \frac{2}{3} = -\frac{4}{3}$$

THE QUOTIENT RULE

Just like products, quotients of functions require their own special method of differentiation. Because multiplication and division are so closely related, you can use the Product Rule to develop the Quotient Rule from scratch. In Example 12, your objective will be to create the Quotient Rule.

 Example 12: Use the Product Rule to find the derivative of $y = \frac{f(x)}{g(x)}$.

Solution: You must rewrite the quotient as a product before you can begin: $y = f(x) \cdot (g(x))^{-1}$. Now, apply the Product Rule and simplify completely.

$$y' = f(x) \cdot (-(g(x))^{-2}) \cdot g'(x) + g(x)^{-1} \cdot f'(x)$$

$$y' = -\frac{f(x)g'(x)}{(g(x))^2} + \frac{(f'(x))}{g(x)}$$

Multiply the second term by $\frac{g(x)}{g(x)}$ to get common denominators and then combine the terms.

$$y' = \frac{g(x)f'(x) - f(x)g'(x)}{(g(x))^2}$$

The Quotient Rule: If $y = \frac{f(x)}{g(x)}$, then $y' = \frac{g(x)f'(x) - f(x)g'(x)}{(g(x))^2}$.

Translation: If a fraction is formed by two functions, the derivative is found by multiplying the bottom by the derivative of the top minus the top times the derivative of the bottom, all divided by the bottom squared. Some people use the following verbal device to remember the Quotient Rule:

$$\frac{d}{dx}\left(\frac{\text{Hi}}{\text{Ho}}\right) = \frac{\text{Ho } d\text{Hi less Hi } d\text{Ho}}{\text{Ho Ho}}.$$

 Example 13: Find $p'(t)$ if $p(t) = \frac{3t^2}{\cot 5t}$.

Solution: Applying the Quotient Rule, the derivative is as follows:

$$p'(t) = \frac{\cot 5t \cdot 6t - 3t^2\left(-\csc^2 5t \cdot 5\right)}{\cot^2 5t}$$

$$p'(t) = \frac{6t\cot 5t + 15t^2 \csc^2 5t}{\cot^2 5t}.$$

 Example 14: Find $h'(e)$ if $h(x) = \text{arccot } x \cdot \left(\frac{3^{2x}}{\ln x}\right)$.

Solution: This problem begins with the Product Rule.

$$h'(x) = \text{arccot } x \left(\frac{3^{2x}}{\ln x}\right)' + \left(\frac{3^{2x}}{\ln x}\right)(\text{arccot } x)'$$

Finding the derivative of $\left(\frac{3^{2x}}{\ln x}\right)$ will require the Quotient and Chain Rules (it's gonna get messy, baby).

$$h'(x) = \text{arc cot } x \cdot \frac{(\ln x)2\ln 3 \cdot 3^{2x} - 3^{2x} \cdot \frac{1}{x}}{(\ln x)^2} + \frac{3^{2x}}{\ln}\left(-\frac{1}{1+x^2}\right)$$

That sure isn't pretty. Now find $h'(e)$, and remember that $\ln e = 1$. That will remove some of the grime.

$$h'(x) = \text{arc cot } e \cdot \frac{2\ln 3 \cdot 3^{2e} - 3^{2e} \cdot \frac{1}{e}}{1^2} + \frac{3^{2e}}{1}\left(-\frac{1}{1+e^2}\right)$$

You can find the decimal value of this load using your calculator, but why put yourself through that much agony? If you do a single thing wrong when entering it into your calculator, you could lose those points you earned the hard way; be a heads-up test taker—if simplifying isn't worth it, leave the answer unsimplified.

PROBLEM SET

You may use your calculator only on Number 4.

1. Find $f'(1)$ if $f(x) = \frac{x^2 \ln x}{3x^3 - 1}$.

2. Prove that $\frac{d}{dx}(\tan x) = \sec^2 x$ using the Quotient Rule.

3. Given $y = \frac{\csc 2x}{e^{-x}}$, find $\frac{dy}{dx}$ using *two different methods*.

> **CAUTION**
>
> There are all kinds of ways to screw up the Quotient Rule. Often, people concentrate so hard on getting the numerator of the Quotient Rule correct, they forget to write the denominator. Don't forget to square your bottom!

4. If $f(x)$ and $g(x)$ are continuous and differentiable functions with some values given by the table below, find $h'(3)$ if $h(x) = \dfrac{f(2x)}{3g(x)}$.

x	$f(x)$	$g(x)$	$f'(x)$	$g'(x)$
3	−2.14	1.05	4.13	−.016
6	3.97	4.11	−2.83	3.28

SOLUTIONS

1. The derivative of the numerator will require the use of the Product Rule, so be cautious.

$$f'(x) = \frac{\left(3x^3 - 1\right)\left(x^2 \cdot \frac{1}{x} + 2x\ln x\right) - \left(x^2 \ln x\right)\left(9x^2\right)}{\left(3x^3 - 1\right)^2}$$

$$f'(1) = \frac{\left((3-1)(1 + 2(\ln 1)) - (1 \cdot \ln 1)(9)\right)}{2^2}$$

$$f'(1) = \frac{(2 \cdot 1 - 1 \cdot 0 \cdot 9)}{4} = \frac{1}{2}$$

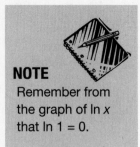

NOTE
Remember from the graph of ln x that ln 1 = 0.

2. You can rewrite $\tan x$ as $\dfrac{\sin x}{\cos x}$ and use the Quotient Rule to find the derivative.

$$\frac{d}{dx}\left(\frac{\sin x}{\cos x}\right) = \frac{\cos x \cdot \cos x - \sin x \cdot (-\sin x)}{\cos^2 x}$$

$$= \frac{\cos^2 x + \sin^2 x}{\cos^2 x}$$

$$= \frac{1}{\cos^2 x} \text{ (by Mamma Theorem)}$$

$$= \sec^2 x$$

3. *Method One: Quotient Rule*

$$\frac{dy}{dx} = \frac{e^{-x}\left(-\csc 2x \cot 2x\right) \cdot 2 - \csc 2x \cdot -e^{-x}}{e^{-2x}}$$

Certainly, simplification is possible, but the answer is quite messy with negative exponents.

Method Two: Product Rule

Rewrite the expression as a product, and things are much easier.

$$y = (\csc 2x)(e^x)$$

$$\frac{dy}{dx} = e^x \cdot \csc 2x + e^x - \csc 2x \cot 2x \cdot 2$$

$$\frac{dy}{dx} = e^x \csc 2x(1 - 2\cot 2x)$$

Both solutions are equivalent, although it is not obvious to the naked eye. Given a choice, Product Rule is clearly the way to go, even though the original problem was written as a quotient.

4. You may begin by factoring $\frac{1}{3}$ out of the expresion to eliminate the 3 in the denominator, but that is not required.

$$h'(3) = \frac{3g(3) \cdot 2f'(2 \cdot 3) - f(2 \cdot 3) \cdot 3g'(3)}{\left(3g(3)\right)^2}$$

$$h'(3) = \frac{3 \cdot 1.05 \cdot 2(-2.83) - 3.97 \cdot 3(-.016)}{(3 \cdot 1.05)^2}$$

$$h'(3) = \frac{-17.829 + .19056}{9.9225}$$

$$h'(3) \approx -1.776$$

A WORD ABOUT RESPECTING VARIABLES

If asked to find the derivative of x, you would probably give an answer of 1. Technically, however, the derivative of x is dx. (Similarly, the derivative of y is dy, etc.) The derivative of x is only 1 when you are differentiating with respect to x. Examine more closely the notation $\frac{dy}{dx}$ that you have unknowingly used all this time. This notation literally means "the derivative of y with respect to x." The numerator contains the variable you are differentiating, and the denominator contains the variable you are "respecting." Using this notation, the derivative of x with respect to x is $\frac{dx}{dx}$, or 1.

Let's look at a more complicated example. You know that the derivative of x^3 (with respect to x) is $3x^2$. Let's be even more careful about the process and use the Chain Rule. The derivative of x^3 with respect to x is $3x^2$ times the derivative of what's inside, x, with respect to x: $\frac{dx}{dx}$. The final answer is $3x^2 \cdot \frac{dx}{dx} = 3x^2 \cdot 1 = 3x^2$. If the variable in the expression matches the variable you are "respecting," you differentiate like you have in previous sections. When the variables don't match, however, things get a little more bizzare.

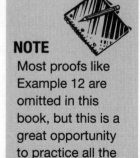

NOTE
Most proofs like Example 12 are omitted in this book, but this is a great opportunity to practice all the methods you've learned thus far.

Example 15: Find $\frac{d}{dx}(5y^4)$, the derivative of $5y^4$ with respect to x.

Solution: Apply the Chain Rule to this expression: the y is the inner function, substituted into x^4. Thus, take the derivative of the outer function leaving y alone,

$$20y^3,$$

and then multiply by the derivative of the inside (y) with respect to x $\left(\frac{dy}{dx}\right)$.

Your final answer is

$$20y^3 \cdot \frac{dy}{dx}.$$

It is very important to notice when you are differentiating with respect to variables that don't match those in the expression. This only happens rarely, but it is important when it happens.

The television show *The Simpsons* used this concept of differentiation in the episode where Bart cheats his way into a school for smart kids. The teacher asks the students to determine what is funny about the derivative of $\frac{r^3}{3}$. The derivative is (technically) r^2dr, or $r \cdot dr \cdot r$ (hardy-har-har). In this problem, dr is the derivative of r because no respecting variable was defined. Had the teacher asked for the derivative of $\frac{r^3}{3}$ with respect to r, the joke would have been less funny, since that derivative is r^2.

Example 16: Find the derivative of each with respect to x:

(a) t

The derivative of t with respect to x is written $\frac{dt}{dx}$.

(b) $7x^3 \sin x$

This expression contains two different functions, so you need to use the Product Rule: $7x^3 \cos x + 21x^2 \sin x$. There is no fraction like the $\frac{dt}{dx}$ from the last problem, because you are taking the derivative with respect to the variable in the expression (they match).

(c) $\cos y^2$

This is a composite function and will require the Chain Rule: The y is plugged into x^2, which is plugged into $\cos x$. The derivative, according to the Chain Rule, is $-\sin y^2 \cdot 2y \cdot \frac{dy}{dx}$. Notice that you leave the y alone until the very end, as it is the innermost function.

    (d) $2xy$

This is the product of $2x$ and y, so use the Product Rule: $2x \cdot \dfrac{dy}{dx} + y \cdot 2$. The derivative of $2x$ with respect to x is simply 2.

PROBLEM SET

Do not use a graphing calculator.

1. What is meant by $\dfrac{dg}{dc}$? Create an expression whose derivative contains $\dfrac{dg}{dc}$.

2. Find the derivative of each with respect to y:

    (a) $\csc 2y$

    (b) e^{x+y}

    (c) $\cos(\ln x) + xy^2$

SOLUTIONS

1. $\dfrac{dg}{dc}$ is the derivative of g with respect to c. This expression appears whenever you differentiate an expression containing g's with respect to c. For example, the derivative of g^2 with respect to c (mathematically, $\dfrac{d}{dc}(g^2)$ is $2g \cdot \dfrac{dg}{dc}$.

2. (a) $-2\csc 2y \cot 2y$: Don't forget to leave the $2y$ alone as you find the derivative of cosecant; finally, multiply by the derivative of $2y$ with respect to y, which is 2.

    (b) $\left(\dfrac{dx}{dy}+1\right)e^{x+y}$: Remember, you are looking for the derivative of x with respect to y in this problem.

    (c) This problem requires the Chain Rule for the first term and the Product Rule for the second:

$$-\sin(\ln x) \cdot \frac{1}{x} \cdot \frac{dx}{dy} + x(2y) + y^2 \frac{dx}{dy}.$$

IMPLICIT DIFFERENTIATION

At this point in your long and prosperous calculus life, you are no longer intimidated by questions such as "Find $\dfrac{dy}{dx}$: $y = 2\sin x + 3x^2$." Without a second thought (hopefully), you would answer $\dfrac{dy}{dx} = 2\cos x + 6x$. Sometimes, however, the questions are not solved for a variable like the above problem was solved for y. In fact, some things you differentiate aren't even functions, like the above example is a function of x. In such circumstances, you will employ *implicit differentiation*. In order to differentiate implicitly, you need to have a basic understanding of what it means to differentiate with respect to a variable (the previous topic in this chapter).

CAUTION
You must be at least 18 to purchase a book that discusses "explicit" differentiation.

Steps for Success in Implicit Differentiation

(These steps assume that you are finding $\frac{dy}{dx}$, as it is your goal 85 percent of the time. If other variables are used, adjust accordingly.)

1. Find the derivative of the entire equation with respect to x.

2. Solve for $\frac{dy}{dx}$.

3. If a specific solution is required, substitute in the corresponding x and y values.

 Example 17: Find the equation of the tangent line to the circle $x^2 + y^2 = 9$ when $x = 1$.

Solution: In order to find the equation of a line, you need a point and a slope. It is very simple to find the point. Substitute $x = 1$ into the formula to find its corresponding y value. This coordinate pair $(1, \sqrt{8})$ marks the point of tangency, as shown in the figure below.

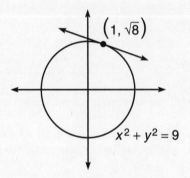

CAUTION
Don't forget to take the derivative of the constant; $\frac{d}{dx}(9) = 0$.

All that remains is to find the slope of the tangent line, which is given by the derivative, $\frac{dy}{dx}$, You need to find this derivative implicitly, however. To do so, first find the derivative of everything with respect to x:

$$2x + 2y\frac{dy}{dx} = 0.$$

To complete the problem, solve for $\frac{dy}{dx}$:

$$2y\frac{dy}{dx} = -2x$$

$$\frac{dy}{dx} = -\frac{2x}{2y} = -\frac{x}{y}.$$

Hence, the derivative for any point (x,y) on the circle is given by the formula $-\frac{x}{y}$. Therefore, the slope of the tangent line at $(1, \sqrt{8})$ is $-\frac{1}{\sqrt{8}}$. Using point-slope form for a line, the equation of the tangent line is

$$y - \sqrt{8} = -\frac{1}{\sqrt{8}}(x - 1).$$

 Example 18: Find the equation of the *normal* line to $\sin(x) + e^{xy} = 3$ when $x = \pi$.

Solution: In order to find the slope of the normal line, you need to take the opposite reciprocal of the slope of the tangent line (the derivative), since they are perpendicular. It is far from easy to solve this equation for y, so you should differentiate implicitly. Again, start by finding the point on the normal line by plugging in $x = \pi$:

$$\sin(\pi) + e^{\pi y} = 3$$

$$0 + e^{\pi y} = 3$$

$$\ln(e^{\pi y}) = \ln(3)$$

$$\pi y = \ln 3$$

$$y = \frac{\ln 3}{\pi} \approx .3496991526.$$

Thus, the tangent and normal lines both pass through $(\pi, .3496991526)$.

Now, find the derivative of the equation with respect to x to get the slope of the tangent line, $\frac{dy}{dx}$:

$$\cos x + e^{xy} \cdot (y + x\frac{dy}{dx}) = 0.$$

Distribute e^{xy} and solve for $\frac{dy}{dx}$:

$$\cos x + ye^{xy} + xe^{xy}\frac{dy}{dx} = 0$$

$$xe^{xy}\frac{dy}{dx} = -\cos x - ye^{xy}$$

$$\frac{dy}{dx} = -\frac{\cos x + ye^{xy}}{xe^{xy}}.$$

Plug in the coordinate $(\pi, .3496991526)$ for (x,y) to get the slope of the tangent line there:

$$\frac{dy}{dx} = -\frac{\cos \pi + .3496991526e^{1.098612289}}{\pi e^{1.098612289}}$$

NOTE
A *normal line* is perpendicular to the tangent line at the point of tangency.

TIP
To keep resulting answers accurate, you shouldn't round to .350 until the problem is completely over.

CAUTION
You must use the Product Rule to differentiate xy.

$$\frac{dy}{dx} = -.0052094021.$$

If this is the slope of the tangent line at $(\pi, .3496991526)$, then the slope of the normal line is the negative reciprocal of $-.0052094021$, or 191.9606091 Therefore, the equation of the normal line is

$$y - .350 = 191.961(x - \pi).$$

(You are allowed to round at the very end of the problem on the AP test.)

PROBLEM SET

You may use a graphing calculator on Number 4 only.

1. Find $\frac{dy}{dx}$: $e^{y^2} + 3y = \tan x$.

2. What is the slope of the tangent line to $\ln(xy) + y^2 = 2y + 3x$ at the point $(e,1)$?

3. Find $\frac{d^2y}{dx^2}$ if $x^2 - y^2 = 16$.

4. Dennis Franz High School ("Home of the Ferocious Prarie Dogs") has had a top-notch track team ever since they installed their elliptical track. Its dimensions, major axis length 536 feet and minor axis length 208 feet, are close to an actual track. Below is a diagram of the Prarie Dogs' track superimposed on a coordinate plane. If the northern boundaries of DF High are linear and tangent to the track at $x = \pm 250$ feet, find the equations of the northern property lines.

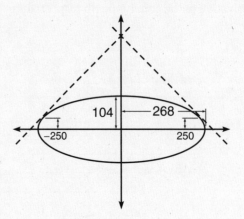

SOLUTIONS

1. Find the derivative with respect to x.

$$e^{y^2} \cdot 2y \frac{dy}{dx} + 3\frac{dy}{dx} = \sec^2 x$$

Now, solve for $\frac{dy}{dx}$. You'll need to factor $\frac{dy}{dx}$ out of both terms on the left side of the equation.

$$\frac{dy}{dx}(2ye^{y^2}+3)=\sec^2 x$$

$$\frac{dy}{dx}=\frac{\sec^2 x}{2ye^{y^2}+3}$$

2. First, find $\frac{d}{dx}$. Remember to use the Product Rule for $\frac{d}{dx}(xy)$:

$$\frac{1}{xy}\cdot(x\frac{dy}{dx}+y)+2y\frac{dy}{dx}=2\frac{dy}{dx}+3.$$

Distribute $\frac{1}{xy}$ and solve for $\frac{dy}{dx}$:

$$\frac{1}{y}\cdot\frac{dy}{dx}+\frac{1}{x}+2y\frac{dy}{dx}=2\frac{dy}{dx}+3$$

$$\frac{1}{y}\cdot\frac{dy}{dx}+2y\frac{dy}{dx}-2\frac{dy}{dx}=3-\frac{1}{x}$$

$$\frac{dy}{dx}(\frac{1}{y}+2y-2)=3-\frac{1}{x}.$$

At this point, you can solve for $\frac{dy}{dx}$, but there is no real need to do so in this problem. You just want the derivative at $(e,1)$, so plug in those values for x and y:

$$\frac{dy}{dx}\left(\frac{1}{1}+2\bullet 1-2\right)=3-\frac{1}{e}$$

$$\frac{dy}{dx}=3-\frac{1}{e}.$$

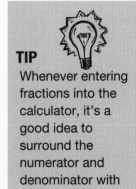

TIP
Whenever entering fractions into the calculator, it's a good idea to surround the numerator and denominator with parentheses.

3. $\frac{d^2y}{dx^2}$ means the second derivative, so begin by finding $\frac{dy}{dx}$, the first derivative:

$$2x-2y\frac{dy}{dx}=0$$

$$\frac{dy}{dx}=\frac{x}{y}.$$

Use the Quotient Rule to find the second derivative:

$$\frac{d^2y}{dx^2}=\frac{y-x\frac{dy}{dx}}{y^2}$$

You already know that $\frac{dy}{dx}=\frac{x}{y}$, so plug it in:

$$\frac{d^2y}{dx^2} = \frac{y - x \cdot \frac{x}{y}}{y^2}$$

Get common denominators for y and $\frac{x^2}{y}$ and simplify:

$$\frac{d^2y}{dx^2} = \frac{\frac{y^2}{y} - \frac{x^2}{y}}{y^2}$$

$$\frac{d^2y}{dx^2} = \frac{y^2 - x^2}{y^3}.$$

Now, the original problem states that $x^2 - y^2 = 16$, so $y^2 - x^2 = -16$. Substitute this, and you're finished:

$$\frac{d^2y}{dx^2} = \frac{16}{y^3}.$$

4. Begin by creating the equation of the ellipse. Remember, standard form of an ellipse is $\frac{x^2}{a^2} + \frac{y^2}{b^2} = 1$, where a and b are half the lengths of the axes. Therefore, the track has the following equation:

$$\frac{x^2}{268^2} + \frac{y^2}{104^2} = 1, \text{ or } \frac{x^2}{71824} + \frac{y^2}{10816} = 1.$$

Now find the y that corresponds to both $x = 250$ and $x = -250$:

$$\frac{62500}{71824} + \frac{y^2}{10816} = 1$$

$$y^2 = (1 - .8701826687) \cdot 10816 = 1404.104255$$

$$y = 37.47137914.$$

Now, you have points $(\pm 250, 37.471379)$. Find $\frac{dy}{dx}$ at these points to get the slope of the tangent lines. (Because the graph of an ellipse is y-symmetric, the slopes at $x = 250$ and $x = -250$ will be opposites.

$$\frac{2}{71824}x + \frac{2}{10816}y\frac{dy}{dx} = 0$$

$$\frac{2}{71824}(250) + \frac{2}{10816}(37.4137914)\frac{dy}{dx} = 0$$

$$\frac{dy}{dx} \approx -1.006$$

The property lines have equations $y - 37.414 = -1.006(x - 250)$ and $y - 37.414 = 1.006(x + 250)$.

HANDS-ON ACTIVITY 4.1: APPROXIMATING DERIVATIVES

Occassionally, the AP writers can be not only tricky but maniacal as well. One such example is their new practice of asking students to find derivatives without even giving them a function. Fear not, weary calculus student! Armed with a little cleverness, a good plan, and a bag of live goldfish, these problems are as easy to conquer as France in a twentieth-century war! (In fact, you may not even need the goldfish.)

1. A function $w(t)$ is continuous; some of its values are listed in this table.

t	0	.5	.7	.9	.95	1
$f(t)$	1	2	3.1	3.4	3.6	3.7

Draw one possible graph of $w(t)$ for $0 \le t \le 1$.

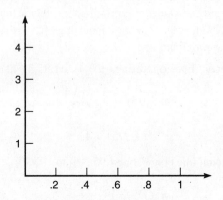

2. Although it likely won't be exact, use a ruler to draw the tangent line to $w(t)$ at $t = .95$.

3. Describe why it is impossible to find the exact value of $w'(t)$.

_____

_____

4. Draw the secant line connecting $t = .9$ and $.95$. Why might this secant line be an important tool? What other secant lines could serve the same purpose?

_____

_____

_____

_____

5. Approximate $w'(.95)$ using three different methods.

_____

_____

_____

6. If a different graph, $f(t)$, were continuous on the interval $[-2,1]$, but you were only given the function values for $t = -2, -\frac{3}{2}, -1, -\frac{1}{2}, \ldots, \frac{3}{2}$, and 2, explain how you would approximate $f'(1)$.

_____

_____

_____

SELECTED SOLUTIONS TO HANDS-ON ACTIVITY 4.1

3. In order to find the exact value of $w'(t)$, you would need to know what rule defines $w(t)$, not just the resulting graph.

4. The secant line has a slope that is very close to that of the tangent line. Using the very simple slope formula from Algebra ($\frac{\Delta y}{\Delta x}$), you can calculate the secant slope and use it as an approximation for the derivative (tangent slope). Other secant lines that are good tangent approximators include the secant connecting $(.95, 3.6)$ to $(1, 3.7)$ and the secant line connecting $(.9, 3.4)$ to $(1, 3.7)$. All of these would be good approximations.

5. (a) Using the secant line connecting $(.9, 3.4)$ to $(.95, 3.6)$:

$$m'(.95) \approx \frac{3.6 - 3.4}{.95 - .9}$$

$$m'(.95) \approx 4.$$

(b) Using the secant line connecting $(.95, 3.6)$ to $(1, 3.7)$:

$$m'(.95) \approx \frac{3.7 - 3.6}{1 - .95}$$

$$m'(.95) \approx 2.$$

(c) Using the secant line connecting $(.9, 3.4)$ to $(1, 3.7)$:

$$m'(.95) \approx \frac{3.7 - 3.4}{1 - .9}$$

$$m'(.95) \approx 3.$$

All of these are approximations, so the answers need not be the same.

6. You could find any of the following secant slopes: $\frac{f(1) - f(.5)}{.5}$, $\frac{f(1.5) - f(1)}{.5}$, or $\frac{f(1.5) - f(.5)}{1}$.

PROBLEM SET

You may use your graphing calculator to answer the following problems.

Problems 1 through 4 refer to $g(x)$ as defined by this graph:

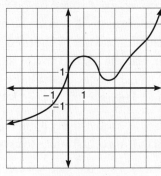

Graph of $g(x)$

1. Give values for $g(2)$, $g(3)$, and $g(4)$. Of these, which can only be an approximation?

2. Approximate $h'(2)$.

3. Approximate the average rate of change of $g(x)$ on the interval $[-2,5]$.

4. Rank the following in numerical order from least to greatest: $h'(-1)$, $h'(1)$, $h'(2)$.

SOLUTIONS

1. The graph makes it pretty clear that $g(2) = 1$ and $g(4) = 2$. You'll have to approximate $g(3)$; according to the graph, $g(3) \approx .6$.

2. The secant line connecting $(1,2)$ and $(2,1)$ is a good approximator; therefore,
   $$g'(2) \approx \frac{(1-2)}{(2-1)} = -1.$$

3. The average rate of change of $g(x)$ is given by
   $$\frac{g(5) - g(-2)}{5 - (-2)}$$
   $$\frac{3 - (-1.7)}{5 - (-2)} \approx .671.$$

4. If you sketch tangent lines at these points, you'll notice that the tangent line has negative slope at $x = 2$, zero slope at $x = 1$, and positive slope at $x = -1$. Although you do not know what the slopes actually are, you still know that *negatives* < 0 < *positives*. Therefore, $h'(2) < h'(1) < h'(-1)$.

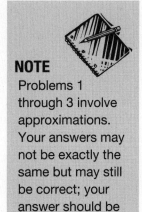

NOTE
Problems 1 through 3 involve approximations. Your answers may not be exactly the same but may still be correct; your answer should be reasonably close.

TECHNOLOGY: FINDING NUMERICAL DERIVATIVES WITH THE GRAPHING CALCULATOR

If you are a studious person and a friend of the environment, you have already read Chapter 1, which discusses when you may and may not use a graphing calculator to justify an answer. You should already know how to graph using the calculator, and you learned how to solve equations using the calculator in Chapter 2. That leaves two topics, and one of these is calculating a numerical derivative.

The TI-83 can evaluate a derivative at any point on its domain (cue bugle fanfare). Unfortunately, it will not actually take the symbolic derivative (cue booing from studio audience). For example, you cannot type $\frac{d}{dx}(\tan x)$ and expect the calculator to respond "$\sec^2 x$" (although some calculators, such as the TI-89, do have this capability). The numerical differentiation function is called "nDeriv" and is found under the [Math] button, option Number 8. To learn this technique, let's revisit an old friend: problem 1 in the Quotient Rule section.

 **Example 19:** Find $f'(1)$ if $f(x) = \dfrac{x^2 \ln x}{3x^3 - 1}$.

Solution: The correct syntax for a numerical derivative is

nDeriv(*function of x*, *x*, *number at which to find derivative*).

Be careful to type the function correctly; this one will require a whole bunch of parentheses.

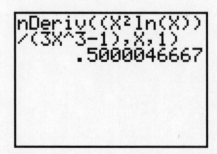

When you did this problem by hand, you got an answer of $\frac{1}{2}$. Note that the calculator's nDeriv result is not the exact answer, as the calculator only approximates it, although it is a pretty darn good approximation. Since the AP test only requires three-decimal accuracy, a truncated or rounded answer of .500 would have been fine, but the complete calculator answer may have been marked *incorrect*, since the actual answer is $\frac{1}{2}$ and not .5000046667.

Always take your calculator's solutions with a grain of salt; they're usually close but not always right.

 Example 20: Use the calculator to find the following numerical derivatives:

(a) $h'(0)$ if $h(x) = 3\sqrt[3]{x} + 2$.

When you type the nDeriv command, you get a very bizzare result. I got 300, as shown below.

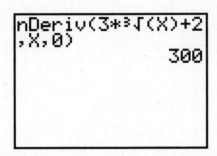

You should check this by hand, because it just feels wrong. You can rewrite h as $h(x) = 3x^{1/3} + 2$, so $h'(x) = 3 \cdot \frac{1}{3}x^{-2/3} = \frac{1}{x^{2/3}}$. Clearly, then, $h'(0)$ is undefined, and the derivative does not exist. (This is because h has a vertical tangent line at $x = 0$, and no derivatives exist at vertical tangent lines.) The calculator has unintentionally lied.

(b) $g'(3)$ if $g(x) = |x - 3| - 1$.

Below is the graph of g and the nDeriv command required to find $g'(0)$.

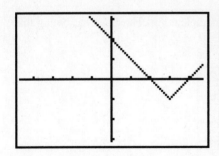 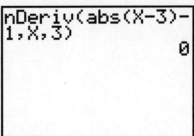

Does something about these images make you uncomfortable? g has a sharp point at $(3,-1)$, but the calculator reports that the derivative is 0. You know better than that—no derivative exists at a sharp point, so the correct answer is $g'(3)$ does not exist.

These examples illustrate a very important point. The calculator is a great estimation tool, but it cannot always replace a good knowledge of where derivatives exist and how to compute them. Be aware of the calculator's limitations, and don't accept the calculator outputs as gospel truth.

PRACTICE PROBLEMS

You may use a graphing calculator only on problems 5 through 12.

1. Find $\dfrac{dy}{dx}$: $y = \sin x \cos \sqrt{x}$.

2. Find $\dfrac{d}{dx}(\sec 3x \ln(\tan x))$.

3. Determine the value of $\dfrac{dy}{dx}$ if $x = 0$ for $e^{xy} + \cos y = 1$.

4. Write the equation of the tangent line to $y = \dfrac{x^2 - 2}{4x + 5}$ at $x = 0$.

5. At what x-values does $(3x^2 + 2x)(x - 1)^{4/5}$ have horizontal tangent lines? On what intervals is the function differentiable?

6. Describe $t(x)$ as completely as you can based on its graph, shown below.

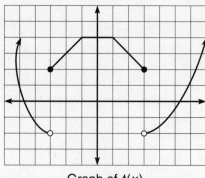

Graph of $t(x)$

7. If $n(x) = (g(x))^2$ and $p(x) = f(x)g(x)$, complete the chart below knowing that $n'(2) = -6$ and $p'(2) = -3$.

x	2
f(x)	9
f'(x)	
g(x)	
g'(x)	1

8. If $b(x)$ has the graph shown below with points of interest x_1 and x_2,

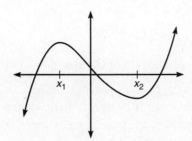

which of the following could be the graph of $b'(x)$?

A.

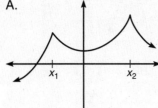

B.

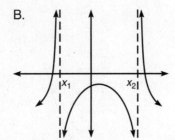

C.

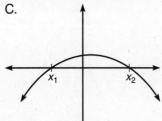

D.

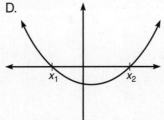

9. If $f(1) = 1, f'(1) = -3, g(1) = 4,$ and $g'(1) = -\dfrac{1}{2},$ find

    (a) $p'(1)$ if $p(x) = \dfrac{f(x)}{2g(x)}$

    (b) $h'(1)$ if $h(x) = (\sqrt{f(x)}\,)^3$

    (c) $j'(1)$ if $j(x) = g(f(1))$

10. If $m(x)$ is an even function differentiable at $x = 4$ and $m'(4) = 0,$ determine which of the following statements *must* be true and briefly justify your answers.

    (a) m is continuous at $x = 4.$

    (b) $\lim\limits_{x \to 4^+} m(x)$ exists.

    (c) $x = 4$ is a root of $m(x).$

    (d) m cannot have a point discontinuity at $x = 4.$

(e) $m'(0) = 4$.

(f) $m'(4) = m'(-4)$.

(g)   The average rate of change of $m(x)$ on $[-4,4] = 0$.

NOTE

The function $s(t)$ in Number 12 is called a *position function*, because its outputs describe where the object in question is at a specified time.

11. Given that the following graph represents an audience member's heart rate (in beats per minute) for the last 16 minutes of the new horror flick *The Bloodening* (starring Whoopi Goldberg and Lassie), approximate the following values and explain what they represent.

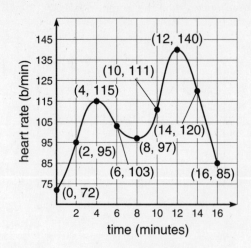

(a) $h(9)$

(b) $h'(6)$

(c) $\dfrac{h(12) - h(8)}{4}$

12. **James' Diabolical Challenge:** A particle moves along the x-axis such that its distance away from the origin is given by $s(t) = t^3 - 9t^2 + 24t - 7$ (where t is in seconds and $s(t)$ is in millimeters).

(a) Find the average velocity of the particle on $[1,3]$.

(b) How quickly is the particle moving at $t = 3$, and in what direction is it moving?

(c) When is the particle at rest?

(d) Graph $s'(t)$ and give the intervals of time during which the particle is moving to the right.

SOLUTIONS

1. The Product Rule should be your first plan of attack (the Chain Rule will follow with $\cos \sqrt{x}$):

$$y' = \sin x \cdot -\sin \sqrt{x} \cdot \frac{1}{2\sqrt{x}} + \cos \sqrt{x} \cdot \cos x.$$

2. This problem is very similar to the last:

$$\sec 3x \cdot \frac{1}{(\tan x)} \cdot \sec^2 x + \ln(\tan x) \cdot 3\sec 3x \tan 3x$$

$$\sec 3x \left(\frac{\sec^2 x}{\tan x} + 3\tan 3x \ln (\tan x) \right).$$

3. First, find the y that corresponds to $x = 0$:

$$e^0 + \cos y = 1$$

$$1 + \cos y = 1$$

$$y = \frac{\pi}{2}.$$

Now, use implicit differentiation, but remember to apply the Product Rule to xy:

$$e^{xy}(y + x \frac{dy}{dx}) - \sin y \frac{dy}{dx} = 0$$

Finally, plug in $\left(0, \frac{\pi}{2} \right)$ and solve for $\frac{dy}{dx}$:

$$e^0 \left(\frac{\pi}{2} + 0 \cdot \frac{dy}{dx} \right) - \sin\left(\frac{\pi}{2} \right) \frac{dy}{dx} = 0$$

$$\frac{\pi}{2} = \frac{dy}{dx}.$$

4. Clearly, the tangent line will intersect the curve at $(0, f(0))$, or $(0, -\frac{2}{5})$. Use the Quotient Rule to find the slope of the tangent line:

$$y' = \frac{(4x + 5)(2x) - (x^2 - 2)(4)}{(4x + 5)^2}$$

$$y'(0) = \frac{0 - (-8)}{25} = \frac{8}{25}.$$

Therefore, the equation of the tangent line is $y + \frac{2}{5} = \frac{8}{25}x$.

5. First, find the derivative using the Product Rule:

$$(3x^2 + 2x)\frac{4}{5}(x - 1)^{-1/5} + (x - 1)^{4/5}(6x + 2).$$

Clearly, the derivative will not exist at $x = 1$, because $(x-1)^{-1/5}$ would cause a zero in the denominator. The function will have a horizontal tangent line when its derivative is 0, so use the calculator to find when the derivative is 0. Horizontal tangent lines should occur when $x = -.366$ and $x = .651$; these results are supported by the graph of the function below:

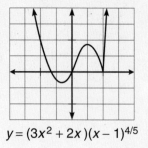

$$y = (3x^2 + 2x)(x-1)^{4/5}$$

TIP

It is clear that t is increasing on (1,3) in Number 6, but to see that t is increasing at an increasing rate, draw a succession of tangent lines to t on (1,3) (remember that the slope of the tangent line is the rate of change at that point). The tangent lines get steeper as you progress from 1 to 3, and a steeper line means a greater slope. Thus, the rate of change is increasing.

6. This is a large list of characteristics of $t(x)$, although there are probably more correct answers: t has jump discontinuities at x = -3 and 3; t has no derivative at x = -3, 3, -1, and 1 (remember that no derivative exists at discontinuities or sharp points); $t'(x) = 0$ on $(-1,1)$; $t'(x) = 1$ on $(-3,-1)$; $t'(x) = -1$ on $(1,3)$; t appears to be an even function; t' is negative on $(-\infty,-3) \cup (1,3)$; t' is positive on $(-3,-1) \cup (3,\infty)$; sketch tangent lines and examine the slopes to see that this is true; t is decreasing at a decreasing rate on $(-\infty,3)$, increasing at a constant rate on $(-3,-1)$, constant on $(-1,1)$, decreasing at a constant rate on $(1,3)$, and increasing at an increasing rate on $(3,\infty)$.

7. First, find the derivative of $n(x)$ and use the fact that $n'(2) = -6$:

$$n'(x) = 2(g(x)) \cdot g'(x)$$
$$n'(2) = 2g(2)g'(2) = -6$$
$$2g(2) \cdot 1 = -6$$
$$g(2) = -3.$$

Now, you only need to find $f'(2)$ to complete the chart. Use the fact that $p'(2) = -3$:

$$p'(2) = f(2)g'(2) + g(2)f'(2) = -3$$
$$9 \cdot 1 + (-3)f'(2) = -3$$
$$(-3)f'(2) = -12$$
$$f'(2) = 4$$

8. Draw a succession of tangent lines to the graph, and examine the slope of the tangent lines, as pictured here:

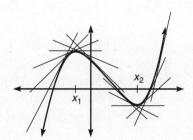

The slopes are positive on $(-\infty, x_1) \bigcup (x_2, \infty)$, so the derivative is positive there also. Similarly, $b'(x)$ will be negative on (x_1, x_2). Furthermore, $b(x)$ will have horizontal tangent lines at x_1 and x_2, so derivatives there will be 0. All these characteristics apply only to graph D.

9. (a) Use the Quotient Rule to find $p'(1)$:

$$p'(1) = \frac{2g(1)f'(1) - f(1) \cdot 2g'(1)}{\left(2g(1)\right)^2}$$

$$p'(1) = \frac{-24 - (-1)}{(2 \cdot 4)^2} = -\frac{23}{64}$$

(b) Rewrite the function as $h(x) = f(x)^{3/2}$, and then apply the Power Rule:

$$h'(1) = \frac{3}{2} f(1)^{1/2} \cdot f'(1)$$

$$h'(1) = \frac{3}{2} \cdot 1 \cdot (-3) = -\frac{9}{2}.$$

(c) Use the Chain Rule to find $j'(1)$:

$$j'(1) = g'(f(1)) \cdot f'(1)$$

$$j'(1) = g'(1) \cdot (-3)$$

$$j'(1) = -\frac{1}{2} \cdot -3 = \frac{3}{2}.$$

10. (a) True: differentiability implies continuity.

(b) True: if continuous at $x = 4$, the limit must exist.

(c) False: $m'(4) = 0$; it doesn't say that $m(4) = 0$.

(d) True: if m is continuous, *no* discontinuity exists.

(e) False: This is wrong for so many reasons…

(f) True: it is also true that $m'(4) = -m'(-4)$ since m is y-symmetric (draw a picture to convince yourself).

(g) True: $m'(4) = m'(-4) = 0$ since the graph is y-symmetric.

NOTE
You may have noticed that when a function increases, its derivative is positive, and when a function decreases, its derivative is negative. This is a very important observation, and it is discussed in depth in Chapter 5.

11. (a) $h(9) \approx 100$ bt/min, which is the audience member's heart rate at $t = 9$ minutes.

(b) $h'(6)$ is the rate of change of the heart rate at $t = 6$. The rate of change of a rate is more commonly known as the acceleration. There is no formula in evidence, so you'll need to approximate the slope of the tangent line. Do so by finding the slope of a nearby secant line. One good approximator is the secant line connecting $(4,115)$ and $(6,103)$ (although it is not the only viable approximation). That slope is given by $\frac{103-115}{6-4} \approx -6 \frac{\frac{bt}{min}}{min}$ or -6 bt/min^2 (the heart rate is decelerating) .

(c) This is the formula for average rate of change on $[8,12]$, or the average rate of change of the heart rate (average heart acceleration). There is no approximation here; your answer will be exact, as all necessary values are given exactly:

$$\frac{140-97}{4} = \frac{43}{4} \text{ bt/sec}^2.$$

12. (a) In order to find velocity, you need to realize that velocity is the rate of change of position. This makes sense; the rate at which an object's position changes tells you how fast that object is moving. So, you are trying to find the average rate of change of the position:

$$\frac{s(3)-s(1)}{3-1}$$

$$\frac{11-9}{2} = 1 \text{ mm/sec.}$$

(b) Here, you are trying to find the velocity (rate of change, or derivative, of position). Thus, the velocity equation will be $v(t) = 3t^2 - 18t + 24$, and $v(3) = 27 - 54 + 24 = -3$ mm/sec. Because velocity is negative, the particle is moving to the left instead of the right.

(c) The particle is not moving when its velocity is 0. So, set $v(t) = 3t^2 - 18t + 24$ equal to 0 and use the calculator to solve the equation. The particle will be at rest when $t = 2$ and $t = 4$ sec. Factoring or using the quadratic formula would also have worked.

(d) $s'(t)$ is the velocity of the particle, and a positive velocity implies that the particle is moving to the right. Thus, you are simply reporting the intervals on which $s'(t)$ is positive, according to the graph.

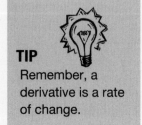

TIP
Remember, a derivative is a rate of change.

NOTE
The rate of change of an object's position (how quickly it is moving) is its *velocity*. The rate of change of an object's velocity (the rate at which it is speeding up or slowing down) is its *acceleration*. More on this in Chapter 6.

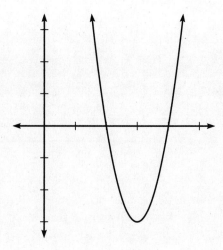

The *t*-intercepts appear to be *t* = 2 and 4, and that is verified by your work in 12(c), so the particle is moving right on $(-\infty,2)\bigcup(4,\infty)$.

Advanced Topics in Differentiation

The Power, Product, Quotient, and Chain Rules allow you to find the derivative of almost any expression or equation you will encounter. However, there are a few exceptions to the rule. This chapter serves to iron out those exceptions and make you invincible in the land of derivatives. In addition, BC students need to know how to differentiate polar, paramtetric, and vector-valued equations, and these topics are included here.

THE DERIVATIVE OF AN INVERSE FUNCTION

Finding the inverse of a function is an important skill in mathematics (although you most certainly never have to use this skill in, say, the french fry department of the fast-food industry). Inverse functions are helpful for so many reasons; the main reason, of course, is their power to cause other functions to disappear. How do you solve the equation $\sec^3 x = 8$? Find the cube root of each side to cancel the exponent and then take the arcsecant of each side to cancel the secant. (The answer, by the way, is   .) Functions this useful are bound to show up on the AP exam, and you should know how to differentiate them.

⊙ Example 1: If $f(x) = (3x + 4)^2$, $x \geq -\dfrac{4}{3}$ find $\left(f^{-1}\right)'(x)$.

Solution: Before you can find the derivative of the inverse, you need to find the inverse function. Use the method outlined in Chapter 2 as follows:

$$x = (3y + 4)^2$$

$$\sqrt{x} - 4 = 3y$$

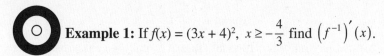

$$f^{-1}(x) = \frac{\sqrt{x} - 4}{3} = \frac{1}{3} \cdot \left(\sqrt{x} - 4\right).$$

Now, find $(f^{-1})'(x)$:

$$\left(f^{-1}\right)'(x) = \frac{1}{3} \cdot \frac{1}{2} x^{-1/2} = \frac{1}{6\sqrt{x}}.$$

That was pretty easy, wasn't it? However, as long as you've been in math classes, you've experienced teachers' uncanny ability to take easy things and make them much more difficult—easy problems tend to leave you this uneasy feeling in the pit of your stomach like you're in a slasher movie, and the killer is right behind you, but you're afraid to turn around. In the case of inverse function differentiation, the killer takes the form of functions for which you cannot easily find an inverse.

How can you find the *derivative* of a function's inverse if you can't even find the inverse? You'll need a formula based on a very simple characteristic of inverse functions: $f(f^{-1}(x)) = x$. Take the derivative of the equation, using the Chain Rule, to get

$$f'(f^{-1}(x)) \cdot (f^{-1})'(x) = 1.$$

Now, solve the equation for $(f^{-1})'(x)$:

$$(f^{-1})'(x) = \frac{1}{f'\left(f^{-1}(x)\right)}.$$

TIP

The formula for inverse derivatives can be remembered as $\frac{1}{\text{P.I.}}$, or "One over Prime/ Inverse." Many students remember that Magnum P.I. was a private *inverse*tigator. Use the mneumonic device to remember the order.

This is an incredibly useful formula to memorize; it's especially handy when you cannot find an inverse function, as in the next example.

 Example 2: Find $(g^{-1})'(3)$ if $g(x) = x^5 + 3x + 2$.

Solution: If you try to find $g^{-1}(x)$ by switching the x and y and solving for y (as in Example 1), you get

$$x = y^5 + 3y + 2$$

in which, despite your best efforts and lots of sweat, you cannot solve for y. Because you cannot easily find $g^{-1}(x)$, you should resort to the Magnum P.I. formula:

$$(g^{-1})'(x) = \frac{1}{g'\left(g^{-1}(x)\right)}, \text{ or in this case:}$$

$$(g^{-1})'(3) = \frac{1}{g'\left(g^{-1}(3)\right)}.$$

How are you going to find $g^{-1}(3)$ if you can't find $g^{-1}(x)$? Good question. Time to be clever. If 3 is an input for $g^{-1}(x)$, then 3 *must be* an output of $g(x)$, since the functions are inverses. Therefore, we should find the domain element x that results in an output of 3 for $g(x)$:

$$x^5 + 3x + 2 = 3$$

$$x^5 + 3x - 1 = 0$$

Solve this equation on your calculator, and the result is $x = .33198902969$. Thus, the point $(.3319890296,3)$ falls on $g(x)$, and resultingly, $(3,.3319890296)$ belongs on $g^{-1}(x)$. Did you miss it? You just found that $g^{-1}(3) = .3319890296$. Return to the formula, and plug in what you know—the rest is easy:

$$(g^{-1})'(3) = \frac{1}{g'\left(g^{-1}(3)\right)}$$

$$(g^{-1})'(3) = \frac{1}{g'(.3319890296)}.$$

Because $g'(x) = 5x^4 + 3$, $g'(.3319890296) = 3.060738622$ and

$$(g^{-1})'(3) = \frac{1}{3.060738622} \approx .327.$$

The process of finding derivatives of inverse functions becomes mechanical with practice. Even after the millionth problem like this, it still excites me to find values of an inverse function I don't even know. What a rush!

PROBLEM SET

You may use your graphing calculator on problem 2 only.

1. If $g(x) = 3\sqrt{x-1} + 2$,

    (a) Evaluate $g^{-1}(4)$ without finding $g^{-1}(x)$.

    (b) Find $g^{-1}(x)$, and use it to verify your answer to part (a).

2. If $k(x) = 4x^3 + 2x - 5$, find $(k^{-1})'(4)$.

3. If $h(x)$ is defined by the graph below, approximate $(h^{-1})'(2)$.

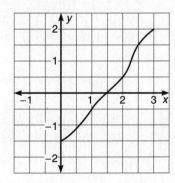

4. If m is a continuous and differentiable function with some values given by the table below, evaluate $(m^{-1})'\left(\frac{1}{2}\right)$.

x	-2	-1	$\frac{1}{2}$	$\frac{2}{3}$	1
$m(x)$	-4	$\frac{1}{2}$	$\frac{2}{3}$	$\frac{3}{2}$	3
$m'(x)$	6	-2	$\frac{5}{4}$	$\frac{5}{4}$	2.5

SOLUTIONS

1. (a)   To find $g^{-1}(4)$, set $g(x) = 4$ and solve for x.

$$3\sqrt{x-1} + 2 = 4$$

$$\sqrt{x-1} = \frac{2}{3}$$

$$x = \frac{4}{9} + 1 = \frac{13}{9}$$

(b)   Switch x and y and solve for y to get $g^{-1}(x)$:

$$x = 3\sqrt{y-1} + 2$$

$$\frac{x-2}{3} = \sqrt{y-1}$$

$$g^{-1}(x) = \frac{(x-2)^2}{3^2} + 1, \text{ so}$$

$$g^{-1}(4) = \frac{(4-2)^2}{3^2} + 1 = \frac{4}{9} + 1 = \frac{13}{9}.$$

2. According to the formula, $(k^{-1})'(4) = \dfrac{1}{k'\left(k^{-1}(4)\right)}$. Find $k^{-1}(4)$ using the method of 1(a):

$$4x^3 + 2x - 5 = 4$$

$$4x^3 + 2x - 9 = 0$$

$$k^{-1}(4) = 1.183617895.$$

Because $k'(x) = 12x^2 + 2$, $\dfrac{1}{k'(1.183617895)} \approx .053$.

3. Again, the formula dictates that $(h^{-1})'(2) = \dfrac{1}{h'\left(h^{-1}(2)\right)}$. From the graph, you can see that $h^{-1}(2) = 3$ (the graph has an output of 2 when $x = 3$), so the formula becomes $\dfrac{1}{h'(3)}$. You will have to approximate $h'(3)$; the secant line connecting $(2.5, 1.5)$ to $(3, 2)$ looks like the best candidate. That secant line slope is $\frac{2-1.5}{3-2.5} = 1$, so a good (and thankfully simple) approximation is $h'(3) = 1$. Therefore, you have

$$(h^{-1})'(2) = \frac{1}{h'(3)} = \frac{1}{1} = 1.$$

4. The Magnum P.I. formula again rears its grotesque head: $(m^{-1})'\left(-\frac{1}{2}\right) =$

$\dfrac{1}{m'\left(m^{-1}\left(-\frac{1}{2}\right)\right)}$. According to the table, $m^{-1}\left(-\frac{1}{2}\right) = -1$ (since $m(-1) = -\frac{1}{2}$).

Also from the table, you see that $m'(-1) = -2$. Therefore, $(m^{-1})'\left(-\frac{1}{2}\right) = -\frac{1}{2}$.

HANDS-ON ACTIVITY 5.1: LINEAR APPROXIMATIONS

Derivatives do so much. They report slopes of tangent lines, describe instantaneous velocities, and illustrate rates of change. Tangent lines, based so closely on derivatives, serve many similar purposes, but one of their most important characteristics is that *tangent lines tend to act like the functions they're tangent to*, at least around the point of tangency.

1. Give the equation of the tangent line to $f(x) = \frac{1}{4}(x-2)^3 + 4$ at $x = 3$.

_____

_____

2. Graph $f(x)$ and its tangent line on your calculator; draw the result below.

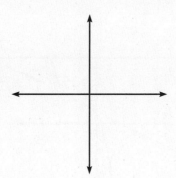

3. Zoom in three times on the point of tangency (using Zoom→Zoom In on the TI-83). What relationship do you notice between the two graphs at this level of magnification?

_____

_____

NOTE

Linear approximations are the first step of a more complicated process known as Euler's Method. Whereas approximations are on both the AB and BC tests, Euler's Method appears only on the BC exam.

4. Return your graph to the standard viewing window (Zoom→standard). Calculate the value of both functions at $x = 3.1$ and $x = 6$. What conclusions can you draw?

_____

_____

_____

5. Describe the special property of tangent lines that is made clear in this activity; make sure to note any drawbacks or weaknesses inherent in this method.

_____

_____

6. Use a tangent line (linear approximation) to estimate the value of $f(5.12)$ if $f(x) = 3x^2 - \dfrac{7}{x^3}$.

_____

_____

_____

SELECTED SOLUTIONS TO HANDS-ON ACTIVITY 5.1

1. The tangent line has equation $3x - 4y = -8$.

2.

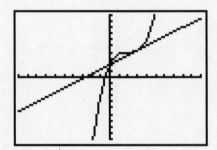

3. Various stages of zooming are shown here:

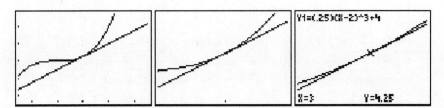

The two graphs have very similar values the closer you zoom in to the point of tangency. In fact, if you keep zooming in, it's hard to tell the two graphs apart.

4. If you call the tangent line $g(x)$, $f(3.1) = 4.33275$ and $g(3.1) = 4.325$. These values are extremely close. However, $f(6) = 20$ and $g(6) = 6.5$. These values aren't even in the same ballpark. The clear conclusion: the tangent line only approximates the function's values near the point of tangency. Further away from this point, all bets are off.

5. I drew this conclusion in number 4; clearly, I am an overachiever.

6. First, find the equation of the tangent line to f at $x = 5$; because 5.12 is close to 5, the function and the tangent line at $x = 5$ will have similar values.

$$f'(x) = 6x + 21x^{-4}$$

$$f'(5) = 30.0336.$$

The tangent line—which passes through $(5, 74.944)$—has slope 30.0336. Thus, tangent line has the following equation:

$$y - 74.944 = 30.0336(x - 5).$$

Now, evaluate the tangent line for $x = 5.12$ for your approximation, and you get 78.548.

PROBLEM SET

You may use your calculator on all of these problems.

1. In the diagram below, the function $f(x)$ meets its tangent line at the point $(a, f(a))$. What is the equation of the linear approximation to f at $x = a$?

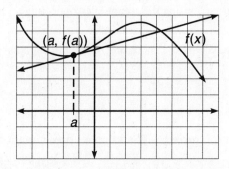

> **NOTE**
>
> The tendency of a graph to look and act like its tangent line immediately surrounding the point of tangency is called *local linearity*. Literally, the graph looks like a line in that localized area, as demonstrated by the zooming exercise in Number 3.

2. Use a linear approximation centered at 2 to estimate $j(2.14)$ if
$j(t) = 2t^3 - \dfrac{1}{t^2} + 4\ln t$.

3. If $h(0) = 4$ and $h'(0) = -5$, use a linear approximation to estimate $h(.25)$.

4. If $g(x)$ is a continuous and differentiable function and contains the values given by the table below, estimate $g(3.2)$ with a linear approximation.

x	$-\dfrac{3}{2}$	$\dfrac{5}{2}$	3	7
$g(x)$	6	$-\dfrac{7}{4}$	$-\dfrac{1}{2}$	14.3

SOLUTIONS

1. The equation of the linear approximation is simply the equation of the tangent line through the point $(a, f(a))$ with slope $f'(a)$: $y - f(a) = f'(a)(x - a)$.

2. The linear approximation is given by the equation of the tangent line to j at $(2, 18.52258872)$. So, find the slope of the tangent line by calcuating the numerical derivative.

$$j'(t) = 6t^2 + \frac{2}{t^3} + \frac{4}{t}$$

$$j'(2) = 24 + \frac{1}{4} + 2 = \frac{105}{4}$$

The equation of the tangent line (linear approximator) is

$$y - 18.52258872 = \frac{105}{4}(x - 2).$$

Substitute $x = 2.14$ to get an approximation of 22.198 for $j(2.14)$.

3. Begin by finding the equation of the tangent line to h when $x = 0$ (since .25 is close to 0, and all the information we have centers around 0). You already know the slope of the line will be -5 and the point of tangency is $(0,4)$, so the equation of the line is $y = -5x + 4$. The linear approximation for $x = .25$ is given by $5(.25) + 4 = 5.25$.

4. The best approximation for $g(3.2)$ will be given by the tangent line constructed at $x = 3$. Clearly, the point of tangency is $(3, -\dfrac{1}{2})$, but you'll have to approximate the derivative at $x = 3$ using the secant line connecting $(\dfrac{5}{2}, -\dfrac{7}{4})$ and $(3, -\dfrac{1}{2})$.

That secant slope $\dfrac{\Delta y}{\Delta x}$ is $\dfrac{5}{2}$. Therefore, the best linear approximation you can make is

$$y + \frac{1}{2} = \frac{5}{2}(x - 3).$$

Evaluate this equation for $x = 3.2$, and your approximation is 0.

NOTE

Problems 2 and 3 are great examples of the usefulness of linear approximation. In the age of calculators, there's rarely a need to approximate a function's value; you can just type it in. However, when only partial information is given (like a table), this tool becomes quite useful.

L'HÔPITAL'S RULE (BC TOPIC ONLY)

L'Hôpital's Rule may sound like the tyrannical rule of a French monarch, but it is actually a way to calculate indeterminate limits. What is an indeterminate limit, you ask? There are numerous forms that an indeterminate limit can take, but you should concentrate on indeterminate forms $\frac{0}{0}$, $0 \cdot \infty$, and $\frac{\infty}{\infty}$. Until now, you were unable to evaluate limits that ended with those results, but now, you will face no such restrictions. Although L'Hôpital's Rule is a BC topic, it is perhaps one of the easiest things in calculus to apply and understand, and AB students are well served to learn it also.

To explore L'Hôpital's Rule, we will revisit an example from the Special Limits section of Chapter 3: $\lim\limits_{x \to 0} \frac{\sin x}{x}$. Substitution resulted in $\frac{0}{0}$, which is indeterminate. None of our other techniques (e.g., factoring, conjugate) applied to this problem either, so we used the graph of $\frac{\sin x}{x}$ to find the limit of 1. If you don't have access to a graphing calculator, however, this problem is a lot more difficult, as the graph is far from trivial to draw. Enter L'Hôpital's Rule.

L'Hôpital's Rule: If $f(x) = g(x) = 0$ or $f(x) = g(x) = \infty$, then $\lim\limits_{x \to a} \frac{f(x)}{g(x)} = \lim\limits_{x \to a} \frac{f'(x)}{g'(x)}$.

Translation: If you are trying to evaluate a limit and you end up with $\frac{0}{0}$ or $\frac{\infty}{\infty}$, you can take the derivatives of the numerator and denominator separately. The new fraction will have the same limit as the original problem, so try and evaluate it. If you get another indeterminate answer, you can repeat the process.

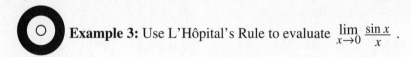 **Example 3:** Use L'Hôpital's Rule to evaluate $\lim\limits_{x \to 0} \frac{\sin x}{x}$.

Solution: As stated previously, substitution results in $\frac{0}{0}$. Apply L'Hôpital's Rule by deriving $\sin x$ and x (with respect to x):

$$\lim\limits_{x \to 0} \frac{\sin x}{x} = \lim\limits_{x \to 0} \frac{\cos x}{1}.$$

Note that $\frac{\sin x}{x}$ and $\frac{\cos x}{1}$ are not *equal*, but they will have the same limit as x approaches 0. The new problem, $\lim\limits_{x \to 0} \cos x$, is very easy by substitution. Thanks to the unit circle, you know $\cos 0 = 1$, which is the answer you memorized "back in the day."

NOTE

An indeterminate result means that the method you chose to evaluate the limit (usually substitution) was ineffectual, and you cannot arrive at an answer. It doesn't mean that no limit exists. You just have to try a different approach.

CAUTION

When applying L'Hôpital's Rule, find the derivatives of the numerator and denominator separately. *Do not use the quotient rule to find the derivative of the entire fraction.*

CAUTION

L'Hôpital's Rule is for use only with indeterminate limits!

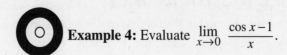

Example 4: Evaluate $\lim\limits_{x \to 0} \dfrac{\cos x - 1}{x}$.

Solution: Remember this one? It's the other limit you memorized. Because substitution results in $\dfrac{0}{0}$, apply L'Hôpital's Rule to get $\lim\limits_{x \to 0} \dfrac{-\sin x}{1}$. Evaluate the new limit by substitution, and you get $\dfrac{0}{1}$, or 0. (Another answer you probably expected. If you didn't expect it, it probably feels like Christmas.)

Example 5: Evaluate $\lim\limits_{x \to \infty} x e^{-x}$.

Solution: If you substitute, you get $\infty \cdot 0$, which is an indeterminate form, but you need a fraction to use L'Hôpital and his fabulous rule. However, you can rewrite the expression as $\dfrac{x}{e^x}$. Substitution now results in $\dfrac{\infty}{\infty}$, and it's time to whip out L'Hôpital; after differentiating, you get

$$\lim\limits_{x \to \infty} \dfrac{1}{e^x}.$$

Substitution results in $\dfrac{1}{e^\infty}$ or $\dfrac{1}{\infty}$, which is 0, according to our special limit rules from Chapter 3.

NOTE

There is no limit to how many times L'Hôpital's Rule can be applied in a problem, as long as each preceding limit is indeterminate.

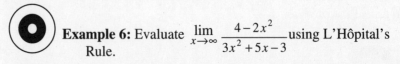

Example 6: Evaluate $\lim\limits_{x \to \infty} \dfrac{4 - 2x^2}{3x^2 + 5x - 3}$ using L'Hôpital's Rule.

Solution: You should be able to get the answer simply by looking at the problem—it's a limit at infinity of a rational function. So, you should compare the degrees of the numerator and denominator. Because the degrees are equal, the limit is the quotient of the leading coefficients: $-\dfrac{2}{3}$. You can verify this with L'Hôpital. (Substitution results in $\dfrac{\infty}{\infty}$, so L'Hôpital's Rule is allowed.) First, find the derivatives as you have done:

$$\lim\limits_{x \to \infty} \dfrac{-4x}{6x + 5}.$$

Substitution *still* results in $\frac{\infty}{\infty}$, so apply L'Hôpital's a second time.

$$\lim_{x \to \infty} \frac{-4}{6} = -\frac{2}{3}.$$

PROBLEM SET

You may not use your graphing calculator on these.

Evaluate the following limits.

1. $\displaystyle\lim_{x \to 0} \frac{\arcsin x}{2x}$

2. $\displaystyle\lim_{x \to 7} \frac{\sqrt{x-3}+x-9}{49-x^2}$

3. $\displaystyle\lim_{x \to 0} \frac{\sqrt{x}+\cos x}{2x+1}$

4. $\displaystyle\lim_{x \to 1} \frac{e^{x-1}-1}{\ln x}$

5. $\displaystyle\lim_{x \to 0} x^2 \cot x$

6. $\displaystyle\lim_{x \to \infty} \left(1+\frac{1}{x}\right)^x$ (**Hint:** use natural log)

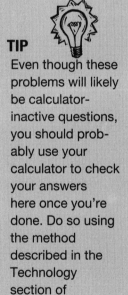

TIP
Even though these problems will likely be calculator-inactive questions, you should probably use your calculator to check your answers here once you're done. Do so using the method described in the Technology section of Chapter 3.

SOLUTIONS

1. Substitution results in $\frac{0}{0}$, so L'Hôpital it to get

    $\displaystyle\lim_{x \to 0} \frac{\frac{1}{\sqrt{1-x^2}}}{2} = \lim_{x \to 0} \frac{1}{2\sqrt{1-x^2}}$. Now, substitution will work, and the answer

    is $\dfrac{1}{2}$.

2. The substitution of $x = 7$ results in $\frac{0}{0}$, so use L'Hôpital's Rule; you should get

    $\displaystyle\lim_{x \to 7} \frac{\frac{1}{2}(x-3)^{-1/2}+1}{-2x}$. At this point, it is no longer illegal to substitute 7; in

    fact, doing so results in the following:

    $$\frac{\frac{1}{2} \cdot \frac{1}{2}+1}{-2 \cdot 7}$$

    $$\frac{\frac{5}{4}}{-14} = -\frac{5}{56}.$$

3. Were you fooled? L'Hôpital's Rule is *not applicable* to this problem, because substitution results in $\frac{1}{1}$, so the limit is equal to 1. Remember to apply L'Hôpital's Rule only in cases of indeterminate limits.

4. Déjà vu all over again! Indeterminate form $\frac{0}{0}$ makes its encore appearance. L'Hôpital is waiting in the wings to tackle this problem like Kevin Costner in *The Bodyguard*:

$$\lim_{x \to 1} \frac{e^{x-1}}{\frac{1}{x}}$$

$$\lim_{x \to 1} x\, e^{x-1}.$$

Substitution results in $1 \cdot 1 = 1$.

5. This limit is of indeterminate form $0 \cdot \infty$. In order to apply L'Hôpital's rule, you'll need a fraction. Because $\tan x$ and $\cot x$ are reciprocals, you can rewrite the limit as $\lim\limits_{x \to 0} \dfrac{x^2}{\tan x}$. Now, the indeterminate form $\frac{0}{0}$ occurs, so you know what to do:

$$\lim_{x \to 0} \frac{2x}{\sec^2 x}.$$

Substitution results in $\frac{0}{1}$, which, of course, is 0.

6. This one is a bit tricky, although you hopefully recognize it from the special limits section; the answer is supposed to be e. In order to find the answer, we make the crazy assumption that there is an answer, and we call it y:

$$y = \lim_{x \to \infty} \left(1 + \frac{1}{x}\right)^x.$$

Here's where the hint comes in. Take the natural log of both sides of this equation:

$$\ln y = \ln \lim_{x \to \infty} \left(1 + \frac{1}{x}\right)^x.$$

You can pull that limit out of the natural log (don't wory so much about why you can) and bring the x exponent down using log properties:

$$\ln y = \lim_{x \to \infty} x \ln \left(1 + \frac{1}{x}\right).$$

This limit is the indeterminate form $\infty \cdot 0$, so you can apply L'Hôpital's Rule as soon as you make it a fraction. If you rewrite x as $\frac{1}{\frac{1}{x}}$ (which is not too obvious to most people), you get the indeterminate form $\frac{0}{0}$, so apply L'Hôpitals Rule:

$$\ln y = \lim_{x \to \infty} \frac{\ln\left(1 + \frac{1}{x}\right)}{\frac{1}{x}}$$

$$\ln y = \lim_{x \to \infty} \frac{\frac{1}{\left(1 + \frac{1}{x}\right)} \cdot \frac{-1}{x^2}}{\left(\frac{-1}{x^2}\right)}$$

$$\ln y = \lim_{x \to \infty} \frac{1}{1 + \frac{1}{x}}$$

Finally, substitution is not illegal, and you can find the limit:

$$\ln y = 1.$$

However, your original goal was to find the limit expressed as y, so you'll have to solve this equation for y by writing e to the power of both sides of the equation:

$$e^{\ln y} = e^1$$

$$y = e.$$

This is the answer we expected, although I don't know that we expected the massive amount of work required.

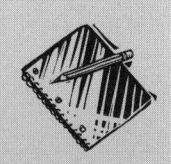

NOTE

The indeterminate form exhibited by Number 6 is 1^∞; while $\frac{0}{0}$ and $\frac{\infty}{\infty}$ are, by far, the most common indeterminate forms on the AP test, there is a slight chance you may see this form or even 0^0 as well.

PARAMETRIC DERIVATIVES (BC TOPIC ONLY)

If you are not yet convinced that derivatives are not only useful but also more fun than your cousin last Thanksgiving when he had the stomach flu, you need only wait until the next chapter. At that point, you will learn that derivatives can serve all kinds of purposes, most of which even have applications in the real world. Because of the extreme handiness of derivatives, it is important that you, the BC student, can differentiate all kinds of equations; specifically, you should be able to find derivatives of polar, parametric, and vector equations and be able to interpret these derivatives. However, one thing at a time; we will start with parametric equations.

In order to find the derivative, $\frac{dy}{dx}$, of a parametrically defined relation, we find the derivatives of the x and y components separately with respect to t and write them as a quotient as follows:

$$\frac{dy}{dx} = \frac{\frac{dy}{dt}}{\frac{dx}{dt}}.$$

NOTE

In essence, the parametric derivative formula is $\frac{dy}{dx} = \frac{y'}{x'}$.

If it's a second derivative you're looking for, the formula is a little different. The second derivative is the derivative of the first derivative with respect to t divided by the derivative of the x component only with respect to t:

$$\frac{d^2y}{dx^2} = \frac{\frac{d}{dt}\left(\frac{dy}{dx}\right)}{\frac{dx}{dt}}.$$

More simply, $\dfrac{d^2y}{dx^2} = \dfrac{\frac{dy'}{dx}}{x'}$.

 Example 7: Find the derivative, $\dfrac{dy}{dx}$, and the second derivative, $\dfrac{d^2y}{dx^2}$, of the parametric equations $x = 3\cos\theta$, $y = 2\sin\theta$.

Solution: The derivatives, $\dfrac{dy}{dt} = -3\sin\theta$ and $\dfrac{dy}{dt} = 2\cos\theta$, are pretty simple to find, and it's only a matter of one more step to get $\dfrac{dy}{dx}$:

$$\frac{dx}{dy} = \frac{\frac{dy}{dt}}{\frac{dx}{dt}} = \frac{2\cos\theta}{-3\sin\theta}$$

$$\frac{dy}{dx} = -\frac{2}{3}\cot\theta.$$

In order to find the second derivative, find the derivative of the expression above, with respect to t and divide it by $\dfrac{dx}{dt}$:

$$\frac{d^2y}{dx^2} = \frac{\frac{2}{3}\csc^2\theta}{-3\sin\theta}$$

$$\frac{d^2y}{dx^2} = -\frac{2}{3}\csc^3\theta.$$

 Example 8: Find the derivative of the parametric equations in Example 7 by a different method.

Solution: You can rewrite this set of parametric equations in rectangular form using the method described in Chapter 2. Because of the Mamma Theorem ($\cos^2x + \sin^2x = 1$), the equivalent rectangular form is $\dfrac{x^2}{9} + \dfrac{y^2}{4} = 1$. In order to find $\dfrac{dy}{dx}$ in this equation, you'll have to use implicit differentiation:

$$\frac{2}{9}x + \frac{1}{2} \cdot y \cdot \frac{dy}{dx} = 0$$

$$\frac{dy}{dx} = -\frac{2x}{9} \cdot \frac{2}{y}$$

$$\frac{dy}{dx} = \frac{-4x}{9y}.$$

Because you know the values of x and y, substitute them into $\frac{dy}{dx}$:

$$\frac{dy}{dx} = -\frac{12\cos\theta}{18\sin\theta}$$

$$\frac{dy}{dx} = -\frac{2}{3}\cot\Theta,$$

which matches the answer we got in Example 7.

NOTE

We converted the parametric equations in Example 8 to rectangular form in the problem set for parametric equations in Chapter 2. Check back if you cannot remember how to convert.

 Example 9: Find the equation of the tangent line to the parametric curve defined by $x = \arcsin t$, $y = e^{3t}$, when $t = \frac{1}{\sqrt{2}}$.

Solution: As always, we need a point and a slope in order to find the equation of a line. When $t = \frac{1}{\sqrt{2}}$, we get the corresponding point $(\arcsin\frac{1}{\sqrt{2}}, e^{\frac{3}{\sqrt{2}}})$, or $(\frac{\pi}{4}, e^{\frac{3}{\sqrt{2}}})$. (You may use your calculator to find the corresponding decimals, but remember not to round anything until the problem is completely over.) To find the slope of the tangent line, you need to calculate $\frac{dy}{dx}$:

$$\frac{dy}{dx} = \frac{3e^{3t}}{\frac{1}{\sqrt{1-t^2}}} = 3e^{3t}\sqrt{1-t^2}.$$

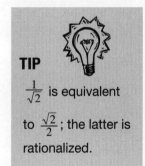

TIP

$\frac{1}{\sqrt{2}}$ is equivalent to $\frac{\sqrt{2}}{2}$; the latter is rationalized.

When $t = \frac{1}{\sqrt{2}}$, the derivative is approximately 17.696, which leads to the tangent line equation of

$$y - e^{\frac{3}{\sqrt{2}}} = 17.696\left(x - \frac{1}{\sqrt{2}}\right), \text{ or}$$

$$y - 8.342 = 17.696\,(x - .785).$$

Vector-valued equations are very similar to parametric equations. Because vector equations are already defined in terms of x and y components, we follow the same procedure to find $\frac{dy}{dx}$. However, vector problems sometimes require us to find *vectors* that represent rate of change. This topic is discussed in detail in the next chapter in the section entitled "Motion in the Plane."

PROBLEM SET

You may use your graphing calculator only on problem 5.

1. What is the derivative of the parametric curve defined by $x = \tan t$, $y = \cot^2 t$?

2. Find the derivative of the parametrically defined curve $x = 1 + 2t$, $y = 2 - t^2$ using two different methods.

3. Find $\frac{d^2 y}{dx^2}$ when $t = 2$ for the parametric curve defined by the equations $x = e^3 t$, $y = t^2 e^{3t}$.

4. If a parametric curve is defined by

$$x = \frac{t^3}{3} + 3t^2 + 5t - 11 \text{ and}$$

$$y = \frac{t^3}{3} - \frac{t^2}{2} - 2t + 5,$$

   (a) At what point(s) does the curve have horizontal tangent lines?

   (b) At what point(s) is the curve nondifferentiable?

5. The position of a crazed lizard, as it runs left and right along the top of a hot brick wall, is defined parametrically as $x = 2e^t$, $y = t^3 - 4t + 7$ for $t \geq 0$ (t is in seconds). Rank the following values of t from least to greatest in terms of how fast the lizard was moving at those moments: $t = 0, .5, 1, 2, 3$.

SOLUTIONS

1. Because $\frac{dy}{dt} = -\csc^2 t$ and $\frac{dx}{dt} = \sec^2 t$,

$$\frac{dy}{dx} = \frac{-\csc^2 t}{\sec^2 t} = -\frac{\cos^2 t}{\sin^2 t}$$

$$\frac{dy}{dx} = -\cot^2 t.$$

2. *Method One:* $\frac{dy}{dx} = \dfrac{\frac{dy}{dt}}{\frac{dx}{dt}}$

$$\frac{dy}{dx} = \frac{-2t}{2} = -t.$$

Method Two: Convert to rectangular form first

Begin by solving either *x* or *y* for *t* and substituting it into the other equation. Because $x = 1 + 2t$, $t = \frac{x-1}{2}$, and you can substitute:

$$y = 2 - t^2$$

$$y = 2 - \frac{1}{4}(x^2 - 2x + 1).$$

Now, take the derivative with respect to *x*:

$$\frac{dy}{dx} = -\frac{1}{2}x + \frac{1}{2}.$$

This answer doesn't seem to match the one above, but it actually does; the difference is that this solution is in terms of *x*, whereas the preceding solution was in terms of *t*. To see that they are equal, remember that $x = 1 + 2t$ and substitute:

$$\frac{dy}{dx} = -\frac{1}{2}(1 + 2t) + \frac{1}{2}$$

$$\frac{dy}{dx} = -\frac{1}{2} - t + \frac{1}{2} = -t.$$

3. First, you need to find $\frac{dy}{dx}$. Don't forget to use the Product Rule when differentiating $t^2 e^{3t}$:

$$\frac{dy}{dx} = \frac{3e^{3t}t^2 + 2te^{3t}}{3e^{3t}}$$

$$\frac{dy}{dx} = \frac{3t^2 + 2t}{3} = t^2 + \frac{2}{3}t.$$

Now, to find the second derivative, divide $\frac{d^2y}{dx^2}$ by $\frac{dx}{dt}$:

$$\frac{d^2y}{dx^2} = \frac{2t + \frac{2}{3}}{3e^{3t}}.$$

The final answer is the derivative evaluated at $t = 2$: $\dfrac{4 + \frac{2}{3}}{3e^6}$. It's ugly, but it's right. More simplification can be done, but it's not necessary and not recommended.

4. (a) First find $\frac{dy}{dx}$ as you have done numerous times already:

$$\frac{dy}{dx} = \frac{t^2 - t - 2}{t^2 + 6t + 5}$$

$$\frac{dy}{dx} = \frac{(t-2)(t+1)}{(t+5)(t+1)}.$$

Horizontal tangent lines have a slope of 0; the derivative will only have a value of 0 when $t = 2$. The point that corresponds with $t = 2$ is $(\frac{41}{3}, \frac{5}{3})$.

(b) $\frac{dy}{dx}$ does not exist when $t = -1$ or -5, because both values cause a 0 in the denominator. While the graphing calculator shows a definite sharp point when $t = -1$,

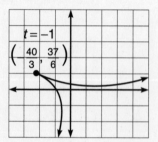

the target line is vertical when t = -5. The points corresponding to these values are $(-\frac{40}{3}, \frac{37}{6})$ (for $t = -1$) and $(-\frac{8}{3}, -\frac{235}{6})$ (for $t = -5$).

5. The derivative of the parametric curve will be

$$\frac{dy}{dx} = \frac{3t^2 - 4}{2e^t}.$$

Rather than substitute each t value in by hand, use your calculator to evaluate each. You should get the following values: $\frac{dy}{dx}(.5) \approx -.986$, $\frac{dy}{dx}(0) = -2$, $\frac{dy}{dx}(1) \approx -.184$, $\frac{dy}{dx}(2) \approx .541$, and $\frac{dy}{dx}(3) \approx .573$. Remember that the derivative is the rate of change, so these numbers are the rate of change of the position of the lizard, or its velocity. Negative velocity indicates that the lizard is running to the left, positive velocity indicates running to the right. However, when speed is the issue, direction does not matter, so the lizard was moving the fastest at $t = 0$. The correct order is: $t = 0$, $t = .5$, $t = 3$, $t = 2$, $t = 1$.

POLAR DERIVATIVES (BC TOPIC ONLY)

One of the defining qualities of polar equations is their similarity to parametric equations. Remember that any polar function $r = f(\theta)$ can be expressed parametrically by $x = r\cos \theta$, $y = r\sin \theta$. Therefore, you differentiate polar equations using essentially the same method outlined for parametric equations:

$$\frac{dy}{dx} = \frac{\frac{dy}{d\theta}}{\frac{dx}{d\theta}} \, .$$

The derivative of the y component divided by the derivative of the x component, both with respect to θ, as both should contain θ's.

The only difference between this and the other formula is that the independent variable is θ instead of t, the typical paramter in parametric equations. While the mathematics involved is not too difficult, there are a lot of places to make mistakes; this is the only thing that makes polar differentiation tricky. Make sure to proceed slowly and cautiously.

 Example 10: Find the slope of the tangent line to $r = 2 + 3\cos\theta$ if $\theta = \frac{\pi}{3}$.

Solution: First, express r parametrically as $x = r\cos\theta$ and $y = r\sin\theta$:

$x = (2 + 3\cos\theta)\cos\theta$ $\qquad\qquad$ $y = (2 + 3\cos\theta)\sin\theta$

$x = 2\cos\theta + 3\cos^2\theta$ $\qquad\qquad$ $y = 2\sin\theta + 3\cos\theta\sin\theta.$

Now, differentiate to get $\frac{dx}{d\theta} = -2\sin\theta - 6\cos\theta\sin\theta$. You'll have to use the Product Rule to find $\frac{dx}{d\theta}$:

$$\frac{dx}{d\theta} = 2\cos\theta + 3(\cos^2\theta - \sin^2\theta)$$

$$\frac{dx}{d\theta} = 2\cos\theta + 3\cos 2\theta.$$

To find the derivative at $\theta = \frac{\pi}{3}$, substitute that into $\frac{dy}{dx}$:

$$\frac{dy}{dx} = r'(\theta) = \frac{2\cos\theta + 3\cos 2\theta}{-2\sin\theta - 6\cos\theta\sin\theta}$$

$$r'\left(\frac{\pi}{3}\right) = \frac{2 \cdot \frac{1}{2} - \frac{3}{2}}{-\sqrt{3} - \left(\frac{3\sqrt{3}}{2}\right)}$$

$$r'\left(\frac{\pi}{3}\right) \approx .115 \, .$$

No particular step of this problem is difficult, but a single incorrect sign could throw off all your calculations.

NOTE

Speed is defined as the absolute value of velocity.

NOTE

$\cos 2\theta = \cos^2\theta - \sin^2\theta$ according to trigonometric double angle formulas. Check Chapter 2 if you don't remember these.

Example 11: At what values of θ, $0 \le \theta \le \frac{\pi}{2}$, does $r = 2\cos$ (3θ) have vertical or horizontal tangent lines?

Solution: Questions regarding tangent lines lead you right to the derivative; begin with parametric representation: $x = 2\cos(3\theta)\cos\theta$, $y = 2\cos(3\theta)\sin\theta$. Now take the derivative of each with respect to θ in order to build $\frac{dy}{dx}$; each will require the Product Rule:

$$\frac{dx}{d\theta} = -2\cos(3\theta)\sin\theta - \cos\theta\, 6\sin(3\theta)$$

$$\frac{dx}{d\theta} = 2\cos(3\theta)\cos\theta - \sin\theta\, 6\sin(3\theta).$$

Remember, horizontal tangents occur when the numerator of the slope is 0 (but the denominator isn't), and vertical tangents occur when the denominator of the slope is 0 (but the numerator isn't). Use your calculator to solve these equations (but you'll have to set it back into rectangular mode first). $\frac{dx}{d\theta}$ equals 0 at $\theta = 0$, .912, and $\frac{\pi}{2}$; the derivative will not exist at these points due to vertical tangent lines. $\frac{dy}{d\theta}$ equals 0 at $\theta = .284$ and 1.103; the derivative will be zero for these values, indicating horizontal tangent lines.

PROBLEM SET

You may use your calculator on problems 3 and 4.

1. If $r(\theta) = 1 + \sin\theta$, where is $r'(\theta)$ defined on $[0,2\pi]$?

2. Find the equation of the tangent line to $r = \tan\theta$ when $\theta = \frac{11\pi}{6}$.

3. If $r(\theta) = 3 - \sin(3\theta)$, at what values of θ is $r'(\theta) = 1$?

4. Find the slopes of the tangent lines to $r = a \cdot \sin(2\theta)$ $(a > 0)$ at the four points furthest from the origin (as indicated in the graph on the following page).

CAUTION

If $\frac{dx}{d\theta}$ and $\frac{dx}{d\theta}$ are 0 at the same time, you cannot draw any conclusions concerning horizontal and vertical asymptotes.

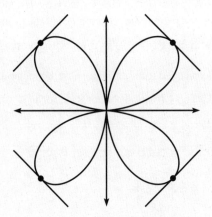

SOLUTIONS

1. First, calculate the derivatives of the x and y components:
$\dfrac{dy}{d\theta} = \cos\theta + 2\sin\theta\cos\theta$ and $\dfrac{dx}{d\theta} = -\sin\theta + \cos 2\theta$. The derivative, $\dfrac{dy}{dx}$, will

be defined wherever $\dfrac{dx}{d\theta} \neq 0$, so set it equal to zero to find these points:

$$-\sin\theta + \cos 2\theta = 0$$

$$-\sin\theta + (1 - 2\sin^2\theta) = 0$$

$$2\sin^2\theta + \sin\theta - 1 = 0$$

$$(2\sin\theta - 1)(\sin\theta + 1) = 0$$

$$\theta = \frac{\pi}{6},\ \frac{5\pi}{6},\ \text{and}\ \frac{3\pi}{2}.$$

The first two values correspond to vertical tangent lines, and the final value corresponds to a sharp point on the graph, (as shown in the graph below).

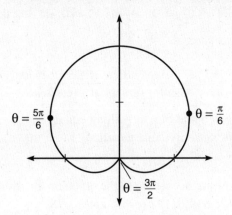

2.   In parametric form, $r = \tan\theta$ becomes $x = \tan\theta\cos\theta = \sin\theta$, and $y = \tan\theta\sin\theta$. Use these to find the rectangular coordinates of the point of tangency (when $\theta = \frac{11\pi}{6}$). If you plug $\theta = \frac{11\pi}{6}$ into both, you get the coordinate $(-\frac{1}{2}, \frac{1}{2\sqrt{3}})$. Now, you need to find the slope of the tangent line. To do so, find the derivtaives of x and y (using Product Rule for y'):

$$x' = \cos\theta$$

$$y' = \tan\theta\cos\theta + \sin\theta\sec^2\theta$$

Therefore, $\dfrac{dy}{dx}\left(\dfrac{11\pi}{6}\right) = \dfrac{y'}{x'}\left(\dfrac{11\pi}{6}\right) =$

$$\frac{-\frac{1}{\sqrt{3}}\cdot\frac{\sqrt{3}}{2} - \frac{1}{2}\cdot\frac{4}{3}}{\frac{\sqrt{3}}{2}}$$

$$\frac{-\frac{1}{2} - \frac{2}{3}}{\frac{\sqrt{3}}{2}}$$

$$-\frac{7}{6}\cdot\frac{2}{\sqrt{3}} = -\frac{7}{3\sqrt{3}}.$$

Therefore, the equation of the tangent line is as follows:

$$y - \frac{1}{2\sqrt{3}} = -\frac{7}{3\sqrt{3}}x + \frac{1}{2}.$$

That was truly an ugly problem, but doing it without your calculator toughened you up some—admit it.

NOTE

In Number 2, the notation x' and y' are used; they refer to $\frac{dx}{d\theta}$ and $\frac{dy}{d\theta}$, respectively.

3.   Once again, it is important to express the polar equation in parametric form: $x = 3\cos\theta - \sin 3\theta\cos\theta$, $y = 3\sin\theta - \sin 3\theta\sin\theta$. Take the derivatives of each to get $\dfrac{dy}{dx} = \dfrac{y'}{x'}$ below:

$$\frac{dy}{dx} = \frac{3\cos\theta - 3\sin\theta\cos 3\theta - \cos\theta\sin 3\theta}{-3\sin\theta + \sin\theta\sin 3\theta - 3\cos\theta\cos\theta}$$

You want to find when that big, ugly thing equals one, so carefully type it into your calculator and solve that equation.

There are four solutions, according to the graph on the following page, and they are $\theta = 1.834, 2.699, 3.142$ (or π), and 5.678.

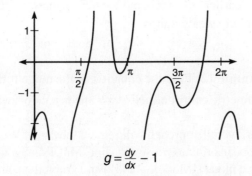

$$g = \frac{dy}{dx} - 1$$

4. You have to decide what points will be the furthest from the origin; in other words, what's the largest $a \cdot \sin(2\theta)$ can be? To start with, $\sin(2\theta)$ can be no larger that 1 and no smaller than –1 (the range of $\sin x$). Therefore, the graph of $a \cdot \sin(2\theta)$ can be no further than $a \cdot 1 = a$ units from the origin. (Note that a distance of $-a$ from the origin is the same, just in the opposite direction.) Set the equation equal to $\pm a$ to find out which values of θ give these maximum distances:

$$a \cdot \sin(2\theta) = \pm a$$

$$\arcsin(\sin(2\theta)) = \arcsin(\pm 1)$$

$$2\theta = \frac{\pi}{2}, \frac{3\pi}{2}, \frac{5\pi}{2}, \frac{7\pi}{2}$$

$$\theta = \frac{\pi}{4}, \frac{3\pi}{4}, \frac{5\pi}{4}, \frac{7\pi}{4}.$$

So, you need to find the derivatives at these values. Note that the derivatives will not change regardless of a's value. To convince yourself of this, you may want to draw a couple of graphs with different a values. There's no shame in using your calculator to evaluate these derivatives since this is a calculator-active question. (If you're not sure how to do that, make sure to read the technology section at the end of the chapter.) The derivatve is 1 for $\theta = \frac{3\pi}{4}$ and $\frac{7\pi}{4}$, and the derivative is –1 for $\theta = \frac{\pi}{4}$ and $\frac{5\pi}{4}$.

CAUTION
Even though your graphs are done in polar mode, all equation solving with the calculator (using x-intercepts) requires that you switch back to rectangular mode first.

TECHNOLOGY: FINDING POLAR AND PARAMETRIC DERIVATIVES WITH YOUR CALCULATOR (BC TOPIC ONLY)

Evaluating polar and parametric derivatives aren't the most difficult topics you'll encounter in AP Calculus. However, when the heat is on during the AP test and you're searching for that needle-in-a-haystack error you committed that is causing your answer to be mortally and inexplicably wrong, these derivatives can be pretty difficult. As you have seen in this chapter, even relatively simple polar equations can have long and yucky derivatives. There is good news—you can use your calculator to find these derivatives, and the process is very simple. To prove it to you, we'll revisit

Example 9 from earlier in the chapter, with our new calculator buddies tucked snugly in our sweaty palms.

 Example 12: Find the equation of the tangent line to the parametric curve defined by $x = \arcsin t$, $y = e^{3t}$, when $t = \frac{1}{\sqrt{2}}$.

Solution: Our overall approach will remain the same. We still need a point and a slope in order to construct a line. Make sure your calculator is in parametric mode ([Mode]→"Par") and graph the equation. (You may want to adjust the graph's [window] to see it better; I used a window of $x = [-2,2]$ and $y = [-1,15]$.) The point of tangency occurs (according to the problem) when $t = \frac{1}{\sqrt{2}}$. To find this point in rectangular coordinates, press [2nd]→[Trace]→"value" and input $t = \frac{1}{\sqrt{2}}$. The point of tangency is (.785,8.342), just as we got previously.

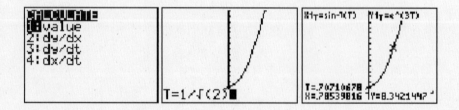

In order to find the derivative when $t = \frac{1}{\sqrt{2}}$, go back to the [Calc] menu and this time select $\frac{dy}{dx}$. Enter $\frac{1}{\sqrt{2}}$ for t as you did last time, and $\frac{dy}{dx} = 17.696$.

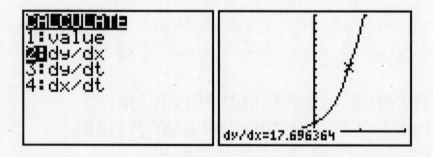

The equation of the tangent line, then, is

$$y - 8.342 = 17.696(x - .785).$$

Polar derivatives are just as easy with the calculator. Who's up for reruns? Let's enjoy Example 10 a second time, this time with our lil' computing buddy by our side.

Example 12: Find the slope of the tangent line to $r = 2 + 3\cos\theta$ if $\theta = \frac{\pi}{3}$.

NOTE

Your calculator can find points of tangency for polar graphs, too. This (like everything else) is also found in the [Calc] menu.

Solution: Make sure to set the calculator to Polar mode ([Mode]→"pol") and graph the equation. Because this graph has trigonometric equations, it's a good idea to choose [Zoom]→"Ztrig" (window settings that are friendly to trigonometric graphs). Much like Example 11, you proceed to the [Calc] menu ([2nd]→[Trace]) and select $\frac{dy}{dx}$, the derivative. The calculator then prompts you to input θ; type in $\frac{\pi}{3}$ (as shown below), and the derivative appears as if by magic: $\frac{dy}{dx} \approx .115$.

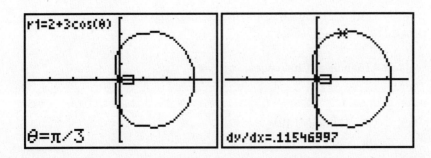

It's worth repeating that you must know how to calculate points of tangency and derivatives without a calculator, although no one would argue that the calculator doesn't make it significantly easier.

PRACTICE PROBLEMS

You may use a graphing calculator on problems 4 through 8 only.

1. Use a linear approximation at $x = \frac{1}{2}$ to estimate $m(.502)$ if $m(x) = \frac{2x^2 - 3x}{x^3}$.

2. Evaluate $\displaystyle\lim_{x \to \infty} \frac{\sqrt{3x^2 + 2x + 5} - 2}{(5x - 7)}$.

3.* Find $\frac{dy}{dx}$ at $\theta = \frac{7\pi}{6}$ for the polar equation $r = \sin\theta \cos\theta$.

4.* At what values of t does $\frac{dy}{dx} = \frac{1}{2}$ for the parametric function defined by $x = 2 + t^2, y = 3 - t$?

5.* If $f(x) = 2\tan x + x^3$ and $g(x)$ is a continuous, origin-symmetric function that contains the following values:

x	−.2	−.1	0
g(x)	.942	.493	0

Estimate $\lim\limits_{x \to 0} \dfrac{f(x)}{g(x)}$.

6.* Find the equation of the vertical tangent line (in rectangular form) to the curve defined by the parametric equations $x = e\,t \sin t$, $y = 2t + 1$, $0 \le t < 2\pi$.

7. If $k(x) = 2x^3 + x + 2$, what is the equation of the tangent line to $k^{-1}(x)$ when $x = 4.3$?

8.* At what rectangular coordinates do the tangent lines to $r = \sin^2\theta$ at $\theta = \dfrac{\pi}{6}$ and

$\theta = \dfrac{2\pi}{5}$ intersect? (*Note:* Answer this as you would a free response question. Show all work, and make sure to include the setup for any answers you give based on calculator work.)

9. **James' Diabolical Challenge:** Let $f(x) = \dfrac{x^2}{e^x}$, $x \le 0$.

    (a)   Evaluate $f^{-1}(0)$.

    (b)   Find $(f^{-1})'(0)$.

    (c)*   What is $\lim\limits_{x \to \infty} f(x)$?

SOLUTIONS

1. We have to begin by finding the equation of the tangent line to m at $x = \dfrac{1}{2}$. We determine the point of tangency by calculating $m\left(\dfrac{1}{2}\right)$:

$$\frac{\dfrac{1}{2} - \dfrac{3}{2}}{\dfrac{1}{8}} = \frac{-1}{\dfrac{1}{8}} = -8.$$

The point of tangency, therefore, is $\left(\dfrac{1}{2}, -8\right)$. The slope of the tangent line will, as always, be given by the derivative, so use the Quotient Rule to find it:

$$m'(x) = \frac{x^3(4x - 3) - (2x^2 - 3x)(3x^2)}{(x^3)^2}$$

---

* a BC-only question.

$$m'(x) = \frac{-2x^4 + 6x^3}{x^6} = \frac{-2x + 6}{x^3}$$

$$m'\left(\frac{1}{2}\right) = \frac{-1 + 6}{\frac{1}{8}} = \frac{5}{\frac{1}{8}} = 40.$$

Therefore, the equation of the tangent line is

$$y + 8 = 40\left(x - \frac{1}{2}\right).$$

Last step: Substitute $x = .502$ into the tangent line to get the linear approximation:

$$y + 8 = 40\,(.502 - .5)$$

$$y + 8 = .08$$

$$y = -7.92.$$

(If you plug the answer into the original equation, $m\left(\frac{1}{2}\right) = -8$, so this approximation is relatively close.)

2. L'Hôpital's Rule is too cumbersome for this problem. Notice that you have a rational function with the same degree $\left(\sqrt{x} = x\right)$ in the numerator and denominator. Thus, the limit at infinity will be the ratio of the numerator's and denominator's leading coefficients: $\frac{\sqrt{3}}{5}$.

3. In order to find $\frac{dy}{dx}$, you first must express the equation parametrically: $x = \sin\theta\cos^2\theta$, $y = \sin^2\theta\cos\theta$. Remember that $\frac{dy}{dx}$ is basically $\frac{y'}{x'}$. You'll need to use the Product Rule to get x' and y':

$$x' = \sin\theta\,(\cos^2\theta)' + \cos^2\theta\,(\sin\theta)'$$

$$x' = -2\cos\theta\sin^2\theta + \cos^3\theta$$

$$y' = \sin^2\theta\,(\cos\theta)' + \cos\theta\,(\sin^2\theta)'$$

$$y' = -\sin^3\theta + 2\cos^2\theta\sin\theta$$

$$\frac{dy}{dx} = \frac{-\sin^3\theta + 2\cos^2\theta\sin\theta}{-2\cos\theta\sin^2\theta + \cos^3\theta}.$$

Now, evaluate $\frac{dy}{dx}$ for $\theta = \frac{7\pi}{6}$:

$$\frac{dy}{dx} = \frac{\frac{1}{8} + 2\cdot\frac{3}{4}\cdot-\frac{1}{2}}{-2\cdot\frac{-\sqrt{3}}{2}\cdot\frac{1}{4} - \frac{\sqrt{27}}{8}}.$$

Of course, there is no need to simplify this answer, and doing so drastically increases your chances of making a mistake. If you simplifed further and want to know if you got it right, use a calculator to evaluate your expression; it should equal 2.887.

4. First, find $\dfrac{dy}{dx}$:

$$\frac{dy}{dx} = \frac{y'}{x'}$$

$$\frac{dy}{dx} = \frac{-1}{2t}.$$

Clearly, the derivative will equal $\frac{1}{2}$ whenever $t = -1$.

5. If you try to evaluate the limit via substitution, you will get indeterminate form $\frac{0}{0}$, so you should default to L'Hôspital's Rule and find $\lim\limits_{x \to 0} \dfrac{f'(x)}{g'(x)}$. Simple differentiation yields $f'(x) = 2\sec^2 x + 3x^2$, so $f'(0) = 2 \cdot 1 + 0 = 2$. To complete the problem, however, you'll also need $g'(0)$, and that will require some estimation. Because g is origin symmetric, you automatically know that $g(.1) = -.492$ and $g(.2) = -.942$. You should use one of the following secant lines to estimate the derivative: from $x = -.1$ to $.1$, from $x = -.1$ to 0, or from 0 to $.1$. If you choose the secant line from $-.1$ to 0, the secant slope is

$$\frac{0 - .493}{0 - (-.1)} = -4.93.$$

Therefore, $\lim\limits_{x \to 0} \dfrac{f'(x)}{g'(x)} = \dfrac{2}{-4.93} \approx -.406.$

6. The slopes of all tangent lines, vertical or not, are furnished by $\dfrac{dy}{dx}$, so begin by finding it:

$$\frac{dy}{dx} = \frac{2}{e^t \cos t + e^t \sin t}.$$

A vertical asymptote will occur when the denominator is 0 but the numerator is not. This numerator is always 2, so you need only determine where the denominator is 0. You can use your calculator to solve this, but the problem is pretty easy without the calculator:

$$e^t(\cos t + \sin t) = 0$$

$$\cos t + \sin t = 0$$

$$\cos t = -\sin t$$

$$t = \frac{3\pi}{4}, \frac{7\pi}{4}$$

NOTE
You need only worry about (cos t + sin t) being equal to 0 in Number 6 because e^t never equals 0 (the range of e^x is $y > 0$).

However, this is not the answer to the question. Instead, you need to give the rectangular equations of the vertical tangent lines that occur at these values of t. To do so, use the $[2^{nd}]\rightarrow[\text{Trace}]\rightarrow$ "value" function of your calculator. When $t = \dfrac{3\pi}{4}$, the corresponding rectangular coordinate is $(7.460, 5.712)$; when $t = \dfrac{7\pi}{4}$, the corresponding coordinate is $(-172.641, 11.996)$. Therefore the equations of the vertical tangent lines are $x = 7.460$ and $x = -172.641$.

7. You'll need, as always, a point and a slope to form a line. You cannot find $k^{-1}(x)$ algebraically, so you'll need to use the Magnum P.I. formula to find the derivative of the inverse function and, hence, the slope of the tangent line:

$$\left(k^{-1}\right)'(4.3) = \frac{1}{k'\left(k^{-1}(4.3)\right)}$$

$$\left(k^{-1}\right)'(4.3) = \frac{1}{k'(.8900107776)}$$

$$\left(k^{-1}\right)'(4.3) \approx .174.$$

Now that you have the point and the slope, it is a trivial pursuit to write the equation of the line (although I always have trouble with those entertainment questions):

$$y - .890 = .174(x - 4.3).$$

8. Use the calculator's $[2^{nd}]\rightarrow[\text{Trace}]\rightarrow$ "value" and $[2^{nd}]\rightarrow[\text{Trace}]$ "$\dfrac{dy}{dx}$" functions to calculate the points of tangency and derivatives for each line. The tangent lines should be

$$y - .125 = 1.0392309\,(x - .21650635), \text{ for } \theta = \frac{\pi}{6}, \text{ and}$$

$$y - .8602387 = -1.235663(x - .2795085), \text{ for } \theta = \frac{2\pi}{5}.$$

To find the intersection of these lines, solve both for y and set them equal to each other:

$1.0392309(x - .21650635) + .125 = -1.235663(x - .2795085) + .8602387.$

You can now solve the equation on your calculator by setting the above equation equal to 0 and finding the root:

$1.0392309(x - .21650635) + .125 + 1.235663(x - .2795085) - .8602387 = 0$

$$x = .57392439.$$

Substitute this value into either line's equation to get the corresponding y-value, and the resulting coordinate is $(.574, .496)$.

CAUTION
You *cannot* use the [2ⁿᵈ]→[Trace]→ "intersection" feature of your calculator on the free-response portion of the test, so use the equation solving feature of the calculator instead.

9. (a) To find f^{-1} at a specific x, set f equal to x and solve

$$\frac{x^2}{e^x} = 0$$

$$x^2 = 0$$

$$x = 0.$$

Therefore, $f^{-1}(0) = 0$.

(b) This requires the inverse function derivative formula. Apply the information from 9(a), and the formula becomes

$$\left(f^{-1}\right)'(0) = \frac{1}{f'(0)}.$$

If you find $f'(x)$ with the Quotient Rule, you get

$$f'(x) = \frac{2xe^x - x^2e^x}{e^{2x}}$$

$$f'(x) = \frac{2x - x^2}{e^x}.$$

NOTE
It makes sense that $(f^{-1})'(0) = 0$. If a function has a horizontal tangent line at 0, its inverse must have a vertical tangent line there, since the graphs are reflections about the line $y = x$.

Therefore, $f'(0) = 0$; substitute this into the inverse function derivative formula:

$$\left(f^{-1}\right)'(0) = \frac{1}{0}$$

$(f^{-1})'(0)$ is undefined; the inverse function is not differentiable there due to a vertical tangent line.

(c) To find this limit, you must apply L'Hôpital's Rule twice (due to presence of indeterminate form $\frac{\infty}{\infty}$:

$$\lim_{x \to \infty} \frac{2x}{e^x}$$

$$\lim_{x \to \infty} \frac{2}{e^x} = \frac{2}{\infty} = 0.$$

Chapter

Applications of the Derivative

By this time, you are beginning to form an opinion about derivatives. Either you like them or you don't. Hopefully, the two of you are at least on speaking terms. If you are, you are going to be very impressed by the things derivatives can do. They can even help you *in the real world*, which sometimes surprises math students. Before we get into these topics, it's time to decide whether or not you are a derivative fan, à la Jeff Foxworthy's classic and oft-used "You Might be a Redneck if…" routine.

You Might Love Derivatives if…

· You have an oversized foam hand that reads "Derivatives #1!"
· When your friend flipped a coin the other day, you said, "That's what I call a rate of *change*." And no one laughed but you.
· You love it when people go off on tangents.
· You took up skiing so you could learn more about slopes.
· You loudly commented at the grocery store the other day that, "This express line looks more like a linear *approximation*."
· You always respond to chain letters because they remind you of the Chain Rule.
· You have a giant tattoo of the Quotient Rule on your back.

RELATED RATES

You already know that a function or equation shows a clear relationship between the variables involved. For example, in the linear equation $y = 3x - 2$, each ordered pair has an x that is 2 less than 3 times as large as y. What you may not know is that when you find the derivative of such an equation with respect to *time*, you find another relationship—one between the rates of change of the variables. Back to our equation: If you find the derivative of $y = 3x - 2$ with respect to t, you get

$$\frac{dy}{dt} = 3\left(\frac{dx}{dt}\right).$$

CAUTION

Of all the topics in AP Calculus, students often forget how to do related rates by test time. Take some extra time and make sure that your understanding is complete.

NOTE

$\frac{dy}{dt}$ is interpreted as the rate of change of *y* with respect to time, and $\frac{dx}{dt}$ is the rate of change of *x* with respect to time.

This means that *y* is changing at a rate 3 times faster than *x* is changing. This makes sense, because the derivative, $\frac{dy}{dx}$, is $\frac{3}{1}$. For every one unit you travel to the right, you must travel up 3 to stay on the graph. These types of problems are called *related rates* (for obvious reasons).

As we progress through the following examples, we will be closely following the plan below. Get used to the chronology of these steps—the method of solving related rates problems always follows the same pattern.

Steps to Success with Related Rates:

· Identify which rate you are trying to find and what information is given to you.

· Find an equation that relates the variables to one another if you're not given one.

· Eliminate extra variables, if at all possible, by substituting in for them (see Example 3).

· Find the derivative of the equation with respect to *t*.

· Plug in what you know, and solve for the required rate.

Example 1: My brother, Dave, and I recently went golfing. After a promising start, he landed 3 consecutive balls in the lake in front of the second green. As the first ball entered the water, it caused a multitude of ripples in the form of concentric circles emanating from the point of impact at a steady rate of $\frac{3}{4}$ ft/sec.

(a) What was the rate of change of the area of the outermost ripple when its radius was 3 feet?

This question concerns area and radius; both of these elements are contained by the formula for area of a circle: $A = \pi r^2$. To find the realtionship between the rates, find the derivative with respect to *t*:

$$\frac{dA}{dt} = 2\pi r \frac{dr}{dt}.$$

You are trying to find $\frac{dA}{dt}$, the rate of increase of *A*. You know that $\frac{dr}{dt} = \frac{3}{4}$ according to the given information, and the problem prompts you that $r = 3$ in this instance. Substitute these values into the equation to solve:

$$\frac{dA}{dt} = 2\pi(3) \cdot \frac{3}{4} = \frac{9\pi}{2} \text{ ft}^2/\text{sec}.$$

(b) What was the rate of change of the golf club he threw with a mighty initial velocity after drowning the third ball?

No one knows, but everyone knew to stay out of the way.

As Example 1 illustrates, make sure that you include the correct units in your final answer when units are included in the problem. Following are the most commonly requested rates (assuming that the problem includes meters and seconds): area (m^2/sec), volume (m^3/sec), length or velocity (m/sec), and acceleration (m/sec^2). If the problem contains units other than meters and seconds, the format is still the same.

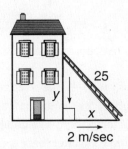 **Example 2:** While painting my house and atop a 25-foot ladder, I was horrified to discover that the ladder began sliding away from the base of my home at a constant rate of 2 ft/sec (don't ask me how I knew that, I just did). At what rate was the top of the ladder carrying me, screaming like a 2-year-old girl, toward the ground when the base of the ladder was already 17 feet away from the house?

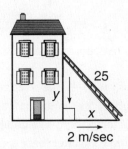

2 m/sec

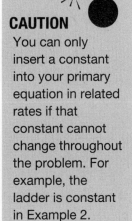

Solution: You first need to set up a relationship that contains your given information and what you need to find. The right triangle made by the ladder and my house contains all of this information (although you wouldn't have to have the same variables, of course), so by the Pythagorean Theorem:

$$y^2 + x^2 = 25^2.$$

Notice that the values of x and y will change as the ladder slides, but the ladder will always be 25 feet long, so I can use this constant rather than a third variable.

Now, find the derivative with respect to t:

$$2y \cdot \frac{dy}{dt} + 2x \cdot \frac{dx}{dt} = 0.$$

The base of the ladder is sliding away from the house at 2 ft/sec, so $\frac{dx}{dt} = 2$; the problem also states that $x = 17$. You'll have to use the Pythagorean Theorem to find the value of y for this specific value of x:

$$y^2 + x^2 = 25$$

$$y^2 + (17)^2 = 25^2$$

$$y = \sqrt{336} \cdot$$

Now, you have all the variables in question except for $\frac{dy}{dt}$, the rate that you are trying to find. Substitute all your values into the derivative you found earlier to find $\frac{dy}{dt}$:

$$2\left(\sqrt{336}\right)\frac{dy}{dt} + 2(17)(2) = 0$$

$$\frac{dy}{dt} \approx -1.855 \text{ ft/s.}$$

Notice that the length of y is decreasing, since the ladder is sliding downward; therefore, $\frac{dy}{dt}$ must be negative.

A few years ago, I had to undergo massive nasal surgery, the focus of which was to scrape out all of my sinus cavities to remove disgusting "mucous cysts" that had gathered there like college students waiting for a party. I wrote the next problem soon after that experience. It was, at the time, the worst thing that could happen. (This problem appeared on my Web site as it was just getting started.)

NOTE
Example 3 is quite difficult when compared to the others. A problem of this difficulty in related rates is relatively rare on the AP test.

Example 3: The nightmare has come to pass. All of Kelley's extensive surgeries and nasal passage scrapings have (unfortunately) gone awry, and he waits in the ear, nose, and throat doctor's office waiting area spewing bloody nose drippings into a conical paper cup at the rate of 2.5 in³/min. The cup is being held with the vertex down and has a height of 4 inches and a base of 3 inches. How fast is the mucous level rising in the cup when the "liquid" is 2 inches deep?

Solution: You should first establish what you know: a cone's volume is $V = \frac{1}{3}\pi r^2 h$, $\frac{dV}{dt} = 2.5$, height of the cone is 4, and the diameter of the base is 3, which makes the radius of the base, r, equal to $\frac{3}{2}$. (You also know that the mucous itself will be in the shape of a cone since it is in a conical container.) What's even more important is what you don't know. You don't know the radius of the mucous, and you don't know its rate of change. Therefore, you should try to eliminate the variable r from the volume equation. Why include a variable you know nothing about? To do so is the most complicated part of this problem. You need to use similar triangles. Below is a cross-section of the cup.

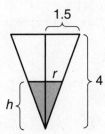

Two similar right triangles can be formed. Look at the set of overlapping triangles on the right. The smaller triangle (representing mucous) has unknown height and radius, whereas the larger triangle (the cup) has height 4 and radius 1.5. This allows you to set up a proportion, since corresponding sides of similar triangles are in proportion.

$$\frac{4}{h} = \frac{\frac{3}{2}}{r}$$

Solve this proportion for r and you get $r = \frac{3h}{8}$. If you substitute this for r into the volume equation, the problem of knowing nothing about r is completely solved.

$$V = \frac{1}{3}\pi r^2 h$$

$$V = \frac{1}{3}\pi \left(\frac{3h}{8}\right)^2 h$$

$$V = \frac{3}{64}\pi h^3$$

Now, find the derivative with respect to t to get rolling

$$\frac{dV}{dt} = \frac{9}{64}\pi h^2 \frac{dh}{dt}$$

and substitute in all the information you know to solve for $\dfrac{dh}{dt}$, the value requested by the problem:

$$2.5 = \frac{9\pi}{64}(2)^2 \frac{dh}{dt}$$

$$\frac{dh}{dt} \approx 1.145 \text{ in/min.}$$

My cup, it overfloweth.

PROBLEM SET

You may use a graphing calculator on problems 2 through 4.

1. A particle moves along the path $y = x^3 - 3x^2 + 2$. If the particle's horizontal rate of change when $x = 4$ seconds is -3 ft/sec, what is its vertical rate of change at that instant?

2. If a spherical balloon is being deflated at a rate of 5 in³/sec, at what rate is the radius of the balloon decreasing when $r = 5$ in?

3. Last week, I accidentally dropped a cube into a vat of nuclear waste, setting off a chain reaction of events that eventually caused the cube to possess super powers, among them the ability to eat rocks. As the cube amassed these super powers, it swelled at a rate of 7 in/sec. At what rate was the surface area of Super Cube changing when one of its sides was 2 *feet* long?

4. During a *Circus of the Stars* presentation, Angela Lansbury displays her athletic prowess by skydiving out of a hovering helicopter 100 feet away from a cliff. However, the chutes unfortunately fail, and the proud actress plummets to certain disaster (*Tragedy, She Wrote*). If her position, in feet, is given by $s(t) = -16t^2 + 15840$, find the rate of change of the angle of depression (no pun intended) in degrees/sec at $t = 30.9$ seconds for a viewer standing at the edge of the cliff, assuming that his head is 600 feet above the floor of the valley below.

NOTE
A fall from a helicopter 3 miles in the sky is not only dangerous but nearly impossible. It should only be attempted by a trained professional or someone *really* famous.

SOLUTIONS

1. The equation is already given, so find the derivative with respect to t and substitute:

$$\frac{dy}{dt} = 3x^2 \frac{dx}{dt} - 6x \frac{dx}{dt}$$

$$\frac{dy}{dt} = 3(4)^2(-3) - 6(4)(-3) =$$

$$\frac{dy}{dt} = -72 \text{ ft/sec.}$$

BC students note: $\dfrac{\frac{dy}{dt}}{\frac{dx}{dt}} = \dfrac{-72}{-3} = 24$, which is also $\dfrac{dy}{dx}$ when $x = 4$.

2. You need to know the volume of a sphere. Find its derivative with respect to t and plug in the given information:

$$V = \frac{4}{3}\pi r^3$$

$$\frac{dV}{dt} = 4\pi r^2 \frac{dr}{dt}$$

$$-5 = 4\pi(5)^2 \frac{dr}{dt}$$

$$\frac{dr}{dt} = -.016 \text{ in/sec.}$$

Notice that you have to make $\dfrac{dV}{dt}$ negative because the volume is *decreasing*.

3. The surface area of a cube is the sum of the areas of its sides. The sides are all squares, so the surface area is $S = 6x^2$, where x is the length of a side. Now find the derivative and substitute:

$$\frac{dS}{dt} = 12x \frac{dx}{dt}$$

$$\frac{dS}{dt} = 12(24)(7)$$

$$\frac{dS}{dt} = 2016 \text{ in}^2/\text{sec.}$$

Note that you use 24 inches instead of 2 feet for x, since the rest of the problem to be given in terms of inches.

NOTE

The surface area of a sphere is given by $4\pi r^2$; it's the derivative of the volume. A similar relationship exists for circles: area = πr^2 and circumference = $2\pi r$.

4. Begin by drawing a picture.

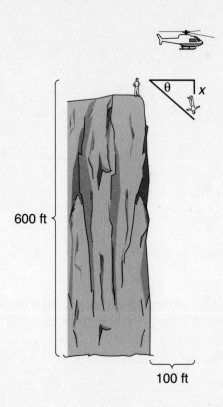

600 ft

100 ft

The horizontal distance between the spectator and Angela will remain fixed, but the vertical distance will change dramatically. You need to use a variable to label any length that can change; here we used x. You'll need an equation that contains θ, since your goal is to find $\frac{d\theta}{dt}$. The perfect choice is tangent, and the equation should be $\tan \theta = \frac{x}{100}$. Now, find its derivative with respect to t:

$$\sec^2 \theta \frac{d\theta}{dt} = \frac{1}{100} \frac{dx}{dt}.$$

You still need to find θ and $\frac{dx}{dt}$ to finish this problem. To find x, you first must find $s(t)$ when $t = 30.9$, so plug 30.9 into the position equation:

$$s(t) = -16(30.9)^2 + 15,840 = 563.04.$$

You can tell by looking at the diagram that $x = 600 - s(t)$, so at the instant that $t = 30.9$, $x = 600 - 563.04 = 36.96$. Finally, we can find θ (in degrees as asked):

$$\tan \theta = \frac{x}{100} = \frac{36.96}{100}$$

$$\theta = 20.28431249°.$$

Now you need to find $\frac{dx}{dt}$. We just said $x = 600 - s(t)$, so find the derivative with respect to t when $t = 30.9$:

$$\frac{dx}{dt} = -s'(t)$$

$$\frac{dx}{dt} = -(-32t) = 32(30.9)$$

$$\frac{dx}{dt} = 988.8 \text{ ft/sec.}$$

All of the required information is finally available, so substitute and finish this problem:

$$\sec^2(20.28431249)\frac{d\theta}{dt} = \frac{1}{100}(988.8)$$

$$\frac{d\theta}{dt} = 8.700 \text{ deg/sec.}$$

HANDS-ON ACTIVITY 6.1: ROLLE'S AND MEAN VALUE THEOREMS

The following activity will help you uncover two of the most foundational theorems of differential calculus. Rolle's Theorem is one of those rare calculus theorems that makes a lot of sense right away. The Mean Value Theorem is not very difficult either, and it usually appears on the AP test pretty frequently. With your deep conceptual understanding of derivatives, you should have no problem at all understanding.

CAUTION
Some students confuse the Mean Value Theorem with the Intermediate Value Theorem. They are similar only in that they are both *existence* theorems, in that they guarantee the existence of something.

1. Let's say that you have the continuous and differentiable function $f(x) = x^3 - 6x^2 + 12x - 5$. Draw the portion of the graph indicated by the axes below ($x = [0,4]$).

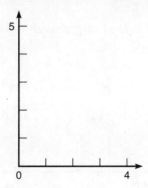

2. Draw the secant line that connects $x = 1$ to $x = 3$ on the graph of f, and calculate the slope of the secant line.

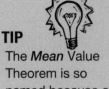

TIP
The *Mean* Value
Theorem is so
named because of
the large role the
average or *mean*
rate of change
plays in the
theorem.

3. Are there any places on the graph of *f* where the tangent line to the graph appears to be parallel to the secant line you've drawn? How many times does this happen on [0,3]? Draw these tangent lines that appear to be parallel to the secant line.

_____

_____

4. What does it mean geometrically if those tangent lines are parallel to the secant line?

_____

_____

5. Write your conclusion to Number 4 as a mathematical formula. If you solve this for *x* (you'll have to use your calculator), you will find the *x* values for the tangent lines. What are they?

_____

_____

_____

6. Time to be generic. Given a continuous, differentiable function $g(x)$ on the interval $[a,b]$, complete the following:

    • Give the slope of the secant line from $x = a$ to $x = b$.

    _____

    • Give the slope of the tangent line to $g(x)$ at any point on its domain.

    _____

    • Use your answers above to fill in the blanks and complete the Mean Value Theorem:

    Mean Value Theorem: Given a function $g(x)$ that is continuous and differentiable on a closed interval $[a,b]$, there exists at least one *x* on $[a,b]$ for which _____, the slope of the secant line, equals _____, the slope of the tangent line.

7. Illustrate the Mean Value Theorem graphically using the graph below of *g* on $[a,b]$.

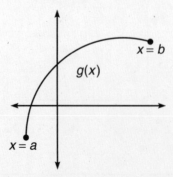

8. Translate the Mean Value Theorem into a statement about rates of change.

---

9. Rolle's Theorem is a specific case of the Mean Value Theorem, which applies whenever $g(a) = g(b)$. What is the slope of the secant line for such a function? What is guaranteed by the Mean Value Theorem as a result, and what does that mean geometrically?

10. Fill in the blanks to complete the theorem:

   Rolle's Theorem: Given a function $f(x)$ that is continuous and differentiable on the closed interval $[a,b]$ and _____ = _____, then there exists at least one x on $[a,b]$ such that _____.

11. Draw a function $g(x)$ on the axes below that satisfies Rolle's Theorem on $[a,b]$ but satisfies it more than one time.

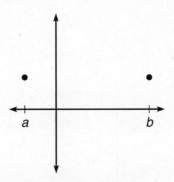

SELECTED SOLUTIONS TO HANDS-ON ACTIVITY 6.1

2. The slope of the secant line is $\dfrac{f(3) - f(1)}{3-1} = \dfrac{4-2}{3-1} = 1$.

3. There are two places on $[0,3]$ where the tangent lines appear to be parallel—one falls nearly midway between $x = 1$ and 2 and another between $x = 2$ and 3.

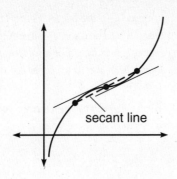

secant line

4. It means that they share the same slope.

5. The formula should state that the secant and tangent slopes are the same:

$$\frac{f(3)-f(1)}{3-1} = f'(x).$$

The left-hand formula is the secant slope, and the derivative on the right represents the tangent slope. You already know the secant slope, so $f'(x) = 1$. Find $f'(x)$ and solve using your calculator; you should get $x \approx 1.423$ and $x \approx 2.577$.

6. The secant slope is $\dfrac{g(b)-g(a)}{b-a}$, and the tangent slope is $g'(x)$.

Mean Value Theorem: …there exists at least one x on $[a,b]$ for which $\dfrac{g(b)-g(a)}{b-a}$ equals $g'(x)$.

7. Estimate on the graph where the secant and tangent lines *appear* to be parallel.

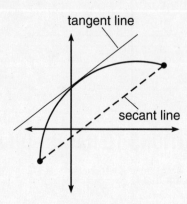

8. At least once on the interval $[a,b]$, the instantaneous rate of change equals the average rate of change for that interval.

9. The slope of the secant line is 0, so the Mean Value Theorem guarantees that somewhere on $[a,b]$, there will be a horizontal tangent line. Geometrically, this means that the graph reaches a maximum or a minimum somewhere between a and b, assuming, of course, that the function is not merely a horizontal line connecting $(a,f(a))$ to $(b,f(b))$.

10. **Rolle's Theorem:** …$[a,b]$ and $f(a)=f(b)$, then there exists at least one x on $[a,b]$ such that $f'(x) = 0$.

11.

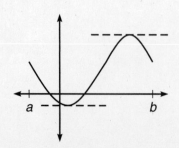

NOTE
The fact that a derivative of 0 indicates a possible maximum or minimum on the graph is an incredibly important fact as this chapter progresses.

PROBLEM SET

You may use a graphing calculator on problem 5 only.

1. Determine if each of the following statements is true or false. If true, justify your answer. If false, provide a counter example.

    If $f(x)$ is a continuous function on $[a,b]$, and $f(a) = f(b)$, then …

    (a) f has an absolute maximim and an absolute minimum.

    (b) There exists a c, $a \le c \le b$, such that $f'(c) = 0$.

2. During vacation, Jennifer is en route to her brother's house. Unbeknownst to her, two policemen are stationed two miles apart along the road, which has a posted speed limit of 55 mph. The first clocks her at 50 mph as she passes, and the second measures her speed at 55 mph but pulls her over anyway. When she asks why she got pulled over, he responds, "It took you 90 seconds to travel 2 miles." Why is she guilty of speeding according to the Mean Value Theorem?

3. Given $g(x)$ as defined by the graph below, and $g(b) = g(c)$.

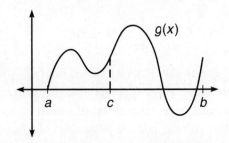

    (a) How many times does $g(x)$ satisfy the Mean Value Theorem on $[a,b]$?

    (b) What conclusions, if any, can be drawn about $g(x)$ using Rolle's Theorem?

4. At what value(s) of x is the Mean Value Theorem satisfied for $m(x) = x^3 - \frac{5}{2}x^2 - 2x + 1$ on $[-2,4]$?

5. If $h'(0)$ satisfies the Mean Value Theorem for $h(x) = \ln(\sin x + 1)$ on $[-\frac{\pi}{4}, b]$, find the smallest possible value of b.

SOLUTIONS

1. (a) True: This value is guaranteed by the Extreme Value Theorem; the function need only be continuous on a closed interval.

   (b) False: It may sound like Rolle's Theorem, but Rolle's Theorem includes the guarantee that the function is differentiable on $[a,b]$; the diagram below shows one possible graph of f for which there is no horizontal tangent line on the interval.

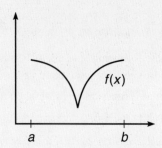

2. You can figure out Jenny's average speed for the two miles between the police; think back to the old, familiar formula $d = rt$ (distance = rate · time). Her rate will be given by $r = \frac{d}{t}$. Your units are hours, so 90 seconds has to be rewritten in terms of hours. Because there are 3600 seconds in an hour, 90 seconds represents $\frac{90}{3600}$ of an hour, or $\frac{1}{40}$. Clearly, $d = 2$ from the information given. Therefore, you can find the average rate of speed:

$$r = \frac{2}{\frac{1}{40}} = 80 \text{ mph.}$$

The Mean Value Theorem says that on a closed interval, your instantaneous rate of change (in this case, velocity) must equal the average rate of change (average speed) at least once. Therefore, ol' leadfoot Jenny had to have traveled 80 mph at least once, and the policeman can ticket her.

3. (a) The tangent line must be parallel to the secant line (shown dotted in the diagram below) four times in that interval at approximately the places marked below.

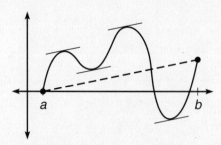

(b) Rolle's Theorem ensures that $g'(x) = 0$ at least once between $x = c$ and $x = b$. In fact, $g'(x) = 0$ twice on the interval, although you may not be able to justify that visually until later in this chapter.

4. The Mean Value Theorem states that at some point c between -2 and 4,

$$m'(c) = \frac{m(4) - m(-2)}{4 - (-2)}.$$

Once you find $m'(c)$ and evaluate the fraction, you can solve for c:

$$3c^2 - 5c - 2 = \frac{17 - (-13)}{6}$$

$$3c^2 - 5c - 2 = 5$$

$$3c^2 - 5c - 7 = 0.$$

This doesn't factor, so you need to resort to the quadratic formula. The solutions are $c = \frac{5 \pm \sqrt{109}}{6}$. However, only the solutions on the interval $[-2,4]$ count. Without a calculator, how can you tell if either of these fall in the interval? Well, $\sqrt{109}$ is between $\sqrt{100}$ (10) and $\sqrt{121}$ (11). If you evaluate the solutions with these estimations in place, both answers fall within the interval, so both values of c are correct.

5. If the Mean Value Theorem is satisfied at $h'(0)$, then the average rate of change on the interval *must* be equal to that value; that value is

$$h'(0) = \frac{\cos 0}{\sin 0 + 1} = 1.$$

Therefore, all you need to do is to find the average value and set it equal to 1 in order to find b:

$$\frac{h(b) - h\left(-\frac{\pi}{4}\right)}{b - \frac{-\pi}{4}} = 1$$

$$\frac{\ln(\sin b + 1) - \ln\left(\sin\left(-\frac{\pi}{4}\right) + 1\right)}{b + \frac{\pi}{4}} - 1 = 0.$$

Once you set this gigantic equation equal to zero, you can use your calculator to solve, and $b \approx 1.07313$.

HANDS-ON ACTIVITY 6.2: THE FIRST DERIVATIVE TEST

You might suspect that there is some kind of relationship between functions and their derivatives. Sometimes, late at night, they show up holding hands, insisting that they are "just friends." It's just as you suspect—derivatives can always tell what a function is doing—they have this connection that other people envy and yearn for. You'll learn more about this connection in the following activity.

1. Graph the function $f(x) = x^3 - \frac{5}{2}x^2 - 2x + 1$ on the axes below.

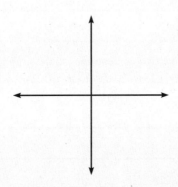

TIP
The relationships between a function and its derivatives are all over the AP test. This is one of the most essential concepts in AP Calculus, so make sure to understand it — it's not that hard, actually.

2. Your goal in this exercise will be to describe where f is increasing and decreasing without depending on its graph. In other words, you want to describe the *direction* of f. **Important fact:** If a graph changes direction, the change will occur at a *critical number*, a number at which the derivative either equals zero or is undefined. What are the critical numbers for f?

_____

_____

3. What relationship do you see between the critical numbers and the direction of f?

_____

_____

4. The line graph below is called a *wiggle graph*. It is used in conjuction with critical numbers to describe a function's direction. Label the graph below by marking off the critical numbers you found in Number 2 above. They should break the wiggle graph into three separate segments.

$$\longleftarrow \qquad\qquad\qquad \longrightarrow f'$$

5. Pick a number from each of the three segments of the wiggle graph and plug each separately into the derivative. If the result is negative, write a "−" above the corresponding segment of the wiggle graph; if the result is positive, denote it with a "+".

6. What relationship do you see between the wiggle graph you have constructed and the graph of f?

_____

_____

7. Complete this statement based on your results: If a function f is increasing on an interval, then its derivative f' will be _____ there. However, if a function is decreasing, it's derivative will be _____.

8. Consider the simple function $g(x) = x^2$. Construct a wiggle graph for it, and give the intervals for which g is increasing.

_____

_____

_____

9. Graph $y = x^2$ and draw two tangent lines to the graph — one on the interval $(-\infty, 0)$ and one on $(0, \infty)$. How do the slopes of these tangent lines support your wiggle graph?

_____

_____

10. How could you tell from the wiggle graph of $p(x) = x^2$ that $x = 0$ was a relative minimum?

_____

_____

11. In general, how can you tell where a relative maximum or minimum occurs using only a wiggle graph?

_____

_____

12. Form the wiggle graph for $h(x) = \dfrac{1}{x^2}$. What are the relative extrema (relative maximims or minimums) for the graph? Why?

_____

_____

SELECTED SOLUTIONS TO HANDS-ON ACTIVITY 6.2

NOTE
It makes sense to call maximums and minimums *extrema* points, because the graph takes on its most extreme values there.

2. To find critical numbers, you must first find the derivative: $f'(x) = 3x^2 - 5x - 2$. Critical numbers occur wherever this is zero or is undefined.

   By factoring, you can see that $f'(x) = 0$ when $x = -\dfrac{1}{3}$ and $x = 2$.

3. At the x-values that represent the critical numbers, the graph seems to change direction.

4.

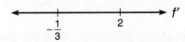

5. You can choose any number from each of the intervals; the numbers $-1, 0$, and 3 are good, simple choices. Plug each into the *derivative*: $f'(-1) = 6$, $f'(0) = -2$, and $f'(3) = 10$. Therefore, mark the intervals left to right as positive, negative, and positive:

Let the meaning of the diagram sink in. You can easily calculate that $f'(-1)$ is positive. That means that *any* value on the interval $(\infty, -\frac{1}{3})$ will return a positive value if substituted into the derivative. Likewise, any number in the interval $(-\frac{1}{3}, 2)$ will return a negative value when substituted into the derivative, according to the wiggle graph.

6. Whenever the wiggle graph is positive, f is increasing, and when the graph wiggles negative, f is decreasing.

7. positive, negative

8. g will be increasing on $(0,\infty)$, since g' is positive on that interval:

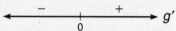

9. Whenever the graph is decreasing (in this case $(-\infty,0)$), the tangent lines on the graph will have a negative slope. When the graph is increasing, the tangent lines have a positive slope. The wiggle graph is based on the derivative, which is defined by the slope of the tangent line.

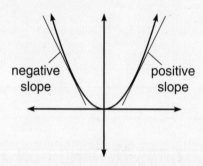

negative slope   positive slope

10. The wiggle graph shows that the derivative changes from negative to positive at $x=0$, which means that g changes from decreasing to increasing to $x=0$. If a graph suddenly stops decreasing and begins increasing, the point at which it stopped must be a minimum — draw it!

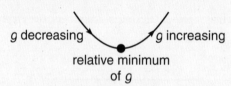

g decreasing g increasing
relative minimum
of g

11. When the derivative changes sign, the function changes direction. Therefore, if a wiggle graph changes from $+$ to $-$ or vice versa, a relative extrema point has occurred.

12. You can rewrite h as x^{-2}; therefore, $h'(x) = -2x^{-3}$, or $\frac{-2}{x^3}$. The derivative never equals zero; however, h' is undefined when $x=0$. Therefore, $x=0$ is a critical number for h. The wiggle graph looks like the following:

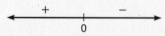

It looks like $x=0$ should be a relative maximum, because the function changes from increasing to decreasing there. However, $x=0$ is *not in the domain* of the original function h! $x=0$ cannot be a maximum on a graph if it's not even in the domain. This is because $h(x)$ has a vertical asymptote at $x=0$. Thus, $x=0$ is not an extrema point. Because it was the only critical number (and all extrema must occur at critical numbers), there are no extrema on h.

PROBLEM SET

Do not use a graphing calculator on these problems.

1. Draw the graph of a function, $b(x)$, that has a relative minimim at $x = -3$ when $b'(-3)$ does not exist.

2. Give the x-values at which each of the following functions have relative extrema, and classify those extrema. Justify your answers with wiggle graphs.

    (a) $y = \dfrac{2x^2 - 9x - 5}{x + 3}$

    (b) $y = \dfrac{e^{2x+1}}{\ln x}$

3. Given the graph below of $m'(x)$, the derivative of $m(x)$, describe and classify the relative extrema of $m(x)$.

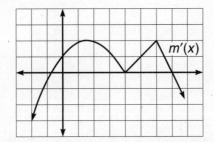

4. Below are two graphs, $g(x)$ and $g'(x)$. Which is which, and why?

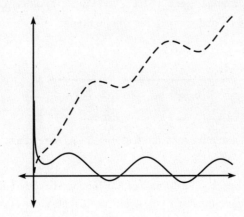

5. If $f(x)$ is defined as follows:

    $$f(x) = \begin{cases} \sin x \cos x, \ 0 \le x < \pi \\ \sin^2 x, \ \pi \le x \le 2\pi \end{cases}$$

    (a) Find all points of discontinuity on $f(x)$.

    (b) Determine if $f'(\pi)$ exists.

    (c) On what intervals is f increasing?

SOLUTIONS

1.

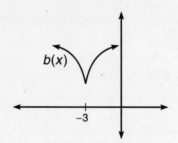

2. (a)   The derivative, using the Quotient Rule, is

$$\frac{(x+3)(4x-9)-\left(2x^2-9x-5\right)}{(x+3)^2}$$

$$y'=\frac{4x^2+3x-27-2x^2+9x+5}{(x+3)^2}$$

$$y'=\frac{2x^2+12x-22}{(x+3)^2}.$$

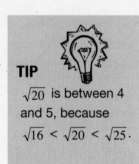

TIP

$\sqrt{20}$ is between 4 and 5, because

$\sqrt{16}<\sqrt{20}<\sqrt{25}$.

The critical numbers occur where the numerator and denominator equal 0. Clearly, the denominator is 0 when $x=-3$. Set the numerator equal to 0, and use the quadratic formula to get $x=-3\pm\sqrt{20}$. Plug test points from each interval into the derivative, and you get this wiggle graph:

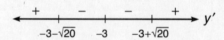

This graph has a relative maximum at $x=-3-\sqrt{20}$ and a relative minimum at $-3+\sqrt{20}$.

(b)   First, realize that the domain of this function is $x>0$, because that is the domain of the denominator. Once again, critical numbers occur wherever the numerator or denominator equals zero. The numerator is the natural exponential function, which never equals zero (its range is $(0,\infty)$). However, the denominator equals 0 when $x=1$. Therefore, $x=1$ is the only critical number.

The resulting wiggle graph looks like the following:

However, there is no relative minimum at $x = 1$, even though the direction of the function changes. Because $x = 1$ is not in the domain of the function, it cannot be an extrema point. In fact, $x = 1$ is a vertical asymptote for this function. There are no extrema points for this function.

3. The derivative changes from negative to positive at $x \approx -.8$, meaning that the original function m changes from decreasing to increasing. Therefore, m has a relative minimum there. Using similar reasoning, m has a relative maximum at $x = 7$. There are no extrema points when $x = 4$, even though the derivative equals 0 there (making it a critical number); the derivative doesn't actually change signs around $x = 4$, so the function does not change direction.

4. The dotted function is $g(x)$. Note that whenever the dotted function reaches a relative maximum or minimum, the other graph has a value of 0 (an x-intercept occurs). Furthermore, whenever the dotted function is increasing, the solid function is positive (above the x-axis) and vice versa.

5. (a) The only possible point of discontinuity of f is at $x = \pi$—if the two graphs do not meet at that point, there will be a jump discontinuity. If you substitute $x = \pi$ into each of the two pieces of the function, they both result in an output of 0, making f continuous on its entire domain $[0,2\pi]$.

(b) These two functions could meet in a sharp point, causing a cusp, and no derivative would exist. To determine if a cusp occurs, you undertake a process similar to 5(a). To be continuous, both pieces had to have the same function value. To be differentiable, both pieces must have the same derivative at $x = \pi$. The derivatives are as follows:

$$\frac{d}{dx}(\sin x \cos x) = \cos^2 x - \sin^2 x$$

$$\frac{d}{dx}(\sin^2 x) = 2\sin x \cos x.$$

When $x = \pi$, the first rule has a derivative of -1, and the second has a derivative of 0. Because these derivatives do not match, no derivative exists on f when $x = \pi$.

(c)   In order to determine direction, you need to set the derivatives found in 5(b) equal to zero and complete a wiggle graph. Because the first rule of the piecewise-defined function pertains only to $[0,\pi)$, the wiggle graph on the same interval will be based on its derivative. Therefore, the interval $[0,\pi)$ has critical numbers $\frac{\pi}{4}$ and $\frac{3\pi}{4}$. Similarly, you set $2\sin x \cos x = 0$ and plot the resulting critical numbers only on the interval $[\pi,2\pi]$. These critical numbers are π, $\frac{3\pi}{2}$, and 2π. The final wiggle graph looks like the following:

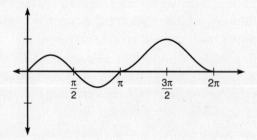

Therefore, f is increasing on $(0,\frac{\pi}{4}) \bigcup (\frac{3\pi}{4},\frac{3\pi}{2})$. In case you are skeptical, here's the graph of f:

CONCAVITY

Two of the major characteristics used to describe graphs are direction and concavity. You have already used the sign of the first derivative to determine the direction of a function, and in this section, you will use the sign of the second derivative to determine the concavity of the original function. In 1955, prison wardens all over the world introduced flouride into the drinking water of their prisons in a coordinated effort to reduce "con" cavities, but their efforts proved to be in vain. Thus, concavity still pervades functions worldwide.

Concavity describes the curviness of a curve. Consider the smiles on the faces drawn below:

The smile on a *happy* face is described as *concave up*, whereas the *frown* is *concave down*. It rhymes. It is also said that milk poured into a concave up curve stays there, whereas milk poured on a concave down curve will splatter on your mom's clean floor and make her angry. In this mnemonic device, a *cup* should be *concave up*.

Notice the signs that constitute the eyes of the faces; these signs remind us of the most important fact concerning concavity: *If a function, f(x), is concave down on an interval, then the second derivative, f″(x), will be negative there. Similarly, if a function is concave up, its second derivative will be positive.*

This is hauntingly similar to our work with direction and the first derivative. In that case, the sign of the *first* derivative indicated direction. Concavity, on the other hand, is dictated by the sign of the *second* derivative. This relationship is explored by the diagram below:

direction   concavity
information   information

The sign of a function describes the direction of the function one step "above" it, and that same sign describes the concavity of the function two steps "above."

 Example 4: On what intervals is the graph of the function $g(x) = x^3 - 2x^2 - 4x + 2$ concave down?

Solution: The concavity of a function is based on the sign of its second derivative, so you need to begin by finding $g''(x)$:

$$g'(x) = 3x^2 - 4x - 4$$

$$g''(x) = 6x - 4.$$

Just like when you found direction, you need to find critical numbers again. This time, the critical numbers occur when the *second* derivative is either zero or undefined. The only critical number is $x = \frac{2}{3}$. Use this to create a *second derivative* wiggle graph, and make sure to label it g''. Choose test numbers from both of the intervals and make sure to plug them into the *second* derivative, as it is the sign of the second derivative that provides the information you are seeking. The wiggle graph looks like the following when you are finished:

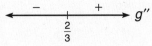

The graph of $g(x)$ is concave up on $\left(-\infty, \frac{2}{3}\right)$.

NOTE
Some psychologists suggest that just about anything that rhymes is usually true. They cite the example "In 1492, Columbus sailed the ocean blue." These psychologists are not highly regarded by their peers.

TIP
Just as you used the wiggle graph to describe direction and direction changes, you'll use a second derivative wiggle graph to describe concavity and concavity changes.

In the preceding example, the graph changed from concave down to concave up at the point $(\frac{2}{3}, -\frac{34}{27})$. The change of concavity makes this a *point of inflection*, much as a change of direction caused points to become extrema points.

 Example 5: Given the below graph of $h'(x)$, the derivative of $h(x)$, describe the concavity of $h(x)$.

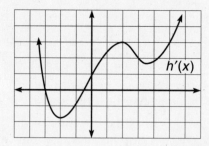

Solution: Not only does the $h''(x)$ describe the concavity of $h(x)$, it also describes the *direction* of the $h'(x)$. (Since $h'(x)$ is one step "above" $h''(x)$.) From the graph, you can tell that $h'(x)$ is increasing on $(-2,2) \cup (3.5,\infty)$ and decreasing on $(-\infty,-2) \cup (2,3.5)$. Therefore, $h''(x)$ will be negative on $(-\infty,-2) \cup (2,3.5)$ and positive on $(-2,2) \cup (3.5,\infty)$, as indicated on the concavity wiggle graph below:

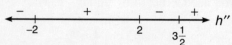

Without having seen a graph of $h(x)$, we can draw from the concavity wiggle graph that h is concave up on $(-2,2) \cup (3.5,\infty)$ and concave down on $(-\infty,-2) \cup (2,3.5)$.

PROBLEM SET

You may use a graphing calculator on problems 3 and 4.

1. Describe the concavity of the generic linear function $y = ax + b$, and interpret your answer.

2. On what intervals is the function $g(x) = \frac{\sin 2x \cos x}{\sin x}$ concave up on the interval $[0,2\pi]$?

TIP
Do you see why it's so important to label your wiggle graph? Without the h'' label in Example 5, it wouldn't be immediately obvious which derivative we were referencing.

3. Given the differentiable graph of $f''(x)$ below, answer the following questions:

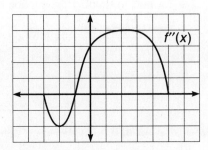

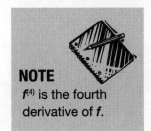

NOTE
$f^{(4)}$ is the fourth derivative of f.

   (a)   Describe the concavity of $f(x)$.

   (b)   At what x-values will $f'(x)$ have its absolute maximum and absolute minimum values?

   (c)   Will $f'''(0)$ be positive or negative? What about $f^{(4)}(0)$?

   (d)   On what intervals will $f'(x)$ be concave down?

4. Given the below graph of $h'(x)$, describe the concavity of $h(x)$, given $h(x)$ is continuous.

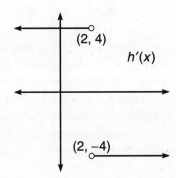

SOLUTIONS

1. The concavity of a function is described by its second derivative, so find that first:

$$y' = a$$

$$y'' = 0.$$

   It is difficult to interpret a value of zero in the second derivative. In this case, it is because a line by itself does not have any concavity. However, a second derivative of 0 does not always mean that no concavity exists (see problem 4 below).

2. It's necessary, again, to find the second derivative. Instead of using the quotient rule, use a double angle formula to simplify the fraction first:

$$g(x) = \frac{2\sin x \cos x \cdot \cos x}{\sin x} = 2\cos^2 x$$

$$g'(x) = -4\cos x \sin x$$

$$g''(x) = -4(\cos^2 x - \sin^2 x).$$

Set the second derivative equal to 0, and solve to get critical numbers of $\frac{\pi}{4}$, $\frac{3\pi}{4}$, $\frac{5\pi}{4}$, and $\frac{7\pi}{4}$. Use these and test points from each interval to construct the following wiggle graph for concavity:

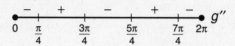

Therefore, $g(x)$ will be concave up on the intervals $(\frac{\pi}{4}, \frac{3\pi}{4})$ and $(\frac{5\pi}{4}, \frac{7\pi}{4})$.

3. (a) $f''(x)$ is negative on $(-3,-1)$, so $f(x)$ will be concave down there. However, $f(x)$ will be concave up on $(-1,5)$, since $f''(x)$ is positive on that interval.

(b) Because $f''(x)$ is negative on $(-3,-1)$, you know that $f'(x)$ is decreasing on that interval. However, $f'(x)$ will be increasing on the interval $(-1,5)$ by the same reasoning. Notice that the graph changes from decreasing to increasing at $x=-1$, so a relative minimum will occur there. That relative minimum will also be the absolute minimum of the graph, because the graph only increases once you pass that point. Because the graph increases for the remainder of its domain, the absolute maximum of the graph will occur at $x=5$, which is the last defined point on the graph.

(c) The third derivative of f describes the direction of the second derivative of f. You are given the graph of the second derivative, and at $x=0$, $f''(x)$ is increasing, so $f'''(0)$ will be positive. On the other hand, $f^{(4)}(x)$ describes the concavity of $f''(x)$, and the graph of $f''(x)$ is concave down at $x=0$. Therefore, $f^{(4)}(0)$ will be negative.

(d) The signs of $f'''(x)$ will describe the concavity of $f'(x)$. How do you determine the signs for $f'''(x)$? Remember that it describes the direction of $f''(x)$. Because $f''(x)$ decreases on $(-3,-2)\bigcup(2,5)$ and increases on $(-2,2)$, you can construct the following wiggle graph of $f'''(x)$:

Therefore, $f'(x)$ will be concave down on $(-\infty,-2)\bigcup(2,\infty)$.

4. This requires some thinking. The graph tells you that $g(x)$ is some function whose derivative is consistently 4 until $x=2$, and then the derivative suddenly changes to -4. What kind of a function has a constant derivative? A linear function. For example, if $y=4x+3$, then $y'=4$. However, what sort of linear

TIP
A *relative extrema point* is the highest or lowest value in a small interval of the graph. An *absolute extrema point* is the highest or lowest point on the entire domain of the graph. Absolute extrema can occur at relative extrema or endpoints, if the function has them.

function suddenly changes derivative? Consider the absolute value graph pictured below:

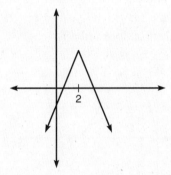

The slope of the curve is 4 until $x = 2$, at which point the slope turns into its opposite, –4. The graph of $h(x)$ is clearly concave down on its entire interval.

MOTION

Calculus has its long, threatening talons in just about every aspect of day-to-day life; luckily, most of us are blissfully ignorant of it and unaware of it stalking us, waiting until we go to sleep, and then messing with our stuff — like putting CDs in the wrong cases and breaking all the points off your pencils on test day. But not even calculus can hide its influence in the topic of motion. Because a derivative describes a rate of change, we have already seen its influence many times and alluded to this very moment: describing how a derivative affects a position equation.

Important Facts about Position Equations

- A position equation is typically denoted as $s(t)$ or $x(t)$; for any time t, its output is the object's position relative to something else. For example, output may represent how far a projectile is off the ground or how far away a particle is from the origin.

- The derivative of position, $s'(t)$ or $v(t)$, gives the *velocity* of the object. In other words, $v(t)$ tells how fast the object is moving and in what direction. For example, if we are discussing a ball thrown into the air and $v(3 \text{ seconds}) = -4$ ft/sec, when time equals 3, the ball is traveling at a rate of 4 ft/sec downward.

- The derivative of velocity, $v'(t)$ or $a(t)$, gives the *acceleration* of the object. This ties in directly to the section you just completed. If an object has positive acceleration, then the position equation (two derivatives "above" $a(t)$) must be concave up, and the velocity equation (one derivative "above" $a(t)$) must be increasing.

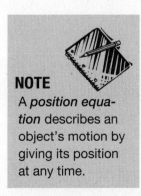

NOTE
A *position equation* describes an object's motion by giving its position at any time.

The most common motion questions on the AP test focus on the motion of a particle on a line, usually horizontal (although the direction of the line doesn't matter). For example, consider a particle moving along the x-axis whose position at any time t is given by $s(t) = t^3 - 10t^2 + 25t - 1$, $t > 0$. The graph of the position equation looks like

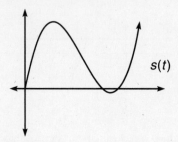

but the particle itself never leaves the x-axis. Let's look at this problem in depth to better understand a typical particle motion problem.

Example 6: If the position (in feet) of a particle moving horizontally along the x-axis is given by the equation $s(t) = t^3 - 10t^2 + 25t$, $t > 0$ seconds, answer the following questions.

(a) Evaluate $s(1)$, $s(4)$, and $s(5)$, and interpret your results.

By simple substitution, $s(1) = 16$, $s(4) = 4$, and $s(5) = 0$. In other words, when 1 second has elapsed, the particle is 16 feet to the right of the origin, but 3 seconds later at $t = 5$, the particle is back to the origin.

(b) At what time(s) is the particle temporarily not moving, and why?

The particle will be temporarily stopped when its velocity equals 0 — this makes a lot of sense, doesn't it? Since the derivative of position is velocity, take the derivative and set it equal to 0:

$$v(t) = s'(t) = 3t^2 - 20t + 25 = 0.$$

Now, factor the quadratic equation to complete the solution:

$$(3t - 5)(t - 5) = 0$$

$$t = \frac{5}{3}, 5.$$

Therefore, at these two moments, the particle is stopped because it is in the process of changing direction. (Remember, part (a) showed you that it changed direction between $t = 1$ and $t = 4$.)

(c) On what interval of time is the particle moving backward?

The particle moves backward when it has negative velocity. Therefore, we will draw a velocity (first derivative) wiggle graph. We already know the critical numbers from part (b), so all that remains is to choose

CAUTION
Some students don't believe me when I say that it is necessary to briefly stop when changing direction, and in demonstrating their point, many of these students end up with neck injuries.

some test points from among the intervals. Because $s(.5)$ is positive, $s(3)$ is negative, and $s(6)$ is positive (of course you wouldn't have to pick the same test points), you get the following wiggle graph:

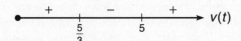

Therefore, the particle is moving backward on $(\frac{5}{3},5)$. This makes sense if you consider the graph of $s(t)$. Remember, this position graph tells how far the particle is away from the origin. On the interval $(\frac{5}{3},5)$, the particle's distance from the origin is *decreasing*, indicating backward movement. It should be no surprise that the velocity is negative then, since velocity is the derivative of that graph, and derivatives have a nasty habit of describing the direction of things.

(d) How far does the particle travel in its first 4 seconds of motion?

You may be tempted to answer 4 feet, since $s(4) = 4$; however, that is what's called the *displacement* of the particle. The displacement is the net change in position. Because $s(0) = 0$ and $s(4) = 4$, no matter what happened in between, the particle ended up a total of 4 units from where it started. However, the problem doesn't ask for displacement — it asks for *total distance traveled*. We need to measure how far it swung out to the right of the origin when it changed direction at $t = \frac{5}{3}$ and then how far back toward the origin it came. We already know $s(0) = 0$, but it is essential to know that $s(\frac{5}{3}) \approx 18.518518518$, because it tells us that the particle traveled 18.518518518 feet in the first $1\frac{2}{3}$ seconds. At this point, the particle changes direction and ends up 4 feet from the origin. In the return trip, then, it traveled $18.518518518 - 4 = 14.518518518$ feet. The total distance it traveled was $18.518518518 + 14.518518518 \approx 33.037$ feet.

The other type of motion probelm the AP test enjoys inflicting upon you is the dreaded trajectory problem. Did you know that anything thrown, kicked, fired, or otherwise similarly propelled follows a predetermined position equation on the earth? It's true, neglecting air resistance of course. The generic projectile position equation is

$$s(t) = -\frac{g}{2}t^2 + v_0t + h_0,$$

where g is the gravitational acceleration constant (32 ft/sec² in the English system and 9.8 m/sec² in the metric), v_0 is the object's initial velocity, and h_0 is the object's initial height. It is probably a good idea to memorize this equation in case you ever need it, although the questions typically asked for this sort of problem are extremely similar to those asked in Example 6.

NOTE
Air resistance has been neglected for so long in theoretical mathematics that it is rumored to have joined a 12-step program.

PROBLEM SET

You may use a graphing calculator on both of these problems.

1. A very neurotic particle moves up and down the y-axis according to the position equation $y = (t^2 - 6t + 8) \cdot \sin t$, $t > 0$, where position is in centimeters and time is in seconds. Knowing this, answer the following questions:

   (a) When is the particle moving down on the interval [0,5]?

   (b) At what values of t is the particle moving at a rate equal to the average rate of change for the particle on the interval [0,5]?

   (c) At what time(s) is the particle exactly 2 cm away from the origin on the interval [0,5]?

   (d) What is the acceleration of the particle the first time it comes to rest?

2. The practice of shooting bullets into the air — for whatever purpose — is extremely dangerous. Assuming that a hunting rifle discharges a bullet with an initial velocity of 3000 ft/sec from a height of 6 feet, answer the following questions (neglecting wind resistance):

   (a) How high will the bullet travel at its peak?

   (b) How long will it take the bullet to hit the ground?

   (c) At what speed will the bullet be traveling when it slams into the ground, assuming that it hits nothing in its path?

   (d) What vertical distance does the bullet travel in the first 100 seconds?

SOLUTIONS

1. (a) The particle is moving down when its position equation is decreasing— when the velocity is negative. You should make a first derivative wiggle graph, so begin by finding the critical numbers of the derivative:

$$v(t) = (t^2 - 6t + 8)(\cos t) + (2t - 6)(\sin t) = 0.$$

It's best to solve this using your graphing calculator. The solutions are $t =$ .738, 2.499, and 3.613. The wiggle graph is

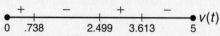

Therefore, the particle is moving down on $(.738, 2.499) \cup (3.613, 5)$.

   (b) The average rate of change of the particle will be $\frac{y(5) - y(0)}{5} =$ $-.5753545648$ cm/sec. To determine when the particle travels this speed, set the velocity equal to this value and solve with your calculator. This is actually the Mean Value Theorem in disguise; we know at least one t will satisfy the requirements in the question, but it turns out that the instantaneous rate of change equals the average rate of change *three* times, when $t = .813, 2.335,$ and 3.770 sec.

(c)   The particle will be two cm away from the origin when its position is 2 (two cm above) *or* –2 (2 cm below). So, you need to solve both the equations $(t^2 - 6t + 8)\sin t = 2$ and $(t^2 - 6t + 8)\sin t = -2$ with your calculator. The solutions are $t = .333$, 1.234, and 4.732 sec.

(d)   The particle first comes to rest at its first critical number, $t = .73821769$. To find acceleration, you need to differentiate the velocity and substitute in the critical number. Rather than doing this by hand, why not use the graphing calculator and the nDeriv function? You would type the following on your TI-83: nDeriv$((x^2 - 6x + 8)(\cos(x)) + (2x - 6)(\sin(x)),x,.73821769)$. The resulting acceleration is -8.116 cm/sec^2.

2.   (a)   You will need to apply the projectile position equation. The initial height and velocity are stated by the problem, and since the question uses English system units (feet), you should use $g = 32$ ft/sec^2 as the acceleration due to gravity. You put your left foot in, you take your left foot out, you put your left foot in, shake it all about, and the position equation is

$$s(t) = -16t^2 + 3000t + 6$$

The bullet will reach its peak at the maximum of the position equation (since it is an upside-down parabola, there will be only one extrema point). To find the t value at which the peak occurs, find the derivative and set it equal to 0 (since the bullet will have a velocity of 0 at its highest point before it begins to fall toward the ground):

$$v(t) = s'(t) = -32t + 3000 = 0$$

$$t = 93.75 \text{ seconds.}$$

This, however, is not the answer. The *height* the bullet reaches at this point is the solution: 140,631 feet, or 26.635 miles.

(b)   The bullet will hit the ground when $s(t) = 0$—literally, when the bullet is 0 feet off of the ground. So, set the position equation equal to 0, and solve (if you use the calculator, you'll have to ZoomOut a few times before the graph appears—these are big numbers). The bullet will remain in the air 187.502 seconds, or 3.125 minutes.

(c) $| \, s'(187.502) \, | = 3000.064$ ft/sec (since speed is the absolute value of velocity, the answer is not negative). The bullet will hit at a speed slightly greater than that at which it was fired. Therefore, being hit by a bullet that was fired into the air will have the same impact as being hit by a bullet at point-blank range.

(d)   You already know that the bullet travels 140,631 feet in the first 93.75 seconds. Because $s(100) = 140,006$, the bullet falls $140,631 - 140,006 = 625$ feet between $t = 93.75$ and $t = 100$. Therefore, the bullet travels a total vertical distance of 140,631 feet up + 625 feet down = 141,256 feet.

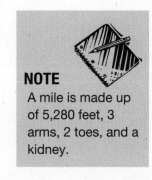

NOTE
A mile is made up of 5,280 feet, 3 arms, 2 toes, and a kidney.

MOTION IN THE PLANE (BC TOPIC ONLY)

Although these questions are less frequent, the AP test sometimes contains questions concerning movement along a parametrically defined or vector path. Vector defined functions are quite easy to differentiate, making this topic relatively simple in the grand scheme of BC topics. Before you read on, you should look back at the introduction to vector functions in Chapter 2.

In order to differentiate a vector function $\mathbf{s} = f(t)\mathbf{i} + g(t)\mathbf{j}$, find the derivatives of the components separately:

$$\frac{d}{dt}(\mathbf{s}) = f'(t)\mathbf{i} + g'(t)\mathbf{j}.$$

Vector position equations work similarly to the position equations we just discussed. The velocity vector is given by the first derivative of the position equation, and the acceleration vector is given by the second.

Example 7: A particle moves in the plane such that the position vector from the origin to the particle is $\mathbf{s} = (\cos t \sin 2t)\mathbf{i} + (2t + 1)\mathbf{j}$, for all t on the interval $[0,2\pi]$.

(a) Find the velocity and acceleration vectors for any time t.

To find the velocity vector, take the derivative of the x and y components separately:

$$\mathbf{v} = (2\cos t \cos 2t - \sin t \sin 2t)\mathbf{i} + 2\mathbf{j}.$$

In order to find the acceleration vector, take the derivative again:

$$\mathbf{a} = (-4\cos t \sin 2t - 2\sin t \cos 2t - 2\sin t \cos 2t - \cos t \sin 2t)\mathbf{i}.$$

(b) What is the velocity vector when $t = \frac{\pi}{2}$, and what is the speed of the particle there?

Plug $t = \frac{\pi}{2}$ into the velocity vector to get the specific answer for that value of t:

$$\mathbf{v} = (0 \cdot -1 - 1 \cdot 0)\mathbf{i} + 2\mathbf{j}$$

$$\mathbf{v} = 2\mathbf{j}.$$

The graph and its velocity vector at $t = \frac{\pi}{2}$ are shown below. The velocity vector has no horizontal component because the graph is changing direction at that point (kind of like a sideways extrema point).

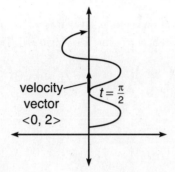

velocity vector <0, 2>

The speed of the particle is given by the norm of the velocity vector. Remember, you find the norm with the equation

$$\|\mathbf{v}\| = \sqrt{\left(\frac{dx}{dt}\right)^2 + \left(\frac{dy}{dt}\right)^2} \; .$$

In this case $\|\mathbf{v}\| = \sqrt{0 + 2^2} = 2$.

PROBLEM SET

You may use your graphing calculator for part (b) of problem 1 and all of problems 2 and 3.

1. A particle moves along the graph defined by $x = \cos t$, $y = \cos 3t$, for t on the interval $[0, 2\pi]$.

   (a) What is the velocity vector, $\mathbf{v}$, when $t = \frac{7\pi}{4}$?

   (b) Draw the path of the particle on the coordinate plane, and indicate the direction the particle moves. Draw the velocity vector for $t = \frac{7\pi}{4}$.

   (c) What is the magnitude of the acceleration when $t = \frac{7\pi}{4}$?

2. A particle moves along a continuous and differentiable path that includes the coordinates (x,y) below for the corresponding values of t:

t	2	2.5	3
x	1.52	4.91	5.02
y	3.64	4.35	4.73

Approximate the speed of the particle at $t = 3$.

3. Create a position equation in vector form for a particle whose speed is 15 when $t = 1$.

SOLUTIONS

1. (a)   The general position vector is given by $\mathbf{s} = (\cos t)\mathbf{i} + (\cos 3t)\mathbf{j}$. Therefore, the velocity vector will be $\mathbf{v} = (-\sin t)\mathbf{i} + (-3\sin 3t)\mathbf{j}$. The velocity vector when

$t = \frac{7\pi}{4}$ will be $\mathbf{v} = \left(\frac{\sqrt{2}}{2}\right)\mathbf{i} + \left(\frac{3\sqrt{2}}{2}\right)\mathbf{j}$, since $3\left(\frac{7\pi}{4}\right) = \frac{21\pi}{4}$, which is

coterminal with $\frac{5\pi}{4}$.

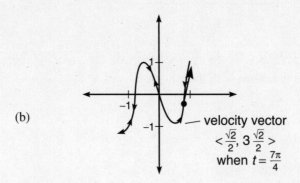

(b)

velocity vector
$< \frac{\sqrt{2}}{2}, 3\frac{\sqrt{2}}{2} >$
when $t = \frac{7\pi}{4}$

The graph proceeds from (1,1) to (−1,−1) on [0,π] and then returns to (1,1) on [π,2π]

(c)   The magnitude of the acceleration is found by calculating the norm of the acceleration, much like the speed is the magnitude (norm) of the velocity. Therefore, you should begin by finding the acceleration vector by differentiating the velocity vector:

$$\mathbf{a} = (-\cos t)\mathbf{i} + (-9\cos 3t)\mathbf{j}.$$

The magnitude of the acceleration is the norm of that vector, so calculate it when $t = \frac{7\pi}{4}$:

$$\|\mathbf{a}\| = \sqrt{\left(-\cos\frac{7\pi}{4}\right)^2 + \left(-9\cos\frac{21\pi}{4}\right)^2}$$

$$\|\mathbf{a}\| = \sqrt{\frac{1}{2} + 81 \cdot \frac{1}{2}}$$

$$\|\mathbf{a}\| = \sqrt{41}.$$

2. Recall the generic formula for speed as the norm of velocity:

$\|\mathbf{v}\| = \sqrt{\left(\frac{dx}{dt}\right)^2 + \left(\frac{dy}{dt}\right)^2}$. Because you do not have the position vector,

you cannot find $\mathbf{v}$. However, you can use slopes of secant lines to approximate $\frac{dx}{dt}$ and $\frac{dy}{dt}$:

$$\frac{dx}{dt} \approx \frac{5.02 - 4.91}{3 - 2.5} = .22$$

$$\frac{dx}{dt} \approx \frac{4.73 - 4.35}{3 - 2.5} = .76$$

Therefore, $\|\mathbf{v}\| = \sqrt{.22^2 + .76^2} \approx .791$.

3. From the given information, we know that

$$\sqrt{\left(\frac{dx}{dt}\right)^2 + \left(\frac{dy}{dt}\right)^2} = 15,$$

so therefore,

$$\left(\frac{dx}{dt}\right)^2 + \left(\frac{dy}{dt}\right)^2 = 225.$$

There are numerous approaches to take, but we will discuss the easiest. Because we can do anything we like (as long as it works), we'll set $\frac{dx}{dt} = 15$ and $\frac{dy}{dt} = 0$. Notice that the sum of their squares equals our goal. Next, we need an expression for x whose derivative, evaluated at 1, is 15. What about $x = \frac{15}{2}t^2$? The trick is taking half of the number you want to end up with and using it as the coefficient, since the Power Rule dictates that you will multiply by 2: $x' = 2 \cdot \frac{15}{2}t = 15t$. Clearly, this has a value of 15 when $t = 1$. You can use any constant for your y component, since its derivative will be 0. One possible answer for this problem, therefore, is $\mathbf{s} = (\frac{15}{2}t^2)\mathbf{i} + 19\mathbf{j}$. Check it to convince yourself that it's right.

OPTIMIZATION

Optimization, like related rates, is one of the most useful topics in calculus because of its direct tie to real-world applications. However, just like related rates, it is one of the topics many students forget about by test time. There are many theories that could account for this forgetfulness in students. One theory is that related rates and optimization, unlike many other calculus topics, require students to follow a strict algorithm in order to arrive at a correct answer (you must proceed from one step to the next, and there are fewer alternative solutions possible than in other calculus topics).

Another less widely held theory is that the concepts of related rates and optimization are imprinted on smaller "memory molecules" than other calculus topics. These smaller molecules are, then, shaken loose from the brain and escape through small lesions in its surface every time you sneeze. I, myself, am torn as to which is actually true, but it does explain the common student phrase "Good luck on the AP Calculus test, and try not to sneeze," which has puzzled scholars for decades.

NOTE
The entire concept of optimization is based on our ability to find max's and min's using critical numbers and wiggle graphs.

Optimization is the process of finding an optimal value, either maximum or minimum, under strict conditions. You may be asked to minimize area, maximize volume, minimize cost, or maximize profit, just to name a few applications. But, we will start out with a simpler example.

TIP
You can tell that this is an optimization problem because it asks for the "smallest possible" number. If the problem is asking for an extreme value (whether large or small), it's an optimization problem.

Example 8: What two positive real numbers give the smallest possible product if one number is two less than three times the other?

Solution: The first step in an optimization problem is to design an equation that represents what you are actually trying to optimize. In this case, you want the minimum product of two numbers. We will set up the equation

$$P = xy,$$

which simply means that some product P is equal to two different numbers, x and y, multiplied. We want a minimum value for P. One problem stands in our way. *Optimization problems require a single variable in the expression.* We have two variables, so we need to go back to the original problem for more information. There is another relationship between the variables: one (it doesn't matter which) is two less than three times the other, so we can write:

$$y = 3x - 2.$$

Now, substitute this y value into our original equation for P to get

$$P = x \cdot (3x - 2) = 3x^2 - 2x.$$

We now have an equation for the product in a single variable! Do you realize how wonderful that is? We already know how to find a maximum or minimum—take the derivative and construct a wiggle graph:

$$P' = 6x - 2 = 0$$

$$x = \frac{2}{6} = \frac{1}{3}$$

From the wiggle graph, it is simple to see that the function P will have a minimum when $x = \frac{1}{3}$. What is the corresponding y? Plug it back into our secondary equation $y = 3x - 2$:

$$y = 3\left(\frac{1}{3}\right) - 2$$

$$y = 1 - 2 = -1.$$

Therefore, the numbers -1 and $\frac{1}{3}$ have the smallest possible product given our initial defining condition of one being two less than three times the other $\left(-1 \text{ is } 2 \text{ less than } 3 \cdot \frac{1}{3}\right)$.

To be honest, this looks like a complicated process, but the method is really quite straightforward and repetitive.

Steps for Success with Optimization Problems

- Write an equation that represents what you are trying to maximize or minimize (this is called the *primary equation*).
- If more than one variable is present, use other information in the problem (in the form of secondary equations like $y = 3x + 2$ in Example 8) to eliminate the excess variables.
- Find the derivative of your primary equation so that you can identify the critical numbers and draw a wiggle graph to find the answer.

If you remember these steps, you are well on your way to succeeding at optimization problems.

Example 9: What point on the graph of $y = \sin x$ is the closest to $(0,1)$?

Solution: You are trying to minimize distance, so use the distance formula as your primary equation:

$$D = \sqrt{\left(x_2 - x_1\right)^2 + \left(y_2 - y_1\right)^2}.$$

Although it doesn't matter which is which, we'll set (x_1, y_1) equal to the stationary point $(0,1)$ and set (x_2, y_2) equal to the point on the graph of sine $(x, \sin x)$. Substitute these points in to get

$$D = \sqrt{x^2 + \left(\sin x - 1\right)^2}$$

$$D = \sqrt{x^2 + \sin^2 x - 2\sin x + 1}.$$

Time to find the derivative and set it equal to zero to find critical numbers:

$$D' = \frac{1}{2} \frac{2x + 2\sin x \cos x - 2\cos x}{\sqrt{x^2 + \sin^2 x - 2\sin x + 1}} = 0.$$

CAUTION
Because optimzation problems have numerous steps, students sometimes stop too early and don't actually answer the question posed to them. When you're done, reread the question and make sure you've answered it.

The critical number will be $x = .4787224241$. You have to graph the derivative to find this value, so while you have D' on your calculator screen, notice that it is negative before the critical number and positive after. Therefore, $x = .4787224241$ is a minimum. (This is a quick calculator shortcut for constructing wiggle graphs without the tedious test point substitution.) In order to finish the problem, however, we need to give the *coordinate* that is closest to sin x, so plug x into sin x to get the y value. The final answer is $(.479,.461)$.

Example 10: You have invented a new and delicious beverage called Schwop!, which tastes inexplicably like cotton balls. Deemed the least refreshing drink in the known universe, it is nonetheless flying off the shelves. If you wish each cylindrical can of Schwop! to contain 100 in³ of beverage, what height must each can be to minimize the amount of aluminum you use to manufacture the cans?

Solution: Each can is made up of a rectangle and two circles (you can't forget the top and bottom of the can!), as shown below:

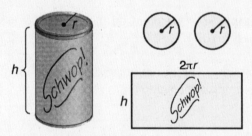

The construction of a can of Schwop!

You first need to write an equation representing what you want to minimize (aluminum), which in this case is the surface area of the can plus the area of the top and bottom of the can:

$$A = 2\pi rh + 2\pi r^2.$$

Danger! There are two variables present. To rectify this, use the other information given by the problem—the volume of Schwop! in each can:

$$V = \pi r^2 h = 100.$$

You can solve this for h to eliminate that pesky extra variable:

$$h = \frac{100}{\pi r^2}.$$

Substitute back into the original equation to get

$$A = 2\pi r\left(\frac{100}{\pi r^2}\right) + 2\pi r^2$$

$$A = \frac{200}{r} + 2\pi r^2.$$

As is typical of optimization problems, the hardest part was finding the equation; now, find the derivative and set up a wiggle graph:

$$A' = -\frac{-200}{r^2} + 4\pi$$

$$r = 2.515397996 \text{ inches.}$$

This is the radius of the smallest possible can. Be careful — the problem does not ask for this value. It asks for the height that corresponds to it. That height is

$$h = \frac{100}{\pi(2.515397996)^2} = 5.031 \text{ inches.}$$

PROBLEM SET

You may use a graphing calculator on problems 2 through 4.

CAUTION
Make sure to include units in your answer to optimization problems, just like you did with related rates.

1. *(A classic maximization problem)* Farmer Rogendoger breeds cows that can't swim. "Nonbuoyant cows just taste better, dadgum it," he insists (and it's best not to question him further). Because the cows are landlocked, it saves him fence costs. If Rogendoger sets up a rectangular pasture that is bordered on one side by a river (which requires no fence, for obvious reasons), what is the maximum area he can enclose with 1000 feet of fence? (Food for thought: How does he prevent cattle theft by riverboat-riding marauders?)

2. What point(s) on the graph of $y = \cot x$ is closest to the coordinate pair $(\pi, 0)$?

3. Find the maximum area of a rectangle that has two vertices on the x-axis and two vertices on the graph of $y = x^2 - 8$.

4. To celebrate our first anniversary, I am commissioning the construction of a four-inch-tall box made of precious metals to give to my bride, Lisa. The jewelry box will have rectangular sides and an open top. The longer sides of the box will be made of gold, at a cost of \$300/in²; the shorter sides will be made of platinum, at a cost of \$550/in². (Let's call it *practical*, not *cheap*, that the shorter sides are more expensive). The bottom will be made of plywood, at a cost of \$.02/in². What dimensions provide me with the lowest cost if I am adamant that the box have a volume of 50 in³?

SOLUTIONS

1. You should begin this problem with a diagram, including all relevant information, like so:

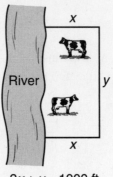

$$2x + y = 1000 \text{ ft.}$$

You want to maximize the *area*, so your primary equation must be the area of the rectangle: $A = xy$. However, there are two variables, so consider the secondary equation based on the limited amount of fence: $2x + y = 1000$ (it's not $2x + 2y$ because you don't need to fence the river). Solve the secondary equation for y to get $y = 1000 - 2x$ and substitute that into the area equation:

$$A = x(1000 - 2x) = 1000x - 2x^2.$$

Now, find the maximum of this equation through the usual channels:

$$A' = 1000 - 4x = 0$$

$$4x = 1000$$

$$x = 250.$$

The optimum value for x is 250, and the corresponding y will be

$$y = 1000 - 2(250) = 500.$$

Therefore, the maximum area is $500 \cdot 250 = 125{,}000 \text{ ft}^2$. As far as the riverboat marauders, answers may vary.

2. This is similar to Example 9, so there's no need to go into a great deal of detail here. Use the same process with points $(\pi, 0)$ and $(x, \cot x)$:

$$D = \sqrt{(x - \pi)^2 + (\cot x)^2}$$

$$D' = \frac{1}{2} \cdot \frac{2(x - \pi) - 2 \cot x \csc^2 x}{\sqrt{(x - \pi)^2 + (\cot x)^2}}$$

CAUTION
250 is almost guaranteed to be an answer choice if this is a multiple-choice question, but it doesn't answer the correct question — so be careful!

There are two critical numbers to worry about, since only two of them are close to $(\pi,0)$: 2.163306396 and 4.119878911. (Don't forget that π is also a critical number since cot x is undefined there.)

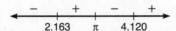

In order to figure out which is closer, you'll need to plug them both into the distance formula. Remember, we have designed this formula to tell us how far away something is from $(\pi,0)$ merely by substituting in the x value. It turns out that they *both* are a distance of 1.187534573 units away, so both are correct answers. The problem does ask for *points* however, so plug both x values (don't round until the very end!) into cot x to get your final answer of $(2.163,-.673)$ and $(4.120,.673)$, shown below.

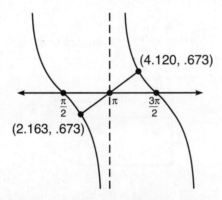

3. You should again begin by drawing a picture:

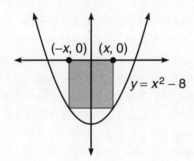

Because the parabola is y-symmetric, our rectangle should be, too, in order to maximize area. The width of the rectangle in the diagram is $2x$, and its length is $-f(x)$, if we set $f(x) = x^2 - 8$. This is because $f(x)$ will be negative, and you don't want a negative length. Your overall goal is to maximize area, so your equation should be area of a rectangle:

$$A = l \cdot w$$

$$A = -(x^2 - 8)(2x) = -2x^3 + 16x.$$

Only one variable is in the equation, so proceed as planned, and critically wiggle:

TIP
Often, the largest rectangle in a restricted space turns out to be a square, as in problem 3.

$$A' = -6x^2 + 16 = 0$$

$$x = \pm\sqrt{\frac{8}{3}}.$$

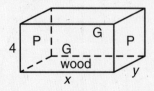

Therefore, $\sqrt{\frac{8}{3}}$ is the optimum value for x, so the maximum width of the rectangle is $2 \cdot \frac{8}{3} = \frac{16}{3}$. The corresponsing length will be $l = -\left[\left(\sqrt{\frac{8}{3}}\right)^2 - 8\right] = \frac{16}{3}$. Finally, the maximum area will be

$$A = \frac{16}{3} \cdot \frac{16}{3} = \frac{256}{9} \text{ square units.}$$

4. Below is a graphic representation of the box in question:

As drawn, the side toward you and in the back are the larger, gold sides, and the left and right sides are the smaller, platinum sides. In addition, x represents the length of the gold side and y represents the length of the platinum side. As stated in the problem, the height of the box is 4 inches. Remember that your overall goal is to find the minimum cost, so you need to design a cost equation based on how expensive the box will be to create. To do so, find the area of each side of the box, and multiply it by how expensive the materials would be for that side. For example, the side facing you is made of gold and has an area of $4x$. Therefore, the cost of the side will be $4x \cdot 300$. There are two of those sides, so you multiply by two to get $2400x$. Finally, add in the platinum sides and the bottom to get

$$C = 2400x + 4400y + xy(.02).$$

Something's wrong—there are two variables. Time to use the last bit of information from the problem: the box must have a volume of 50 in³. That is written as $4xy = 50$. If you solve for y, you get $y = \frac{25}{2x}$. Substitute this back into the cost equation to get

$$C = 2400x + 4400 \cdot \frac{25}{2x} + \frac{25}{2}\left(\frac{2}{100}\right)$$

$$C = 2400x + 55,000x^{-1} + \frac{1}{4}.$$

Take the derivative and do the wiggle thing:

$$C' = 2400 - \frac{55,000}{x^2} = 0$$

$$\frac{55,000}{x^2} = 2400$$

$$2400x^2 = 55,000$$

$$x = \pm\sqrt{\frac{275}{12}}.$$

Only the positive answer makes sense (your box cannot have a negative length without ripping apart time and space, creating a black hole, and swallowing everyone within 2 light years in its diabolical gaping open top).

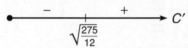

Therefore, the optimal value for x will be 4.787135539, which makes the optimal value for y

$$y = \frac{25}{2 \cdot 4.787135539} \approx 2.611.$$

Therefore, the dimensions providing the lowest cost are 4.787 in $\times$ 2.611 in $\times$ 4 in. (By the way, the lowest cost will be nearly \$23,000. Do you think I still need to buy a card?)

CAUTION
Don't write a book about this evil jewelry box— Stephen King is almost finished one and you'll look like a hack.

TECHNOLOGY: MODELING A PARTICLE'S MOVEMENT WITH A GRAPHING CALCULATOR

The TI-83 includes a neat feature that allows you to view particle motion problems quite simply. If you are an AB student, you'll need to delve into the world of parametric equations (just for a second, and it won't hurt a bit — I promise). Although a lot of information can be gathered from a particle's position equation, nothing beats seeing the particle running back and forth across the x-axis, working up particle sweat and checking its particle pulse. Let's revisit Example 6 from our not-too-distant past to see how this works.

Example 11: If a particle moves along the x-axis according to position equation $s(t) = t^3 - 10t^2 + 25t$, how many times does the particle change direction for $t \geq 0$?

Solution: Switch your calculator to parametric mode. This is done by pressing [Mode], arrowing down to the "Par" option, and pressing [Enter]. Notice that the "Y=" screen has changed; every equation now requires an x and a y component to graph. In the "X$_{1T}$=" line, type the position equation, pressing [x,t,Θ] for each variable. In the "Y$_{1T}$=" line, type 0—this ensures that the particle will always have a height of and never leave the x-axis. Finally, use the arrows to make your way to the little "\" symbol next to "X$_{1T}$" =. This chooses what the graph will look like. Pressing [Enter] twice on this symbol changes it to "–O". Your calculator screen should look like the graphic below before you continue:

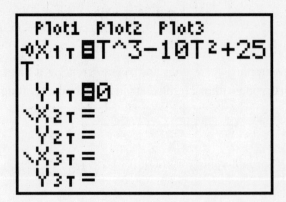

Now, press the [Window] button. Make sure your "Tmin" value is 0, since the problem requires that $t \geq 0$. A good value for "Tmax" is 10. Now, press [Graph]. The little particle should zip off to the right, turn around, and then move off to the right again—this time forever. Thus, it changed direction twice. If the particle moved off the screen, you can always increase the "Xmin" and "Xmax" values of the graph.

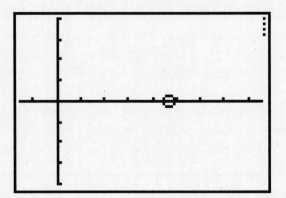

You may be wondering how mind-bogglingly useful this trick is in the grand scheme of things. The answer is "not very." However, if you've had difficulty imagining the movement of a particle on a horizontal axis, this exercise can be very enlightening. If the particle is moving on the y-axis, reverse the values you typed, making the "X$_{1T}$" value 0 and the "Y$_{1T}$" the equation. If the particle's movement is not restricted to a line, the process is even easier, as demonstrated in the next example.

Example 12: Cleverly instruct your calculator to model the motion of a cannonball whose position equation is $s(t) = -16t^2 + 20t$.

Solution: Set your calculator back to "Func" mode by pressing the [Mode] button and type the above equation on the "Y=" screen. (This time, the variables will be x's instead of t's, but that won't affect the graph one bit.) Make sure to arrow over to the "\" symbol, and change it to a "–O". Go ahead and [Graph] the equation. You may want to adjust the window a little bit. The graph below (with the cannonball in mid-air) has the following settings: Xmin = 0, Xmax = 1.5, Ymin = 0, Ymax = 7.

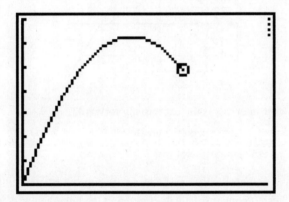

PRACTICE PROBLEMS

You may use a graphing calculator on problems 7 through 12.

1. Draw the graph of a function, g, that satisfies the following conditions: g has domain $[-5,8]$, g has relative minima at $x = -3$ and $x = 4$, g has relative maxima at $x = -1$ and $x = 6$, g has its absolute maximum at $x = -5$, and g's absolute minimum occurs at $x = 4$.

2. The graphs of $h(x)$, $h'(x)$, and $h''(x)$ are given below. Determine which graph is which, and justify your answer.

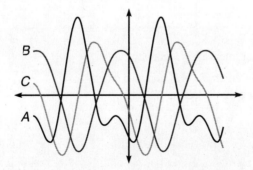

3. Draw the graph of $f(x)$ based on the following chart of the signs of f's derivatives, given that f is continuous, $a < b < c < d < e$, and $f(a) = f(e) = 0$. (Note: DNE means "does not exist.")

	$a<x<b$	b	$b<x<c$	c	$c<x<d$	d	$d<x<e$
$f'(x)$	$-$	DNE	$+$	0	$+$	DNE	$-$
$f''(x)$	$-$	DNE	$-$	0	$+$	DNE	$+$

4. Below is a graph of a car's *velocity*. Answer the following questions based on the graph.

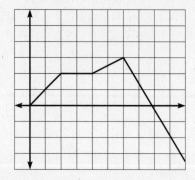

   (a) On what intervals is the car moving forward?

   (b) When is the car's speed the greatest?

   (c) When is the car decelerating?

5. A particle moves along the y-axis according to the continuous, differentiable curve $s(t)$, which contains the values given in the table below.

t	0	1	2	3	5	10
$s(t)$	-2	3	6	9	11	15

What is the approximate velocity of the particle at $t = 5$?

6. Below is a graph of $h''(x)$. Answer the following questions based on the graph.

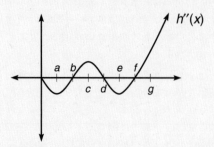

   (a) On what intervals is h concave up?

   (b) On what intervals is h' concave down?

   (c) What are the inflection points of h'?

7. If a particle moves along the *x*-axis according to the position equation $s(t) = t^4 - 4t^2 + 3$ (for $t \geq 0$). If *s* is measured in feet and *t* in seconds, answer the following questions:

   (a) What is the value of *t* guaranteed by the Mean Value Theorem for the interval [0,3]?

   (b) What is the particle's velocity and acceleration when $t = 3$?

   (c) When is the particle moving forward?

8. (For fans of the movie *Willie Wonka and the Chocolate Factory*): Naughty children. No one will listen to Mr. Wonka's instructions. Now Violet has gone and chewed the three-course-meal gum, and she's begun to turn blue. In fact, she's swelling to the size of a giant blueberry! If her torso swells such that her radius is increasing at a constant rate of 2 in/sec, at what rate is the surface area of her berry body increasing when her radius is 3 feet?

9. What is the volume of the largest right circular cylinder that can be inscribed in a sphere of radius 4 feet?

10. Two cars move on straight roads that are at a 50° angle to each other, as pictured below.

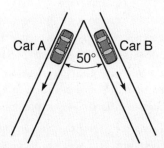

If car A moves at a constant rate of 40 mph, car B moves at a constant rate of 35 mph, and they started from the intersection at the same time, what is the rate of change of the distance between the cars once car A has traveled 25 miles? (**Hint:** the Law of Cosines)

11. *A particle moves along the path defined by $x = \cos t$, $y = \sin 2t$.

    (a) Find the acceleration vector for the particle at $t = \frac{7\pi}{6}$.

    (b) When is the speed of the particle $\frac{1}{2}$?

12. **James' Diabolical Challenge:** You are contracted to build an animal cage for the Discovery Channel. One of the sides of the cage will be a river, with the remaining boundaries being constructed of fencing. You are given 400 feet of fence and are required to incorporate at least 100 feet of coastline. You can build either a rectangular fence or a semi-circular one. Find the dimensions of the cage that give the greatest area, and justify your answer mathematically.

*a BC-only problem

SOLUTIONS

1. There are numerous possible solutions to this problem, but they all look very similar. Remember, relative extrema are little "hills and valleys" in the graph, and absolute extrema (guaranteed by the Extreme Value Theorem) are the highest and lowest points on a closed interval. Also, remember that absolute extrema must occur either at a relative extrema (like $x = 4$) or at an endpoint (like $x = -5$).

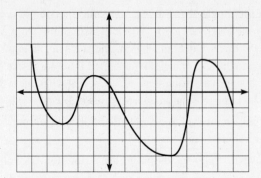

2. Graph B is $h(x)$, graph C is $h'(x)$, and graph A is $h''(x)$. Notice that C has a value of 0 each time B has a relative extrema point. Furthermore, C is positive whenever B is increasing and negative when B is decreasing. Notice that A is negative whenever B is concave down and positive whenever B is concave up. (Therefore, A has a value of 0, or an x-intercept, each time B has an inflection point, and each time C has an extrema point.) You don't need to apply every one of these connections—just enough to differentiate (no pun intended) among the three.

3. Let's take the interval $a < x < b$ as an example. On that interval, both $f'(x)$ and $f''(x)$ are negative. Therefore, $f(x)$ will be decreasing and concave down. There's only one way to draw such a curve, and it looks like the following:

    You know that f is continuous, but no derivative exists at $x = b$ and $x = d$. As you draw the graph, you can tell that no derivative exists because both values of x result in a cusp. A correct graph looks something like the following:

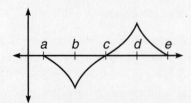

4. (a) The car's moving forward whenever its velocity is positive. We are given the graph of velocity, and it is positive on $(0,8)$. The car's velocity is clearly decreasing on $(6,10)$, but the velocity is not negative until $t > 8$.

    (b) Because speed is the absolute value of velocity, the direction of the car is irrelevant. Therefore, the speed of the car is greatest at $t = 10$. The speed there is a little more than 3, whereas the top speed the car reached while traveling forward was at $t = 6$, when the speed and velocity were both 3.

(c)   Acceleration is based on the second derivative. The car will decelerate when the position function is concave down and when the *velocity* function is decreasing. Therefore, the car is decelerating on (6,10). This may confuse you, since the car is traveling its fastest at $t = 10$. The magnitude (absolute value) of the car's acceleration there is actually pretty high, but the car still is traveling backward, so the acceleration is becoming more and more negative (technically decelerating).

5.   The question is asking you to approximate $s'(5)$. To do so, calculate the slope of the secant line connecting the points (3,9) and (5,11)—this is the best approximation we can use. Therefore, $s'(5) \approx 1$.

6.   (a) h is concave up whenever h'' is positive, so the answer is $(b,d) \bigcup (f,\infty)$.

(b)   In order to determine the concavity of h', you need to move "down" two derivatives to h''' (since the signs of the second derivative of a function describe its concavity). You also know that the signs of h''' will describe the direction of h''. The question is essentially asking you where h''' is negative, and that will happen wherever h'' is decreasing. Therefore, the answer is $(0,a) \bigcup (c,e)$.

(c)   The inflection points of h' occur whenever h''' equals 0 and the concavity actually changes (much like an extrema point is where the derivative equals 0 and the direction changes). You also know that h''' describes the direction of d''. Therefore, whenever d'' changes direction, h' will have an inflection point: $x = a$, $x = c$, and $x = e$.

7.   (a)   The Mean Value Theorem states that there exists some c on the interval such that

$$s'(c) = \frac{s(3) - s(0)}{3 - 0};$$

in other words, there exists a tangent line at some c that has the same slope as the secant line over the entire interval.

$$4c^3 - 8c = \frac{48 - 3}{3}$$

$$4c^3 - 8c = 15$$

$$c \approx 1.975$$

(b)   The velocity is the first derivative of position, and acceleration is the second:

$$v(t) = s'(t) = 4t^3 - 8t$$

$$v(3) = 84 \text{ ft/sec}$$

$$a(t) = v'(t) = 12t^2 - 8$$

$$a(3) = 100 \text{ ft/sec}^2.$$

(c)   The particle is moving forward whenever its first derivative is positive, so construct a wiggle graph for v:

$$v(t) = 4t^3 - 8t = 0$$

$$4t(t^2 - 2) = 0$$

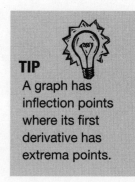

TIP

A graph has inflection points where its first derivative has extrema points.

Critical numbers: $t = 0, \pm\sqrt{2}$.

$$\xrightarrow[\quad 0 \qquad\qquad \sqrt{2} \qquad]{\quad - \qquad\qquad + \qquad} s'(t)$$

According to the wiggle graph, the particle is moving forward on the interval $(\sqrt{2}, \infty)$. Don't forget that the original problem stipuates $t \geq 0$. Without that restriction, the answer would also have included $(-\sqrt{2}, 0)$.

8. This is a related rates problem. Begin with the formula for surface area of a sphere, and take the derivative with respect to time:

    $$S = 4\pi r^2$$

    $$\frac{dS}{dt} = 8\pi r \frac{dr}{dt}.$$

Now, plug in the given information to solve for $\frac{dS}{dt}$:

$$\frac{dS}{dt} = 8\pi(36)(2) = 576\pi \text{ in}^2/\text{sec.}$$

9. Your ultimate goal is to maximize the volume of a cylinder, so your primary equation should be the following formula:

    $$V = \pi r^2 h$$

However, you have two variables, and it's going to require some cleverness to eliminate one of them. Below is a diagram of our situation. On it is drawn a triangle, which connects the center of the sphere to the intersection point of the cylinder and sphere to a point on the cylinder at height $\frac{h}{2}$.

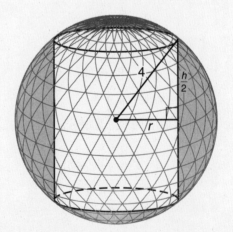

By the Pythagorean Theorem, you have

$$r^2 + \frac{h^2}{4} = 16.$$

Solve this equation for r^2 to eliminate the r^2 in the primary equation:

$$r^2 = 16 - \frac{h^2}{4}.$$

Substitute this into the primary equation, and you can use the familiar methods to maximize the volume:

$$V = \pi \left(16 - \frac{h^2}{4} \right) h$$

$$V = 16\pi h - \frac{\pi}{4} h^3$$

$$V' = 16\pi - \frac{3\pi}{4} h^2 = 0$$

$$h = 4.618802154 \text{ ft.}$$

The wiggle graph verifies that 4.618802154 is the maximum height. The corresponding radius will be

$$r = \sqrt{16 - \frac{4.618802154^2}{4}}$$

$$r = 3.265986324,$$

so the maximum volume is $V = \pi r^2 h \approx 154.778 \text{ ft}^3$.

10. The first question to be answered in this related rates dilemma is how much time has passed if the first car has traveled 25 miles. A simple proportion helps you to figure this out. If a car travels 40 miles in an hour (60 minutes), then how many minutes, m, does it require to travel 25 miles?

$$\frac{40}{60} = \frac{25}{m}$$

$$40m = 1500$$

$$m = 37.5 \text{ minutes}$$

Now you can figure out how far car B has traveled using the same method:

$$\frac{35}{60} = \frac{d}{37.5}$$

$$60d = 1312.5$$

$$d = 21.875 \text{ miles.}$$

With all this fabulous information, you can construct the diagram below.

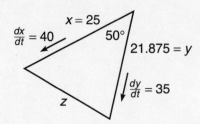

It is quite easy to figure out z using the Law of Cosines:

$$z^2 = 25^2 + 21.875^2 - 2(25)(21.875)\cos 50$$

$$z = 20.01166352 \text{ miles.}$$

To tie this problem up into a nice bundle, you should use the Law of Cosines (since the triangle is oblique—not a right triangle) and derive with respect to t:

$$z^2 = x^2 + y^2 - 2xy \cos 50$$

$$2z\frac{dz}{dt} = 2x\frac{dx}{dt} + 2y\frac{dy}{dt} - 2x(\cos 50)\frac{dy}{dt} - (y\cos 50)2\frac{dx}{dt}$$

$$2(20.01166352)\frac{dz}{dt} = 2(25)(40) + 2(21.875)(35) - 2(25)(\cos 50)(35) -$$

$$(21.875)(\cos 50)(2)(40)$$

$$\frac{dz}{dt} \approx 32.019 \text{ mph.}$$

11. (a) The position vector is given by $\mathbf{s} = (\cos t)\mathbf{i} + (\sin 2t)\mathbf{j}$. Take two derivatives to get the acceleration vector: $\mathbf{a} = (-\cos t)\mathbf{i} + (-4\sin 2t)\mathbf{j}$. The acceleration vector at $t = \frac{7\pi}{6}$ will be

$$\mathbf{a} = (-\cos \frac{7\pi}{6})\mathbf{i} + (-4\sin \frac{7\pi}{3})\mathbf{j}$$

$$\mathbf{a} = \frac{\sqrt{3}}{2}\mathbf{i} - 2\sqrt{3}\,\mathbf{j}.$$

(b) First, find the velocity vector by taking the derivative of position:
$$\mathbf{v} = (-\sin t)\mathbf{i} + (2\cos 2t)\mathbf{j}.$$

The speed of the particle is given by the norm of the velocity vector, so set its norm equal to $\frac{1}{2}$ and solve for t:

$$\|v\| = \sqrt{\sin^2 t + 4\cos^2 2t} = \frac{1}{2}$$

$$\sin^2 t + 4\cos^2 2t = \frac{1}{4}.$$

If you try to solve this equation by graphing, you'll see that the graph never crosses the x-axis, so there are no solutions. The particle never travels at a speed of $\frac{1}{4}$.

12. This problem asks you to find the maximum area, with the condition that a non-fenced coastline be used (ignore the 100 feet of coastline requirement for now). Below are the two cages whose area you want to maximize:

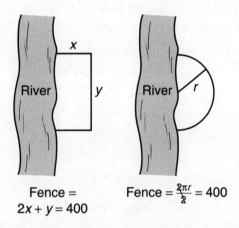

Here are the optimization problems, one at a time:

Rectangular cage:

You want to maximize area, so it is your primary equation:

$$A = xy.$$

You also know that $2x + y = 400$, so use that information to eliminate a variable by solving for x or y:

$$2x + y = 400$$

$$y = 400 - 2x.$$

Now, substitute back into the area equation to find the maximum:

$$A = x(400 - 2x)$$

$$A = 400x - 2x^2$$

$$A' = 400 - 4x = 0$$

$$x = 100$$

$$y = 400 - 2(100) = 200$$

Therefore, the maximum dimensions of the rectangular cage are 100 ft × 200 ft, for an area of 20,000 ft². (By the way, this meets the 100 ft coastline requirement without any trouble at all. See? I told you not to worry about it!)

Semicircular fence:

This problem does not require the process of optimization at all. If you know that the circumference of the semicircle is 400, then $\pi r = 400$ and $r = \frac{400}{\pi}$. Therefore, the enclosed area will be

$$A = \frac{\pi r^2}{2}$$

$$A = \frac{160000\pi}{2\pi^2}$$

$$A \approx 25464.791 \text{ ft}^2.$$

You should definitely go with the semicircular fence. You get more than 5000 ft² extra space.

CAUTION
Remember that this area is a *semicircle*, so use all of the circle formulas divided by 2.

Integration

Now that you know just about everything there is to know about taking derivatives, it's time to pull the rug out from under you. The third major topic of calculus (limits and derivatives being the first two) is *integration*, or *antidifferentiation*. That's right, Mr. Prefix; *anti-* means "the opposite of," so it's time to explore the process of derivatives reversed. Previously, you would be asked to find $\frac{d}{dx}(x^3)$; clearly, the answer is $3x^2$. Now, you'll be given $3x^2$ and required to come up with the *antiderivative*, x^3.

But, it's never that easy, is it? As a matter of fact, $x^3 + 1$ is also an antiderivative of $3x^2$! So is $x^3 - 14$. Therefore, we say that the antiderivative of $3x^2$ is $x^3 + C$, where C can be any number at all. But, we're getting ahead of ourselves. Let's jump right in—the water's fine.

BASIC ANTIDERIVATIVES

Just as the notation $\frac{dy}{dx}$ or y' indicated to you that differentiation was necessary, the notation

$$\int \cos x \, dx$$

indicates the same for integration. The above is read "the integral (or antiderivative) of $f(x)$ with respect to x." Respecting variables in differentiation was sometimes a complicated procedure. Remember implicit differentiation and related rates? Luckily, respecting variables is not nearly as difficult in integration; you just have to make sure the dx gets multiplied by everything in the integral. But, enough talk—let's get down to business.

Think back for a moment: The Power Rule (one of your earliest and dearest calculus friends) dealt with deriving simple expressions—a single variable to a constant power. There is an equivalent rule for integrating, so we'll call it (get this) the Power Rule for Integration. Clever, eh?

The Power Rule for Integration: If a is a constant, $\int x^a \, dx = \frac{x^{a+1}}{a+1} + C$.

Translation: In order to find the integral of x^a, add 1 to the exponent and divide the term by the new exponent.

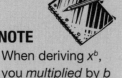

 Example 1: Evaluate $\int x^3 dx$.

Solution: Add one to the exponent ($1 + 3 = 4$), and divide by the new exponent, 4: $\frac{x^4}{4} + C$.

More about that weird C now. It is called the *constant of integration*. It is simply a real number, and we have no idea exactly what that number is (for now). However, $\frac{x^4}{4} + 2$, $\frac{x^4}{4} + 113.4$, and $\frac{x^4}{4} - \pi$ all have a derivative of x^3 (since the derivative of the constant is 0). Therefore, when we write "$+ C$" at the end of an antiderivative, we are admitting that there may have been a constant there, but we do not know it.

Now, let's discuss the two major properties of integrals; both of them are very similar to derivatives:

1. $\int a \cdot f(x) dx = a \int f(x) dx$

If a constant or coefficient is present in your integral, you may ignore it, like you did with derivatives. In fact, you may pull the constant completely out of the integral.

2. $\int (f(x) \pm g(x)) dx = \int f(x) dx \pm \int g(x) dx$

If an integral contains numerous terms being added or subtracted, then these terms can be split apart into separate integrals. In differentiation, given the problem $\frac{d}{dx}(x^3 - 5x)$, you could find the derivatives of the terms separately: $3x^2 - 5$. The same goes for integration. For example, $\int (x^3 + x^5) dx = \int x^3 dx + \int x^5 dx = \frac{x^4}{4} + \frac{x^6}{6} + C$.

In Example 2, we'll apply these properties of integration to some more complex integration problems.

 Example 2: Evaluate the following antiderivatives:

(a) $\int \frac{5}{3x^2} dx$

This expression can be rewritten as

$$\int \frac{5}{3} x^{-2} dx.$$

The $\frac{5}{3}$ is merely a coefficient, so we can apply the first rule of antiderivatives and pull it out of the integral:

$$\frac{5}{3}\int x^{-2}dx.$$

Now, apply the power rule for integrals, but make sure to *add* 1 to the original power of –2.

$$\frac{5}{3}\cdot\frac{x^{-1}}{-1}+C$$

$$-\frac{5}{3x}+C$$

(b) $\int\left(\sqrt{x}-3x^4\right)dx$

This integral must first be rewritten as

$$\int(x^{1/2}-3x^4)dx.$$

Because the two terms are being added, we can split the above into two separate integrals (and pull out the coefficients):

$$\int x^{1/2}dx-3\int x^4dx.$$

Now, apply the Power Rule for Integration:

$$\frac{x^{3/2}}{\frac{3}{2}}-\frac{3x^5}{5}+C$$

$$\frac{2}{3}x^{3/2}-\frac{3}{5}x^5+C.$$

The $\frac{3}{2}$ power is the result of adding 1 to the original exponent ($\frac{1}{2}$).

(c) $\int\frac{7x^2+5x}{x^{5/3}}dx$

Again, rewriting the integral is the first order of business. Instead of one fraction, rewrite as the sum of two fractions with the same denominator. Also, apply the integration properties:

$$7\int\frac{x^2}{x^{5/3}}dx+5\int\frac{x}{x^{5/3}}dx.$$

CAUTION
You can only pull coefficients out of integrals. For example, it would be incorrect to rewrite

$\int\left(x^4+2x\right)dx$ as

$x\int\left(x^3+2\right)dx.$

NOTE
Notice how everything in the integral $\int(x^4+2x)dx$ gets multiplied by *dx*; this is important notation. Although the AP readers have not made it a point to deduct points if you forget to write the "*dx*," that's no guarantee that they will continue to do so.

CAUTION
Once you integrate, make sure to stop writing the $\int$ and *dx* symbols. They only hang around until you're done integrating. When their work is done, they vanish.

CAUTION
Although the Power Rule for Integrals is relatively easy, it is also easy to make mistakes when the exponents are fractions or have negative powers. Be careful.

Remember way back to algebra and exponent properties: $\dfrac{x^a}{x^b} = x^{a-b}$.

Therefore, $\dfrac{x^2}{x^{5/3}} = x^{2-5/3} = x^{1/3}$, and $\dfrac{x}{x^{5/3}} = \dfrac{x^1}{x^{5/3}} = x^{1-5/3} = x^{-2/3}$. Use this to rewrite the problem as

$$7\int x^{1/3}\,dx + 5\int x^{-2/3}\,dx,$$

and apply the Power Rule for Integrals to get

$$7\cdot\frac{3}{4}x^{4/3} + 5\cdot\frac{3}{1}x^{1/3} + C$$

$$\frac{21}{4}x^{4/3} + 15x^{1/3} + C.$$

Do you see the shortcut for integrating fractional exponents? When you integrate $x^{1/3}$, instead of writing the step $\dfrac{x^{4/3}}{\frac{4}{3}}$, remember that dividing by $\frac{4}{3}$ is the same as multiplying by $\frac{3}{4}$. Therefore, the answer is $\frac{3}{4}x^{4/3} + C$.

Well, the Power Rule for Integrals is all well and good, but there is one instance in which it is completely useless. Consider the integral:

$$\int \frac{1}{x}\,dx.$$

This can be rewritten as $\int x^{-1}\,dx$, but if you try to integrate, you get $\frac{x^0}{0} + C$, and the zero in the denominator spoils everything. How, then, are you to integrate $\frac{1}{x}$? Believe it or not, you already know the answer to this[md]you just have to dig it out of your long-term memory. Remember, integration is the opposite of differentiation, so the expression that has the *derivative* of $\frac{1}{x}$ will be the *integral* of $\frac{1}{x}$. You learned in Chapter 4 that $\dfrac{d}{dx}(\ln x) = \dfrac{1}{x}$. Therefore,

$$\int \frac{1}{x}\,dx = \ln |x| + C.$$

(You need to use the absolute value signs since $\ln x$ has domain $(0,\infty)$—the function wouldn't know what to do with negative inputs.)

If you have forgotten the large list of derivatives you were to have memorized, it's time to refresh your memory. Only two of the integrals look a little different from their derivatives. We have already looked at the first:

$\int \dfrac{1}{x}\,dx$ (its integral has that unexpected absolute value). One other problem shows a slight difference in its absolute values:

$$\int \frac{dx}{x\sqrt{x^2-1}} = \operatorname{arc\,sec}|x| + C.$$

You see arcsec x so infrequently on the test, it's hardly worth mentioning, but it is important. In addition, we will take a closer look at inverse trigonometric and exponential integrals a little later in this chapter. Here are a few problems to get you brushed up on the throwback integrals.

 Example 3: Evaluate the following integrals:

(a) $\int -\sin x\, dx$

This problem asks, "What has a derivative of $-\sin x$?" The answer is, of course, $\cos x + C$, since $\frac{d}{dx}(\cos x + C) = -\sin x$. If the problem had been $\int \sin x\, dx$, the answer would have been $-\cos x + C$, since $\frac{d}{dx}(-\cos x + C) = -(-\sin x) = \sin x$.

(b) $\int \dfrac{1}{\sin^2 x}\,dx$

First, rewrite this problem as $\int \csc^2 x\, dx$. What has a derivative of $\csc^2 x$? Well, $\frac{d}{dx}(\cot x) = -\csc^2 x$, and that's only off by a negative sign. Therefore, add a negative sign to $\cot x$ to account for the missing sign, and the answer is

$$\int \csc^2 x\, dx = -\cot x + C.$$

(c) $\int \dfrac{1}{\sqrt{1-x^2}}\,dx$

This is simply the derivative of arcsin x, so the answer is arcsin $x + C$.

(d) $\int e^x\, dx$

If the derivative of e^x is e^x, then $\int e^x\, dx = e^x + C$.

PROBLEM SET

You may not use a graphing calculator on any of these problems.

Evaluate each of the following integrals.

1. $\int\left(\dfrac{2}{3}x^3 - \dfrac{5}{x^7} + \sqrt[3]{x^2}\right)dx$

2. $\int bx^a\,dx$, if a and b are real numbers

3. $\int\left(x^{2\pi} + \dfrac{4}{7x^{2/3}}\right)dx$

4. $\int(x^2 - 1)(x + 2)\,dx$

5. $\int\dfrac{2m^4 + 3m^2 - m + 3}{m^3}\,dm$

6. $\int\dfrac{\sin x\,dx}{\cos^2 x}$

7. $\int\left(-\tan^2 x - 1\right)dx$

8. $\int\dfrac{x\sqrt{x^2-1}+1}{x\sqrt{x^2-1}}$

SOLUTIONS

1. Begin by splitting the integral into pieces and rewriting it so that you can apply the Power Rule for Integrals:

$$\frac{2}{3}\int x^3\,dx - 5\int x^{-7}\,dx + \int x^{2/3}\,dx$$

It's ready to be power ruled, so go to it:

$$\frac{2}{3}\cdot\frac{x^4}{4} - 5\cdot\frac{x^{-6}}{-6} + \frac{3}{5}x^{5/3} + C.$$

$$\frac{x^4}{6} + \frac{5}{6x^6} + \frac{3x^{5/3}}{5} + C.$$

2. Because b is a coefficient, it can be pulled out of the integral.

$$b \int x^a dx$$

$$b \cdot \frac{x^{a+1}}{a+1} + C$$

$$\frac{b}{a+1} x^{a+1} + C$$

3. Rewrite this integral before starting, and remember that 2π is just a constant, so the Power Rule for Integrals still applies (just like it did to the a exponent in problem 2.

$$\int x^{2\pi} dx + \frac{4}{7} \int x^{-2/3} dx$$

$$\frac{x^{2\pi+1}}{2\pi+1} + \frac{12}{7} x^{1/3} + C$$

4. Before you can integrate, you need to multiply the binomials together. There is no Product Rule for Integration (which makes things tricky later) but for now, we can avoid the problem by multiplying.

$$\int (x^3 + 2x^2 - x - 2) dx$$

$$\frac{x^4}{4} + \frac{2x^3}{3} - \frac{x^2}{2} - 2x + C$$

5. You can begin by writing each of the terms of the numerator over the denominator. This is a long step, and if you can do it in your head, you are encouraged to do so—carefully! So that you can see exactly what's happening, the step is included:

$$\int \left(\frac{2m^4}{m^3} + \frac{3m^2}{m^3} - \frac{m}{m^3} + \frac{3}{m^3} \right) dx$$

$$\int (2m + 3m^{-1} - m^{-2} + 3m^{-3}) dx$$

$$m^2 + 3\ln |x| + \frac{1}{m} - \frac{3}{2m^2} + C.$$

6. This problem looks pretty complicated, but if you are clever (and who doesn't like being clever now and again?), it becomes quite easy. The trick is to rewrite the fraction as follows:

$$\int \frac{\sin x}{\cos x} \cdot \frac{1}{\cos x} dx.$$

NOTE

In some solutions, integrals are written as $\frac{2x^3}{5}$, and in others, the solutions are written $\frac{2}{5} x^3$. Neither is more correct, and both are 100 percent acceptable.

If you multiply those two fractions together, you still get $\frac{\sin x}{\cos^2 x}$, so we haven't actually changed anything's value. However, now we can rewrite $\frac{\sin x}{\cos x}$ as $\tan x$ and $\frac{1}{\cos x}$ as $\sec x$:

$$\int \tan x \sec x \, dx.$$

Perhaps you'll remember it better if it's written this way:

$$\int \sec x \tan x \, dx.$$

You know that is the derivative of $\sec x$, so the final answer is

$$\sec x + C.$$

7. That negative sign looks like it's just beggin' to get factored out, so we'll oblige it (and bring it out of the integral as the coefficient -1):

$$-\int (\tan^2 x + 1) dx.$$

Now that looks familiar. In fact, it is most of the Pappa Theorem! (Remember your Pappa: $\tan^2 x + 1 = \sec^2 x$.) Therefore, we'll use a Pappa substitution to rewrite it:

$$-\int \sec^2 x \, dx.$$

Because $\tan x$ has a derivative of $\sec^2 x$, the answer is

$$-\tan x + C.$$

8. Even though this looks ugly, begin the same way you did with problem 5—write each term of the numerator over the denominator:

$$\int \left(\frac{x\sqrt{x^2-1}}{x\sqrt{x^2-1}} + \frac{1}{x\sqrt{x^2-1}} \right) dx.$$

The first gigantic fraction simplifies to 1, making things much, much happier in the world:

$$\int \left(1 + \frac{1}{x\sqrt{x^2-1}} \right) dx.$$

The integral of 1 is simply x (since $\frac{d}{dx}(x) = 1$), and the other term is the derivative of arcsec x (don't forget the absolute value signs we discussed earlier in this section):

$$x + \text{arcsec} \, |x| + C.$$

HANDS-ON ACTIVITY 7.1: APPROXIMATING AREA WITH RIEMANN SUMS

There is an essential calculus tie between the integral and the area that is captured beneath a graph. You may, in fact, already know what it is. If you do, well, pin a rose on your nose. Those of you who don't know will be kept in the dark for a couple of sections so that some suspense will build (I am nothing if not a showman). For now, let's focus on using archaic and simplistic means to estimating the area "under" a curve. (The means are so simplistic that some students are actually disappointed. "This is calculus?" they ask, brows furrowed and tears forming in the corners of their eyes. My advice: embrace the easy. Just because a lot of calculus is tricky, not all of it has to be.) By the way, you may use your calculator freely on this activity.

1. Draw the graph of $f(x) = -\dfrac{1}{10}x^2 + 3$ on the axes below.

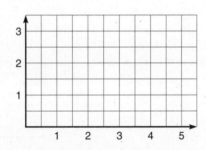

2. We are going to approximate the area between f and the x-axis from $x = 0$ to $x = 4$ using rectangles (the method of Riemann sums). This is not the entire area in the first quadrant, just most of it. Draw four *inscribed* rectangles of width 1 on the interval [0,4] on your graph above.

3. What are the heights of each of the four rectangles? What is the total area of the rectangles? This area, although not the same as the area beneath the curve, is an approximation for that area called the *lower sum*.

_____

_____

4. The actual area between f and the x-axis on the interval [0,4] is $\dfrac{28}{3}$. Why is one area greater?

_____

_____

5. How could you get a better approximation for the area beneath the curve if you still used inscribed rectangles?

_____

NOTE
We are not really finding the area "under" a curve. For most graphs, there is infinite area under the curve. Instead, we will be calculating the area between the given curve and the *x*-axis. Keep that in the back of your mind.

NOTE
Consider only the endpoints' heights when drawing the inscribed and circumscribed rectangles because you can compare their heights very easily. Choose the lower of the two heights for the inscribed and the larger of the two heights to draw cirsumscribed rectangles.

6. On the axes below, graph *f* again. This time, draw four rectangles of width 1 that circumscribe the graph. Use the area of the circumscribed rectangles to approximate the area beneath the curve. This approximation is called the *upper sum*.

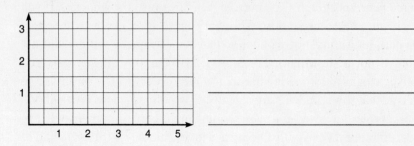

7. Compare the approximation you got in number 6 to the actual area.

_____

_____

8. On the axes below, graph *f* again, and this time draw four rectangles of width one such that the height of each rectangle is given by the midpoint of each interval. (The approximate area is called the *midpoint sum*.)

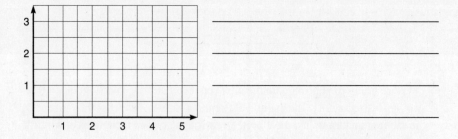

9. Now you have found an inscribed sum, a circumscribed sum, and a midpoint sum. Explain what is likely meant by each of the remaining sums, and draw a sample rectangle on [*a*,*b*] that you would use with the technique to approximate the area beneath *g*.

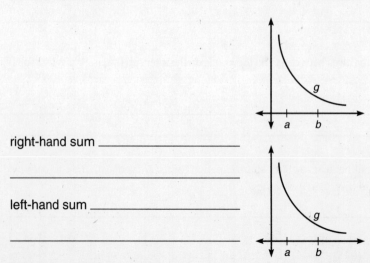

right-hand sum _____

_____

left-hand sum _____

_____

10. Use 5 rectangles of equal width and the technique of midpoint sums to approximate the area beneath the curve $h(x) = x^3 - 2x^2 - 5x + 7$ on [0,4].

Hint: To figure out the width, Δx, of n rectangles on the interval $[a,b]$, use the formula $\Delta x = \frac{b-a}{n}$.

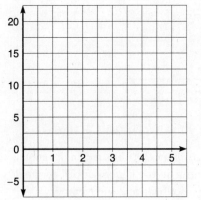

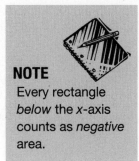

NOTE
Every rectangle *below* the x-axis counts as *negative* area.

SELECTED SOLUTIONS TO HANDS-ON ACTIVITY 7.1

2. The rectangles cannot cross the graph since they are inscribed. Thus, look at the endpoints of each interval, and choose the lower of the two endpoints' heights. That will be the height of the inscribed rectangle for that interval.

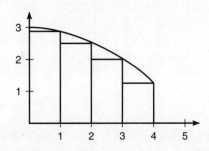

3. The heights of the four rectangles are $f(1) = 2.9$, $f(2) = 2.6$, $f(3) = 2.1$, and $f(4) = 1.4$. The area of each rectangle is height · width, and each of the widths is 1. Therefore, the total combined area is $2.9 \cdot 1 + 2.6 \cdot 1 + 2.1 \cdot 1 + 1.4 \cdot 1 = 9$.

4. The actual area is $9\frac{1}{3}$. Our area approximation is less than the actual area because it excludes little slivers of area between the curve and the rectangles. The most area is omitted on the interval [3,4]. Thus, the approximation is lower, hence the term *lower sums*.

5. If you used more rectangles, the approximation would be much better. In fact, the more rectangles you used, the less space would be omitted and the closer the approximation.

6. In order to draw rectangles that circumscribe the graph, look at each interval separately and choose the higher endpoint height—it will give the height for that rectangle.

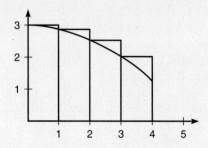

The rectangles' total area will be $3 \cdot 1 + 2.9 \cdot 1 + 2.6 \cdot 1 + 2.1 \cdot 1 = 10.6$

7. This area is too large, which was expected (*upper sum*). The rectangles contain more area than the curve. In fact, the error in the circumscribed rectangle method was greater than the error in the inscribed rectangle method.

8. In this case, the heights of the rectangles will be given by the function values of the midpoints of the following intervals: $f\left(\frac{1}{2}\right)$, $f\left(\frac{3}{2}\right)$, $f\left(\frac{5}{2}\right)$, and $f\left(\frac{7}{2}\right)$.

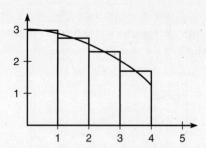

The widths of the rectangles are still 1, so the total Riemann midpoint sum is $1 \cdot 2.975 + 1 \cdot 2.775 + 1 \cdot 2.375 + 1 \cdot 1.775 = 9.9$.

9. The rectangle in a right-hand sum has the same height as the function at the right-hand endpoint of each interval. Similarly, a left-hand sum rectangle has the height of the function at the left-hand endpoint on each interval.

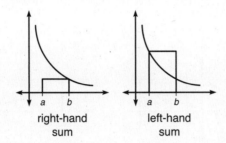

right-hand   left-hand
sum   sum

10. According to the given formula, each rectangle will have width $\Delta x = \frac{4-0}{5} = \frac{4}{5}$, as pictured in the below graph.

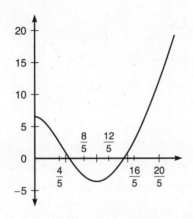

The five intervals in this graph are $(0, \frac{4}{5})$, $(\frac{4}{5}, \frac{8}{5})$, $(\frac{8}{5}, \frac{12}{5})$, $(\frac{12}{5}, \frac{16}{5})$, and $(\frac{16}{5}, 4)$. Because you're doing midpoint sums, you should draw rectangles whose heights are given by the function values of the midpoints of those intervals: $h(\frac{2}{5})$, $h(\frac{6}{5})$, $h(2)$, $h(\frac{14}{5})$, and $h(\frac{18}{5})$. The total area of these rectangles (and it's no sin to use your calculator to help you out here) is

$$\frac{4}{5} \cdot 4.744 + \frac{4}{5} \cdot -.152 + \frac{4}{5} \cdot -3 + \frac{4}{5} \cdot -.728 + \frac{4}{5} \cdot 9.736 = 8.48.$$

PROBLEM SET

You may use your graphing calculator on all of these problems.

1. What type of Riemann sum is being applied in each of the following diagrams? If there is more than one correct answer, give both.

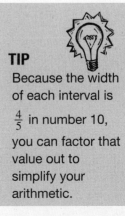

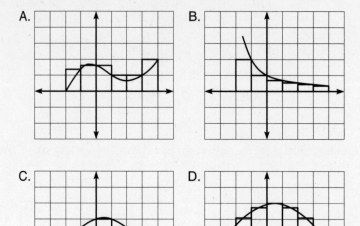

2. Approximate the area bounded by $f(x) = \sin x$ and the x-axis on the interval $[0,\pi]$ using 4 rectangles and upper sums.

3. If $g(x)$ is a continuous function that contains the values in the following table,

x	0	1	2	3	4	5	6	7	8
g(x)	2.1	3.6	5.1	6.3	6.9	7.2	7.3	4.0	3.2

approximate the area bounded by $g(x)$ and the x-axis on $[0,8]$ using

(a) 8 rectangles and right-hand sums

(b) 4 rectangles and midpoint sums

SOLUTIONS

1. (a) Upper sums, circumscribd rectangles: On the first two rectangles, the right-hand endpoint is used to determine height, whereas the left-hand endpoint is being used for the third and fourth rectangles. Thus, it cannot be right- or left-hand sums.

(b) Lower sums, right-hand sums, inscribed rectangles: All these descriptions apply to this diagram since the right-hand endpoint of each interval forms the inscribed rectangles.

(c) Lower sums, inscribed rectangles: The lower of the two endpoints' heights is chosen each time, not the right- or left-hand endpoint on a consistent basis.

(d) Midpoint sums: That one's pretty clear from the diagram. The function value at each interval midpoint dictates the height of the rectangle there.

2. (a) If 4 rectangles are used, the width of each interval will be $\Delta x = \frac{\pi - 0}{4} = \frac{\pi}{4}$. Therefore, the intervals will be $(0, \frac{\pi}{4})$, $(\frac{\pi}{4}, \frac{3\pi}{4})$,

$(\frac{3\pi}{4}, \frac{5\pi}{4})$, and $(\frac{5\pi}{4}, \frac{7\pi}{4})$. Because upper sums (circumscribed rectangles) are specified, the heights of the rectangles, from left to right, will be $\sin \frac{\pi}{4}$, $\sin \frac{\pi}{2}$, $\sin \frac{\pi}{2}$, and $\sin \frac{3\pi}{4}$, as these are the higher of the two endpoint function values for each interval.

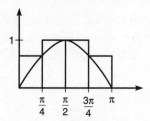

The upper sum is $\frac{\pi}{4}\left(\frac{\sqrt{2}}{2} + 1 + 1 + \frac{\sqrt{2}}{2}\right) \approx 2.682$.

3.   (a) If 8 rectangles are used, the width of each will be 1. Because right-hand sums are specified, the function value at the right endpoint of each interval dictates the height. Thus, the right-hand sum will be

$$1 \cdot (g(1) + g(2) + g(3) + g(4) + g(5) + g(6) + g(7) + g(8)).$$

$$3.6 + 5.1 + 6.3 + 6.9 + 7.2 + 7.3 + 4.0 + 3.2 = 43.6$$

   (b) Each rectangle will have width $\Delta x = 2$, so the intervals are (0,2), (2,4), (4,6), and (6,8). The midpoints of these intervals are simple, and the heights come from their function values, so the midpoint sum is

$$2 \cdot (g(1) + g(3) + g(5) + g(7))$$

$$2(3.6 + 6.3 + 7.2 + 4.0) = 42.2.$$

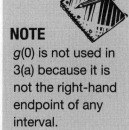

NOTE
$g(0)$ is not used in 3(a) because it is not the right-hand endpoint of any interval.

THE TRAPEZOIDAL RULE

Sure, Riemann sums give some approximation of the area under a curve, but they sure as heck don't give a terrific approximation. In fact, they're as clumsy as your cousin Irene that time she fell down the stairs at your parents' barbeque. There are other, more accurate methods, and this section focuses on one of them. (The other major method, Simpson's rule, is no longer on the AP test. D'oh! It used intervals with little parabolas on the end to approximate area.)

Consider the following graph and the method of upper sums used to approximate the area it bounds.

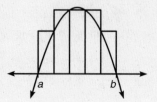

There's no avoiding it or ignoring it any longer: Those rectangles just contain way too much extra space. Enter (to thunderous applause) the Trapezoidal Rule. Instead of little rectangles, we will use little trapezoids to approximate the area.

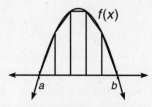

A Cunning Trapezoidal
Rule Problem

These trapezoids are formed by marking the function values at both endpoints of an interval and connecting the two dots. See how all that extra area disappears in just one application, returning the original showroom shine to the car's finish? You also may have noticed that the two "trapezoids" at the edges are really triangles. That's okay. Believe it or not, a triangle is just a special kind of trapezoid (a trapezoid with one base that has length 0). These aren't the pretty trapezoids you're used to seeing, but remember, a trapezoid is a quadrilateral with exactly one pair of parallel sides, and the vertical sides of these trapezoids are their bases.

According to geometry (and the voices echoing in my head), the area of a trapezoid is

$$A = \frac{1}{2}h(b_1 + b_2),$$

where h is the height of the trapezoid and b_1 and b_2 are the lengths of its parallel sides, or bases.

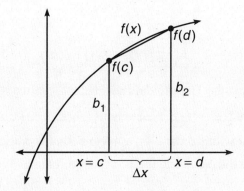

The bases in the above trapezoid have length $f(c)$ and $f(d)$, and the height, h (the distance between the bases), is $d - c$, or Δx.

Let's return to the diagram above, entitled, "A cunning Trapezoidal Rule problem," and determine the area according to the Trapezoidal Rule. In that problem, we know there are 5 trapezoids, but let's pretend we don't and say that there are n trapezoids. This way, we can develop the Trapezoidal Rule from scratch.

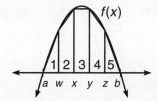

The area of the first trapezoid (conveniently labeled trapezoid 1) will be $= \frac{h}{2}\big(f(a) + f(w)\big)$. To find the height, remember the formula from the section on Riemann sums:

$$\Delta x = h = \frac{b - a}{n};$$

therefore,

$$\frac{h}{2} = \frac{b - a}{2n}.$$

The area of the second trapezoid is $\frac{1}{2}h\big(f(w) + f(x)\big)$. (The h will have the same value, since our intervals will always be of equal measure.) So far the area is

$$\frac{b - a}{2n}\big(f(a) + f(w) + f(w) + f(x)\big).$$

Let's skip ahead to the sum of all the trapezoids.

$$\frac{b-a}{2n}\big(f(a)+f(w)+f(w)+f(x)+f(x)+f(y)+f(y)+f(z)+f(z)+f(b)\big)$$

$$\frac{b-a}{2n}\big(f(a)+2f(w)+2f(x)+2f(y)+2f(z)+f(b)\big)$$

This, in essence, is the Trapezoidal Rule, but let's define it carefully:

The Trapezoidal Rule: If f is a continuous function on $[a,b]$ divided into n equal intervals of width $\Delta x = \frac{b-a}{n}$, as pictured in the diagram below, then the area between the curve and the x-axis is approximately

$$\frac{b-a}{2n}\big(f(a)+2f(x_1)+2f(x_2)+\ldots+2f(x_{n-1})+f(b)\big).$$

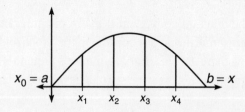

Translation: In order to approximate the area beneath the curve $f(x)$, find the width of one interval $\Delta x = \dfrac{b-a}{n}$ and divide it by 2. Then multiply by the sum of all the function values doubled, but don't double $f(a)$ or $f(b)$, the beginning and ending function values.

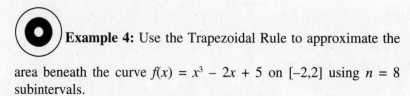 **Example 4:** Use the Trapezoidal Rule to approximate the area beneath the curve $f(x) = x^3 - 2x + 5$ on $[-2,2]$ using $n = 8$ subintervals.

Solution: If 10 subintervals are used, then the width of each will be

$$\Delta x = \frac{b-a}{n} = \frac{2-(-2)}{8} = \frac{4}{8} = \frac{1}{2}.$$

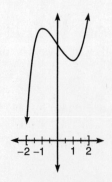

You can then apply the Trapezoidal Rule easily, using the x values $-2, -\dfrac{3}{2}$,

$-1, -\dfrac{1}{2}, 0, \dfrac{1}{2}, 1, \dfrac{3}{2}$, and 2:

$$Area \approx \frac{2-(-2)}{2 \cdot 8}\left(\begin{array}{l} f(-2)+2f\left(-\dfrac{3}{2}\right)+2f(-1)+2f\left(-\dfrac{1}{2}\right)+ \\ 2f(0)+2f\left(\dfrac{1}{2}\right)+2f(1)+2f\left(\dfrac{3}{2}\right)+f(2) \end{array}\right)$$

$$A \approx \frac{1}{4}(80) = 20.$$

In fact, this is the exact area beneath the curve, although you don't know how to verify this yet—that's in the next section.

PROBLEM SET

You may use a graphing calculator only on problem 3.

1. The Trapezoidal Rule can give the exact area beneath what types of functions?

2. Use the Trapezoidal Rule to aproximate the area beneath the curve $h(x) = x^2 + 1$ on the interval [0,4] using

  (a)   2 subintervals

  (b)   4 subintervals

  (c)   3 subintervals

3. (Based on the Stephen King book *The Girl who Loved Tom Gordon*) Nine-year-old Trisha McFarland is hopelessly lost in the woods, and the efforts of those looking for her have turned up nothing. The graph below shows the woods in which she is lost and measurements of the width of the woods (in miles) at regular intervals.

  (a) Use the Trapezoidal Rule to approximate the area of the woods.

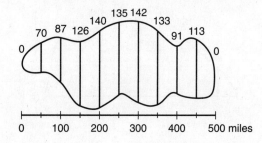

  (b) Assuming that one person can search 15 mi² in one day, how many people will it take to scour the entire woods in a single day?

TIP
The number of terms you add in the Trapezoidal Rule is always one more than the *n* you are using.

SOLUTIONS

1. The Trapezoidal Rule gives the exact area beneath linear functions, since the tops of the trapezoids used to make the approximations are linear.

2.   (a) $\Delta x = 2$

$$\frac{4-0}{2 \cdot 2}\big(h(0)+2h(2)+h(4)\big)$$

$$1 \cdot (1 + 10 + 17) = 28$$

    (b) $\Delta x = 1$

$$\frac{4-0}{(2 \cdot 4)}\big(h(0)+2h(1)+2h(2)+2h(3)+h(4)\big)$$

$$\frac{1}{2}(1 + 4 + 10 + 20 + 17) = 26$$

    (c) $\Delta x = \dfrac{4-0}{3} = \dfrac{4}{3}$

$$\frac{(4-0)}{2 \cdot 3}\left(h(0)+2h\left(\frac{4}{3}\right)+2h\left(\frac{8}{3}\right)+h(4)\right)$$

$$\frac{2}{3}\left(1 + \frac{50}{9} + \frac{146}{9} + 17\right)$$

$$\frac{2}{3}\left(\frac{358}{9}\right) = \frac{716}{27}$$

By the way, $\dfrac{716}{27}$ is approximately 26.519.

3. It might confuse you that the woods wiggle around as they do. If it helps, you can smoosh the woods against the *x*-axis to look something like this:

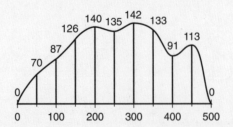

Either way, it is clear that $f(100) = 87$, $f(350) = 133$, and so on. The Trapezoidal Rule is actually quite easy to apply; you don't even have to plug into a function to get its values, as you did in number 2. The graph clearly is defined on the interval [0,500], and there are 10 subintervals, each with a length 50 miles.

$$Area \approx \frac{500 - 0}{2 \cdot 10} \big[(f(0) + 2f(50) + 2f(100) + 2f(150) + 2f(200) + 2f(250) +$$
$$2f(300) + 2f(350) + 2f(400) + 2f(450) + f(500)) \big]$$

$$Area \approx 25(140 + 174 + 252 + 280 + 270 + 284 + 266 + 182 + 226) =$$
$$41,480 \text{ mi}^2$$

(b) Even though 15mi^2 is a generous estimate of how much area one person can search, it would take

$$\frac{51,850}{15} = 3456.667$$

at least 3,457 people to conduct the search simultaneously.

THE FUNDAMENTAL THEOREM OF CALCULUS

I once had a Korean professor in college named Dr. Oh. He once said something I remember to this day: "Fundamental theorems are like the beginning of the world. Yesterday, not very interesting. Today, interesting." This is quite accurate, if not a little understated. In this theorem lies the fabled connection between the antiderivative and the area beneath a curve. In fact, the fundamental theorem has two major parts. Mathematicians can't seem to agree which is the more important part and, therefore, number them differently. Some even refer to one as the Fundamental Theorem and the other as the Second Fundamental Theorem. I love them both equally, as I would my own children.

The first part of the Fundamental Theorem deals with *definite integrals*. These are slightly different from the integrals we've been dealing with for two reasons: (1) they have boundaries, and (2) their answers are not functions with a "+ *C*" tacked on to the end—their answers are numbers. These are, indeed, two giant differences, but you'll be surprised by how much they actually have in common with our previous integrals, which we will now refer to by their proper name, *indefinite integrals*.

The Fundamental Theorem, Part One:

If *f(x)* is a continuous function on [*a,b*] with antiderivative *g(x)*, then

$$\int_a^b f(x)dx = g(b) - g(a).$$

Translation: In order to evaluate the definite integral $\int_a^b f(x)dx$, find the antiderivative of *f(x)*. Once you've done that, plug the upper bound, *b*, into the *anti*derivative. You should get a number. From that number, subtract the result of plugging the lower bound into the antiderivative.

What is the purpose of definite integrals? They give the *exact* area beneath a curve. Let's return to a problem from the Trapezoidal Rule section. You

NOTE
Remember, you can't spell Fundamental Theorem without "fun"!

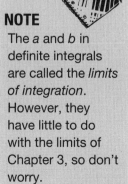

NOTE
The *a* and *b* in definite integrals are called the *limits of integration*. However, they have little to do with the limits of Chapter 3, so don't worry.

used 2, 3, and 4 subintervals to approximate the area beneath $y = x^2 + 1$ on [0,4]. Let's find out what the exact area is.

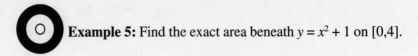 **Example 5:** Find the exact area beneath $y = x^2 + 1$ on [0,4].

Solution: The specified area is the result of the following integral:

$$\int_0^4 \left(x^2 + 1\right)dx.$$

So, you need to find the antiderivative of $x^2 + 1$. When you do, drop the integration sign and dx, as you did before.

$$\left(\frac{x^3}{3} + x\right)\Big|_0^4$$

The problem is not yet over, and you signify that the boundaries of integration still must be evaluated with the vertical line (or right bracket, if you prefer) and the boundaries next to the antiderivative. To finish the problem, then, you plug the upper limit of integration into the expression (both x's!) and then subtract the lower limit plugged in:

$$\left(\frac{4^3}{3} + 4\right) - \left(\frac{0^3}{3} + 0\right)$$

$$\frac{64}{3} + 4 = \frac{76}{3} = 25.333 \text{ units squared.}$$

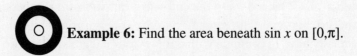 **Example 6:** Find the area beneath $\sin x$ on $[0,\pi]$.

Solution: This area is found by evalating the definite integral

$$\int_0^\pi \sin x \, dx.$$

Integrate $\sin x$ to get

$$-\cos x \Big|_0^\pi.$$

Now, substitute in π and 0 and subtract the two results:

$$-\cos \pi - (-\cos 0)$$

$$-(-1) + 1 = 2 \text{ units squared.}$$

Example 7: Evaluate $\int_0^\pi \cos x\, dx$, and explain the answer geometrically.

Solution: To begin, apply the Fundamental Theorem.

$$\sin x \Big|_0^\pi$$

$$\sin \pi - \sin 0 = 0 - 0 = 0$$

How can a curve have no area beneath it? Consider the graph of $y = \cos x$ on $[0,\pi]$:

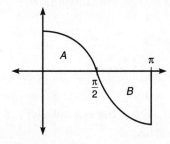

The area is made up of two separate areas, marked *A* and *B* on the diagram. Let's find those two separate areas:

$$Area\ A = \int_0^{\pi/2} \cos x\, dx$$

$$\sin x \Big|_0^{\pi/2} = \sin \frac{\pi}{2} - \sin 0 = 1 - 0$$

$$Area\ A = 1$$

$$Area\ B = \int_{\pi/2}^{\pi} \cos x\, dx$$

$$\sin x \Big|_{\pi/2}^{\pi} = \sin \pi - \sin \frac{\pi}{2} = 0 - 1$$

$$Area\ B = -1$$

These answers should not be too surprising. They are the same, although one is located under the *x*-axis, so its *signed area* is the opposite of the other. When you add these two areas together, you get 0. So, the geometric explanation is that the areas are opposites, and the resulting sum is 0.

Example 7 is a great segue to a few definite integral properties that are essential to know:

NOTE
You may argue that no area is technically negative. That's true. However, any area that falls beneath the *x*-axis is considered negative. To avoid this logical dilemma, the area bounded by definite integrals is often referred to as *signed area*—area that is positive or negative based on its position with relation to the *x*-axis.

- $\int_a^c f(x)dx = \int_a^b f(x)dx + \int_b^c f(x)dx$, if $a \leq b \leq c$.

Translation: You can split up an integral into two parts and add them up separately. Instead of integrating from a to c, you can integrate from a to b and add the area from b to c. We did this in Example 7. In that case, $a = 0$, $c = \pi$, and $b = \frac{\pi}{2}$.

- $\int_a^a f(x)dx = 0$

Translation: If you start and end at the same x value, you are technically not covering any area. Therefore, if the upper and lower limits of integration are equal, the resulting area and definite integral have a value of 0.

- $\int_a^b f(x)dx = -\int_b^a f(x)dx$

Translation: In a typical definite integral, the upper bound, b, is greater than the lower bound, a. If you switch them, the answer you get will be the opposite of your original answer. For example, let's redo Example 6 with the limits of integration switched:

$$\int_\pi^0 \sin x \, dx = -\cos x \Big|_\pi^0$$

$$-\cos 0 - (-\cos \pi)$$

$$-1 - 1 = -2.$$

In essence, switching the boundaries of integration has the effect of commuting (switching the order of) the subtraction problem dictated by the Fundamental Theorem, making it $g(a) - g(b)$ instead of $g(b) - g(a)$. This causes the sign change.

- *The definite integral represents accumulated change.*

Translation: In the same way that derivatives expressed a rate of change, the integral goes in the other direction and reports accumulated change. For example, consider the graph below of a car's velocity:

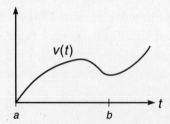

At any time t, the graph tells how fast the car was going. (This is the graph of the rate of change of position, velocity.) However, definite integrals give accumulated change. Therefore, $\int_a^b v(t)dt$ actually gives the distance the car traveled between time a and b. If the graph represented the rate of sale of socks over time, then the definite integral represents the number of socks sold over that time. More appropriate for me, if the graph represents the rate of hair loss over time, then the definite integral represents the actual amount of hair lost over the interval of time. Get the picture? More on this in Chapter 9.

Now that you know quite a bit about definite integrals, it's time to spring Part Two of the Fundamental Theorem on you. In essence, this theorem shows that differentiation and integration are opposites of one another.

The Fundamental Theorem, Part Two:

If c is a constant and t and x are variables,

$$\frac{d}{dx}\left(\int_c^x f(t)dt\right) = f(x).$$

Translation: This theorem is very specific in its focus and purpose. It applies only if (1) you are finding the derivative of a definite integral, (2) you are differentiating with respect to the same variable that is in the upper limit of integration, and (3) the lower limit of integration is a constant. If all these conditions are met, the derivative of the integral is simply the function inside the integral (the derivative and integral cancel each other out) with the upper bound plugged in.

Although this part of the Fundamental Theorem may sound awfully complicated (and some books make it sound nearly impossible), it is really quite easy. The following example will lead you through the process.

 Example 8: Evaluate the following derivatives:

(a) $\dfrac{d}{dx}\left(\displaystyle\int_3^x \cos t\, dt\right)$

You are taking the derivative of an integral with respect to the variable in the upper limit of integration. In addition, the lower limit is a constant. Because all these things are true, you may apply Part Two. In order to do so, simply plug the upper bound, x, into the function to get $\cos x$. That's all there is to it.

If you forgot this handy trick, you can still integrate as in the past:

NOTE
If the upper bound is something other than a single variable, according to the Chain Rule, you must multiply by its derivative.

NOTE
I will refer to Part Two of the Fundamental Theorem as "Part Two" to avoid repeating myself. However, refer to it as merely the Fundamental Theorem if justifying an answer on the AP test.

TIP
If you forget the Fundamental Theorem, Part Two, you're okay. You can still evaluate most of these problems by using regular definite integrals, as shown in Example 8(a). However, it is quite a handy shortcut.

$$\frac{d}{dx}\left(\int_3^x \cos t\, dt\right) = \frac{d}{dx}\left(\sin t\Big|_3^x\right)$$

$$\frac{d}{dx}(\sin x - \sin 3)$$

$$\cos x.$$

Because sin 3 is a constant, its derivative is 0. This is why the lower boundary must be a constant for the theorem to work and also why it doesn't matter what that lower boundary is.

(b) $\dfrac{d}{dh}\left(\displaystyle\int_5^{h^2} (x^2 - 2x)dx\right)$

The upper bound is h^2 rather than just h, but the variable you are deriving with respect to still matches, so you can still apply Part Two. Simply plug the upper bound into the function to get

$$(h^2)^2 - 2(h^2).$$

However, this is not the final answer. Because the upper bound is not merely h, you have to multiply by its derivative according to the Chain Rule. In this case, just multiply your previous answer by $\frac{d}{dh}\left(h^2\right) = 2h$.

$$(h^4 - 2h^2) \cdot 2h$$

$$2h^5 - 4h^3$$

If you like, you can verify this by using Part One of the Fundamental Theorem.

(c) $\dfrac{d}{dt}\displaystyle\int_{2x}^{x^3} \sqrt{t}\, dt$

Bad news: The lower bound of this integral is not a constant. Therefore, we cannot apply Part Two. So, we default back to Part One of the Fundamental Theorem. It helps to rewrite $\sqrt{t}$ as $t^{1/2}$ before you begin.

$$\frac{d}{dt}\left(\frac{2}{3}t^{3/2}\,\Big|_{2x}^{x^3}\right)$$

$$\frac{d}{dt}\left[\frac{2}{3}\left((x^3)^{3/2} - (2x)^{3/2}\right)\right]$$

$$\frac{d}{dt}\left[\frac{2}{3}\left(x^{9/2} - (2x)^{3/2}\right)\right]$$

Don't forget to take the derivative once you're finished integrating.

$$\frac{2}{3}\left(\frac{9}{2}x^{7/2} - \frac{3}{2}\sqrt{2x}\cdot 2\right)$$

$$3x^{7/2} - 2\sqrt{2x}$$

PROBLEM SET

You may use a graphing calculator on problems 3 and 4 only.

1. Evaluate the following definite integrals:

   (a) $\displaystyle\int_{2}^{6} x^{1/3}\left(2x^4 - \frac{1}{\sqrt{x}}\right)dx$

   (b) $\displaystyle\int_{0}^{3\pi/4} \sec x\left(\sec x - \tan x\right)dx$

   (c) $\displaystyle\int_{-1}^{2} |x|dx$

2. If $f(x)$ is defined by the graph below and consists of a semicircle and numerous line segments, evaluate the following:

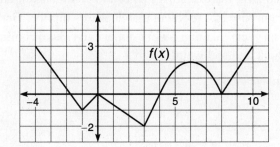

   (a) $\displaystyle\int_{0}^{3} f(x)dx$

   (b) $\displaystyle\int_{4}^{4} f(x)dx$

   (c) $\displaystyle\int_{0}^{10} f(x)dx$

   (d) $\displaystyle\int_{-4}^{0} f(x)dx$

3. Find the vertical line $x = c$ that splits the area bounded by $y = \sqrt{x}$, $y = 0$, and $x = 8$ exactly in half.

4. Evaluate the following derivatives:

(a) $\dfrac{d}{dx}\left(\displaystyle\int_{3x^2}^{9} \sin t \cos t \, dt\right)$

(b) $\dfrac{d}{dy}\left(\displaystyle\int_{y-1}^{3y} m \, dm\right)$

SOLUTIONS

1. (a) Begin by rewriting the expression and distributing the $x^{1/3}$.

$$\int_{2}^{6}\left(2x^{13/3} - x^{1/3-1/2}\right)dx$$

$$\int_{2}^{6}\left(2x^{13/3} - x^{-1/6}\right)dx$$

$$\frac{3}{8}x^{16/3} - \frac{6}{5}x^{5/6}\Big|_{2}^{6}$$

$$\left(\frac{3}{8}(6)^{16/3} - \frac{6}{5}\cdot 6^{5/6}\right) - \left(\frac{3}{8}(2)^{16/3} - \frac{6}{5}(2)^{5/6}\right)$$

Without a calculator, there's really no need to continue simplifying. What was the purpose of a problem that doesn't simplify? Get used to getting weird answers that don't work out evenly. Have confidence in answers that look and feel weird.

(b) Distribute the sec x to get

$$\int_{0}^{3\pi/4}\left(\sec^2 x - \sec x \, \tan x\right)dx.$$

You can integrate each of those terms.

$$(\tan x - \sec x)\Big|_{0}^{3\pi/4}$$

$$\left(-1 + \frac{2}{\sqrt{2}}\right) - (0 - 1)$$

$$\frac{2}{\sqrt{2}}\left(\text{or } \sqrt{2}, \text{ if you rationalize}\right)$$

(c) Even though we never discussed absolute value definite integrals, the answer is as simple as looking at the graph:

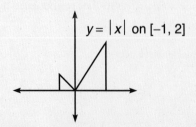

$y = |x|$ on $[-1, 2]$

The area beneath the graph is composed of two right triangles, and all you need to find the area of a triangle is $\frac{1}{2}bh$. Therefore, the area of both triangles is

$$\frac{1}{2} \cdot 1 \cdot 1 + \frac{1}{2} \cdot 2 \cdot 2 = \frac{1}{2} + 2 = \frac{5}{2}.$$

Not bad, eh? Look at the problems at the end of the chapter to practice nonlinear absolute value definite integrals.

2. Your work on 1(c) should make this easier. In order to calculate the definite integrals, use geometric formulas for triangles and semicircles.

    (a) –3: This is a triangle with base 3 and height 2. It is also below the x-axis, so its signed area is negative.

    (b) 0: One of the properties of definite integrals stated that an integral with equivalent upper and lower limits of integration has zero value, as no area is accumulated.

    (c) $2\pi - 1$: The area from 0 to 4 is –4. The area from 4 to 8 is a semicircle of radius 2, which has area $\frac{1}{2}\pi(2)^2 = 2\pi$. The area from 8 to 10 is right triangle of area 3. The sum is $2\pi - 1$.

    (d) $\frac{5}{2}$: In order to do this problem, you first need to find the x-intercept of the line segment from $(-4,3)$ to $(-1,-1)$. To do this, find the equation of the line by finding the slope

$$\text{slope} = \frac{-1-3}{-1-(-4)} = \frac{-4}{3}$$

and substituting a point.

$$y - (-1) = \frac{-4}{3}\left(x - (-1)\right)$$

$$y + 1 = -\frac{4}{3}x - \frac{4}{3}$$

$$3y + 3 = -4x - 4$$

Then, set the y equal to 0 in order to find the x-intercept:

$$3 = -4x - 4$$

$$x = -\frac{7}{4}.$$

Therefore, the area from –4 to 0 consists of two right triangles: one with positive area (height = 3 and base = $4 - \frac{7}{4} = \frac{9}{4}$) and one with negative area (height = 1 and base = $\frac{7}{4}$). Therefore, the total area will be

$$\frac{1}{2} \cdot \frac{9}{4} \cdot 3 - \frac{1}{2} \cdot \frac{7}{4} \cdot 1$$

$$\frac{27}{8} - \frac{7}{8} = \frac{5}{2}.$$

3. The region bounded by all those graphs is simply

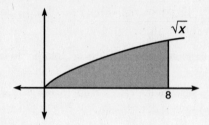

and this area is given by

$$\int_0^8 \sqrt{x}\,dx$$

$$\frac{2}{3} x^{3/2} \Big|_0^8$$

$$\frac{2}{3} \cdot 8^{3/2}.$$

We are looking for a c between 0 and 8 that has exactly half of the area, or an area of $\frac{1}{3} \cdot 8^{3/2}$. In other words,

$$\int_0^c \sqrt{x}\,dx = \frac{1}{3} \cdot 8^{3/2}.$$

Integrate the left side to get

$$\frac{2}{3} x^{3/2} \Big|_0^c = \frac{1}{3} \cdot 8^{3/2}$$

$$\frac{2}{3} c^{3/2} = \frac{1}{3} \cdot 8^{3/2}$$

$$2c^{3/2} = 8^{3/2}.$$

Square both sides to solve

$$4c^3 = 8^3$$

$$c^3 = \frac{512}{4} = 128$$

$$c = \sqrt[3]{128} = 4\sqrt[3]{2}.$$

4.   (a) You cannot apply Part Two right away since the lower bound isn't a constant. However, one of the properties of definite integrals says that you can switch the order of the bounds if you take the opposite of the integral:

$$-\frac{d}{dx}\left(\int_9^{3x^2} \sin t \, \cos t \, dt\right).$$

Apply Part Two and be finished. Don't forget to multiply by the derivative of the upper bound.

$$-6x(\sin 3x^2 \cos 3x^2)$$

(b) No Part Two here, as the lower bound again isn't a constant, and this time you can't do much about it. Just use the Fundamental Theorem Part One:

$$\frac{d}{dy}\left(\frac{m^2}{2}\bigg|_{y-1}^{3y}\right)$$

$$\frac{d}{dy}\left(\frac{9y^2}{2}-\frac{(y-1)^2}{2}\right)$$

$$\frac{d}{dy}\left(\frac{9}{2}y^2-\frac{1}{2}\left(y^2-2y+1\right)\right).$$

Again, don't forget to take the derivative to finish!

$$9y-\frac{1}{2}(2y-2)$$

$$9y-(y-1)=8y+1$$

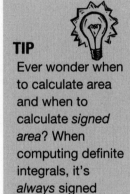

TIP
Ever wonder when to calculate area and when to calculate *signed area*? When computing definite integrals, it's *always* signed area.

HANDS-ON ACTIVITY 7.2: ACCUMULATION FUNCTIONS

Accumulation functions are a new emphasis on the AP test. Not all books refer to them by this name, but the leaders in calculus reform have begun to use this terminology. Therefore, I will bend to peer pressure and use it as well. Accumulation functions are natural (and neato) extensions to definite integrals.

1. Given a function $f(x) = \int_0^x g(t)\,dt$ and the function $g(t)$ defined by the graph

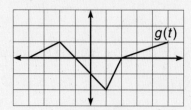

   evaluate $f(1)$. Plug 1 into $f(x)$, even though it may feel strange, and evaluate the definite integral.

   _____

   _____

2. How did the function f get its value?

   _____

   _____

3. Evaluate $f(-1)$. Why is your answer positive?

   _____

   _____

4. For what integral value(s) of c does $f(c) = -1$? Complete the chart below to decide.

x	−4	−3	−2	−1	0	1	2	3	4	5
$f(x)$			0						$-\dfrac{11}{6}$	

5. Based on the table you created in number 4, graph $f(x)$.

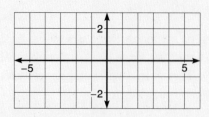

6. Find $f'(x)$, and justify your answer.

   _____

   _____

7. What relationship do you notice between $f(x)$ and $g(t)$, and why?

_____

_____

8. Evaluate $f''(-3)$.

_____

SELECTED SOLUTIONS TO HANDS-ON ACTIVITY 7.2

1. $f(1) = \int_0^1 g(t)dt$. In other words, $f(1)$ is the area beneath $g(t)$ on the interval [0,1]. That area is a small trapezoid with height 1 and bases of length 1 and 2. (You could also calculate the area as the sum of the square and triangle that make up the trapezoid.) Either way, the signed area is $-\frac{3}{2}$ (negative because it is below the x-axis).

2. f got its value from the amount of area "above" $g(t)$ on the interval 0 to whatever the input was (in this case, 1). In other words, by accumulating signed area, f is an accumulation function.

3. $f(-1) = \int_0^{-1} g(t)dt$. Notice that the lower number is on top, which is weird, since it is usually on the bottom. Therefore, you should change it so that the lower limit of integration is on the bottom, but remember that doing so makes the definite integral its opposite (since $\int_a^b f(x)dx = -\int_b^a f(x)dx$.

    $$-\int_{-1}^0 g(t)dt$$

    The triangle between -1 and 0 is below the x-axis, so it should have area $-\frac{1}{2}$, so the integral equals:

    $$-\left(-\frac{1}{2}\right) = \frac{1}{2}.$$

4. Use geometric formulas to calculate the areas. Don't forget that $f(4) = \int_0^4 g(t)dt$, which is *all* of the accumulated area from $x = 0$ to $x = 4$. The area from $x = 0$ to $x = 1$ is $-\frac{3}{2}$, as we've said. The area from $x = 1$ to $x = 2$ is a right triangle of area -1. The area from $x = 2$ to $x = 4$ is a right triangle with base 2 and height $\frac{2}{3}$ (since the slope of the line from $x = 2$ to $x = 5$ is $\frac{1}{3}$). Therefore, all of the accumulated area from $x = 0$ to $x = 4$ is $-\frac{3}{2} - 1 + \frac{1}{2} \cdot 2 \cdot \frac{2}{3} = -\frac{11}{6}$.

x	−4	−3	−2	−1	0	1	2	3	4	5
f(x)	−1	$-\frac{3}{4}$	0	$\frac{1}{2}$	0	$-\frac{3}{2}$	$-\frac{5}{2}$	$-\frac{7}{3}$	$-\frac{11}{6}$	−1

The correct answers are $c = -4$ and 5.

5.

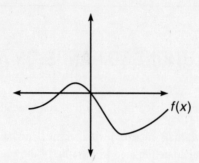

6. $\frac{d}{dx}\left(\int_0^x g(t)dt\right) = g(x)$, according to Fundamental Theorem Part Two.

Don't miss the importance of this: the graph of $f'(x)$ is the graph you were given in the first place, $g(t)$. (The fact that there's a t instead of an x makes no difference in the graph, but you should use $g(x)$ to keep your notation consistent.)

7. The graph of $g(x)$ acts like any first derivative graph. $g(x)$ is positive when $f(x)$ is increasing (look at the interval $(-4,-1)$ on each). $g(x)$ is 0 whenever $f(x)$ has a relative extrema point (look at $x = -1$ and -2).

8. $f''(-3)$ is the slope of the tangent line to $f'(x)$ at $x=-3$. Since $g(x)$ is $f'(x)$, you are trying to find the slope of the tangent line to $g(x)$ at $x=-3$. Luckily, $g(x)$ is linear at $x=-3$, and the slope of the line segment is $\frac{1}{2}$. Remember, the slope of a tangent line to a linear graph is simply the slope of that linear graph. Therefore, $f''(-3) = \frac{1}{2}$.

PROBLEM SET

Do not use a graphing calculator on any of these problems.

1. If $f(x) = \int_0^3 g(x)dx$, why is f not an accumulation function?

2. One famous function gets its value by accumulating area beneath the graph of $y = \frac{1}{x}$.

   (a) If we call the famous function $f(x)$ and define it as $f(x) = \int_1^x \frac{1}{t}dt$, find $f(2)$.

   (b) What is the name of the famous function?

(c) Evaluate $f(10)$, and verify that the result matches the output of the famous function.

3. If $g(x) = \int_0^{x/4} \sec^2 t \, dt$, evaluate $g(3\pi)$, $g'(3\pi)$, and $g''(3\pi)$.

4. If $h(t)$ is defined by the below graph and $m(x) = \int_{1/2}^{x} h(t) \, dt$,

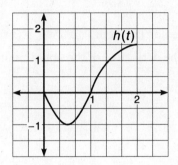

(a) Approximate $m(\dfrac{3}{2})$.

(b) Evaluate $m'(2)$.

(c) Describe the concavity of $m(x)$ for $[0,2]$.

(d) Write the following function values in order from least to greatest:

$m(\dfrac{1}{2})$, $m(1)$, and $m(2)$.

SOLUTIONS

1. An accumulation function has an x in its limits of integration.

2. (a) $f(2) = \int_1^2 \dfrac{1}{t} dt$. You know that the integral of $\dfrac{1}{t}$ is $\ln |t|$, so you get

$$\ln |t| \Big|_1^2$$

$$\ln 2 - \ln 1,$$

and since $\ln 1 = 0$, the answer is $\ln 2$. Notice, $f(2) = \ln 2$.

(b) The name of this famous function is the *natural logarithm!*

$f(x) = \ln x$. It makes sense since $f'(x) = \dfrac{1}{x}$.

(c) $f(10) = \int_1^{10} \dfrac{1}{t} dt$. So,

$$f(10) = \ln 10 - \ln 1 = \ln 10.$$

This works every time, since the lower limit of integration is always 1 and $\ln 1$ is 0.

3. Let's do one at a time and start with $g(3\pi)$.

$$g(3\pi) = \int_0^{3\pi/4} \sec^2 t \, dt$$

$$g(3\pi) = \tan t \Big|_0^{3\pi/4}$$

$$g(3\pi) = \tan \frac{3\pi}{4} - \tan 0 = -1 - 0 = -1$$

According to Fundamental Theorem Part Two, $g'(x) = \frac{1}{4} \sec^2(\frac{x}{4})$. (Don't forget to multiply by the derivative of the upper limit of integration, which is $\frac{1}{4}$.) Therefore,

$$g'(3\pi) = \frac{1}{4} \sec^2\left(\frac{3\pi}{4}\right)$$

$$g'(3\pi) = \frac{1}{4} \cdot \left(\frac{-2}{\sqrt{2}}\right)^2$$

$$g'(3\pi) = \frac{1}{2}.$$

Finally, use the Chain Rule to find $g''(x)$:

$$g''(x) = \frac{1}{2} \sec \frac{x}{4} \cdot \sec \frac{x}{4} \tan \frac{x}{4} \cdot \frac{1}{4}$$

$$g''(x) = \frac{1}{8} \sec^2 \frac{x}{4} \tan \frac{x}{4}$$

$$g''(3\pi) = \frac{1}{8} \sec^2\left(\frac{3\pi}{4}\right) \tan\left(\frac{3\pi}{4}\right)$$

$$g''(3\pi) = \frac{1}{8} \cdot 2 \cdot -1 = -\frac{1}{4}.$$

4. (a) $m\left(\frac{3}{4}\right) = \int_{1/2}^{3/2} h(t)dt$. Because there is no function, you cannot find this exact area. However, you do know that the area from $x = \frac{1}{2}$ to $x = 1$ is negative, and the area from $x = 1$ to $x = \frac{3}{2}$ is positive. You could use Riemann sums or the Trapezoidal Rule to approximate; however, these methods are mostly used when you happen to know the function. Since there is no function, why not make it a total approximation and *count boxes of area*? Each box on the graph of the function is one square unit. Therefore, the negative area is approximately −1.9 (your answer may be different but should be relatively close), and the positive area should be 2.

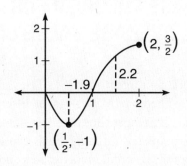

Therefore, $m\left(\dfrac{3}{2}\right) \approx 0.3.$

(b)   According to Fundamental Theorem Part Two, $m'(x) = h(x)$. Therefore, $m'(2) = h(2) = \dfrac{3}{2}$, according to its graph.

(c)   The concavity of $m(x)$ is described by the signs of $m''(x)$. In addition, $m''(x)$'s signs describe the direction of $m'(x)$, or $h(x)$. Therefore, whenever $h(x)$ is increasing, $m''(x)$ will be positive and vice versa. To summarize, $m(x)$ will be concave up whenever $h(x)$ is increasing: $\left(\dfrac{1}{2},2\right)$, and $m(x)$ will be concave down wherever $h(x)$ is decreasing: $\left(0,\dfrac{1}{2}\right)$.

(d) Using the method of counting boxes, we have already determined that $m(1) \approx -1.9$. We also said that $m\left(\dfrac{3}{2}\right) \approx .3$, so it makes sense to say that $m(2) > m\left(\dfrac{3}{2}\right)$, since m accumulates a lot of additional *positive* area from $x = \dfrac{3}{2}$ to $x = 2$. Therefore, we know that $m(2)$ is positive (if you count boxes, $m(2) \approx 4$). Finally, $m\left(\dfrac{1}{2}\right)$ is 0, since

$m\left(\tfrac{1}{2}\right) = \displaystyle\int_{1/2}^{1/2} h(t)$ (according to a property of definite integrals). With

all this in mind, we know that $m(1) < m\left(\dfrac{1}{2}\right) < m(2)$.

NOTE
Counting boxes to approximate area may feel inaccurate, but it's often more accurate than Riemann sums, and since you'd be approximating function values to use the Trapezoidal Rule in this case anyway, why bother? Sure it's a guess, but the directions do say *approximate*.

THE MEAN VALUE THEOREM FOR INTEGRATION, AVERAGE VALUE OF A FUNCTION

The Mean Value Theorem for Integration (MVTI) is an existence theorem, just like the Mean Value Theorem (MVT) was for differentiation. The MVT guaranteed the existence of a tangent line parallel to a secant line. The MVTI guarantees something completely different, but because it involves integration, you can guess that the theorem involves area and, therefore, definite integrals.

Although the MVTI is a very interesting theorem (and I'm not lying just to try and keep you interested), it is not widely used. I call the MVTI the flour theorem, because it has everything to do with making cookies, a necessary precursor to one of my favorite hobbies, eating cookies.

The Mean Value Theorem for Integration: If $f(x)$ is a continuous function on the interval $[a,b]$, then there exists a real number c on that interval such that $\int_a^b f(x)dx = f(c)(b-a)$.

Translation: You can create a rectangle whose base is the interval and whose height is one of the function values in that interval. This is a very special rectangle because its area is exactly the same as the area beneath the curve on that interval.

Look at the graph below of $f(x)$. If you count the boxes of area beneath it on the interval [1,9], you will get approximately 36.

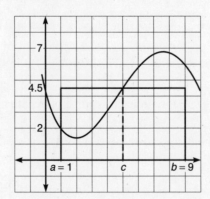

Also on the graph is a rectangle with length $9 - 1 = 8$. It stretches across the interval at a height of 4.5, and its area is 36. Notice that $f(c) = 4.5$; this is the value promised by the MVTI. Let's break down the parts of the theorem:

$$\underbrace{\int_a^b f(x)dx}_{\substack{\text{area beneath} \\ \text{the curve from} \\ a \text{ to } b}} = \underbrace{f(c)}_{\substack{\text{height} \\ \text{of the} \\ \text{rectangle}}} \cdot \underbrace{(b-a)}_{\substack{\text{width} \\ \text{of the} \\ \text{rectangle}}} = \underbrace{}_{\substack{\text{area} \\ \text{of the} \\ \text{rectangle}}}$$

The height of the rectangle, $f(c)$, is also called the *average value of $f(x)$*. In the example above, the average value is $f(c) = 4.5$. Imagine that the graph of $f(x)$ represents the flour in a measuring cup whose width is the interval $[a,b]$:

If you shake the measuring cup back and forth, the flour will level out to its average height or average value. The resulting flour will have the same volume as the hillier version of it—it's just flattened out. The same thing happens in two dimensions with the MVTI.

The AP test loves to ask questions about the average value of a function. Because of this, it helps to have the MVTI written in a different way—a way that lets you get right at the average value with no hassle. You get this formula quite easily; just multiply both sides of the MVTI by $\frac{1}{b-a}$, and you've got it.

The Average Value of a Function: If $f(x)$ is continuous on $[a,b]$, the average value of the function, $f(c)$, is given by

$$f(c) = \frac{1}{b-a} \cdot \int_a^b f(x)dx.$$

 Example 9: Find the average value of the function $g(x) = \sqrt{x} + x$ on $[4,9]$.

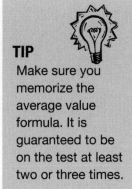

Solution: The average value, $g(c)$, will be given by

$$g(c) = \frac{1}{b-a} \cdot \int_a^b g(x)dx$$

$$g(c) = \frac{1}{5}\int_4^9 \left(x^{1/2} + x\right)dx$$

$$g(c) = \frac{1}{5}\left(\frac{2}{3}x^{3/2} + \frac{x^2}{2}\right)\Bigg|_4^9$$

$$g(c) = \frac{1}{5}\left(\left(\frac{2}{3}\cdot 27 + \frac{81}{2}\right) - \left(\frac{2}{3}\cdot 8 + 8\right)\right)$$

$$g(c) = \frac{1}{5}\left(\frac{36}{2} + \frac{81}{2}\right) - \left(\frac{16}{3} + \frac{24}{3}\right)$$

$$g(c) = \frac{1}{5}\left(\frac{117}{2} - \frac{40}{3}\right) = \frac{271}{30}.$$

CAUTION

$\frac{271}{30}$ is a tiny bit more than $\frac{270}{30} = 9$, so in this case (although it was close), the average value is not halfway between the absolute maximum (12) and absolute minimum (6) for the closed interval. Some students assume that the average value always falls exactly in the middle.

One final note before a parting example: Students sometimes get the average value of a function confused with the average rate of change of a function. Remember, the average value is based on definite integrals and is the "flour flattening" height of a function. The *average rate of change* is the slope of a secant line and describes rate over a period of time. Problem 4 following this section addresses this point of confusion.

 Example 10: (a) Use your graphing calculator to find the average value of $h(x) = x \cos x$ on the interval $[0,\pi]$.

We have no good techniques for integrating $x \cos x$, and we'll need to do so to find the average value of the function. The method is the same, but the means will be different:

$$h(c) = \frac{1}{\pi - 0}\int_0^\pi x \cos x \, dx.$$

Type the above directly into your calculator, using the "fnInt" function found under the [Math] menu. If you do not know how to use your calculator to evaluate definite integrals, immediately read the technology section at the end of this chapter. The average value turns out to be $h(c) = -.6366197724$.

In fact, the actual answer is $-\frac{2}{\pi}$.

(b) Find the c value guaranteed by the Mean Value Theorem for Integration.

The MVTI guarantees the existence of a c whose function value is the average value. In this problem, there is one input c whose output is $-\dfrac{2}{\pi}$; in other words, the point $(c, -\dfrac{2}{\pi})$ is on the graph of $h(x)$. To find the c, plug the point into $h(x)$:

$$h(c) = c \cos c$$

$$-\frac{2}{\pi} = c \cos c.$$

Solve this using your calculator, and you get: $c = 1.911$.

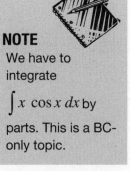

NOTE
We have to integrate

$\int x \cos x \, dx$ by

parts. This is a BC-only topic.

PROBLEM SET

You may use your calculator on problems 3 and 4 only.

1. Find the average value of $g(x) = \sec^2 x$ on the closed interval $\left[\dfrac{3\pi}{4}, \dfrac{5\pi}{4}\right]$.

2. If $\displaystyle\int_2^5 f(x)dx = 10$ and $\displaystyle\int_{14}^2 f(x)dx = -29$, what is the average value of $f(x)$ on the interval $[5,14]$?

3. Find the value c guaranteed by the Mean Value Theorem for Integrals for the function $h(x) = x^2 - \dfrac{1}{x}$ on the interval $[2,6]$.

4. A particle travels along the x-axis according to the position function $s(t) = \sin\frac{t}{3}\cos t$.

   (a) What is the particle's average velocity from $t = \frac{\pi}{2}$ to $t = 2\pi$?

   (b) What is the velocity of the particle at any time t?

   (c) Find the average value of the velocity function you found in part (b) on the interval $[\frac{\pi}{2}, 2\pi]$, and verify that you get the same result you did in part (a).

NOTE
The MVTI guarantees that at least one such c will exist, but multiple c's could be lurking around.

SOLUTIONS

1. The average value, $g(c)$, is given by

$$g(c) = \frac{1}{\frac{\pi}{2}} \int_{3\pi/4}^{5\pi/4} \sec^2 x\, dx$$

$$g(c) = \frac{2}{\pi} (\tan x)\Big|_{3\pi/4}^{5\pi/4}$$

$$g(c) = \frac{2}{\pi}(1 - -1) = \frac{4}{\pi}.$$

2. First of all, you can rewrite the second definite integral as

$$\int_2^{14} f(x)dx = 29, \text{ so}$$

Using another property of definite integrals, you know that

$$\int_2^5 f(x)\,dx + \int_5^{14} f(x)\,dx = \int_2^{14} f(x)\,dx$$

$$10 + \int_5^{14} f(x)\,dx = 29.$$

Therefore, $\int_5^{14} f(x)dx = 19$. With this value, you can complete the average value formula:

$$f(c) = \frac{1}{14-5}\int_5^{14} f(x)dx$$

$$f(c) = \frac{1}{9} \cdot 19 = \frac{19}{9}.$$

3. First, you should find the average value of the function:

$$h(c) = \frac{1}{6-2}\int_2^6 \left(x^2 - \frac{1}{x}\right)dx$$

$$h(c) = \frac{1}{4} \cdot \left(\frac{x^3}{3} - \ln|x|\right)\Bigg|_2^6$$

$$h(c) = \frac{1}{4} \cdot \left((72 - \ln 6) - \left(\frac{8}{3} - \ln 2\right)\right).$$

Although this is the average value of the function, it is not the c guaranteed by the MVTI. However, if you plug c into $h(x)$, you should get that result. Therefore,

$$c^2 - \frac{1}{c} = 17.05868026.$$

Solve this with your graphing calculator to find that $c \approx 4.159$.

4. (a) The average velocity is given by the slope of the secant line to a *position function*, just as the tangent lines to *position functions* give instantaneous velocity. The slope of the secant line is

$$\frac{\frac{\sqrt{3}}{2}}{2\pi - \frac{\pi}{2}} = .1837762985.$$

(b) Since s is the position function, the velocity, $v(t)$, is the derivative. Use the Product Rule (and the Chain Rule) to get

$$v(t) = -\sin\frac{t}{3}\sin t + \frac{1}{3}\cos t \cos\frac{t}{3}.$$

(c) Use your calculator's fnInt function to find the average value of $v(t)$:

$$\frac{1}{3\pi/2} \cdot \int_{\pi/2}^{2\pi}\left(-\sin\frac{t}{3}\sin t + \frac{1}{3}\cos t \cos\frac{t}{3}\right)dt = .1837762985.$$

Therefore, you can find the average rate of change of a function two ways: (1) calculate the slope of the secant line of its original function, or (2) find a function that represents the rate of change and then calculate the average value of it. Pick your favorite technique. Collect 'em and trade 'em with your friends!

U-SUBSTITUTION

Until this point, you have been able to integrate painfully few things. For example, you can solve $\int \cos x\, dx$, but not $\int \cos 5x\, dx$. If you thought $\int \cos 5x\, dx = \int \sin 5x\, dx$, all you have to do is take the derivative (to check):

$$\frac{d}{dx}(\sin 5x + C) = \frac{1}{5}\cos 5x$$

Since the result was not cos 5x, our antiderivative of sin5x doesn't check out. However, we were pretty close.

In order to integrate things like cos 5x or e^{2x}, you need to employ a method called *u-substitution*. This method allows you to integrate composite functions, sort of like the Chain Rule allowed you to differentiate composite functions. Here's a good rule of thumb: If you would use the Chain Rule to take the derivative of an expression, you should use u-substitution to integrate it. How important is u-substitution? It is all over the AP test, and it is an essential skill you'll use for the remainder of the year. Let's begin by integrating the above examples using u-substitution.

 Example 11: Integrate the following:

(a) $\int \cos 5x\, dx$

If this were cos x, you could integrate it. Therefore, we will introduce a new variable, u, like so: $u = 5x$. This way, the expression will become cos u, and we know the integral of cos u—it is sin u. However, before we get ahead of ourselves, we need to find the derivative of u (with respect to x); this is the all-important second step.

NOTE
Remember, there is no *u*-substitution without *you*. Isn't that nice? Do you feel all fuzzy inside? No? Me neither, so get used to it.

$$du = 5dx$$

We want to solve this for dx. Why? We will be replacing x's with u's in the integral, so we want to replace dx's with du's so all the variables match. Solving for dx is very easy:

$$\frac{du}{5} = dx.$$

Now, substitute $u = 5x$ and $\frac{du}{5} = dx$ into the original integral to get

$$\int \cos u \frac{du}{5}.$$

At this point, you can pull the $\frac{1}{5}$ out of the integral since it's just a constant and the result is beautiful:

$$\frac{1}{5} \int \cos u\, du.$$

The integral of $\cos u$ is $\sin u$, so you get

$$\frac{1}{5} \sin u + C.$$

However, your final answer can't include u's since the original problem didn't include u's. To finish, substitute in the original value for u, $u = 5x$.

$$\frac{1}{5} \sin 5x + C$$

(b) $\int e^{2x} dx$

We can integrate e^x, but not e^{2x}. Because the $2x$ is not a single variable (which we want), we replace it with a single variable, u:

$$u = 2x$$

$$du = 2dx.$$

Again, solve for dx so we can replace it in the integral and make the variables match.

$$\frac{du}{2} = dx$$

Now, substitute into the integral, make those x's a bad memory, and factor out that $\frac{1}{2}$.

$$\int e^u \frac{du}{2}$$

$$\frac{1}{2} \int e^u\, du$$

Nothing is easier to integrate than e^u! $\int e^u = e^u + C$.

$$\frac{1}{2}e^u + C$$

Get everything back to x's and you're finished.

$$\frac{1}{2}e^{2x} + C$$

If you don't feel confident that these answers are correct, take their derivatives and check them.

So far, you've seen that u-substitution can take the place of the "inner function" of a composite function and make the integral simpler by replacing that troublesome inner function with a single variable. But, this is only one good use of u-substitution. Often, you'll see a complex integral problem with this specific characteristic: part of the integral's derivative is also in the problem. For example, consider this integral: $\int \frac{\ln x}{x} dx$. Of the two pieces in the integral, whose derivative is present? The answer is $\ln x$. We can rewrite the integral to look like this:

$$\int \ln x \cdot \frac{1}{x} dx.$$

The derivative of $\ln x$ is $\frac{1}{x}$, and it's in the integral! Why is this important? Watch and see.

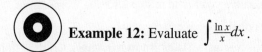 **Example 12:** Evaluate $\int \frac{\ln x}{x} dx$.

Solution: Because, as we said above, $\ln x$ has its derivative in the integral, we set it equal to u (and take the derivative as we did each time in Example 11).

$$u = \ln x$$

$$du = \frac{1}{x} dx$$

That's why it's important that the derivative was also present. Because we could rewrite the original integral expression as

$$\int \ln x \cdot \frac{1}{x} dx.$$

(as we showed above), we can now replace $\frac{1}{x}dx$ with du. Why? Because we found out they were equal when we differentiated the statement $u = \ln x$. By substitution, our original integral becomes

$$\int u \, du.$$

This is really easy to integrate:

$$\frac{u^2}{2} + C$$

Simply replace the u's with x's, and you're finished. The final answer is

$$\frac{(\ln x)^2}{2} + C.$$

The hardest part of u-substitution is deciding what the u should be. If you try a few things and they don't work, don't get discouraged. Try other things. Eventually, you'll find something that works.

NOTE

If you can't integrate by simple means (like the Power Rule for Integration), immediately try u-substitution. Make it your on-deck batter.

Steps to Success with U-Substitution

1. Choose part of the integral expression to be your u. This is the diabolical part. If you're dealing with a composite function, you might want to set u equal to the inner function. If an expression $f(x)$ and its derivative $f'(x)$ are both in a function, set u equal to $f(x)$. There are no hard and fast rules. Practice is the key.

2. Find the derivative of the u expression with respect to x (or whatever the variable present is).

3. If necessary, solve the derivative you found in step 2 for dx.

4. Substitute back into the original integral and integrate.

5. Replace your u's using your original u statement from number 1.

 Example 13: Solve the following integrals using u-substitution.

(a) $\displaystyle\int \frac{x^2 + 1}{x^3 + 3x} dx$

This looks complicated. You might try a couple of things for u, but if you set $u = x^3 + 3x$, watch what happens:

$$du = \left(3x^2 + 3\right)dx$$

$$\frac{du}{3} = \left(x^2 + 1\right)dx.$$

The entire numerator and the dx get replaced with $\frac{du}{3}$, the original integral expression becomes

$$\frac{1}{3}\int \frac{1}{u}du,$$

and you can integrate this quite easily.

$$\frac{1}{3}\ln|u| + C$$

$$\frac{1}{3}\ln\left|x^3 + 3x\right| + C$$

(b) $\int \dfrac{\tan x}{\cos^2 x} dx$

There are at least two good ways to solve this, both using u-substitution.

Method One: Rewrite tan x *as* $\dfrac{\sin x}{\cos x}$.

$$\int \frac{\sin x}{\cos^3 x}$$

$$u = \cos x$$

$$du = -\sin x \, dx$$

$$-du = \sin x \, dx$$

$$-\int \frac{1}{u^3}du$$

$$-\int u^{-3} du$$

$$\frac{1}{2}u^{-2} + C$$

$$\frac{1}{2\cos^2 x} + C$$

Method Two: Rewrite $\dfrac{1}{\cos^2 x}$ *as sec^{2}x.*

$$\int \tan x \, \sec^2 x \, dx$$

$$u = \tan x$$

$$du = \sec^2 x \, dx$$

TIP
If you cannot figure out what u should be and the integral expression is a fraction, try and set u equal to the denominator, as in Example 13(a).

$$\int u \, du$$

$$\frac{u^2}{2} + C$$

$$\frac{\tan^2 x}{2} + C$$

Although those answers do not immediately look the same, they are equivalent. If you need proof, here it is:

$$\frac{1}{2\cos^2 x} + C$$

$$\frac{1}{2}\sec^2 x + C.$$

Use Pappa to substitute $1 + \tan^2 x$ for $\sec^2 x$.

$$\frac{1}{2}\left(1 + \tan^2 x\right) + C$$

$$\frac{1}{2} + \frac{\tan^2 x}{2} + C$$

Realize that $\frac{1}{2} + C$ is simply another constant, or a different C, and the expressions are equal. There is never a guarantee that two C's are equal, even though we use the same variable to represent them—C is an arbitrary constant.

$$\frac{\tan^2 x}{2} + C$$

You can also use u-substitution in definite integrals. The only adjustment you must make is that your limits of integration (which are x limits) must become u limits. It's very easy to accomplish this.

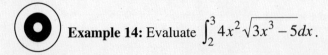 **Example 14:** Evaluate $\int_{2}^{3} 4x^2 \sqrt{3x^3 - 5} \, dx$.

Solution: Begin by pulling the constant 4 out of the integral to get

$4\int_{2}^{3} x^2 \sqrt{3x^3 - 5} \, dx$. Now, set u equal to the value inside the radical.

$$u = 3x^3 - 5$$

$$du = 9x^2 dx$$

$$\frac{du}{9} = x^2 dx$$

At this point, it is incorrect to write $\frac{4}{9} \int_2^3 u^{1/2}$. The boundaries 2 and 3 are x boundaries. We have to make all the variables match (which is why dx has to become du.) Therefore, plug each of these x values into the u expression $u = 3x^3 - 5$ to get the corresponding u boundary:

$$u = 3(2)^3 - 5 = 19$$

$$u = 3(3)^3 - 5 = 76.$$

Therefore, we can rewrite the original integral as

$$\frac{4}{9} \int_{19}^{76} u^{1/2} du.$$

Integrate and apply the Fundamental Theorem to finish.

$$\frac{4}{9} \cdot \frac{2}{3} u^{3/2} \bigg|_{19}^{76}$$

$$\frac{8}{27} \left(76^{3/2} - 19^{3/2} \right) \approx 171.773$$

The last important topic of this section (and it was a long one, wasn't it?) is integrating trigonometric functions. You already know the integrals of $\sin x$ and $\cos x$, but you don't know the other four, and it's important that you know all six. It is actually quite easy to find $\int \tan x \, dx$. If you rewrite it in terms of sine and cosine, you can integrate using u-substitution:

$$\int \tan x \, dx = \int \frac{\sin x}{\cos x} dx$$

$$u = \cos x, \, du = -\sin x \, dx$$

$$-du = \sin x \, dx$$

$$-\int \frac{1}{u} du$$

$$-\ln |\cos x| + C.$$

You can integrate cotangent in a similar way to get

$$\int \cot x \, dx = \ln |\sin x| + C.$$

CAUTION

When you change the x boundaries to u boundaries, the lower bound might end up higher than the upper bound. If this happens, don't panic and don't change them! Leave them as is, and complete the problem.

TIP

It sometimes is useful to set u equal to the part of your integral that is raised to a power. In Example 14, $3x^3 - 5$ is raised to the $\frac{1}{2}$ power.

Integrating sec x and csc x are a little more difficult and require a trick or two, so we won't get into that. However, it is important that you know what their integrals are, if not where they came from:

$$\int \sec x \, dx = \ln|\sec x + \tan x| + C$$

$$\int \csc x \, dx = -\ln|\csc x + \cot x| + C.$$

Just like you memorized the derivatives of these functions, it's equally important to memorize their integrals. Except for sine and cosine, all of the trig integrals contain "ln." To help memorize $\int \sec x \, dx$, remember its derivative. $\frac{d}{dx}(\sec x) = \sec x \, \tan x$, and

$$\int \sec x \, dx = \ln|\sec x + \tan x| + C.$$

One multiplies the terms, and the other adds them. The same goes for csc x. You really shouldn't need a trick to memorize the integrals for tan x and cot x; as you've seen with tan x, they are very easy to find with u-substitution.

PROBLEM SET

Do not use a graphing calculator on any of these problems.

Evaluate the following integrals.

1. $\int_0^3 \sqrt{3x+14} \, dx$

2. $\int \frac{\sin e^{-x}}{e^x} dx$

3. $\int \cos^2 5x \, \sin \, 5x \, dx$

4. $\int_{\pi/4}^{\pi/3} \csc \frac{x}{2} \cot \frac{x}{2} \, dx$

5. $\int e^{\sin 3x} \cos 3x \, dx$

6. $\int \frac{x}{(\ln 3)(x^2 + 4)} dx$

SOLUTIONS

1. This is a composite function, so set $u = 3x + 14$, the inner function. (This also follows the earlier tip that something to a power can be the u.)

$$u = 3x + 14$$

Therefore, the derivative is $du = 3dx$ and $\frac{du}{3} = dx$. The new u boundaries will be

$$u = 3(0) + 14 = 14 \text{ and}$$

$$u = 3(3) + 14 = 23.$$

The integral then becomes

$$\frac{1}{3}\int_{14}^{23} u^{1/2}\, du$$

$$\frac{1}{3} \cdot \frac{2}{3} u^{3/2} \Big|_{14}^{23}$$

$$\frac{2}{9}\left(23^{3/2} - 14^{3/2}\right)$$

2. This is a fraction that contains a composite function. The inner function is e^{-x}, so make that your u:

$$u = e^{-x}$$

$$-du = e^{-x} dx$$

$$-du = \frac{dx}{e^{x}}$$

The $-du$ will replace the denominator as well as dx so you can integrate.

$$-\int \sin u\, du$$

$$-(-\cos u) + C$$

$$\cos e^{-x} + C$$

3. The best u in this problem is $u = \cos 5x$. Therefore, $du = -5\sin 5x\, dx$, and $-\frac{1}{5} du = \sin 5x\, dx$.

$$-\frac{1}{5}\int u^2 du$$

$$-\frac{1}{15} u^3 + C$$

$$-\frac{1}{15}\cos^3 5x + C$$

4. Another composite function: Set $u = \dfrac{x}{2}$, so $du = \frac{1}{2} dx$ and $2du = dx$. The new u boundaries will be

$$u = \frac{\frac{\pi}{3}}{2} = \frac{\pi}{6}$$

$$u = \frac{\frac{\pi}{4}}{2} = \frac{\pi}{8},$$

and the new integral will be

$$2\int_{\pi/8}^{\pi/6} \csc u \cot u \, du$$

$$-2\csc u \Big|_{\pi/8}^{\pi/6}$$

$$-2\csc\frac{\pi}{6} + 2\csc\frac{\pi}{8}.$$

5. You could set $u = \sin 3x$, but if you set $u = e^{\sin 3x}$, watch what happens:

$$du = 3e^{\sin 3x} \cdot \cos 3x \, dx$$

$$\frac{du}{3} = e^{\sin 3x} \cdot \cos 3x \, dx.$$

All of the integral disappears, to be replaced with $\frac{du}{3}$.

$$\frac{1}{3}\int du$$

Of course, the integral of du is u, so the answer is

$$\frac{1}{3}u + C = \frac{1}{3}e^{\sin 3x} + C.$$

This is a great technique for integrating *any* kind of exponential function; it works like magic.

6. The $\frac{1}{\ln 3}$ is merely a constant we can pull out of the integral; next, set $u = x^2 + 4$. This results in $du = 2x\,dx$, and $\frac{du}{2} = x\,dx$. When you substitute back you get

$$\frac{1}{2 \cdot \ln 3}\int \frac{du}{u}$$

$$\frac{1}{2 \cdot \ln 3}\ln\left(x^2 + 4\right) + C.$$

There's no need to use absolute values for the natural log, since $x^2 + 4$ will always be positive. You might also apply log properties and write the answer as

$$\frac{1}{\ln 9} \ln\left(x^2 + 4\right) + C.$$

INTEGRATING INVERSE TRIGONOMETRIC FUNCTIONS

You already know a lot about integrating inverse trigonometric functions. In fact, you know so much that your parents pull pictures of you from their wallet and drone on and on about your intimate knowledge of arcsin. In fact, they wish you were going out with arcsin instead of that no-good you're currently dating. Well, it's time to give them even more to brag about.

You already know that $\int \dfrac{1}{\sqrt{1 - x^2}} \, dx = \arcsin x + C$. However, a radical in the denominator does not have to contain a 1, and the squared term does not have to be x^2. You can integrate any expression of the form

$$\frac{1}{\sqrt{a^2 - u^2}}$$

to get $\arcsin \dfrac{u}{a}$ (where a is a constant and u is a function of x).

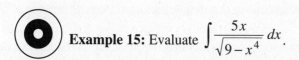 **Example 15:** Evaluate $\displaystyle\int \dfrac{5x}{\sqrt{9 - x^4}} \, dx$.

Solution: First of all, pull that 5 out of the numerator. The numerator *has* to be 1, according to the formula above. Don't worry about the presence of the x for now… that will take care of itself later.

$$5 \int \frac{x}{\sqrt{9 - x^4}} \, dx$$

We still have the arcsin form $\sqrt{a^2 - u^2}$ in the denominator; $a^2 = 9$ and $u^2 = x^4$. Therefore, $a = 3$ and $u = x^2$. The presence of a u reminds you to do u-substitution.

$$u = x^2$$

$$du = 2x \, dx$$

Therefore, $\frac{du}{2}$ will replace the xdx in the denominator when we substitute. (See? I told you the x would take care of itself.)

$$\frac{5}{2}\int \frac{du}{\sqrt{a^2 - u^2}}$$

According to the formula above, this equals $\arcsin \frac{u}{a}$, which is

$$\frac{5}{2}\arcsin\frac{x^2}{3} + C.$$

That's all there is to it. Recognize the pattern of $\sqrt{\text{number}^2 - \text{variable}^2}$ in the denominator, and it's a good clue to try and integrate using $\arcsin x$. However, there are two other patterns you want to memorize as well (not five, as you might have feared). Here are the remaining two (and they look remarkably similar to the derivatives you found earlier in the book).

$$\int \frac{1}{a^2 + u^2}du = \frac{1}{a}\arctan\frac{u}{a} + C$$

Pattern to look for: The sum of a number and a variable to a power in the denominator.

$$\int \frac{1}{u\sqrt{u^2 - a^2}}\,du = \frac{1}{a}\text{arcsec}\frac{|u|}{a} + C$$

Pattern to look for: A radical in the denominator and the difference of a variable to a power and a number. Arcsec is very close to arcsin, but the order of the subtraction is reversed—not to mention the presence of that extra x in the arcsec formula.

Both of these formulas have a $\frac{1}{a}$ in front of the inverse trigonometric formula, whereas $\arcsin x$ does not. It does not change the procedure at all; just don't forget it.

 Example 16: Integrate the following:

(a) $\int \frac{2x^2}{x^3 \left(x^6 - 4\right)}\,dx$

This is a job for arcsec. The $u\sqrt{u^2 - a^2}$ pattern is evident in the denominator: $u = x^3$ and $a = 2$. Don't forget—the u reminds you to do u-substitution. If $u = x^3$, then $du = 3x^2$ and $\frac{du}{3} = x^2$. This gives you

$$\frac{2}{3}\int \frac{du}{u\sqrt{u^2 - a^2}}$$

$$\frac{2}{3} \cdot \frac{1}{a}\operatorname{arc\,sec}\frac{|u|}{a} + C$$

$$\frac{1}{3}\operatorname{arc\,sec}\frac{|x^3|}{2} + C.$$

(b) $\int \frac{((\sin x))}{\cos^2 x + 3}dx$

This one may not look like arctan, but it is. The denominator has the form $u^2 + a^2$, where $u = \cos x$ and $a = \sqrt{3}$. The constant does not have to be a perfect square. This works exactly the same way. Don't forget about u-substitution, though. If $u = \cos x$, $du = -\sin x$, and $-du = \sin x$.

$$-\int \frac{du}{u^2 + a^2}$$

$$-\frac{1}{a}\arctan\frac{u}{a} + C$$

$$-\frac{1}{\sqrt{3}}\arctan\frac{\cos x}{\sqrt{3}} + C$$

You can rationalize this if you wish to get

$$-\frac{\sqrt{3}}{3}\arctan\frac{\sqrt{3}\cos x}{3} + C.$$

PROBLEM SET

Evaluate each of the following without a calculator.

1. $\int_0^{\pi/6} \frac{1}{4x^2 + 7}$

2. $\int \frac{dx}{x + 7}$

3. $\int \frac{4dx}{\sqrt{64 - 16x^2}}$

4. $\int \frac{\tan x}{\sin^2 x\sqrt{\cot^2 x - 16}}dx$

SOLUTIONS

1. This is a clear arctan problem with $u = 2x$ and $a = \sqrt{7}$. Therefore, $du = 2dx$ and $\frac{du}{2} = dx$.

$$\frac{1}{2}\int_0^{\pi/3} \frac{du}{u^2 + a^2}$$

$$\left(\frac{1}{2\sqrt{7}}\arctan\frac{u}{\sqrt{7}}\right)\Big|_0^{\pi/3}$$

$$\frac{1}{2\sqrt{7}}\arctan\frac{\pi}{3\sqrt{7}}$$

2. You cannot use inverse trig formulas to solve this. If you tried, you would have set $u = \frac{1}{\sqrt{x}}$, but the resulting u-substitution would have been impossible—you'd need another $\frac{1}{\sqrt{x}}$ in the problem. Instead, this is a simple u-substitution problem. Set $u = x + 7$ and $du = dx$. You can then rewrite the integral as

$$\int \frac{du}{u},$$

which is simply

$$\ln|u| + C$$
$$\ln|x+7| + C.$$

3. This is definitely an arcsin problem, but it's much easier if you factor out a 16 from the denominator and simplify first.

$$4\int \frac{dx}{\sqrt{16}\sqrt{4-x^2}}$$

$$\int \frac{dx}{4-x^2}$$

This could hardly be more straightforward now. Set $u = x$ and $a = 2$, so $du = dx$ and the integral becomes

$$\int \frac{du}{\sqrt{a^2 - u^2}}$$

$$\arcsin\frac{x}{2} + C.$$

4. Your instinct should tell you that this is an arcsec problem, since there is a radical in the denominator and the order of subtraction is *variable – constant*. However, if $u = \cot x$, shouldn't there be a cot x in front of the radical to match

the correct denominator form of $x\sqrt{x^2 - u^2}$? Watch what happens when you rewrite the trig functions using the reciprocal identities:

$$\int \frac{\csc^2 x \, dx}{\cot x \sqrt{\cot^2 x - 16}}.$$

Now, if $u = \cot x$ (and $a = 4$), $du = -\csc^2 x \, dx$, so $-du = \csc^2 x \, dx$. Rewrite the integral to get

$$-\int \frac{du}{u\sqrt{u^2 - a^2}},$$

which is the exact formula for arcsec, and everybody's happy.

$$-\frac{1}{4}\operatorname{arc\,sec}\frac{|\cot x|}{4} + C$$

TECHNOLOGY: EVALUATING DEFINITE INTEGRALS WITH YOUR GRAPHING CALCULATOR

The final of the four calculator skills (also known as the four skills of the apocalypse by those lacking technological know-how) you are required to know for the AP test is how to calculate a definite integral. As was the case with derivatives, the TI-83 cannot find symbolic integrals. In other words, it does not know that $\int x^2 dx = \dfrac{x^3}{3} + C$. However, it can find a darn good approximation for the area beneath a curve. Always remember and never forget that you'll have to be able to solve definite integrals without your graphing calculator, so don't come to rely so much on the tool that you forget the Fundamental Theorem in all its glory.

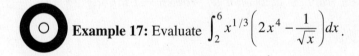 **Example 17:** Evaluate $\displaystyle\int_2^6 x^{1/3}\left(2x^4 - \frac{1}{\sqrt{x}}\right)dx$.

Solution: This was problem 1(a) from the Fundamental Theorem problem set; although it wasn't very hard, there are lots of places to make a mistake. To use your calculator, press [Math], [9], or just press [Math] and arrow down to the 9th option, "fnInt." The correct syntax for evaluating definite integrals on the TI-83 is

fnInt(*integral,x,lower limit,upper limit*).

Type this into your calculator,

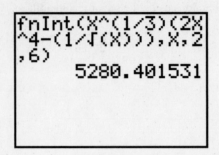

and you get 5280.402. Are you convinced that this is the same thing as the answer we got back in the pre-calculator day?

$$\left(\frac{3}{8}(6)^{16/3} - \frac{6}{5} \cdot 6^{5/6}\right) - \left(\frac{3}{8}(2)^{16/3} - \frac{6}{5}(2)^{5/6}\right)$$

To make sure they are the same (if you feel the need, are masochistic, or are, by nature, dubious of others), you can type the above "number" into your calculator, and see what happens.

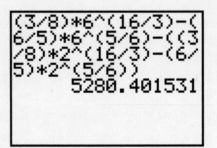

Oh, ye of little faith—they're the same.

PRACTICE PROBLEMS

You may use a graphing calculator on problems 11 through 13 only.

1. $\int \tan^2 x \, dx$

2. $\int \frac{e^{\tan x}}{1 - \sin^2 x} dx$

3. $\int_0^3 f(x)dx = 10$ and $\int_3^0 g(x)dx = 12$, evaluate the following:

    (a) $\int_0^3 (g(x) - 3f(x))dx$

    (b) $\int_0^3 (f(x) + 2)dx$

4. Evaluate $\displaystyle\int \frac{\sin \frac{1}{x^4}}{x^5}dx$.

5. What is the average value of the function $v(x) = 4^{2\sec x} \cdot \dfrac{\sin x}{\cos^2 x}$ on $[0, \frac{\pi}{3}]$?

6. Set up, but do not evaluate, $\displaystyle\int_{-3}^{10} \left| x^2 - 4x - 5 \right| dx$.

7. If $j(x) = \displaystyle\int_{3}^{\cos x} \ln y^3 dy$, $j'\left(\dfrac{\pi}{4}\right)$?

8. What expression has an integral of $3\ln\left| m(3x) \right| + C$, if m is a function of x?

9. $\displaystyle\int \tan x \, \sec\left(\ln|\cos x|\right)dx$

10. The following graph, $r(t)$, represents the rate of sales of the Furby toy (in hundreds of thousands of toys per month) from January 1998 to June 1999.

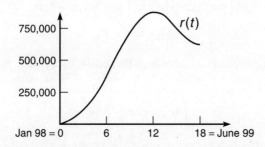

(a) Write a definite integral that represents total sales from February 1998 to March 1999.

(b) Write, but do not evaluate, an expression that represents the average rate of sales over the entire period of time.

(c) Where will the graph of total sales be concave up?

11. If $\displaystyle\int_{2}^{b}\left(x^{1/3} - \dfrac{a}{x^2}\right)dx = 11$ (a and b are real numbers), find a in terms of b.

12. If it takes NASCAR driver Dale Jarrett 1 minute to complete a lap around a track and his speed is measured every 10 seconds (in mph) as indicated below, answer the following.

time	0	10	20	30	40	50	60
speed(mph)	0	98	117	225	233	228	241

(a) According to midpoint sums and $n = 3$ rectangles, approximately how long is one lap on the track?

(b) What estimation of Dale's distance traveled is given by the Trapezoidal Rule with $n = 6$ trapezoids?

(c) Using your results from part (b), approximate Dale's average speed.

13. **James' Diabolical Challenge:** A particle's velocity over time (in inches/sec) as it moves along the x-axis is governed by the function $v(t) = 3t^2 - 10t + 15$.

(a) If the particle's position at $t = 1$ second is 8 inches, find the exact position equation of the particle, $s(t)$.

(b) What is the distance traveled by the particle from $t = 0$ sec to $t = 5$ sec?

(c) At what time(s) from $t = 0$ sec to $t = 5$ sec is the particle traveling its average velocity?

SOLUTIONS

1. If you use the Pappa Theorem to rewrite $\tan^2 x$ as $\sec^2 x - 1$, you can easily integrate:

$$\int \left(\sec^2 x - 1\right) dx = \int \sec^2 x\, dx - \int dx$$

$$\tan x - x + C.$$

2. Another trigonometric substitution is needed in this problem. According to the Mamma Theorem, $1 - \sin^2 x = \cos^2 x$. Therefore, the integral can be rewritten as

$$\int \sec^2 x \cdot e^{\tan x} dx.$$

This is especially useful since the derivative of $\tan x$ is $\sec^2 x$, which prods you to use u-substitution. If $u = \tan x$, $du = \sec^2 x\, dx$:

$$\int e^u du = e^u + C$$

$$e^{\tan x} + C.$$

3. First of all, get the boundaries to match up. According to definite integral properties, $\int_0^3 g(x)dx = -12$.

(a) $\int_0^3 \left(g(x) - 3f(x)\right)dx = \int_0^3 g(x)dx - 3\int_0^3 f(x)\,dx = -12 - 3 \cdot 10 = -42$

(b) Think about the graph of $f(x) + 2$ when compared to $f(x)$. The graph is the same, only moved up two units. The effect is demonstrated by the diagram below:

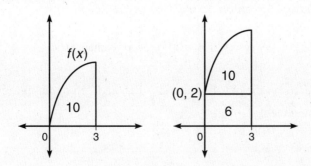

The new graph has an additional rectangle of area 6 (length 3 and height 2) beneath the original area of 10. Therefore, $\int_0^3 (f(x) + 2)dx = 16$.

4. If $u = x^{-4}$, then $du = -4x^{-5}dx$, so $-\dfrac{du}{4} = \dfrac{1}{x^5}\,dx$. This u-substitution makes the integral

$$-\frac{1}{4}\int \sin u\,du$$

$$\frac{1}{4}\cos u + C = \frac{1}{4}\cos\frac{1}{x^4} + C.$$

5. In order to find the average value, you'll need $\int_0^{\pi/3} v(x)dx$. Hopefully, your instinct is pushing you toward u-substitution. When integrating exponential functions, it's best to set u equal to the entire exponential function, so

$$u = 4^{2\sec x}$$

$$du = 4^{2\sec x} \cdot 2\sec x\,\tan x \cdot \ln\,4 \cdot dx$$

$$\frac{du}{2\ln\,4} = 4^{2\sec x}\sec x\tan x\,dx.$$

That seems all well and good, but where is the "$\sec x\,\tan x$" in the original problem? Do you see it? Rewrite $\dfrac{\sin x}{\cos^2 x}$ as $\dfrac{\sin x}{\cos x} \cdot \dfrac{1}{\cos x}$, and the original function becomes

$$v(x) = 4^{2\sec x}\sec x\tan x.$$

Now it's time to actually find the average value of v:

$$v(c) = \frac{1}{b-a} \cdot \int_a^b v(x)dx$$

$$v(c) = \frac{1}{\frac{\pi}{3}} \cdot \int_0^{\pi/3} 4^{2\sec x}\sec x\tan x\,dx.$$

Now, do the *u*-substitution, as outlined above, to get

$$v(c) = \frac{3}{\pi} \cdot \frac{1}{2\ln 4} \int_{16}^{256} du.$$

Don't forget to get the new boundaries of 16 and 256 by plugging 0 and $\frac{\pi}{3}$ into the *u* statement.

$$v(c) = \frac{3}{\pi \ln 16} \cdot u \Big|_{16}^{256}$$

$$v(c) = \frac{3}{\pi \ln 16}(240)$$

6. The graph of $x^2 - 4x - 5$ without the absolute values is a concave-up parabola. To find the roots, factor and set each equal to zero. The roots are, therefore, -1 and 5.

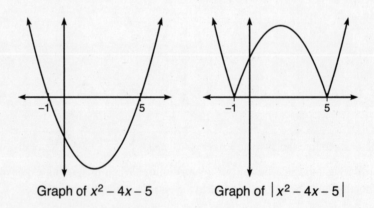

Graph of $x^2 - 4x - 5$ Graph of $\left| x^2 - 4x - 5 \right|$

The graph of $\left| x^2 - 4x - 5 \right|$ is the same, except that its negative portion flips above the *x*-axis, as the absolute value cannot have a negative output. Therefore, to find the area beneath the right graph above, we will find the area beneath the left graph and take the opposite of the area between -1 and 5 (since that area will be negative as it is below the *x*-axis).

$$\int_{-3}^{10} \left| x^2 - 4x - 5 \right| dx = \int_{-3}^{-1} \left(x^2 - 4x - 5 \right) dx - \int_{-1}^{5} \left(x^2 - 4x - 5 \right) dx + \int_{5}^{10} \left(x^2 - 4x - 5 \right) dx$$

7. You should use Fundamental Theorem Part Two. Plug in the upper bound, and multiply by its derivatite to get $j'(x)$:

$$j'(x) = \ln(\cos^3 x) \cdot (-\sin x)$$

$$j'\left(\frac{\pi}{4}\right) = -\frac{\sqrt{2}}{2} \cdot \ln\left(\frac{\sqrt{2}}{2}\right)^3.$$

NOTE

If number 6 had asked for

$$\int_{3}^{10} (x^2 - 4x - 5)dx,$$

you would not have taken the opposite of the negative signed area between $x = -1$ and $x = 5$. The absolute value signs mandate that all of the signed area must be turned positive.

8. Find $\frac{d}{dx}\left(3\ln\left|m(3x)\right|+C\right)$; whatever you get must integrate to get the original expression. You don't have to worry about the absolute value signs at all—they're only there for the benefit of the ln function.

$$3\cdot\frac{1}{m(3x)}\cdot m'(3x)\cdot 3$$

$$\frac{9m'(3x)}{m(3x)}$$

9. This problem was much easier if you memorized your trig integrals. First of all, you'd notice that tan x and its integral's opposite, $\ln\left|\cos x\right|$, are both in the expression. Therefore, you should do a u-substitution with $u=\ln\left|\cos x\right|$. That makes $du=-\tan x\,dx$, and $-du=\tan x\,dx$. The integral now becomes

$$-\int\sec u\,du.$$

You're not out of the woods yet! You still have to remember what the integral of sec u is. Are you thinking that you should have memorized those silly things? Do it now!

$$-\ln\left|\sec u+\tan u\right|+C$$

$$-\ln\left|\sec\left(\ln\left|\cos x\right|\right)+\tan\left(\ln\left|\cos x\right|\right)\right|+C$$

10. (a) Because the definite integral represents accumulated change, it gives you total sales (not total rate of sales or anything weird like that). If January 1998 = 0, then February 1998 = 1 and March 1999 = 15, and the correct definite integral is $\int_{1}^{15}r(t)dt$.

   (b) The average rate of change over the interval [0,18] is

   $$\frac{1}{18}\cdot\int_{0}^{18}r(t)dt.$$

   (c) $r(t)$ is the rate of change, or *derivative*, of the total sales function. Therefore, the total sales function will be concave up whenever $r(t)$ is increasing, which is on the interval [0,12].

11. Use the Fundamental Theorem in order to evaluate the definite integral:

$$\int_{2}^{b}\left(x^{1/3}-ax^{-2}\right)dx=11$$

$$\left(\frac{3}{4}x^{4/3}+\frac{a}{x}\right)\Big|_{2}^{b}=11$$

$$\left(\frac{3}{4}b^{4/3}+\frac{a}{b}\right)-\left(\frac{3}{4}\cdot 2^{4/3}+\frac{a}{2}\right)=11$$

All that remains is to solve the equation for a:

$$a\left(\frac{1}{b}-\frac{1}{2}\right)=11-\frac{3}{4}\left(b^{4/3}-2^{4/3}\right)$$

$$a=\frac{11-\frac{3}{4}\left(b^{4/3}-2^{4/3}\right)}{\frac{1}{b}-\frac{1}{2}}$$

12. (a) We can assume that Dale stuck to the inner lanes as much as possible to cut down on his time, so you are basically finding the length of the innermost lane of the track, since the definite integral of velocity is distance traveled. Before you begin, you must standardize the units. Since the speed is in mph, you should transform the seconds into hours to match. For example, since there are $60^2 = 3600$ seconds in an hour, 10 seconds is equal to $\frac{10}{3600}=\frac{1}{360}$ hours. If you convert all the times, the chart becomes

time(hrs)	0	$\frac{1}{360}$	$\frac{1}{180}$	$\frac{1}{120}$	$\frac{1}{90}$	$\frac{1}{72}$	$\frac{1}{60}$
speed(mph)	0	98	117	225	233	228	241

Using midpoint sums, the width of each interval will be

$\Delta x=\frac{b-a}{n}=\frac{\frac{1}{60}}{3}=\frac{1}{180}$. The midpoints of the intervals occur at $t=10, 30$, and 50 seconds, or $\frac{1}{360}$, $\frac{1}{120}$, and $\frac{1}{72}$ hours. Therefore, the midpoint sum is given by

$$\frac{1}{180}(98+225+228)\approx 3.061 \text{ miles}.$$

(b) The Trapezoidal Rule estimation for this problem is

$$\frac{\frac{1}{60}}{12}(0+2\cdot98+2\cdot117+2\cdot225+2\cdot233+2\cdot228+241)$$

$$\frac{1}{720}\cdot2043=2.8375 \text{ miles}.$$

(c) Dale's average speed is the average value of the velocity function. The fact that speed is the absolute value of velocity is irrelevant in this problem, as Dale's velocity is always positive.

$$\textit{Average speed} = \frac{1}{\frac{1}{60}}\int_0^{1/60} v(t)dt$$

In part (b), you found that $\int_0^{1/60} v(t)\,dt \approx 2.8375$. Plug this value into the average speed formula.

$$Average\ speed = 60 \cdot 2.8375 \approx 170.250\ mph$$

13. (a) Since position is the integral of velocity, you know that

$$s(t) = t^3 - 5t^2 + 15t + C.$$

But, you also know that $s(1) = 8$, so plug that into the function to find C.

$$s(1) = 1^3 - 5 \cdot 1^2 + 15 \cdot 1 + C = 8$$

$$1 - 5 + 15 + C = 8$$

$$C = -3$$

Therefore, the exact position equation is

$$s(t) = t^3 - 5t^2 + 15t - 3.$$

(b) The distance traveled is not just $\int_0^4 v(t)dt$! That gives you the total

displacement of the function. If the particle changes direction and comes back toward the origin, your answer will be wrong. For example, if the particle moves right 15 inches until $t = 3$ and then moves left 8 inches from $t = 3$ to $t = 5$, the definite integral above will give a result of 7 inches, whereas the particle really moved $15 + 8 = 23$ inches. So, you have to make sure the particle does not change direction on [0,5]. How do you do that? Dust off the wiggle graph.

$$s'(t) = v(t) = 3t^2 - 10t + 15 = 0$$

Luckily, $v(t) > 0$, and the particle never stops or turns backward (according to its graph). Therefore, the total distance traveled *will* be

$$\int_0^5 \left(3t^2 - 10t + 15\right) dt = 75\ inches.$$

In fact, because the particle never turns around, you could say that the total distance traveled is $s(5) - s(0)$, where it stopped minus where it started, which is $78 - 3 = 75$.

NOTE
Even though you are only given velocity values and not the actual velocity function, you are still approximating the area beneath the velocity function in 12(a) and (b). Therefore, we call that mystery function $v(t)$ in 12(c), even though we don't know what it is.

NOTE
Doesn't it feel good to finally unmask one of those C's? In order to do so, we needed just one value of the function. That value is sometimes called the *initial condition*.

NOTE
You can also find average velocity by calculating the slope of the secant line of the *position* function from $t = 0$ to $t = 5$: $\dfrac{78-3}{5-0} =$ 15 in / sec.

(c) The average velocity is the average value of the velocity function (most of which you've already figured out):

$$\frac{1}{5} \cdot \int_0^5 \left(3t^2 - 10t + 15\right) dt$$

$$\frac{1}{5} \cdot 75 = 15 \text{ in / sec}.$$

When is the particle actually traveling at a rate of 15 in/sec? Set the velocity equation equal to 15, and solve.

$$3t^2 - 10t + 15 = 15$$

$$3t^2 - 10t = 0$$

$$t = 0, \frac{10}{3}$$

Advanced Methods of Integration

Occasionally, you'll encounter an integration problem that smells like trouble. AB students need only complete the first section of this chapter, whereas BC students have to plod all the way through it. Happy trails.

MISCELLANEOUS METHODS OF INTEGRATION

Most of the integration on the AP test is done using the Power Rule and *u*-substitution. Occasionally, the test writers (while deviously twisting their thin moustaches) will throw in a tricky integral or two and cackle uproariously. Integration is unlike differentiation in a fundamental way: Using the Power, Product, Quotient, and Chain Rules, you can differentiate just about anything that comes your way. Integration requires many more methods, some of which only work in very specific circumstances. However, don't be discouraged. Below are five things you can try if all else has failed and you simply cannot integrate the problem at hand. One of these will help you if nothing else can.

1. **Use a trigonometric substitution**

Although the problem $\int \cot^2 x \, dx$ looks just about as impossible as can be, you can use the Baby Theorem to rewrite $\cot^2 x$ and change the integral to $\int \left(\csc^2 x - 1 \right) dx$. This is substantially easier because it is now possible. In the same way, $\int \tan x \, dx$ was impossible until we rewrote it as $\int \frac{\sin x}{\cos x} dx$ and used *u*-substitution. If the problem is trigonometric, you've got options.

2. **Split up the integral**

If an integral looks too complicated, rewrite it in pieces, if possible. In the case of a fraction, rewrite each term of the numerator over the denominator, as in the following example.

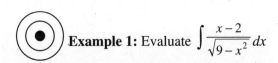 **Example 1:** Evaluate $\int \frac{x-2}{\sqrt{9-x^2}} dx$

Solution: Although the denominator certainly looks like an arcsin is in your future, the numerator makes things too complicated. However, split this into two separate fractions, and the work is half done.

$$\int \frac{x}{\sqrt{9-x^2}}\,dx - 2\int \frac{dx}{\sqrt{9-x^2}}$$

Let's do the left integral first by u-substitution. If $u = 9 - x^2$, $du = -2x\,dx$, and $-\frac{du}{2} = x\,dx$.

$$-\frac{1}{2}\int u^{-1/2}\,du$$

$$-\frac{1}{2}\cdot 2\cdot u^{1/2} + C$$

$$-\sqrt{9-x^2} + C$$

The second integral is an arcsin problem with $a = 3$ and $u = x$. Since $du = dx$, you have

$$-2\arcsin\frac{x}{3} + C.$$

Therefore, the final answer is

$$-\left(\sqrt{9-x^2} + 2\arcsin\frac{x}{3}\right) + C.$$

3. **Long division**

If the integral at hand is a fraction made up of polynomials, *and the degree of the numerator is greater than or equal to the degree of the denominator*, you can use long division on the problem before you begin to simplify the integral:

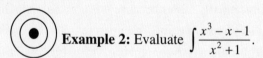 **Example 2:** Evaluate $\displaystyle\int \frac{x^3 - x - 1}{x^2 + 1}$.

Solution: Because the degree of the numerator is greater than (or equal to) the degree of the denominator, you can simplify the problem by long division first:

TIP

If none of these techniques work, there's always weeping, cursing, and breaking things. Although they won't help you solve the problem, you'll feel a whole lot better when you're through.

NOTE

Don't forget you are subtracting the entire second integral. That's why it is negative.

$$x^2 + 0x + 1 \overline{)x^3 + 0x^2 - x - 1} \quad \begin{array}{l} x + 0 \end{array}$$

$$\begin{array}{l} -\ x^3 - 0x^2 - x \\ \hline \qquad\qquad -2x - 1 \\ \qquad\qquad\qquad\quad 0 \\ \hline \qquad\qquad -2x - 1 \end{array} \quad \begin{array}{l} = x + \frac{-2x - 1}{x^2 + 1} \\[4pt] = x - \frac{2x + 1}{x^2 + 1} \end{array}$$

We are not finished by any means, but our integral can now be rewritten as $\int \left(x - \frac{2x + 1}{x^2 + 1} \right) dx$. In order to finish this problem, you'll have to separate it into pieces, just as you did Example 1. Once separated, you get

$$\int x\,dx - 2\int \frac{x}{x^2 + 1}\,dx - \int \frac{1}{x^2 + 1}\,dx.$$

You'll use the Power Rule for Integrals, u-substitution, and arctan, respectively, to solve this, and the final answer will be

$$-\frac{x^2}{2} - \ln\!\left(x^2 + 1\right) - \arctan x + C.$$

4. Complete the square

This technique is useful when you have quadratic polynomials in the denominator of your integral and, typically, only a constant in the numerator. Once you complete the square, you are able to default back to an inverse trig formula.

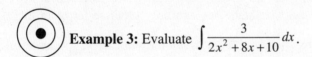 **Example 3:** Evaluate $\int \dfrac{3}{2x^2 + 8x + 10}\,dx$.

Solution: The quadratic in the denominator and no variables in the numerator alert us to complete the square. To do so, you'll have to factor 2 out of the terms in the denominator (since the x^2 must have a coefficient of 1).

$$\frac{3}{2}\int \frac{1}{x^2 + 4x + 5}\,dx$$

When you complete the square, you'll add $\left(\frac{1}{2}\cdot 4\right)^2$ and subtract it in the denominator to ensure that the value of the fraction does not change.

NOTE

When you add $C + C$, you do not get $2C$. Since each C is "some number," when you add them, you'll get some other number, which we also call C. Handy, eh?

CAUTION

When using long division, remember to use place holders of 0 for terms that are not present, like $0x^2$ or $0x$.

$$\frac{3}{2}\int \frac{1}{x^2 + 4x + 4 + 5 - 4}dx$$

$$\frac{3}{2}\int \frac{1}{(x+2)^2 + 1}dx$$

This is now a pretty easy arctan function with $u = x + 2$, $du = dx$, and $a = 1$. The final answer is

$$\frac{3}{2}\arctan(x+2) + C.$$

5. Add and subtract (or multiply and divide) the same thing

This exercise might sound fruitless. If you add and subtract the same thing, you get zero. What's the point? In the above example, you saw how adding and subtracting 4 allowed you to complete the square. So, it's not a complete waste of time. This method is used most often when you are trying to do a u-substitution and the problem won't cooperate with you. If you need something in the problem that isn't there to finish a u-substitution, why not just add it right in (as long as you remember to subtract it as well).

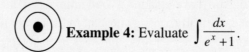 **Example 4:** Evaluate $\int \frac{dx}{e^x + 1}$.

Solution: We can't complete the square (not a quadratic denominator), we can't long divide (that's just crazy), we can't do a trig substitution, we can't separate (we can only separate terms in the numerator—the expression $\frac{1}{e^x} + \frac{1}{1}$ is *not* the same as $\frac{1}{e^x + 1}$), and u-substitution comes up short. What are we to do? Try u-substitution again, and force it.

Let's set $u = e^x + 1$, so $du = e^x dx$. We've got a problem: There is no $e^x dx$ present in the numerator—only the dx is there. So, we will add and subtract e^x in the numerator like so:

$$\int \frac{1 + e^x - e^x}{e^x + 1}dx.$$

(Don't forget that there was a 1 in the denominator to start with. It wasn't $0dx$ up there.) If we split this integral up, something magical transpires.

$$\int \frac{1 + e^x}{e^x + 1}dx - \int \frac{e^x}{e^x + 1}dx$$

The first fraction cancels out, since the numerator and denominator are equal. The second fraction is integrated with a simple u-substitution of $u = e^x + 1$. After taking these steps, you get

$$\int 1\,dx - \int \frac{du}{u},$$

and the final answer will be

$$x - \ln\left(e^x + 1\right) + C.$$

PROBLEM SET

Do not use a graphing calculator to integrate the following.

1. $\int \dfrac{x}{x^2 + 5x + 9}\,dx$

2. $\int \dfrac{4}{\sqrt{-x^2 - 6x + 12}}\,dx$

3. $\int \dfrac{x^3 + 4x + 3}{x - 2}\,dx$

4. Each of the following integrals varies just slightly from the others. However, each requires a completely different integration method. Discuss the method you would use to begin each.

   (a) $\int \dfrac{1}{x^2 + 3x + 10}\,dx$

   (b) $\int \dfrac{x}{x^2 + 3x + 10}\,dx$

   (c) $\int \dfrac{x^3}{x^2 + 3x + 10}\,dx$

SOLUTIONS

1. (This one's pretty tough.) If you set $u = x^2 + 5x + 9$, $du = (2x + 5)\,dx$. In order to use u-substitution, the numerator needs to be $2x + 5$. First, get the $2x$ up there by multiplying by 2 and $\frac{1}{2}$ at the same time:

$$\frac{1}{2}\int \frac{2x}{x^2 + 5x + 9}\,dx.$$

Now, you can add and subtract 5 to get that required $2x + 5$:

$$\frac{1}{2} \int \frac{2x+5-5}{x^2+5x+9}\, dx.$$

Split the integral up now

$$\frac{1}{2} \int \frac{2x+5}{x^2+5x+9}\, dx - \frac{5}{2} \int \frac{1}{x^2+5x+9}\, dx,$$

and the first piece can be solved by u-substitution (now that you have arranged it). The second integral requires you to complete the square.

$$\frac{1}{2} \int \frac{1}{u}\, du - \frac{5}{2} \int \frac{1}{x^2+5x+\frac{25}{4}+9-\frac{25}{4}}$$

$$\frac{1}{2} \ln\left(x^2+5x+9\right)+C - \frac{5}{2} \int \frac{1}{\left(x+\frac{5}{2}\right)^2+\frac{11}{4}}\, dx$$

$$\ln\sqrt{x^2+5x+9} - \frac{5}{\sqrt{11}} \arctan \frac{2\left(x+\frac{5}{2}\right)}{\sqrt{11}}+C$$

2. This is a completing-the-square question. Begin by factoring the negative out of the first two terms in the denominator so that the coefficient of x^2 is 1:

$$4\int \frac{1}{\sqrt{(-1)\left(x^2+6x\right)+12}}\, dx.$$

Now, complete the square in the denominator:

$$4\int \frac{1}{\sqrt{(-1)\left(x^2+6x+9\right)+12+9}}\, dx.$$

Even though it looks like you are adding 9 twice, remember that the 9 in parentheses gets multiplied by that -1, so it's really -9.

$$4\int \frac{1}{\sqrt{21-(x+3)^2}}\, dx$$

This is the arctan form with $a = \sqrt{21}$ and $u = x+3$. Since $du = dx$, you can rewrite the integral as

$$4\int \frac{1}{\sqrt{a^2-u^2}}\, du,$$

and the answer is

$$4\arcsin\frac{x+3}{\sqrt{21}}+C.$$

3. The numerator degree is larger, so use long division (or even synthetic division since the denominator is a linear binomial).

$$
\begin{array}{r}
x^2+2x+8 \\
x-2\overline{)x^3+0x^2+4x+3} \\
-x^3+2x^2 \\
\hline
2x^2+4x \\
-2x^2+4x \\
\hline
8x+3 \\
-8x+16 \\
\hline
19
\end{array}
\qquad
\begin{array}{r}
2\,|\ 1\quad 0\quad 4\quad 3 \\
2\quad 4\ 16 \\
\hline
1\quad 2\quad 8\ 19
\end{array}
$$

Either way, the quotient is $x^2+2x+8+\frac{19}{x-2}$. So, the integral can be rewritten as

$$\int\left(x^2+2x+8\right)dx+19\int\frac{1}{x-2}\,dx,$$

which you can integrate using the Power Rule and a u-substitution of $u=x-2$ to get

$$\frac{x^3}{3}+x^2+8x+19\ln\big|\,x-2\,\big|+C.$$

4. (a) Because the denominator is a quadratic with a constant numerator, you will complete the square in the denominator to integrate.

(b) You will use u-substitution to integrate, with $u=x^2+3x+10$. Since $du=(2x+3)dx$, you'll have to make the numerator match it, as you did in number 1.

(c) Because the degree of the numerator is greater than that of the denominator, long division will be your first step. I have a sinking feeling that there will be other methods required before that one is done, though.

PARTS (BC TOPIC ONLY)

Integration by parts is a technique based completely on the Product Rule. However, it's unlikely that you'll recognize that familiar and happy rule once we're done mangling it. This integrating method was made famous in the movie *Stand and Deliver*, when Jaime Escalante stands in front of the chalkboard and chants "Come on, it's tic-tac-toe." That small cameo role catapulted integration by parts to fame, and it eventually ended up on *General Hospital* playing a handsome gangster doctor. However, one day, everyone noticed it was just a math formula, and it was immediately fired. The formula is still very bitter.

We'll begin by using the Product Rule to find the derivative of the expression uv, where both u and v are functions:

$$d(uv) = u \cdot v' + v \cdot u', \text{ or}$$

$$d(uv) = u\,dv + v\,du.$$

If you integrate both sides of the equation, you get

$$uv = \int u\,dv + \int v\,du.$$

Finally, solve this for $\int u\,dv$, and you get the formula for integration by parts:

$$\int u\,dv = uv - \int v\,du.$$

The focus of this method is splitting your difficult integral into two parts: a u and a dv like the left side of the above equation. The u portion must be something you can differentiate, whereas the dv must be something you can integrate. After that, it's all downhill.

 Example 5: Evaluate $\int x\,\cos x\,dx$.

Solution: None of the methods we've discussed so far can handle this baby. We'll use parts instead. First of all, set $u = x$ (because you can easily find its derivative) and $dv = \cos x\,dx$ (because you can easily find its integral). It's true that you could have set $u = \cos x$ and $dv = x\,dx$, but if at all possible, you should choose a u whose derivative, if you kept taking it again and again, would eventually equal 0. The derivative of $\cos x$ will jump back and forth between $\cos x$ and $\sin x$ without ever becoming 0.

NOTE
You'll see why it's important to pick a u that, through differentiation, eventually becomes 0 in Example 6.

Since $u = x$, $du = dx$, and if $dv = \cos x\,dx$, $v = \int \cos x\,dx = \sin x$. (Don't worry about "$+ C$"s for now—we'll take care of them later.) According to the parts formula,

$$\int u\,dv = uv - \int v\,du,$$

and we know what u, du, v, and dv are, so plug them in.

$$\int x\,\cos x\,dx = x\,\sin x - \int \sin x\,dx$$

Our original integral, on the left, equals (and can be rewritten as) the expression on the left, which contains a very simple integral in $\int \sin x\,dx$. The final answer is

$$x\,\sin x + \cos x + C.$$

If you don't believe that this is the answer, take its derivative, and you get

$$x \cos x + \sin x - \sin x = x \cos x,$$

which is the original integral.

That wasn't so bad, was it? Sometimes, however, it's less tidy. For example, if you are integrating $\int x^2 \cos x \, dx$ by parts, you'd set $u = x^2$ and $dv = \cos x \, dx$. Therefore, $du = 2xdx$ and $v = \sin x$. According to the formula,

$$\int x^2 \cos x \, dx = x^2 \sin x - 2 \int x \sin x \, dx.$$

Do you see what's troubling in this equation? You cannot integrate $x \sin x \, dx$ easily. In fact, guess what method you'll have to use? Parts! You'll have to set aside the $x^2 \sin x$ portion for now and expand the integral using the parts formula again, this time setting $u = x$ and $dv = \sin x \, dx$. These sorts of things happen when the du term isn't something pretty and frilly like 1, as it was in Example 5. But don't give up hope—there's a handy chart you can use to integrate by parts that feels like no work at all. The only limitation it has is that the u term must eventually differentiate to 0.

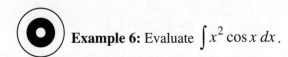 **Example 6:** Evaluate $\int x^2 \cos x \, dx$.

Solution: To set up the chart, make a u column, a dv column, and a column labeled "+/– 1". In the first row, list your u, your dv, and a "+1". In the second row, list du, v, and a "–1". In the third row, take another derivative, another integral, and change the sign again. Continue until the u column becomes 0, but take the signs column one row further than that. (You'll see why in a second).

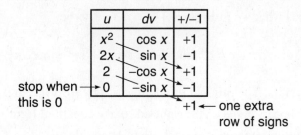

Now draw diagonal arrows beginning at the x^2 and continuing down and to the right. Do this until you get to the 0 term. Multiply each of the terms along the arrow (for example, in the first arrow, you multiply $x^2 \cdot \sin x \cdot$ "+1") to get a term in the answer ($x^2 \sin x$). You will make three arrows in this chart, so the answer has three terms. (Multiplying anything by 0 results in 0, so there's no need for a fourth arrow.) The final answer is

$$x^2 \sin x + 2x \cos x - 2 \sin x + C.$$

This method is preferred by students. In fact, look how easy Example 5 is if you use the chart:

u	dv	+/-1
x	cos x	+1
1	sin x	-1
0	-cos x	+1
		-1

$$x \sin x + \cos x + C.$$

However, as wonderful as the chart is, it's not so handy when the u term's derivative does not eventually become 0. The final example is about as complicated as integration by parts gets.

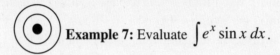 **Example 7:** Evaluate $\int e^x \sin x \, dx$.

Solution: It's wise to choose $u = \sin x$ and $dv = e^x dx$, since the integral of dv is very easy. If you do so, $du = \cos x \, dx$ and $v = e^x$.

$$\int e^x \sin x \, dx = e^x \sin x - \int e^x \cos x \, dx$$

This is unfortunate. We have to use parts again to evaluate the new integral. Stick with it, though—it will pay off. For now, we'll ignore the $e^x \sin x$ (even though it will be part of our eventual answer) and focus on $\int e^x \cos x \, dx$. As before, we set $u = \cos x$ and $dv = e^x \, dx$; so, $du = -\sin x \, dx$ and $v = e^x$.

$$\int e^x \cos x \, dx = e^x \cos x + \int e^x \sin x \, dx$$

Watch carefully now. The original integral, $\int e^x \sin x \, dx$, is equal to $e^x \sin x$ minus what we just found $\int e^x \cos x \, dx$ to be:

$$\int e^x \sin x \, dx = e^x \sin x - \left(e^x \cos x + \int e^x \sin x \, dx \right).$$

It looks hopeless. The original problem was $\int e^x \sin x \, dx$, and that same expression appears *again* on the right side of the equation! Here's what you do: distribute that negative sign on the right-hand side and add $\int e^x \sin x \, dx$ to both sides of the equation.

$$2 \int e^x \sin x \, dx = e^x \sin x - e^x \cos x$$

To get your final answer, just divide by 2.

$$\int e^x \sin x \, dx = \frac{e^x \sin x - e^x \cos x}{2}$$

PROBLEM SET

Do not use a graphing calculator on any of these problems.

1. $\int x^2 \sin 4x \, dx$

2. $\int 10x \sec x \, \tan x \, dx$

3. $\int e^x \cos x \, dx$

4. $\int x^3 e^x dx$

5. $\int \ln x \, dx$ (**Hint:** So far, you know of no easy integral for ln x.)

SOLUTIONS

1. If $u = x^2$ and $dv = \sin 4x \, dx$, the derivative of u will eventually become 0, so you can use a chart to find the integral.

u	dv	$+/-1$
x^2	$\sin 4x$	$+1$
$2x$	$\dfrac{-\cos 4x}{4}$	-1
1	$\dfrac{-\sin 4x}{16}$	$+1$
0	$\dfrac{+\cos 4x}{64}$	-1
		$+1$

$$-\frac{1}{4}x^2 \cos \ 4x + \frac{1}{8}x \sin \ 4x + \frac{1}{64}\cos \ 4x + C$$

2. It's smart to put $dv = \sec x \tan x \, dx$, since the resulting v is sec x. Therefore, $u = 10x$ and $du = 10dx$. Use a chart or the formula; either works fine.

$$\int 10x \ \sec x \ \tan x \, dx = 10x \ \sec x - \int 10 \sec x \, dx$$

$$= 10x \ \sec x - 10\ln \ | \ \sec x + \tan x \ | + C$$

3. This one is tricky, like Example 7. Let $u = \cos x$ and $dv = e^x dx$; therefore, $du = -\sin x \, dx$ and $v = e^x$.

$$\int e^x \ \cos x \, dx = e^x \ \cos x + \int e^x \ \sin x \, dx$$

The rightmost integral must be evaluated using parts again, this time with $u = \sin x$ ($du = \cos x \, dx$) and $dv = e^x \, dx$ ($v = e^x$).

$$\int e^x \ \sin x \, dx = e^x \ \sin x - \int e^x \ \cos x \, dx$$

Therefore, the original integral becomes

$$\int e^x \ \cos x \, dx = e^x \ \cos x + e^x \ \sin x \ - \int e^x \ \cos x \, dx.$$

Add $\int e^x \cos x \, dx$ to both sides of the equation to get

$$2\int e^x \cos x \, dx = e^x \cos x + e^x \sin x$$

$$\int e^x \cos x \, dx \, \frac{e^x \cos x + e^x \sin x}{2}.$$

4. This baby is a prime candidate for the chart, with $u = x^3$ and $dv = e^x dx$.

u	dv	+/−1
x^3	e^x	+1
$3x^2$	e^x	−1
$6x$	e^x	+1
6	e^x	−1
0	e^x	+1
		−1

$$e^x(x^3 - 3x^2 + 6x - 6) + C$$

5. If you don't know an integral for $\ln x$, then you cannot set it equal to dv. So, $u = \ln x$ and $du = \frac{1}{x} dx$. Therefore, the dv must be dx; it's the only thing left! If $dv = dx$, then $v = x$.

$$\int \ln x \, dx = x \, \ln \, x - \int x \cdot \frac{1}{x} dx$$

$$= x \, \ln \, x - \int dx$$

$$= x \, \ln \, x - x + C$$

POWERS OF TRIGONOMETRIC FUNCTIONS (BC TOPIC ONLY)

If we've done anything, we've done a lot of trigonometric integration, so here's a little more to throw on the top of the pile. In this section, you learn a few more coping strategies for when all of our other methods fail. These methods, like others we've covered, often help out when u-substitution does not *quite* work out. The first of these is something I call the *Odd Man Out Rule*, and it works for sine and cosine.

Odd Man Out Rule: If an integral contains positive powers of sine and cosine, and only one of the powers is odd, keep one of the odd-powered factors and convert the rest to the other trigonometric expression using the Mamma Theorem. Is there anything your Mamma can't do?

NOTE
This is not universally called the "Odd Man Out Rule," so don't refer to it as such on the test; this rule has no universally accepted name.

 Example 8: Evaluate $\int \sin^2 x \, \cos^3 x \, dx$.

Solution: In this integral, $\cos x$ is the odd man out, since it has the odd power of the two factors. Therefore, we want to leave behind only one $\cos x$ and convert the other cosines to sines (truly making $\cos x$ the odd man out).

$$\int \sin^2 x \, \cos^2 x \, \cos x \, dx$$

The Mamma Theorem tells you that $\cos^2 x = 1 - \sin^2 x$, so replace the $\cos^2 x$ to get

$$\int \sin^2 x \left(1 - \sin^2 x\right) \cos x \, dx.$$

Distribute the $\sin^2 x$ and the $\cos x$ to both terms, and split the integral to get

$$\int \sin^2 x \cos x \, dx - \sin^4 x \cos x \, dx.$$

Now, you can perform u-substitution in each expression with $u = \sin x$ to get

$$\frac{u^3}{3} - \frac{u^5}{5} + C$$

$$\frac{\sin^3 x}{3} - \frac{\sin^5 x}{5} + C.$$

What if you have *only* sines *or* cosines in the problem and not both? In this case, you cannot count on the odd man out to help with u-substitution. If this occurs, you will have to resort back to the power-reducing formulas from way back in Chapter 2. Once applied, they make the problem almost a trivial pursuit.

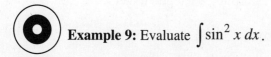 **Example 9:** Evaluate $\int \sin^2 x \, dx$.

Solution: The power-reducing formula for $\sin^2 x$ is $\dfrac{1 - \cos 2x}{2}$, so you can substitute that into the integral and factor out $\frac{1}{2}$.

$$\frac{1}{2} \int (1 - \cos 2x) dx$$

Split the integral up,

$$\frac{1}{2} \int dx - \frac{1}{2} \int \cos 2x \, dx$$

and use u-substitution (with $u = 2x$) to integrate $\cos 2x$.

$$\frac{1}{2} x - \frac{1}{4} \sin 2x + C$$

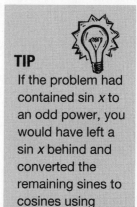

TIP
If the problem had contained sin *x* to an odd power, you would have left a sin *x* behind and converted the remaining sines to cosines using Mamma—the same procedure.

The final rule for trigonometric powers works for secants and tangents in the same integral. I call it the *Steven and Todd* Rule, since it has to do with *secant*s being *even* and *tangent*s being *odd*.

Steven and Todd Rule: If an integral contains positive powers of secant and tangent and the power of secant is even, save a $\sec^2 x$ and convert the remaining secants to tangents using Pappa. If, however, the power of tangent is odd, save a $\sec x \tan x$ and convert the remaining tangents to secants using Pappa.

This might be the opposite of your first instincts. If you are focusing on an even power of secant, you are preparing a $\sec^2 x$ term, which is the derivative of tangent. If you are focusing on the odd power of tangent, then you are preparing a $\sec x \tan x$ term, the derivative of secant. However backward it may seem, it works like a charm.

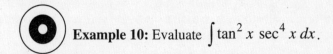 **Example 10:** Evaluate $\int \tan^2 x \, \sec^4 x \, dx$.

Solution: Steven is in this problem, waving to you, eating a hotdog, and waiting until you see him (since secant is even). Therefore, you prepare a $\sec^2 x$ term and convert the remaining $\sec^2 x$ term to $1 + \tan^2 x$ using Pappa.

$$\int \tan^2 x \, \sec^2 x \, \sec^2 x \, dx$$

$$\int \tan^2 x \left(1 + \tan^2 x\right) \sec^2 \, dx$$

Distribute the $\tan^2 x \, \sec^2 x$, and split the integral to get

$$\int \tan^2 x \, \sec^2 x \, dx + \int \tan^4 x \, \sec^2 x \, dx.$$

Use *u*-substitution with $u = \tan x$ to finish.

$$\frac{u^3}{3} + \frac{u^5}{5} + C$$

$$\frac{\tan^3 x}{3} + \frac{\tan^5 x}{5} + C$$

This answer looks hauntingly similar to Example 8, and that's no real coincidence. Odd Man Out and Steven-Todd are techniques that set you up for a simple *u*-substitution and differ only in very minor ways.

There may be instances in which you cannot use any of the techniques outlined here. Do not panic. Try some old-fashioned elbow grease, and experiment until something works. Break the integral into smaller pieces, and bring Mamma, Pappa, and Baby into the picture.

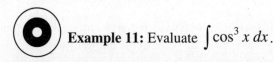 **Example 11:** Evaluate $\int \cos^3 x \, dx$.

Solution: No u-substitution is possible yet. If you set $u = \cos x$, there's no sine to help out on the du. So, let's introduce a sine (or two) into the problem with Mamma.

$$\int \cos^3 x \, dx = \int \cos^2 x \, \cos x \, dx$$

$$\int \left(1 - \sin^2 x\right) \cos x \, dx$$

$$\int \cos x \, dx - \int \sin^2 x \, \cos x \, dx$$

The left integral is easy, and the right integral is a simple u-substitution ($u = \sin x$).

$$\sin x - \frac{\sin^3 x}{3} + C$$

PROBLEM SET

Evaluate the following without a graphing calculator.

1. $\int \sqrt[3]{\cos^2 x} \, \sin^3 x \, dx$

2. $\int \tan^3 x \, \sec^5 x \, dx$

3. $\int \tan x \, \sec^6 x \, dx$

4. $\int \cos^4 x \, dx$

5. $\int \csc^3 x \, \cos^3 x \, dx$

SOLUTIONS

1. You need to use the Odd Man Out Rule for this integral (even though the cosine is to a weird power). Save a sin x, and transform the remaining sin^2x to $(1 - \cos^2 x)$:

$$\int \cos^{2/3} x \left(1 - \cos^2 x\right) \sin x \, dx$$

$$\int \cos^{2/3} x \, \sin x \, dx - \int \cos^{8/3} x \, \sin x \, dx.$$

Both of these integrals require u-substitution with $u = \cos x$ (don't forget that $du = -\sin dx$).

$$-\frac{3}{5} \cos^{5/3} x + \frac{3}{11} \cos^{11/3} x + C$$

2. It's Todd (tangent is odd), so save a sec x tan x term and transform all the tangents to secants using Pappa.

$$\int (\sec x \, \tan x) \tan^2 x \, \sec^4 x \, dx$$

$$\int (\sec x \, \tan x)(\sec^2 x - 1) \sec^4 x \, dx$$

$$\int \sec^6 x \, \sec x \, \tan x \, dx - \int \sec^4 x \, \sec x \, \tan x \, dx$$

Now, use u-substitution with $u = \sec x$ to finish.

$$\frac{\sec^7 x}{7} - \frac{\sec^5 x}{5} + C$$

3. Both Steven and Todd are in this problem, so you can do the problem either way. However, Todd provides the easier way (since saving a sec x tan x term leaves no other tangents to transform to secants). Bring it home, Todd.

$$\int \sec^5 x \, \sec x \, \tan x \, dx$$

Use u-substitution with $u = \sec x$ (just as in number 2).

$$\frac{\sec^6 x}{6} + C$$

4. There is no sine in this problem to help with u-substitution. So, do we use the technique of Example 9 or Example 11? Because the power is even, we'll use the technique of Example 9, where the power also was even.

$$\int \cos^4 x \, dx = \int \left(\cos^2 x\right)^2 dx$$

Now, use the power-reducing formula for cos^2x.

$$\int \frac{(1 + \cos 2x)^2}{2^2} \, dx$$

$$\frac{1}{4}\int(1+\cos 2x)^2\,dx$$

$$\frac{1}{4}\int\left(1+2\cos 2x+\cos^2 2x\right)dx$$

The first two integrals are easy, but you have to use *another* power-reducing formula for $\cos^2 2x$. Let's focus on that for a moment. (Don't forget the $\frac{1}{4}$ that needs to be distributed to each integral—easy and hard alike.)

$$\frac{1}{4}\int\cos^2 2x\,dx$$

$$\frac{1}{4}\int\frac{1+\cos 2(2x)}{2}\,dx$$

$$\frac{1}{8}\int(1+\cos 4x)\,dx$$

All together (don't forget the two easy integration terms above), the answer is

$$\frac{1}{4}x+\frac{1}{4}\sin 2x+\frac{1}{8}x+\frac{1}{32}\sin 4x+C$$

$$\frac{3}{8}x+\frac{1}{4}\sin 2x+\frac{1}{32}\sin 4x+C.$$

5. Holy smokes, this doesn't match a single one of the techniques we've covered! To begin, rewrite as sine and cosine to see if things get any easier.

$$\int\frac{\cos^3 x}{\sin^3 x}\,dx$$

$$\int\cot^3 x\,dx$$

That looks a little more compact, if nothing else. This, however, looks a lot like Example 11; in fact, it differs by only a single letter. Try that technique on a whim, and see what pans out.

$$\int\cot x\,\cot^2 x\,dx$$

$$\int\cot x\left(\csc^2 x-1\right)dx$$

$$\int\cot x\,\csc^2 x\,dx-\int\cot x\,dx$$

The left integral is a *u*-substitution, and you should have the right integral memorized.

$$-\frac{\cot^2 x}{2}-\ln\left|\sin x\right|+C$$

PARTIAL FRACTIONS (BC TOPIC ONLY)

Integration by partial fractions is a method used to simplify integrals based on a very cool trick. The trick is so unique that it tends to stick with you, so you shouldn't have any trouble remembering how it's done come test time. That's good news, especially since the test is such a high-pressure situation that most people forget important things, such as what their name is, when the Magna Carta was signed, what the current exchange rate is between major world currencies, and what exactly that little symbol is that became The Artist Formerly Known as Prince's new name.

Partial fraction decomposition allows you to break a rational (fractional) expression into the sum of a couple of smaller fractions. Here's the great thing: The denominators of the smaller fractions are the factors of the original denominator. The numerators of those smaller fractions are just constants. It's your job to find out what they are exactly.

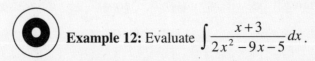 **Example 12:** Evaluate $\int \dfrac{x+3}{2x^2-9x-5}\,dx$.

Solution: You may be tempted to try a u-substitution with $u = 2x^2 - 9x - 5$ and to try and force the numerator into $du = (4x - 9)dx$ (as we did earlier in this chapter). However, you should use partial fractions *because the denominator is factorable*.

$$\int \frac{x+3}{(2x+1)(x-5)}\,dx$$

Partial fraction decomposition tells us that

$$\frac{x+3}{(2x+1)(x-5)} = \frac{A}{2x+1} + \frac{B}{x-5}.$$

NOTE

When you integrate by partial fractions on the AP test, the denominators will always have linear factors. This technique is slightly modified when the factors have higher degrees, but you don't have to worry about that for the AP test.

In other words, we can find two constants A and B such that the sum of the two right fractions equals the larger fraction on the left. How do you do that? First, eliminate the fractions by multiplying both sides of the above equation by $(2x + 1)(x - 5)$.

$$x + 3 = A(x - 5) + B(2x + 1)$$

Now, distribute the constants.

$$x + 3 = Ax - 5A + 2Bx + B$$

You're almost finished; factor the x out of the Ax and $2Bx$ terms.

$$x + 3 = (A + 2B)x - 5A + B$$

Stop and look at that for a second. If the two sides are equal, then $A + 2B$ (the coefficient of the x on the right side) must be equal to 1 (the coefficient of the x on the left side). Similarly, $-5A + B$ must equal 3. Therefore, you get the system of equations:

$$A + 2B = 1$$

$$-5A + B = 3.$$

Use whatever technique you want to simultaneously solve these equations (linear combination, substitution, matrices, etc.) to figure out that $A = -\frac{5}{11}$ and $B = \frac{8}{11}$. Therefore,

$$\frac{x+3}{(2x+1)(x-5)} = -\frac{\frac{5}{11}}{2x+1} + \frac{\frac{8}{11}}{x-5},$$

and instead of integrating the ugly left side, we can integrate the slightly less ugly right side.

$$-\frac{5}{11} \int \frac{1}{2x+1}\, dx + \frac{8}{11} \int \frac{1}{x-5}\, dx$$

Use u-substitution in each integral with $u =$ the denominator of each and the result is

$$-\frac{5}{22} \ln|2x+1| + \frac{8}{11} \ln|x-5| + C;$$

or if you feel like getting common denominators and going nuts with log properties, you can rewrite as

$$\ln \left(\frac{\dfrac{x-5^{16}}{|2x+1|^5}}{22} \right) + C$$

I have no idea why you would ever want to do that, but hey, whatever floats your boat.

In conclusion, you should integrate by partial fractions if you can factor the denominator. Create a sum of new fractions such that the denominators of the new fractions are the factors of the original denominator and the numerators of the new fractions are constants. Once you determine what those constants must be, all that remains is to integrate the string of smaller fractions, which is typically very easy.

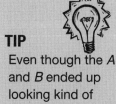

TIP
Even though the A and B ended up looking kind of gross, integrating was still quite easy. That's why integrating by partial fractions is so great.

PROBLEM SET

Do not use a calculator when you integrate the following.

1. $\displaystyle\int \frac{x-3}{x^2+9x+20}\,dx$

2. $\displaystyle\int \frac{x+2}{x^2-2x-8}\,dx$

3. $\displaystyle\int \frac{4}{3x^2+x-14}\,dx$

4. $\displaystyle\int \frac{2x+1}{x^3+x^2-6x}\,dx$

SOLUTIONS

1. The denominator factors to $(x+4)(x+5)$; to begin partial fractions, you set up the following equation:

$$\frac{x-3}{x^2+9x+20}=\frac{A}{x+4}+\frac{B}{x+5}.$$

Multiply through by $(x+4)(x+5)$ to eliminate fractions, and then find A and B.

$$x-3 = A(x+5)+B(x+4)$$

$$x-3 = (A+B)x+5A+4B$$

This results in the system of equations:

$$A+B=1 \text{ and } 5A+4B=-3$$

$$A=-7,\ B=8.$$

The original integral now becomes

$$-7\int \frac{1}{x+4}\,dx+8\int \frac{1}{x+5}\,dx$$
$$-7\ln|x+4|+8\ln|x+5|+C \cdot$$

2. This problem does not require partial fractions. If you factor the denominator, the fraction simplifies.

$$\int \frac{x+2}{(x-4)(x+2)}\,dx$$

$$\int \frac{1}{x-4}\,dx$$

This is an easy *u*-substitution problem if $u=x-4$.

$$\ln|x-4|+C$$

3. You can pull the 4 out of the integral if you want (and replace it with a 1), but that doesn't make the problem significantly easier. Because there is no x in the numerator, its coefficient must be 0 (that'll be important in a few seconds). Factoring the denominator may be the hardest part of this problem.

$$\frac{4}{3x^2 + x - 14} = \frac{A}{3x+7} + \frac{B}{x-2}$$
$$4 = A(x-2) + B(3x+7)$$
$$4 = (A+3B)x + (-2A+7B)$$

This leads to the system $A + 3B = 0$ and $-2A + 7B = 4$, whose solution is $A = -\frac{12}{13}$ and $B = \frac{4}{13}$.

$$-\frac{12}{13}\int \frac{1}{3x+7}\,dx + \frac{4}{13}\int \frac{1}{x-2}\,dx$$

$$-\frac{4}{13}\ln|3x+7| + \frac{4}{13}\ln|x-2| + C$$

4. The denominator has three factors this time, but the technique stays exactly the same.

$$\frac{2x+1}{x^3 + x^2 - 6x} = \frac{A}{x} + \frac{B}{x+3} + \frac{C}{x-2}$$
$$2x+1 = A(x+3)(x-2) + Bx(x-2) + Cx(x+3)$$
$$2x+1 = A(x^2 + x - 6) + B(x^2 - 2x) + C(x^2 + 3x)$$
$$2x+1 = (A+B+C)x^2 + (A-2B+3C)x - 6A$$

The only constant term on the right is $-6A$, so $-6A = 1$ and $A = -\frac{1}{6}$. There is no squared term on the left, so (after substituting in the value of A) you have the system

$$-\frac{1}{6} + B + C = 0 \quad \text{and} \quad -\frac{1}{6} - 2B + 3C = 2$$

$$B + C = \frac{1}{6} \quad \text{and} \quad -2B + 3C = \frac{13}{6}.$$

After solving the system, you get $A = -\frac{1}{6}$, $B = -\frac{1}{3}$, and $C = \frac{1}{2}$, making the integral

$$-\frac{1}{6}\int \frac{1}{x}\,dx - \frac{1}{3}\int \frac{1}{x+3} + \frac{1}{2}\int \frac{1}{x-2}$$

$$-\frac{1}{6}\ln|x| - \frac{1}{3}\ln|x+3| + \frac{1}{2}\ln|x-2| + C.$$

CAUTION
Don't forget to use *u*-substitution when integrating in the final step — even if you just do it mentally.

$$\int \frac{1}{3x+7}\,dx \neq$$

$$\ln|3x+7| + C;$$

you have to multiply by $\frac{1}{3}$.

IMPROPER INTEGRALS (BC TOPIC ONLY)

Improper integrals are bizarre integrals that have at least one of the three following qualities: (1) one of the limits of integration is infinity, (2) the curve being integrated has an infinite discontinuity between the limits of integration, or (3) the integral has an obsession for fruit-scented candles. As the final condition is often difficult to measure, most mathematicians are satisfied with the first two.

Consider the integral $\int_0^3 \frac{1}{\sqrt{9-x^2}}$. The upper limit of integration causes the expression to be undefined. How can you find the area beneath a curve when $x = 3$ if the curve doesn't exist at 3? This is an interesting question. To counter the dilemma, we will evaluate the *limit as x approaches 3*. To be completely mathematically accurate, we will evaluate the limit as x approaches 3 *from the left* (since those left-hand values are in our interval of [0,3] but the right-hand values are not).

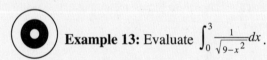 **Example 13:** Evaluate $\int_0^3 \frac{1}{\sqrt{9-x^2}} dx$.

Solution: Since the integration limit of 3 causes trouble, you substitute a constant (we will use b since it is the upper limit) for it and rewrite the integral as follows:

$$\lim_{b \to 3^-} \int_0^b \frac{1}{\sqrt{9-x^2}}.$$

For now, ignore the limit and evaluate the integral.

$$\lim_{b \to 3^-} \left(\arcsin \frac{x}{3} \Big|_0^b \right)$$

$$\lim_{b \to 3^-} \left(\arcsin \frac{b}{3} - \arcsin \frac{0}{3} \right)$$

Since $\arcsin 0 = 0$, the integral is simply $\arcsin \frac{b}{3}$. It's time to bring that limit back into the picture (to evaluate it, you just substitute 3 in for b).

$$\lim_{b \to 3^-} \arcsin \frac{b}{3} = \arcsin \frac{3}{3}$$

$$\arcsin 1 = \frac{\pi}{2}$$

NOTE

If the other limit, 0, had been the trouble, we would have examined the limit as *x* approached 0 from the right.

As you can see, the method was quite easy. The problem limit is replaced by a constant, and you let that constant approach the problem limit. The next example involves another tricky integration boundary, but the resulting limit is a little more complicated.

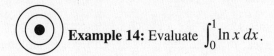 **Example 14:** Evaluate $\int_0^1 \ln x \, dx$.

Solution: This is an improper integral because $\ln x$ has an infinite discontinuity when $x = 0$. Therefore, we replace the 0 boundary with a constant as follows:

$$\lim_{a \to 0^+} \left(\int_a^1 \ln x \, dx \right).$$

We know the integral of $\ln x$ from problem 5 of the integration by parts section. $\int \ln x \, dx = x \ln x - x + C$.

$$\lim_{a \to 0^+} \left(x \ln x - x \right) \Big|_a^1$$
$$\lim_{a \to 0^+} \left((1 \cdot 0 - 1) - (a \ln a - a) \right)$$
$$\lim_{a \to 0^+} \left(-1 - a \ln a + a \right)$$

NOTE
The fact that we were finding the limit *from the left* really didn't affect the integral at all.

Now it's time to let a approach 0. The only difficult part of the above is $\lim_{a \to 0^+} a \ln a$. Did you notice that this is the indeterminate form of $0 \cdot \infty$? You can rewrite the limit as

$$\lim_{a \to 0^+} \frac{\ln a}{\frac{1}{a}}$$

and apply L'Hôpital's Rule. This is the only difficult step in the problem, but it's a doozy. If you find the derivative of the numerator and denominator, you get

$$\lim_{a \to 0^+} -\frac{\frac{1}{a}}{\frac{1}{a^2}}$$
$$\lim_{a \to 0^+} -a,$$

which is definitely 0. That eliminates the hardest part of the expression.

$$\lim_{a \to 0^+} (-1 - a \ln \ a + a)$$

$$\lim_{a \to 0^+} (-1 - 0 - 0) = -1$$

The final answer is –1. If this seemed like a bunch of smoke and mirrors rather than a real magic trick, evaluate the integral with your graphing calculator to see that we're right—remember that the calculator only gives you an approximation!

We have yet to discuss the other major type of improper integral—an integral that actually has ∞ as one of the limits of integration. You'll be glad to hear that the process used to solve those integrals is identical to the process we have used in the previous two examples.

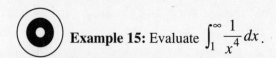 **Example 15:** Evaluate $\int_1^\infty \frac{1}{x^4} dx$.

Solution: You may think that the answer is automatically ∞. We are integrating from $x = 1$ all the way to $x = \infty$, for goodness sakes! Can a shape with an infinite boundary have a finite area? This one will. (I hope I didn't ruin the suspense. If you think I did, the answer is $\frac{1}{3}$—so there, I ruined even more of the suspense.)

The first step to solving this problem is to introduce a constant in place of the troublesome boundary.

$$\lim_{b \to \infty} \left(\int_1^b x^{-4} dx \right)$$

Integrate as usual and then apply the limit.

$$\lim_{b \to \infty} \left(-\frac{1}{3x^3} \Big|_1^b \right)$$

$$\lim_{b \to \infty} \left(-\frac{1}{3b^3} + \frac{1}{3} \right)$$

As b goes to infinity, $3b^3$ will become mega-gigantic, which makes $\frac{1}{3b^3}$ really tiny, or essentially 0.

$$0 + \frac{1}{3}$$

CAUTION

Not all shapes with infinite boundaries have finite area. If the area is not finite, the integral is said to *diverge*, whereas a finite integral is said to *converge*.

Therefore, the definite integral is equal to $\frac{1}{3}$, and I'm sorry I ruined the surprise.

Example 15 highlights a very important characteristic of improper integrals. Any improper integral of the form

$$\int_0^\infty \frac{1}{x^p}\, dx$$

(where the exponent p is a real number) will always converge (result in a finite or numerical answer) if $p > 1$. If, however, $p \leq 1$, the integral will not have a numerical answer (the corresponding area is infinite), and the integral is termed *divergent*. Therefore, you can tell without any work that $\int_0^\infty \frac{1}{x^2}\, dx$ converges (although you'll have to work it out to determine the actual value), whereas $\int_0^\infty \frac{1}{\sqrt{x}}\, dx$ will diverge. This is sometimes called the *p-test*.

The final thing the AP test expects you to do with improper integrals is to compare them with other improper integrals. In such cases, the test will not ask you to evaluate an integral but rather ask you to determine its convergence or divergence. Rather than spending the time to integrate these, it's best to find an integral that is extremely similar to those that are much easier to work with.

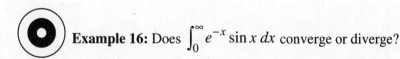 **Example 16:** Does $\int_0^\infty e^{-x} \sin x\, dx$ converge or diverge?

Solution: Compare this to $\int_0^\infty e^{-x}\, dx$ (which is similar but much easier to integrate). Because the range of $\sin x$ is $[-1,1]$, multiplying e^{-x} by $\sin x$ will *never* give you a result larger than e^{-x}. The most you are multiplying e^{-x} by is 1, and that will return the same value (of course). Otherwise, you are multiplying e^{-x} by a number smaller than one, and the result will be less than e^{-x}. Therefore, we can unequivocally say that

$$\int_0^\infty e^{-x} \sin x\, dx \leq \int_0^\infty e^{-x}\, dx.$$

Why is this important? We can show that $\int_0^\infty e^{-x}$ is a finite area, so if the original area is less than a finite area, it must also be a finite area and, therefore, converge.

All that remains is to actually show that $\int_0^\infty e^{-x}$ is finite through improper integration.

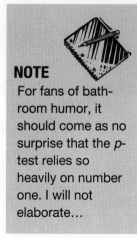

NOTE
For fans of bathroom humor, it should come as no surprise that the *p*-test relies so heavily on number one. I will not elaborate…

TIP

Finding the right integral to compare with is a skill that must be practiced. You will receive much more practice in Chapter 11, Sequences and Series. The comparison of improper integrals is just a taste of what's to come.

$$\lim_{b\to\infty}\int_0^b e^{-x}dx$$

$$\lim_{b\to\infty}\left(-e^{-x}\Big|_0^b\right)$$

$$\lim_{b\to\infty}\left(-e^{-b}+e^0\right)$$

$-e^{-\infty}$ is essentially 0 (to visualize this, graph $-e^x$ and let x approach $-\infty$), but e^0 equals 1, so the limit is

$$0+1=1.$$

Because $\int_0^\infty e^{-x}dx$ converges, then $\int_0^\infty e^{-x}\sin x\,dx$ (being smaller) must converge.

PROBLEM SET

Use a graphing calculator to check your work on these problems only.

For problems 1 through 4, determine whether or not the improper integrals converge; if they do, evaluate them.

1. $\displaystyle\int_{-1}^0 \frac{1}{\sqrt[3]{x+1}}dx$

2. $\displaystyle\int_{-2}^0 \frac{1}{x^2+6x+8}dx$

3. $\displaystyle\int_{-2}^3 \frac{1}{x+1}dx$

4. $\displaystyle\int_0^\infty \frac{dx}{x^{2/3}}$

5. Determine whether or not $\displaystyle\int_1^\infty \frac{dx}{e^x+6}$ converges by comparing it to another, simpler improper integral.

SOLUTIONS

1. The lower limit of integration, –1, is the troublemaker here. Replace it with a constant and evaluate the limit:

$$\lim_{a\to-1^+}\int_a^0 \frac{1}{\sqrt[3]{x+1}}dx.$$

To integrate, use *u*-substitution with $u = (x + 1)$:

$$\lim_{a \to -1^+} \int_{a+1}^{1} u^{-1/3} du$$

$$\lim_{a \to -1^+} \left(\frac{3}{2} u^{2/3} \Big|_{a+1}^{1} \right)$$

Don't forget the *u* boundaries.

$$\lim_{a \to -1^+} \left(\frac{3}{2} - \frac{3}{2}(a+1)^{2/3} \right).$$

When you evaluate the limit, you essentially plug –1 in for *a* to get

$$\frac{3}{2} - 0 = \frac{3}{2}.$$

2. If you factor the denominator into $(x+4)(x+2)$, you can see that the integration limit of –2 will cause the fraction to be undefined, making the integral improper.

$$\lim_{a \to -2^+} \int_{a}^{0} \frac{1}{(x+4)(x+2)} dx$$

You'll have to use partial fractions to integrate.

$$\frac{1}{(x+4)(x+2)} = \frac{A}{x+4} + \frac{B}{x+2}$$
$$A(x+2) + B(x+4) = 1$$
$$x(A+B) + 2A + 4B = 1$$

This creates the system $A + B = 0$ and $2A + 4B = 1$. The solution to the system is $A = -\frac{1}{2}$ and $B = \frac{1}{2}$. Substitute these into the expanded integral and solve to get

$$\lim_{a \to -2^+} \left(-\frac{1}{2} \ln|x+2| + \frac{1}{2} \ln|x+4| + C \right) \Big|_{a}^{0}$$

There is still a problem, however. When you substitute –2 into $\ln|x+2|$, the expression is still undefined. Our method could not correct the problems inherent in the integral, so the integral is divergent.

3. The infinite discontinuity in this integral does not occur at the endpoints but rather when $x = -1$. However, you need the discontinuity to exist at an endpoint to use the method you've practiced and know so well. Therefore, you can rewrite the integral (using a property of definite integrals) as a sum of smaller integrals, each with –1 as a limit of integration:

$$\int_{-2}^{3} \frac{1}{x+1}\,dx = \int_{-2}^{-1} \frac{1}{x+1}\,dx + \int_{-1}^{3} \frac{1}{x+1}\,dx.$$

Each of these integrals needs to be done separately.

$$\int_{-2}^{-1} \frac{1}{x+1}\,dx$$

$$\lim_{b\to-1^-} \int_{-2}^{b} \frac{1}{x+1}\,dx$$

$$\lim_{b\to-1^-} \ln|x+1|\,\Big|_{-2}^{b}$$

As you continue to solve this, you will end up with $\ln(-1+1) = \ln 0$, which is undefined. Therefore, this integral is divergent. (The same problem will occur as you integrate the second integral.)

4. Replace that infinite boundary with a constant and integrate.

$$\lim_{b\to\infty} \int_{0}^{b} x^{-2/3}\,dx$$

$$\lim_{b\to\infty} \left(3x^{1/3}\,\Big|_{0}^{b} \right)$$

$$\lim_{b\to\infty} \left(3b^{1/3} - 0 \right)$$

TIP

You could also have used the *p*-test to do number 4. Since $\frac{2}{3} \le 1$, you know that the integral will diverge.

The function $3b^{1/3}$ (three times the cube root function) will grow infinitely large as $b\to\infty$, so this integral diverges.

5. Compare this integral to $\int_{1}^{\infty} \frac{dx}{e^x}$. Because $e^x + 6 > e^x$, $\frac{1}{e^x+6} < \frac{1}{e^x}$. Therefore, if you can prove $\int_{1}^{\infty} \frac{dx}{e^x}$ has a finite area (and is thus convergent), our original integral must also be convergent.

$$\lim_{b\to\infty} \int_{1}^{b} e^{-x}\,dx$$

$$\lim_{b\to\infty} \left(-e^{-x} \right)\Big|_{1}^{b}$$

$$\lim_{b\to\infty} -e^{-b} + e^{-1}$$

You know from Example 16 that $\lim_{b\to\infty} -e^{-b} = 0$:

$$\lim_{b\to\infty} e^{-1} = \frac{1}{e}.$$

Therefore, the original integral $\int_{1}^{\infty} \frac{dx}{e^x+6}$ must also converge.

TECHNOLOGY: DRAWING DERIVATIVE AND INTEGRAL GRAPHS WITH YOUR CALCULATOR

In past chapters, you were asked to describe characteristics of functions based on the graphs of their first and second derivatives. Your calculator (although slow) is able to draw a function's derivative or integral graph, even if you cannot figure out what it is. This is not an extraordinarily useful tool for the AP test, but it gives you an infinite amount of practice drawing the graphs of derivatives and integrals (and solving each one for you).

 Example 17: Draw the graphs of $\int \cos x \, dx$ and $\frac{d}{dx}(\cos x)$.

Solution: You already know what the integral of $\cos x$ is—it's $\sin x$. However, the process you use with any function will be the same. Before we begin, let's set up a good [Window] for this problem. Press [Zoom]→"Ztrig", and then adjust the [Window] so that "Xmin" is 0 and "Xmax" is 4π. This will set up a very pretty window for $\cos x$. Now, go to the [Y=] screen and assign $Y_1 = \cos(x)$. We want Y_2 to be the graph of the integral, so its equation is

$$Y_2 = \text{fnInt}(Y_1, x, 0, x).$$

This is the area beneath the Y_1 curve with respect to x from $x = 0$ to $x =$ whatever the current x is. In other problems, you may want to adjust the lower limit of integration (in this example 0) to match the "Xmin" value on your graph.

It's also a good idea to darken the graph of the integral so you can tell them apart. Press [Graph], wait a bit, and the graph of the integral slowly scrawls itself across the axes. No big surprise here; it's the graph of $\sin x$, the integral of $\cos x$.

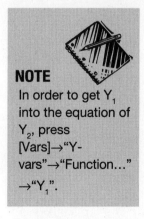

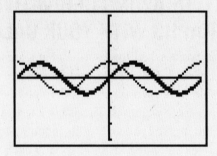

If you want to draw the derivative of cos x, enter

$$Y_2 = nDeriv(Y_1,x,x).$$

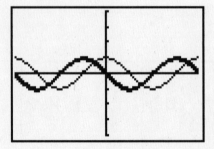

Just as we suspected, the graph of the derivative of cos x is the horizontal reflection of sin x, or $-\sin x$.

If you are still confused about the relationship between a graph and its derivatives, this technique can give you all the practice you need. Make up any function at all, and enter it as Y_1. Predict, based on that graph, what its integral and derivative graphs look like, and use the above commands to check yourself.

Example 18: Show that ln x gets its value by accumulating area beneath $\frac{1}{x}$.

Solution: Enter $Y_1 = \dfrac{1}{x}$. Remember that $\ln x$ gets its value from the accumulation function

$$\ln x = \int_1^x \left(\frac{1}{x}\right) dx.$$

So, $Y_2 = \text{fnInt}(Y_1, x, 1, x)$. Make sure to set your Xmin to a number slightly larger than 0 (since 0 is undefined for $\dfrac{1}{x}$. I chose a value of .001 for Xmin.

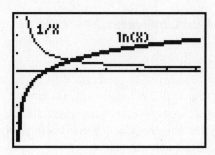

This activity sheds a whole new light on why $\ln x$ is negative from $x = 0$ to $x = 1$: $\ln x$ is based on an integral with inverted limits of integration on that interval. For example,

$$\ln \frac{1}{2} = \int_1^{1/2} \frac{1}{x} dx.$$

Because those limits are "backward" (with the larger limit in the lower position), the result is negative (based on a property of definite integrals).

PRACTICE PROBLEMS

Do not use a graphing calculator on a single, solitary one of these problems.

Evaluate each of the following integrals, if possible.

1. $\displaystyle\int \frac{1}{1+\cos x}\,dx$

2.* $\displaystyle\int \frac{x^2-1}{x^3-7x^2+6x}\,dx$

3.* $\displaystyle\int x\,\ln x\,dx$

4.* $\displaystyle\int_2^6 \frac{\sqrt{6+x}}{\sqrt{6-x}}\,dx$

5. $\displaystyle\int \sec^2 x\,\tan^5 x\,dx$

* *a BC-only question.*

6. $\int \dfrac{x\,dx}{x+3}$

7. $\int \dfrac{x^3+6x^2+2x-4}{x+7}dx$

8.* $\int \cos^4(3x)\,\sin^3(3x)\,dx$

9.* $\int_{1}^{\infty} x\,\sin x\,dx$

10.* **James' Diabolical Challenge:** $\int x^4 \cos 3x\,dx$

SOLUTIONS

1. The best way to integrate this is using a method called the *conjugate*. It hasn't been included until now because it is much easier than the other methods. To apply it, multiply the numerator and denominator by the conjugate of the denominator $(1 - \cos x)$; you will get

$$\int \frac{1-\cos x}{1-\cos^2 x}dx.$$

Use Mamma to change the denominator and split the integral into two parts.

$$\int \frac{1}{\sin^2 x}dx - \int \frac{\cos x}{\sin^2 x}dx$$
$$\int \csc^2 x\,dx - \int \frac{\cos x}{\sin^2 x}dx$$

The first integral is easy, and the second is a simple *u*-substitution (with $u = \sin x$).

$$-\cot x - \int u^{-2}du + C$$
$$-\cot x + \frac{1}{u} + C$$
$$-\cot x + \csc x + C$$

2. This is not a long-division problem because the degree of the denominator is greater than the degree of the numerator. Instead, because the denominator is factorable, you should use partial fractions.

$$\frac{x^2-1}{x(x-1)(x-6)} = \frac{A}{x} + \frac{B}{x-1} + \frac{C}{x-6}$$
$$x^2-1 = A\left(x^2-7x+6\right) + Bx(x-6) + Cx(x-1)$$
$$x^2-1 = x^2\left(A+B+C\right) + x\left(-7A-6B-C\right) + 6A$$

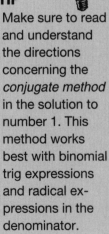

---

* *a BC-only question.*

Therefore, $A = -\frac{1}{6}$, $B = 0$, and $C = \frac{7}{6}$. The integral is now written as

$$-\frac{1}{6}\int \frac{1}{x}dx + \frac{7}{6}\int \frac{1}{x-6}dx$$

$$-\frac{1}{6}\ln|x| + \frac{7}{6}\ln|x-6| + C.$$

3. Integration by parts is the way to go here. It's best to set $u = \ln x$; although you probably have memorized its integral, you'd have to integrate twice if you set $dv = \ln x$, so that rules out the chart. Instead, set $u = \ln x$ and $dv = xdx$, which makes $du = \frac{1}{x}dx$ and $v = \frac{x^2}{2}$.

$$\int x \ln x \, dx = \frac{x^2 \ln x}{2} - \frac{1}{2}\int xdx$$

$$\int x \ln xdx = \frac{x^2 \ln x}{2} - \frac{x^2}{4} + C$$

4. (This is a toughy.) This is an improper integral since the boundary of 6 wreaks havoc in the denominator.

$$\lim_{b\to 6^-}\int_2^b \frac{\sqrt{6+x}}{\sqrt{6-x}}dx$$

The best method to use is the conjugate method (since the denominator is a radical surrounding a binomial). Multiply the top and bottom of the fraction by $\sqrt{6+x}$ to get

$$\lim_{b\to 6^-}\int_2^b \frac{6+x}{\sqrt{36-x^2}}dx.$$

This integral must be broken into two parts. We'll solve them one at a time. The first is

$$\lim_{b\to 6^-}\left(6\int_2^b \frac{1}{\sqrt{36-x^2}}dx\right)$$

$$\lim_{b\to 6^-}\left(6\cdot\arcsin\frac{x}{6}\Big|_2^b\right)$$

$$6\arcsin 1 - 6\arcsin\frac{1}{3}$$

$$6\cdot\frac{\pi}{2} - 6\arcsin\frac{1}{3}$$

$$3\pi - 6\arcsin\frac{1}{3}.$$

The second integral is completed using u-substitution.

$$\lim_{b \to 6^-} \left(\int_2^b \frac{x}{\sqrt{36 - x^2}} dx \right)$$

$$u = 36 - x^2, \ du = -2x dx, \ -\frac{du}{2} = x dx$$

$$\lim_{b \to 6^-} \left(-\frac{1}{2} \int_{32}^{36 - b^2} u^{-1/2} du \right)$$

$$\lim_{b \to 6^-} \left(-u^{1/2} \Big|_{32}^{36 - b^2} \right)$$

$$\lim_{b \to 6^-} \left(-\sqrt{36 - b^2} + \sqrt{32} \right)$$

$$\sqrt{32}$$

The final answer is the sum of the two integrals:

$$3\pi - 6 \arcsin \frac{1}{3} + \sqrt{32}.$$

5. This is a simple u-substitution problem; BC students can also use Steven or Todd (although Steven, in this case, is essentially u-substitution).

$$u = \tan x, \ du = \sec^2 x dx$$

$$\int u^5 du$$

$$\frac{\tan^6 x}{6} + C$$

6. There are two good ways to do this. If you use long division (since the degrees of top and bottom are equal), you get

$$\int \left(1 - \frac{3}{x + 3} \right) dx$$

$$x - 3 \ln |x + 3| + C,$$

which is quite easy. You can also do a bizarre u-substitution. If you set $u = x + 3$, $du = dx$ and the integral becomes

$$\int \frac{x \, du}{u}.$$

How do you get rid of that pesky x in the numerator? Remember that you just set $u = x + 3$. Therefore, $x = u - 3$. If you substitute that value in for x, you get

$$\int \frac{u - 3}{u} du,$$

which can be split up, resulting in

$$\int du - 3\int \frac{1}{u} du$$
$$u - 3\ln|u| + C$$
$$x + 3 - 3\ln|x+3| + C, \text{ or}$$
$$x - \ln\left|(x+3)^3\right| + C$$

(since the 3 is just a constant and can be added to the C). Both methods result in the same answer.

7. You need to begin this problem with long or synthetic division (synthetic division is shown below).

$$\begin{array}{r} -7|\ \ 1\ \ \ 6\ \ \ 2\ \ -4 \\ -7\ \ \ 7\ -63 \\ \hline 1\ -1\ \ \ 9\ -67 \end{array}$$

$$(x^3 + 6x^2 + 2x - 4) \div (x + 7) = x^2 - x + 9 - \frac{67}{x+7}$$

The integral can be rewritten as

$$\int\left(x^2 - x + 9 - \frac{67}{x+7} dx\right)$$
$$\frac{x^3}{3} - \frac{x^2}{2} + 9x - 67\ln|x+7| + C.$$

8. The odd man out in this problem is sin $(3x)$. Therefore, save one sin $(3x)$, and change the others to cosines.

$$\int \cos^4(3x)\ \sin^2(3x)\ \sin(3x)\ dx$$
$$\int \cos^4(3x)\ \left(1 - \cos^2(3x)\right)\ \sin(3x)\ dx$$
$$\int \cos^4(3x)\ \sin(3x)\ dx - \int \cos^6(3x)\ \sin(3x)\ dx$$

Both integrals require u-substitution with $u = \cos(3x)$ and $-\frac{du}{3} = \sin(3x)dx$.

$$-\frac{1}{3}\left(\frac{\cos^5(3x)}{5} - \frac{\cos^7(3x)}{7}\right) + C$$

9. This is an improper integral due to the infinite upper limit.

$$\lim_{b\to\infty}\int_1^b x\ \sin x\ dx$$

You should integrate $x \sin x$ by parts with $u = x$ and $dv = \sin x \, dx$. In fact, you can use the parts chart.

u	dv	+/−1
x	$\sin x$	+1
1	$-\cos x$	−1
0	$-\sin x$	+1
		−1

$$\lim_{b \to \infty} \left(-x \cos x + \sin x \right) \Big|_1^b$$

$$\lim_{b \to \infty} -b \, \cos b + \sin b - (-\cos 1 + \sin 1)$$

As b approaches infinity, neither $\cos b$ nor $\sin b$ approach any one height. They both oscillate infinitely between −1 and 1. Therefore, there is no limit as b approaches infinity, and this integral diverges.

10. This is one mother of an integration by parts problem. If you don't use the chart, you are just masochistic.

u	dv	+/−1
x^4	$\cos 3x$	+1
$4x^3$	$\frac{1}{3} \sin 3x$	−1
$12x^2$	$-\frac{1}{9} \cos 3x$	+1
$24x$	$-\frac{1}{27} \sin 3x$	−1
24	$\frac{1}{81} \cos 3x$	+1
0	$-\frac{1}{243} \sin 3x$	−1
		+1

Multiply along the diabolical diagonals (and simplify fractions, if you feel the urge) to get the final answer:

$$\frac{1}{3} x^4 \sin 3x + \frac{4}{9} x^3 \cos 3x - \frac{4}{9} x^2 \sin 3x - \frac{8}{27} x \cos 3x - \frac{8}{81} \sin 3x + C.$$

Applications of Integration

Learning how to integrate, although a fun adventure in and of itself (yeah right), is just the beginning. Just as derivatives had keen applications such as related rates and optimization, integrating has its own applications, which are primarily concerned with finding area and volume generated by graphs. These topics always remind me of one thing—getting close to the AP test! Although BC students have an extra chapter (and what a fun chapter that is), AB students are nearing the home stretch for the AP test. Don't give up, and keep your focus—you're almost done.

HANDS-ON ACTIVITY 9.1: AREA BETWEEN CURVES

You can already find the area between a curve and the *x*-axis (thanks to the Fundamental Theorem). Now, you will learn how to find the area between any two integrable curves. I can tell that you're excited. Let's get right to it.

1. If a circle is inscribed in a square whose side length is 9, find the area of the shaded region.

_____

_____

2. Describe the technique you used to complete number 1.

_____

_____

3. Below are the graphs of $y = 4$ and $y = x^2$ in the first quadrant. Where do these two graphs intersect?

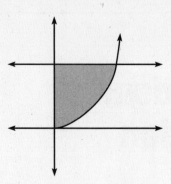

4. Evaluate the definite integrals $\int_0^2 4\,dx$ and $\int_0^2 x^2\,dx$. Using those two integrals, how can you find the shaded area? (Use a technique similar to the circle and square problem from number 1.)

5. Fill in the blanks below to complete the statement:

When finding the area enclosed by two curves that contain x variables, you are adding the sum of the areas of an infinite number of vertical _____s. In practice, you subtract the definite integral of the _____ *from* the definite integral of the _____. The limits of integration for those integrals should be the x-values of the _____.

6. Graph the curves $y = 2 - x^2$ and $y = x^2 - 3$. Lightly shade the total area between the curves (not just in the first quadrant this time). Set up the integral that represents the area, but do not integrate it.

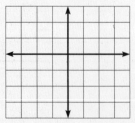

7. If your functions both contain y's, how will your technique for finding the area between them differ?

_____

_____

8. Fill in the blanks below to complete the statement:

When finding the area enclosed by two curves that contain y variables, you are adding the sum of the areas of an infinite number of _____. In practice, you subtract the definite integral of the _____ *from* the definite integral of the _____. The limits of integration for those integrals should be the _____.

9. Find the area *in the first quadrant* enclosed by the curves $x = y^2$ and $x = \sqrt{y}$.

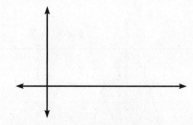

_____

_____

SELECTED SOLUTIONS TO HANDS-ON ACTIVITY 9.1

1. The radius of the circle must be $\frac{9}{2}$ since its diameter is the same as the side of the square. The shaded area is $81 - \frac{81\pi}{4}$.

2. You find the area of the square and subtract the area of the circle.

3. You can find this answer by setting them equal to one another.

$$x^2 = 4$$

$$x = \pm 2$$

They intersect at the point (2,4).

4. $\int_0^2 4\,dx = 8$ and $\int_0^2 x^2\,dx = \frac{8}{3}$, as shown in the following diagrams.

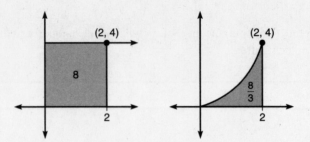

Notice that the left area minus the right area equals the shaded area in the problem. Therefore, the answer is $8 - \dfrac{8}{3} = \dfrac{16}{3}$.

5. When finding the area enclosed by two curves that contain x variables, you are adding the sum of the areas of an infinite number of vertical <u>rectangles</u>. In practice, you subtract the definite integral of the <u>lower curve</u> from the definite integral of the <u>higher curve</u>. The limits of integration for those integrals should be the x-values of the <u>intersection points</u>.

6.

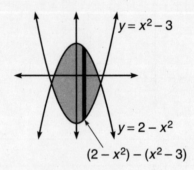

The diagram contains a single, dark vertical rectangle, which is one of the infinite number of rectangles that make up the shaded area. This makes it easy to see that you will integrate the top curve minus the bottom curve, or $(2 - x^2)-(x^2 - 3)$, as that is the length of the rectangle.

Find the intersection values by setting the functions equal.

$$2 - x^2 = x^2 - 3$$

$$2x^2 = 5$$

$$x = \pm\sqrt{\dfrac{5}{2}}$$

Therefore, the area is given by

$$\int_{-\sqrt{5/2}}^{\sqrt{5/2}} \left(\left(2 - x^2\right) - \left(x^2 - 3\right) \right) dx.$$

7. When functions contain y's instead of x's, the *rightmost* function is the greater instead of the higher function. Also, you would use the y-coordinate of the intersection point rather than the x-coordinate.

8. <u>horizontal rectangles</u>, <u>leftmost curve</u>, <u>rightmost curve</u>, <u>y-coordinates of the points of intersection</u>.

9.

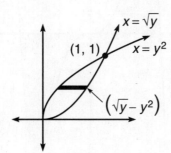

Because these functions both contain y's, you are using horizontal rectangles of length (right − left) = $\sqrt{y} - y^2$. The functions intersect at $(1,1)$ and $(0,0)$. Don't forget that you have to use the y-coordinate (even though it's the same as the x in this case).

$$\int_0^1 \left(\sqrt{y} - y^2 \right) dy$$

$$\left. \left(\frac{2}{3} y^{3/2} - \frac{y^3}{3} \right) \right|_0^1 = \frac{1}{3}$$

TIP

If you have problems drawing functions of y, plot points to construct the graph.

PROBLEM SET

Do not use a graphing calculator on problems 1 and 4.

1. Calculate the area bounded by $f(x) = x^5 - 5x^3$ and the x-axis. What important conclusion can be drawn from this problem?

2. Calculate the area bounded by $y = x^3 - 4x$ and $y = (x + 2)^2$ in the second quadrant.

3. Calculate the area in the first quadrant bounded by $x = 5$ and $y = x^2 + 2$ twice— once with horizontal rectangles and once with vertical ones.

4. Calculate the area bounded by $x = y^2 + 4y - 5$ and $y = -\frac{1}{2}x + 1$.

5. Calculate the area bounded by $y = 3x^3 - 8x$, $y = 10x^2$, $x = 0$, and $x = -1$.

TIP

remember that vertical rectangles relate to *x* variables and horizontal rectangles relate to *y* variables. This will help you immensely in the next section.

SOLUTIONS

1. First, you need to find the points of intersection of the graphs by setting them equal.

$$x^5 - 5x^3 = 0$$

$$x^3(x^2 - 5) = 0$$

$$x = 0, \pm\sqrt{5}$$

NOTE
For any origin-symmetric function $g(x)$

$$\int_{-a}^{a} g(x)dx = 0$$

since the regions will have congruent but opposite signed areas and will cancel one another out. That's why you have to split up number 1. Similarly, if $h(x)$ is even,

$$\int_{-a}^{a} h(x)dx = 2\int_{0}^{a} h(x)dx$$

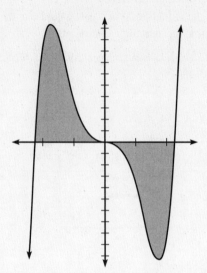

This problem must be split into two separate integrals. On $[-\sqrt{5},0]$, $f(x)$ is the "top" function, whereas on $[0,\sqrt{5}]$, the x-axis ($y = 0$) is the "top" function. However, you only need to calculate *one* of these regions since $f(x)$ is odd and therefore origin-symmetric (both of the regions will have the same area).

We will calculate the area of the region on $[0,\sqrt{5}]$. Don't forget that $y = 0$ is the "top" function.

$$\int_{0}^{\sqrt{5}} \left(0 - \left(x^5 - 5x^3\right)\right)dx$$

$$\int_{0}^{\sqrt{5}} \left(-x^5 + 5x^3\right)dx$$

$$\left(-\frac{x^6}{6}\right) + \frac{5}{4}x^4 \Bigg|_{0}^{\sqrt{5}}$$

$$\left(-\frac{125}{6} + \frac{125}{4}\right) = \frac{125}{12}$$

CAUTION
The definite integral:

$$\int_{-\sqrt{5}}^{\sqrt{5}} \left(x^5 - 5x^3\right)dx = 0$$

Therefore, the entire region between the two curves is $\frac{125}{12} \cdot 2 = \frac{125}{6}$. The important conclusion to be drawn? The Fundamental Theorem of Calculus is just an extension of this method. (Before, our unstated lower curve was always 0, since *upper curve* – 0 = *upper curve*.)

2. These are both functions of x, so you use x-boundaries and integrate *top* – *bottom*. (You also use vertical rectangles with x functions.) Graph them to see which is which.

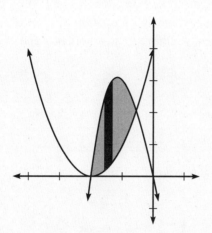

The top function is $x^3 - 4x$. Now, find the intersection points of the graphs.

$$x^3 - 4x = x^2 + 4x + 4$$

$$x^3 - x^2 - 8x - 4 = 0$$

Solve this equation using your calculator to get $x = -2$ and $-.5615528128$ (the other solution corresponds to an intersection point in the first quadrant). The area will be

$$\int_{-2}^{-.5615528128} \left(\left(x^3 - 4x \right) - (x+2)^2 \right) dx.$$

The area is approximately 2.402 square units.

3. Using vertical rectangles is much easier in this problem, although both give you the same solution.

Vertical rectangles (using x's):

The area of the region is found using the Fundamental Theorem.

$$\int_0^5 \left(x^2 + 2 \right) dx = \frac{155}{3}$$

Horizontal rectangles (using y's):

From $y = 0$ to $y = 2$, the right-hand function is $x = 5$, and the left-hand function is $x = 0$. In fact, that area is simply a rectangle of area 10.

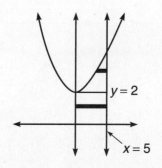

How do you find that upper intersection point? You cannot just set the equations equal, because one is an equation in terms of x. If you want to use horizontal rectangles, you have to go whole hog and change everything to y's.

$$y = x^2 + 2$$

$$y - 2 = x^2$$

$$x = \sqrt{y-2}$$

Now you can set the two equal to find the boundaries.

$$\sqrt{y-2} = 5$$

$$y - 2 = 25$$

$$y = 27$$

The intersection point is (5,27). Notice that the horizontal rectangles from $y = 2$ to $y = 27$ have a right-hand boundary of $x = 5$ and a left-hand boundary of $x = \sqrt{y-2}$. Since we already know the lower rectangular area is 10, let's find the other area.

$$\int_2^{27} \left(5 - \sqrt{y-2}\right) dy = \frac{125}{3}.$$

The total area is $10 + \frac{125}{3} = \frac{155}{3}$, and the answer matches our vertical rectangle answer.

NOTE
The reason we had to split the problem into two integrals was the change of boundaries at $y = 2$. With vertical rectangles, the boundaries stayed the same the whole time. That's why it was easier.

4. You have to use horizontal rectangles here. One of the functions is in terms of y, and there is no changing that one into terms of x easily. Therefore, you need to change the other and put it in terms of y also.

$$y = -\frac{1}{2}x + 1$$

$$y - 1 = -\frac{1}{2}x$$

$$2 - 2y = x$$

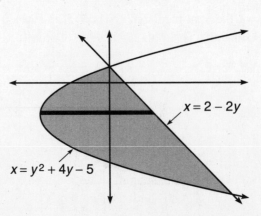

The intersection points will be

$$y^2 + 4y - 5 = 2 - 2y$$

$$y^2 + 6y - 7 = 0$$

$$y = 1, -7.$$

On the entire interval from $y = 1$ to $y = -7$, the line forms the right-hand boundary, so the area between the curves will be

$$\int_{-7}^{1} \left((2 - 2y) - \left(y^2 + 4y - 5 \right) \right) dy$$

$$\int_{-7}^{1} \left(7 - 6y - y^2 \right) dy$$

$$\left(7y - 3y^2 - \frac{y^3}{3} \right) \Big|_{-7}^{1} = \frac{256}{3}.$$

5. Both of these functions are in terms of x, so you should use vertical rectangles and x boundaries for integration.

This graph is stretched to feature the area in question.

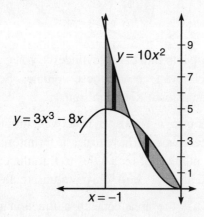

These graphs intersect in the interval $[-1,0]$. Find that point.

$$3x^3 - 8x = 10x^2$$

$$3x^3 - 10x^2 - 8x = 0$$

$$x(3x + 2)(x - 4) = 0$$

$$x = 0, \ -\frac{2}{3}, \text{ and } 4$$

Clearly, $-\frac{2}{3}$ is the value we need. However, we will need two separate integrals. On the interval $[-1, -\frac{2}{3}]$, the quadratic function is on top. However, on $[-\frac{2}{3}, 0]$, the cubic function is on top. Therefore, the total area is

$$\int_{-1}^{-2/3}\left(10x^2-\left(3x^3-8x\right)\right)dx+\int_{-2/3}^{0}\left(3x^3-8x-10x^2\right)dx$$

$$\frac{235}{324}+\frac{52}{81}=\frac{443}{324}\text{ or }1.367.$$

THE DISK AND WASHER METHODS

My students are always surprised when I tell them that calculus can help them find the volume of a butternut squash. However, when they see how it's done, they are often disappointed. "That's not calculus!" they exclaim. "Is too!" I cleverly retort. "Is not!" they protest. Such academic debates are essential to producing enlightened and well-spoken students.

There is no formula in your textbook concerning the volume of produce, so you need to break the squash into smaller, more manageable pieces.

A squash's cross-section

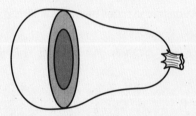

If you slice the squash into thin disks (cylinders), you can find the approximate volume of each and then add those volumes. Sound familiar? It's basically a three-dimensional Riemann sum.

The volume of a disk is $\pi r^2 h$. However, if we want to find the *exact* volume of a three-dimensional shape, we have to use an infinite number of disks (like we used an infinite number of rectangles to find the exact area when we applied the Fundamental Theorem). This formula is called the *Disk Method*.

The Disk Method: If the area beneath a function is rotated in three dimensions and the resulting solid has *no holes or gaps*, its volume is given by

$$V=\pi\int_{a}^{b}\left(r(x)\right)^2dx,$$

where a and b are the endpoints of the original area and $r(x)$ is the radius of the three-dimensional solid.

This sounds mighty complicated, but it is really quite simple. (And that's no typical math teacher mumbo jumbo.) One thing to keep in mind: The Disk Method uses rectangles that are *perpendicular* to the rotational axis. If you remember that, the process is *much* simpler.

Example 1: Find the volume generated if the region bounded by $f(x) = \sin x$ on $[0,\pi]$ is rotated around the x-axis.

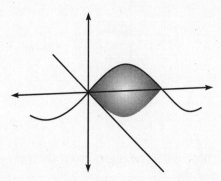

Solution: Draw the original region first (you don't have to be able to picture it in three dimensions to get the problem right.) The x-axis (the rotational axis specified by the problem) is horizontal, so we will have to use rectangles perpendicular to that (vertical rectangles) to complete the problem. Remember from last section that vertical rectangles mean that everything is in terms of x.

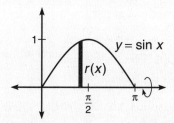

To find $r(x)$, just draw one sample vertical rectangle on the interval. How do you find its height? Use *top – bottom* like we did last section. So, $r(x) = \sin x - 0$. According to our formula, the volume of the resulting solid will be

$$V = \pi \int_0^\pi \sin^2 x \, dx.$$

AB students will have to use their calculators to integrate this, but BC students should be able to do it by hand. The answer is $\frac{\pi^2}{2}$ or approximately 4.935.

 Example 2: Find the volume of the region bounded by $x = -y^2 + 4y - 2$ and $x = 1$ rotated about the line $x = 1$.

Solution: This time, the rotational axis is vertical, so you have to use horizontal rectangles (which should be very easy, since the function is already in terms of y). All you have to do is find a, b, and $r(y)$ and plug them right into the formula. The endpoints are $y = 1$ and $y = 3$ (they have

CAUTION
Don't forget to multiply by the π outside the integral.

to be *y* values). Once again, you can draw a sample rectangle in order to find the radius.

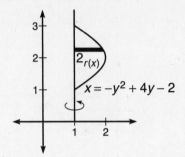

To find length with horizontal rectangles, you take *right – left*, so the radius is given by $(-y^2 + 4y - 2) - 1$.

$$\pi \int_1^3 \left(-y^2 + 4y - 2 - 1\right)^2 dy$$

$$\pi \int_1^3 \left(y^4 - 8y^3 + 22y^2 - 24y + 9\right) dy$$

$$\pi \cdot \left(\frac{y^5}{5} - 2y^4 + \frac{22}{3}y^3 - 12y^2 + 9y\right)\Bigg|_1^3$$

The rotational volume is $\frac{16\pi}{15}$, which is much prettier an answer than you thought it would be, isn't it?

Sometimes, rotational solids aren't all that solid. Consider the region bounded by the horizontal lines $y = 1$ and $y = 2$ on the interval [0,3].

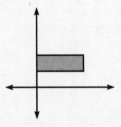

If this figure is rotated about the *x*-axis, the empty space between $y = 0$ and $y = 1$ gets rotated, too, creating a three-dimensional doughnut. The more practical-minded mathematicians of the days of yore thought it looked more like a washer (hence the name *Washer Method*).

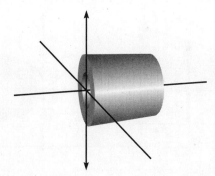

The Washer Method works the same way as carving a pumpkin for Halloween. If you want to find the volume of a jack-o-lantern, what would you have to do? First, you'd have to find the volume of the whole pumpkin (including pumpkin guts). Second, you'd have to find the volume of the hollowed-out space inside the pumpkin and subtract that from the original volume (sort of like the shaded area problem in the beginning of this chapter). In the Washer Method, you use the disk method *twice*, once to find the overall volume and once to find the volume of the hole.

The Washer Method: The rotational volume of the area bounded by two functions, *f*(*x*) and *g*(*x*), on [*a*,*b*] if *f*(*x*) ≥ *g*(*x*) on [*a*,*b*]) is given by

$$V = \pi \int_a^b \left(\left(R(x) \right)^2 - \left(r(x) \right)^2 \right) dx,$$

where $R(x)$ is the *outer radius* (the radius of the outer edge of the region) and $r(x)$ is the *inner radius* (the radius of the hole of the region).

Again, this method looks insanely difficult, but it is not bad if you've been paying attention since the beginning of this chapter.

Example 3: Find the volume generated by revolving the area bounded by the curves $y = \sqrt{x} + 1$, $x = 4$, and $y = 1$ about the *x*-axis.

Solution: If you rotate the given area around the line *y* = 1, there is no hole in the rotational solid. However, since you are revolving around the line *y* = 0, there is a gaping hole there.

NOTE
You can also use the Washer Method with *y*'s by adjusting the same way you did with the Disk Method.

NOTE
Use rectangles perpendicular to the rotational axis with the Washer Method, just like you did with the disk. Wash*er* = perpendicula*r*. Remember that both words growl at the end (just like you do after you eat chili).

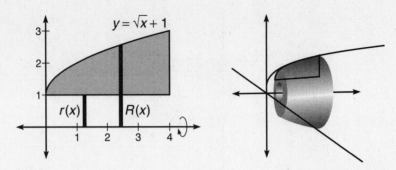

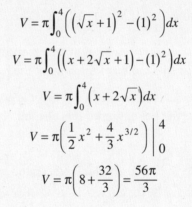

We are using vertical rectanges, since the axis of revolution (the *x*-axis) is horizontal. Notice that the outer radius, $R(x)$, reaches from the axis of rotation to the outer edge of the region. The inner radius, $r(x)$, reaches from the axis of revolution to the outer edge of the gap between the region and the axis. You can find the length of each by subtracting *top – bottom*:

$$R(x) = \sqrt{x} + 1$$

$$r(x) = 1 - 0.$$

The boundaries of the region are $x = 0$ and $x = 4$. Throw all these components together, and you get the volume according to the Washer Method:

$$V = \pi \int_0^4 \left(\left(\sqrt{x} + 1 \right)^2 - (1)^2 \right) dx$$

$$V = \pi \int_0^4 \left(\left(x + 2\sqrt{x} + 1 \right) - (1)^2 \right) dx$$

$$V = \pi \int_0^4 \left(x + 2\sqrt{x} \right) dx$$

$$V = \pi \left(\frac{1}{2}x^2 + \frac{4}{3}x^{3/2} \right) \Big|_0^4$$

$$V = \pi \left(8 + \frac{32}{3} \right) = \frac{56\pi}{3}$$

CAUTION
Don't forget to square the radii in the Disk and Washer Methods; the formula is based on πr2, as you saw at the beginning of the section. I forget to square all the time—don't let yourself get into the habit.

PROBLEM SET

You may use a graphing calculator on problems 4 and 5.

1. Find the volume generated by revolving the area in the first quadrant bounded by $y = x^2$ and $x = 3$ about the line $x = 3$.

2. True or False: If $\int_a^b f(x)dx = \int_c^d g(x)dx$, then the volume generated by revolving each of those regions about the *x*-axis is equal. Give an example that supports your position.

3. If you rotate the line segment below about the *y*-axis, you get a right circular cone of height *h* and radius *r*. Verify that the cone has a volume of $\frac{1}{3}\pi r^2 h$.

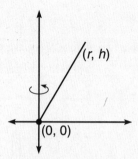

4. Find the volume, *V*, generated by revolving the region bounded by $y = (x+2)^3$, $y = 0$, and $x = 0$ about $x = 1$.

5. Find the volume, *V*, generated by revolving the region bounded by $y = \tan x$, $y = 0$, and $x = \frac{\pi}{4}$ about the *x*-axis. Then, find the value of *c* on $[0, \frac{\pi}{4}]$ such that a plane perpendicular to the *x*-axis at $x = c$ divides *V* exactly in half.

SOLUTIONS

1. Because you are revolving about a vertical axis, you need to use horizontal rectangles. With such rectangles, everything must be in terms of *y*, so you have to rewrite $y = x^2$ as $x = \sqrt{y}$.

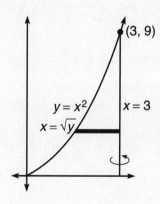

The length of the example rectangle is given by *right – left*, or $3 - \sqrt{y}$. Therefore, the volume is

$$\pi \int_0^9 \left(3 - \sqrt{y}\right)^2 dy$$

$$\pi \int_0^9 \left(9 - 6y^{1/2} + y\right) dy$$

$$\pi \cdot \left(9y - 4y^{3/2} + \frac{y^2}{2}\right) \Big|_0^9$$

$$\pi \cdot \left(81 - 108 + \frac{81}{2}\right)$$

$$\frac{27\pi}{2}.$$

2. The problem proposes that two regions of equal area result in the same volume once rotated about the x-axis. This is false. Consider the two regions pictured below, both of area 1.

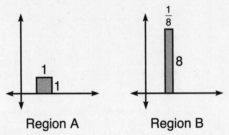

Region A   Region B

Both revolutions are solid disks with known radii and heights, so we can apply the formula $V = \pi r^2 h$ to find their volumes. The solid of revolution generated by region A has height = 1 and radius = 1; the volume is π. Region B generates a disk of height $\frac{1}{8}$ and radius 8; its volume will be $\pi \cdot 8^2 \cdot \frac{1}{8} = 8\pi$. The volumes are very different.

3. To start, you need to find the equation of that line segment. Because it passes through $(0,0)$ and (r,h), it has slope $\frac{h}{r}$ and y-intercept 0. Thus, the equation is

$$y = \frac{h}{r}x.$$

However, you are revolving about a vertical axis, so the rectangles need to be horizontal, and the equation needs to be in terms of y:

$$x = \frac{r}{h}y.$$

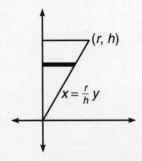

Therefore, the volume generated after revolution will be

$$\pi \cdot \int_0^h \left(\frac{r}{h}y\right)^2 dy$$

$$\pi \cdot \int_0^h \frac{r^2 y^2}{h^2} dy.$$

You can pull out the constants to simplify the integral.

$$\frac{r^2 \pi}{h^2} \int_0^h y^2 dy$$

$$\frac{r^2 \pi}{h^2} \cdot \frac{y^3}{3} \bigg|_0^h$$

$$\frac{r^2 \pi}{h^2} \cdot \frac{h^3}{3}$$

$$\frac{r^2 h \pi}{3}$$

4. This solid of revolution will have a hole in it, so you have to use the Washer Method. Again, a vertical axis of revolution means horizontal rectangles. Therefore, the function must be rewritten as $x = \sqrt[3]{y} - 2$. The inner radius (from the axis of revolution to the outside of the region) is $1 - \left(\sqrt[3]{y} - 2\right)$, and the outer radius (from the axis of revolution to the outside of the hole) is $1 - 0 = 1$.

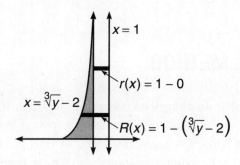

NOTE
The upper boundary of the integral is still 8, even though we are revolving around $x = 1$. The integral boundaries must be the bounds of the region.

Therefore, the volume is

$$V = \pi \int_0^8 \left(\left(1 - \sqrt[3]{y}\right)^2 - (1)^2 \right) dy$$

$$V \approx 35.186.$$

You might as well use your calculator to find the answer. You should know how to use the Fundamental Theorem by now.

5. Finally, a problem in which we don't have to convert the function in terms of y! Vertical rectangles are perpendicular to the x-axis, so the volume, V, will be

$$V = \pi \int_0^{\pi/4} \tan^2 x \, dx.$$

We can integrate $\tan^2 x$ by hand if we replace it with $(\sec^2 x - 1)$ using Pappa. However, save time by using your calculator since it's allowed on this problem.

$$V \approx .6741915533$$

We want to find a revolutionary volume that is exactly half of that, so we should set up this equation:

$$\pi \int_0^c \tan^2 x \, dx = \frac{1}{2} \cdot .6741915533.$$

Unless you have a symbolic integrator (like a TI-89), you'll have to integrate the left side by hand.

$$\pi \int_0^c \left(\sec^2 x - 1 \right) dx = .3370957767$$

$$\pi (\tan x - x) \big|_0^c = .3370957767$$

$$\pi (\tan c - c) = .3370957767$$

$$(\tan c - c) = .1073009183$$

Use your calculator to solve this equation; you get

$$c \approx .645.$$

THE SHELL METHOD

NOTE
There are no places on the AP test in which you *have* to apply the Shell Method, according to the College Board. However, the Shell Method *can* save you time, so it's good to know.

I must confess: I really don't like baseball all that much. I used to love *playing* baseball, but watching it on television (or even in the stadium, I am sad to say) just doesn't cut it for me. A bunch of guys standing around in tight pants sounds like Broadway musical tryouts. At least in football, the guys in tight pants are trying to hurt one another, so we can forgive the "nothing left to the imagination" attire.

One of the more interesting pieces of minutiae in baseball (which is far from trivial to the avid fan) is the designated hitter rule. You see, the American League allows coaches to replace the pitchers (who are generally *very* poor batters) with other players, whose only job is to hit for the pitcher in the lineup. It's a good concept: allow one person good at something to stand in for someone else at the plate.

The Shell Method is the designated hitter for calculus, and there's no debate about whether or not it is good for baseball. When the Disk or Washer Method is hard (or even impossible) to use, the Shell Method steps in and hits one out of the park. The Shell Method doesn't even care if the rotational solid has a hole in it or not (whereas that scares the Disk Method off). The Shell Method uses rectangles that are *parallel* to the rotational axis, so it can even save you time with variable conversions. The rule itself is a little bizzare, though. I find it difficult to visualize, so spend more time on memorizing and applying it than trying to figure out its place in the cosmos.

The Shell Method: If a region is rotated about a horizontal or vertical axis on an interval [*a*,*b*], the resulting volume is given by

$$2\pi \int_{a}^{b} d(x)h(x)\, dx \text{ or}$$

$$2\pi \int_{f(a)}^{f(b)} d(y)h(y)\, dy$$

Now, you probably see why I call this the **D**esignated **H**itter"; it helps you memorize the formula. But what do the two functions stand for?

$d(x)$ is the distance from the rotational axis to a rectangle in the region.

$h(x)$ is the height of that rectangle.

 Example 4: Consider the following region, bounded by functions $f(x)$ and $g(x)$, that intersect at $x = a$ and $x = b$. What is the volume of this region if rotated about the *y*-axis?

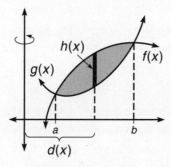

Solution: The key to the Shell Method is drawing the darkened rectangle in the region (which is parallel to the rotational axis). The function $d(x)$ represents how far that rectangle is from the *y*-axis. We could have drawn that rectangle anywhere on the interval, and we can't be sure *exactly* where it is, so we will say that it is a distance of *x* away. Thus, $d(x) = x$. Secondly, what is the height of that rectangle? You find it the same way you have all chapter: *top – bottom*. This is how $h(x)$ gets its value, so $h(x) = f(x) - g(x)$. Therefore, the volume of the indicated region is

$$2\pi \int_a^b d(x)h(x)\,dx$$

$$2\pi \int_a^b x\big(f(x)-g(x)\big)dx.$$

Sometimes, students are confused about how to find $d(x)$. It's not complicated, so don't get frustrated. Think about it this way: If you are given the graph below and asked to give the coordinates of the point, how would you respond?

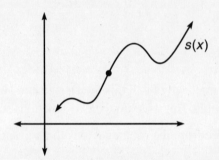

The only correct response would be very generic: $(x,s(x))$. In this instance, you assign an unknown horizontal distance a value of x. That's all you're doing with the Shell Method. However, $d(x)$ is not always just x (or y), as you'll see in the next example.

NOTE

You can use the Washer Method to solve Example 5, but we'll use the Shell Method, since that's the name of this section.

 Example 5: What volume results if you rotate the region in the first quadrant bounded by $y = x^2 + 1$, $x = 1$, and $y = 1$ about the line $y = \frac{1}{2}$?

Solution: To begin, breathe deeply, and draw the region. Then, draw a dark rectangle in the region parallel to the rotational axis.

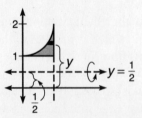

All of the rectangles are horizontal now, so all of the equations must be in terms of y. Therefore, you must rewrite $y = x^2 + 1$ as $x = \sqrt{y-1}$. How far is that darkened rectangle from the x-axis? Because you don't know exactly, you must say that it is *y* units away. However, that's not *d(y)*. To find it, we need to know how far the rectangle is from the *rotational axis*. Because the rectangle is *y* units above the *x* axis and the rotational axis is $\frac{1}{2}$ unit above the rotational axis, $d(y) = y - \frac{1}{2}$. This is much easier to visualize than to explain; look at the graph above for help. All you have left to do is to determine the value of *h(y)*, the height (or length) of the rectangle. Its length is given by *right – left* and is $h(y) = 1 - \sqrt{y-1}$. Therefore, the volume will be

$$2\pi \int_1^2 \left(y - \frac{1}{2} \right)\left(1 - \sqrt{y-1}\right) dy.$$

You can multiply this out and integrate each separately, but let's embrace calculator technology for the time being. The answer is approximately 1.676.

This is the same answer you get using the washer method. If you don't believe me, I'll show you. The outer radius will be $R(x) = (x^2 + 1) - \frac{1}{2}$, and the inner radius is $r(x) = \frac{1}{2}$. Therefore, the volume is

$$\pi \int_0^1 \left(\left(x^2 + \frac{1}{2} \right)^2 - \left(\frac{1}{2} \right)^2 \right) dy = 1.676.$$

PROBLEM SET

You may not use a graphing calculator on these problems.

1. Find the volume generated by rotating the area in the first quadrant bounded by $y = x^2$ and $x = 2$ about the line $x = -2$. Use the Shell Method.

2. Find the volume generated by rotating the area bounded by $y = \sqrt{16 - x^2}$ and $y = 2$ about the *x*-axis. Use the most appropriate method.

3. Find the volume generated by rotating the area bounded by $y = \sqrt{x}$, $x = 4$, and $y = 0$ about the line $y = -3$ using the
   (a) Shell Method.
   (b) Washer Method.

4. Find the volume generated by rotating the area bounded by $y = x^3 + x + 1$, $y = 0$, and $x = 1$ about the line $x = 1$, and explain why the Shell Method *must* be used.

SOLUTIONS

1. First, graph the region and draw your rectangle.

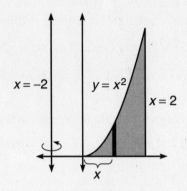

In this case, $d(x)$ is more than just x; in fact, it is exactly two units more, so $d(x) = x + 2$. Even easier, $h(x) = x^2$. Therefore, the volume is

$$2\pi \int_0^2 (x+2)\left(x^2\right)dx$$

$$2\pi \int_0^2 \left(x^3 + 2x^2\right)dx$$

$$2\pi \cdot \left(\frac{x^4}{4} + \frac{2}{3}x^3\right)\bigg|_0^2$$

$$2\pi\left(4 + \frac{16}{3}\right) = \frac{56\pi}{3}.$$

TIP

The best method to use in rotational solid volumes is the method that requires the least amount of conversion.

2. The Shell Method is not your best bet in this problem. You'd have to use horizontal rectangles, so the function would have to be put in terms of y, and the problem would only get more complicated from there.

Because the rotational solid will have a hole, you'll need to use the Washer Method (which requires perpendicular rectangles; in this case, x's). First, you need to find the intersection points of the graphs.

$$\sqrt{16 - x^2} = 2$$

$$16 - x^2 = 4$$

$$x^2 = 12$$

$$x = \pm 2\sqrt{3}$$

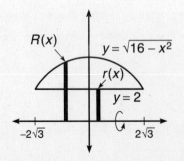

$R(x) = \sqrt{16 - x^2}$ and $r(x) = 2$; therefore, the volume is

$$\pi \int_{-2\sqrt{3}}^{2\sqrt{3}} \left(16 - x^2 - 4\right) dx$$

$$\pi \int_{-2\sqrt{3}}^{2\sqrt{3}} \left(12 - x^2\right) dx$$

$$\pi \left(\left(24\sqrt{3} - 8\sqrt{3}\right) - \left(-24\sqrt{3} + 8\sqrt{3}\right)\right)$$

$$32\pi\sqrt{3}.$$

3. (a) The Shell Method will use horizontal rectangles and requires that you rewrite $y = \sqrt{x}$ as $x = y^2$.

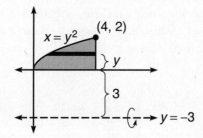

Based on the diagram, $d(y) = y + 3$ and $h(y) = 4 - y^2$. Therefore, the volume is

$$2\pi \int_0^2 (y + 3)\left(4 - y^2\right) dy$$

$$2\pi \int_0^2 \left(4y - y^3 - 3y^2 + 12\right) dy$$

$$2\pi \left(2y^2 - \frac{y^4}{4} - y^3 - 12y\right) \Big|_0^2$$

$$2\pi(8 - 4 - 8 - 24) = 40\pi.$$

(b) The Washer Method uses vertical rectangles and the function in terms of x.

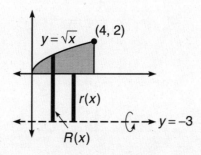

The outer radius is $R(x) = \sqrt{x} - (-3)$ and the inner radius is $r(x) = 0 - (-3)$. Therefore, the volume is

$$\pi \int_0^4 \left(\left(\sqrt{x} + 3 \right)^2 - 9 \right) dx$$

$$\pi \int_0^4 \left(x + 6\sqrt{x} \right) dx$$

$$\pi \left(\frac{x^2}{2} + 4x^{3/2} \right) \Big|_0^4$$

$$\pi(8 + 32) = 40\pi.$$

4. Even though the solid of revolution will have no hole in it, you cannot use the Disk Method. To do so, you would have to use horizontal rectangles and put the $y = x^3 + x + 1$ in terms of y, which is impossible (you cannot solve that equation for x). Therefore, the shell method is the only way to go (since that requires vertical rectangles).

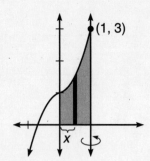

The distance from the axis of revolution to the rectangle in the region is $1 - x$, and the height of the rectangle is $x^3 + x + 1$. So, the volume generated will be

$$2\pi \int_0^1 (1 - x)\left(x^3 + x + 1 \right) dx$$

$$2\pi \int_0^1 \left(-x^4 + x^3 - x^2 + 1 \right) dx$$

$$2\pi \left(-\frac{x^5}{5} + \frac{x^4}{4} - \frac{x^3}{3} + x \right) \Big|_0^1$$

$$2\pi \left(-\frac{1}{5} + \frac{1}{4} - \frac{1}{3} + 1 \right) = \frac{43\pi}{30}.$$

FINDING THE VOLUME OF REGIONS WITH KNOWN CROSS SECTIONS

Until now, all of our rotational solids have had circular cross sections. This section discusses figures whose cross sections are, perhaps, triangles, squares, or semicircles instead. This concept seems incredibly hard until you realize its inherent simplicity. This is just an extension of the Disk Method, and nobody hates the Disk Method! Even many of the most evil people throughout history harbored a fondness for the Disk Method, among them Jesse James, Dracula, and the guy who canceled Star Trek.

In order to find the volume with the disk method, you integrated the area of one cross section. Because the cross section was circular, you integrated the formula for the area of a circle (πr^2). So, if a new problem has squares as cross sections instead (for example), you will integrate the formula for area of a square ($side^2$).

There is one other difference in these types of problems. Noncircular cross sections are *not* the result of a rotation, as every other problem has been so far. (You have been rotating about the *x*- or *y*-axis or a line such as $x = 1$.) Instead, these solids *grow out of a base on the coordinate plane into the third dimension*. It's not as hard as it sounds.

Consider a circular base on the coordinate axes with equation $x^2 + y^2 = 9$. This circle is actually made up of two functions (if you solve for *y*): $y = \sqrt{9 - x^2}$ and $y = -\sqrt{9 - x^2}$.

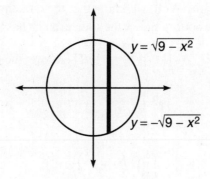

Here's where your imagination and visualization come in to play. This circle is sort of like how a pedestal is analogous to a statue. A three-dimensional form will sit on *top* of it and come out of your paper. Imagine that the darkened rectangle in the figure above is the bottom of a square that is sitting on the base. That is not the only square, however. There are squares all along the circle at every possible vertical rectangle. The resulting three-dimensional shape would look something like this:

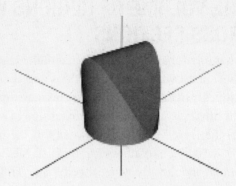

Our job in the next example will be to find the volume of this shape.

Example 6: If a solid has square cross sections perpendicular to the *x*-axis and has a base bounded by $x^2 + y^2 = 9$, what is the volume of that solid?

Solution: We must use vertical rectangles, since they are perpendicular to the *x*-axis. As stated earlier, all we have to do is to integrate the formula for the area of the figure in question. The formula for the area of a square (with side *s*) is s^2. Therefore, the integral we use to find the volume is

$$\int_{-3}^{3} s^2 dx.$$

(The –3 and 3 boundaries are the *x*-boundaries of the region.) How long is one side of the square? Well, we know that the darkened rectangle in the diagram above is the bottom side of the square, and it has length *top – bottom*

$= \sqrt{9-x^2} - \left(-\sqrt{9-x^2}\right) = 2\sqrt{9-x^2}$. Therefore, the total volume is

$$\int_{-3}^{3} \left(2\sqrt{9-x^2}\right)^2 dx$$

$$\int_{-3}^{3} \left(4(9-x^2)\right)dx$$

$$\left(36x - \frac{4}{3}x^3\right)\Big|_{-3}^{3} = 144.$$

This process is a little bizarre, but it is easy to learn.

Steps to Success with Known Cross Sections

1. Draw the graph of the base on the coordinate plane and darken a sample rectangle on it (be careful and draw it as instructed, i.e., perpendicular to the *x*-axis or *y*-axis).

NOTE
This is what is meant by a figure "growing out of a base." The circle $x^2 + y^2 = 9$ is like a fertile field out of which squares whose side length is equal to a vertical rectangle's length.

2. Determine the length of that rectangle and what relation it has to the shape of the cross section. In the previous example, the rectangle was one of the sides of the squares that formed the cross sections.

3. Integrate the formula for the area of the given cross section, inserting the information you have about the darkened rectangle. Make sure the boundaries match the shape of the rectangle (e.g., y-boundaries if the rectangle is horizontal).

Example 7: Find the volume of a solid that has semi-circular cross sections perpendicular to the x-axis whose base is bounded by the graphs of $y = x^2$ and $y = \sqrt{x}$.

Solution: Begin by drawing the base. Because these cross sections are perpendicular to the x-axis, we will use vertical rectangles and x's.

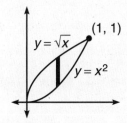

If you are curious, the solid looks like this:

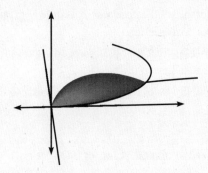

However, it is neither important nor useful to be able to draw this region, so don't worry if you can't. The darkened rectangle on the base will have length $\sqrt{x} - x^2$, but what does that length represent? If the cross sections are semicircles, then that must be the diameter, and the semicircles sprout from there. If $\sqrt{x} - x^2$ is the diameter, then $\frac{1}{2}\left(\sqrt{x} - x^2\right)$ is the radius of those semicircles. That's important because the formula for the area of a semi-circle is $\frac{\pi r^2}{2}$. Put all these pieces together to get the total volume:

$$\int_0^1 \frac{\pi r^2}{2}\, dx$$

$$\frac{\pi}{2} \int_0^1 r^2\, dx$$

$$\frac{\pi}{2} \int_0^1 \left(\frac{1}{2}\left(\sqrt{x}-x^2\right) \right)^2 dx$$

$$\frac{\pi}{8} \int_0^1 \left(x - 2x^{5/2} + x^4 \right) dx$$

$$\frac{\pi}{8} \left(\frac{x^2}{2} - \frac{4}{7}x^{7/2} + \frac{x^5}{5} \right)\Big|_0^1$$

$$\frac{\pi}{8} \left(\frac{1}{2} - \frac{4}{7} + \frac{1}{5} \right) = \frac{9\pi}{560}.$$

PROBLEM SET

You may use a graphing calculator on part (b) of each problem.

1. Find the volume of the solid whose base is the region bounded by $y = -\sqrt{x}$ and $y = -\frac{1}{2}x$ that has cross sections

   (a) that are rectangles of height 3 perpendicular to the x-axis.

   (b) that are equilateral triangles perpendicular to the y-axis.

2. Find the volume of the solid whose base is a circle with radius 5 centered at the origin and that has cross sections

   (a) that are isosceles right triangles perpendicular to the x-axis (such that the hypoteneuse lies on the base).

   (b) that are semiellipses of height 2 perpendicular to the x-axis.

SOLUTIONS

1. (a) First, we need to find the points of intersection.

$$-\sqrt{x} = -\frac{1}{2}x$$

$$x = \frac{x^2}{4}$$

$$4x - x^2 = 0$$

$$x = 0, 4$$

NOTE
The area of an equilateral triangle with side s is $\frac{\sqrt{3}}{4}s^2$. You should memorize that if you don't know it.

NOTE
The area of an ellipse is πab.

Since the cross sections are rectangles of height 3, the volume will be given by

$$\int_0^4 length \cdot width\, dx$$

$$\int_0^4 3\left(-\frac{1}{2}x + \sqrt{x}\right)dx.$$

To find the width of the rectangle, you calculate the length of the darkened rectangle, using *top – bottom*.

$$3\left(-\frac{1}{4}x^2 + \frac{2}{3}x^{3/2}\right)\Big|_0^4$$

$$3\left(-4 + \frac{16}{3}\right) = 4$$

(b) This time, you have to rewrite the equations in terms of *y* (because the rectangles are perpendicular to the *y*-axis and, therefore, horizontal).

The function $y = -\sqrt{x}$ becomes $x = y^2$; $y = -\frac{1}{2}x$ becomes $x = -2y$.

The length of a side of the triangle is given by *right – left* $= -2y - y^2$. Therefore, the volume is

$$\int_{-2}^0 \frac{\sqrt{3}}{4}s^2\, dy$$

$$\int_{-2}^0 \frac{\sqrt{3}}{4}\left(-2y - y^2\right)^2 dy.$$

You can use your calculator to find the volume; it is approximately .462.

2. (a) First of all, we need to figure out how to find the area of an isosceles right triangle based on the length of its hypotenuse.

According to the Pythagorean Theorem, $2s^2 = h^2$. Therefore, $s = \frac{h}{\sqrt{2}}$. The area of the triangle is $\frac{1}{2}bh = \frac{1}{2}s^2$. If we substitute in for s, we get

$$\frac{1}{2}\left(\frac{h}{\sqrt{2}}\right)^2 = \frac{1}{2} \cdot \frac{h^2}{2} = \frac{h^2}{4}.$$

Now, we can integrate this formula to get the volume of the shape, as soon as we figure out what the length of the hypotenuse is.

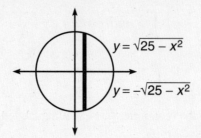

We know that the darkened rectangle represents a hypotenuse, and its length is $h = top - bottom = 2\sqrt{25 - x^2}$. Therefore, the volume is

$$\frac{1}{4}\int_{-5}^{5}\left(2\sqrt{25 - x^2}\right)^2 dx$$

$$\frac{1}{4}\int_{-5}^{5}\left(4\left(25 - x^2\right)\right)dx$$

$$\frac{1}{4}\left(100x - \frac{4}{3}x^3\right)\Big|_{-5}^{5}$$

$$\frac{1}{4}\left(\frac{1000}{3} - \left(-\frac{1000}{3}\right)\right) = \frac{500}{3}.$$

(b) This one is a little tougher to picture. The semiellipse should come out of the base something like this:

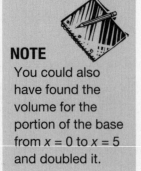

NOTE

You could also have found the volume for the portion of the base from $x = 0$ to $x = 5$ and doubled it.

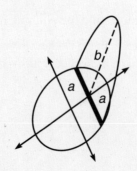

Therefore, the darkened rectangle represents 2*a* (which means $a = \sqrt{25 - x^2}$), and *b* = 2 (the height). The area of a semiellipse is $\frac{\pi ab}{2}$, so the volume of the solid will be

$$\frac{1}{2}\int_{-5}^{5} \pi ab \, dx$$

$$\frac{\pi}{2}\int_{-5}^{5} 2\sqrt{25 - x^2} \, dx$$

$$\pi\int_{-5}^{5} \sqrt{25 - x^2} \, dx .$$

Use your calculator to evaluate this definite integral. The volume will be approximately 123.370.

ARC LENGTH (BC TOPIC ONLY)

The practice of finding arc length has a long and storied history, stretching all the way back to Noah. Of course, he used cubits, but we can use any unit of measure. It is arguable that Noah had more things on his mind, though: the impending destruction of the human race, ensuring his ark was water-tight, and placing layer upon layer of newspaper on the floor. All you have to worry about is passing the AP test, and the good news is that this topic is very easy; all you have to do is memorize a pair of formulas—one for rectangular equations and one for parametric equations.

The Length of a Rectangular Curve (or Arc): If *f*(*x*) is an integrable function on [*a*,*b*], then the length of the curve from *a* to *b* is $\int_{a}^{b} \sqrt{1 + \left(f'(x) \right)^2} \, dx$.

That's all there is to it. Find the derivative and plug it right into the formula—it couldn't be easier.

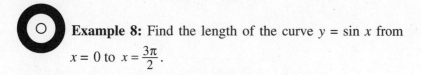 **Example 8:** Find the length of the curve *y* = sin *x* from $x = 0$ to $x = \frac{3\pi}{2}$.

Solution: First, find the derivative (which is very easy).

$$f(x) = \sin x$$

$$f'(x) = \cos x \, dx$$

Now, plug into the formula.

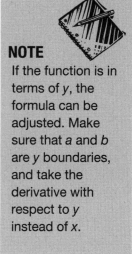

NOTE
If the function is in terms of *y*, the formula can be adjusted. Make sure that *a* and *b* are *y* boundaries, and take the derivative with respect to *y* instead of *x*.

$$\int_0^{\frac{3\pi}{2}} \sqrt{1+\cos^2 x}\; dx$$

This is really not very easy to integrate with our methods, so use the calculator to finish. The arc (or curve) length is approximately 5.730.

It is just as easy to find arc length when you're dealing with parametric equations; in fact, the formula is very similar to rectangular arc length. The only less obvious difference is that the boundaries of the definite integral in parametric arc length are *t*-values, not *x* or *y* values.

CAUTION

Many of the integrals you end up with when finding arc length are difficult and have been handed quite a beating with the ugly stick. Don't expect to be able to solve them all by hand.

The Length of a Parametric Curve (or Arc): If *a* and *b* are *t*-values for a parametric function, its length between those *t*-values is given by

$$\int_a^b \sqrt{\left(\frac{dx}{dt}\right)^2 + \left(\frac{dy}{dt}\right)^2}\; dt.$$

The hardest part of differentiating parametric functions was finding $\frac{dy}{dx}$. Deriving with respect to *t* is so easy that it's just silly.

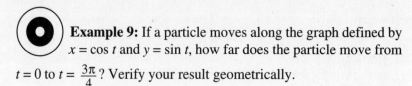

Example 9: If a particle moves along the graph defined by $x = \cos t$ and $y = \sin t$, how far does the particle move from $t = 0$ to $t = \frac{3\pi}{4}$? Verify your result geometrically.

Solution: Since the particle moves along the curve, we are just finding the length of the curve. Begin by finding the derivatives.

NOTE

It's unlikely that you will be asked on the AP text to verify your result geometrically, as this problem does. However, doing so will help you as you learn this topic.

$$x = \cos t;\; \frac{dx}{dt} = -\sin t$$

$$y = \sin t;\; \frac{dy}{dt} = \cos t$$

Now, substitute these values into the arc-length formula.

$$\int_0^{3\pi/4} \sqrt{\left(-\sin t\right)^2 + \left(\cos t\right)^2}\; dt$$

$$\int_0^{3\pi/4} \sqrt{\sin^2 t + \cos^2 t}\; dt$$

According to the Mamma Theorem, the contents of the radical are equal to 1:

$$\int_0^{3\pi/4} dt$$

$$t \Big|_0^{3\pi/4} = 3\pi/4 \approx 2.356.$$

To justify this geometrically, you must realize that the parametric equations result in a circle of radius 1 (the unit circle, actually). The arc length from 0 to $\dfrac{3\pi}{4}$ represents $\dfrac{3}{8}$ of the circumference of the circle.

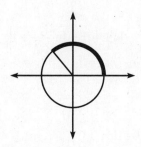

The circumference of the unit circle is

$$2\pi r = 2\pi \approx 6.283185307,$$

and $\frac{3}{8}$ of that total is 2.35619449; this matches our answer above.

PROBLEM SET

You may use a graphing calculator on problems 2 through 5.

1. What is the length of the curve $y = \ln(\sin x)$ from $x = \frac{\pi}{4}$ to $x = \frac{\pi}{2}$?

2. Find the arc length of $y = \sqrt{x - 1}$ from $x = 1$ to $x = 3$.

3. How long is the curve defined by $x = y^3 - 4y + 2$ from $y = -1$ to $y = 1$?

4. Find the length of the parametric curve defined by $x = \sqrt{t + 1}$, $y = t^2 + t$ from $t = 1$ to $t = 5$.

5. Find the perimeter of the ellipse $\frac{x^2}{4} + \frac{y^2}{9} = 1$ using parametric equations.

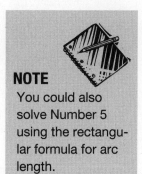

NOTE
You could also solve Number 5 using the rectangular formula for arc length.

SOLUTIONS

1. Begin by finding y':

$$y' = \frac{1}{\sin x} \cdot \cos x = \cot x.$$

The arc length will be

$$\int_{\pi/4}^{\pi/2} \sqrt{1 + \cot^2 x}\, dx.$$

The Pappa Theorem allows you to replace the contents of the radical with $\csc^2 x$.

$$\int_{\pi/4}^{\pi/2} \sqrt{\csc^2 x}\,dx$$

$$\int_{\pi/4}^{\pi/2} \csc x \, dx$$

$$-\ln|\csc x + \cot x|\Big|_{\pi/4}^{\pi/2}$$

$$-\ln|1+0| + \ln\left|\frac{2}{\sqrt{2}} + 1\right|$$

$$\ln\left|\frac{2}{\sqrt{2}} + 1\right|$$

2. Again, find the derivative first.

$$y' = \frac{1}{2\sqrt{x-1}}$$

The arc length will be

$$\int_1^3 \sqrt{1 + \frac{1}{4(x-1)}}\,dx$$

$$\int_1^3 \sqrt{1 + \frac{1}{4x-4}}\,dx.$$

Evaluate this with your graphing calculator to get 2.562.

3. Even though everything is in terms of y, it would be the exact same problem if every variable were an x.

$$\int_{-1}^{1} \sqrt{1 + (3y^2 - 4)^2}\,dy$$

The arc length will be 6.361.

4. First, find the derivatives with respect to t:

$$\frac{dx}{dt} = \frac{1}{2\sqrt{t+1}}$$

$$\frac{dy}{dt} = 2t+1.$$

The arc length will be as follows:

$$\int_1^5 \sqrt{\frac{1}{(4t+4)} + (2t+1)^2} \, dt \approx 28.025.$$

5. The first order of business is to rewrite the ellipse in parametric form. To do this, remember your Mamma.

$$\cos^2 t + \sin^2 t = 1$$

Compare that equation with the ellipse in standard form.

$$\frac{x^2}{4} + \frac{y^2}{9} = 1$$

You can set $\cos t = \frac{x}{2}$ and $\sin t = \frac{y}{3}$, according to Mamma. Now, solve the new equations for x and y to get the parametric form of the ellipse:

$$x = 2\cos t, \; y = 3\sin t.$$

The ellipse is drawn completely from $t = 0$ to $t = 2\pi$, so you can use those values to bound the definite integral. However, exactly half of the integral is drawn from $t = 0$ to $t = \pi$, so let's evaluate that integral and double it (for a change of pace). Both answers will be the same.

$$2 \cdot \int_0^\pi \sqrt{(-2\sin t)^2 + (3\cos t)^2} \, dt$$

The perimeter is $2 \cdot 11.05174608 = 22.103$.

POLAR AREA (BC TOPIC ONLY)

You may have wondered why we didn't discuss polar arc length. Perhaps you answered that question for yourself. Remember, any polar equation is easily expressed parametrically using the formulas $x = r\cos \theta$ and $y = r\sin \theta$. Therefore, finding polar arc length equates to finding parametric arc length. However, you will need to be able to calculate the area enclosed by polar curves. Smile—this is the last you will see of parametric and polar equations for the AP test. Polar area isn't hard, but it is different, and it will take a moment or so to get used to.

No self-respecting calculus topic comes without a formula to despair over and ultimately memorize. The formula for polar area is different from all previous area formulas, because it is not based on rectangles. Instead, polar area uses an infinite number of *sectors* to find area. A *sector* is a hunk of circle; for example, a piece of pie is a sector of the entire pie. The area of a sector of a circle is given by $\frac{\theta r^2}{2}$, but since we are using an infinite number of θ's to calculate exact area, we replace θ with $d\theta$. Therefore, the formula for the area bounded between radial lines $\theta = a$ and $\theta = b$ is

$$\frac{1}{2}\int_a^b r^2 \, d\theta .$$

We'll start with an easy example and dip our toes into the swimming pool. Once you see how good the water feels, you won't mind diving right in.

 Example 10: Show that the area of a quarter circle of radius 2 equals π using polar area.

Solution: A circle is quite easy to express in polar form; a circle of radius 2 is just the equation $r = 2$. Since we are trying to find the area of a quarter circle, you have to find the area of the circle in one of the quadrants. For simplicity's sake, let's choose the first quadrant.

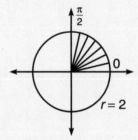

Our area formula sums the areas of the little sectors in the first quadrant with the formula

$$\frac{1}{2}\int_0^{\pi/2} \left(2\right)^2 d\theta .$$

Integrate and apply the Fundamental Theorem to get

$$\frac{1}{2}\cdot\left(4\theta\right)\Big|_0^{\pi/2}$$

$$\left(\frac{1}{2}\right)(2\pi) = \pi .$$

You see, that's not so bad. In fact, you've probably had burps that made you feel worse than that. Let's try a shaded-region problem to up the ante a little. To solve it, you'll have to find the area of the outer portion and subtract the inner. You can use your calculator with no shame on this example.

 Example 11: Find the shaded area in the graph below of $r = 2 + 3\cos\theta$.

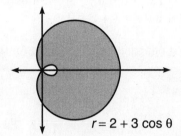

$r = 2 + 3 \cos \theta$

Solution: First of all, we should find where the graph intersects the pole (origin).

$$2 + 3\cos\theta = 0$$

Set your calculator to rectangular mode for a moment to solve this equation. The answers are $\theta = 2.300523983$ and $\theta = 3.982661324$. This is the meat of the problem. As θ goes from 0 to 2.300523983, the top of the graph is drawn *excluding the inner loop*. As θ goes from 2.300523983 to 3.982661324, the inner loop is drawn. Finally, from 3.982661324 to 2π, the bottom of the graph is drawn, again excluding the inner loop.

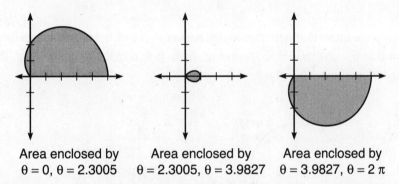

Area enclosed by   Area enclosed by   Area enclosed by
$\theta = 0$, $\theta = 2.3005$ $\theta = 2.3005$, $\theta = 3.9827$ $\theta = 3.9827$, $\theta = 2\pi$

Therefore, we can get the area of the requested region by doubling the left area above and then subtracting the middle area. (The rightmost area is exactly the same as the leftmost area, since the equation is symmetric about the x-axis.) Therefore, the answer will be

$$\frac{1}{2} \cdot 2 \int_0^{2.300523983} \left(2 + 3\cos\theta\right)^2 d\theta - \frac{1}{2}\int_{2.300523893}^{3.982661324} \left(2 + 3\cos\theta\right)^2 d\theta.$$

The area is approximately 25.822.

The final polar area problem you can expect from the AP test concerns finding the area bounded by multiple functions. These problems are no more difficult if you draw a graph first and proceed very carefully.

Example 12: Find the area bounded by the graphs $r = 3 \sin \theta$ and $r = 1 + \cos \theta$.

Solution: First, we should determine the value of θ at which the graphs intersect.

$$3 \sin \theta = 1 + \cos \theta$$

$$3 \sin \theta - \cos \theta - 1 = 0$$

$$\theta = .6435011088$$

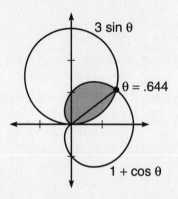

This is the important step: the shaded region above is defined by $3 \sin \theta$ from $\theta = 0$ to $\theta = .6435011088$; however, from $\theta = .6435011088$ to $\theta = \pi$, the shaded region is bounded by $1 + \cos \theta$.

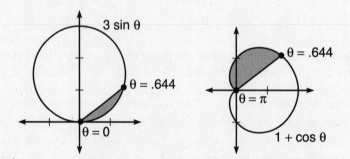

Therefore, the area will be the sum of those two smaller regions.

$$\frac{1}{2} \int_0^{.6435011088} (3 \sin \theta)^2 \, d\theta + \frac{1}{2} \int_{.6435011088}^{\pi} (1 + \cos \theta)^2 \, d\theta$$

The total area (please use a calulator!) is approximately 1.521.

PROBLEM SET

You may use a graphing calculator on problems 4 and 5.

1. Find the area bounded by $r = -2\sin\theta$.

2. Find the area in the second quadrant bounded by $r = \cos\theta + \sin\theta$.

3. Find the area enclosed by the graph of $r = \cos 2\theta$.

4. Find the area of the region bounded by $r = 2 + \cos\theta$ and $r = 2$.

5. Find the area bounded by $r = 3 - 2\sin\theta$ and $r = 4\cos\theta$.

SOLUTIONS

1. The entire graph is drawn from $\theta = 0$ to $\theta = \pi$. If you calculate the area from 0 to 2π, your answer will be two times too big.

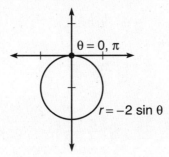

Other than that, the setup is very easy:

$$\frac{1}{2}\int_0^\pi (2\sin\theta)^2 \, d\theta = \pi.$$

This is similar to Example 10, since both problems boil down to finding the area of a circle with radius 1.

2. The radial lines that bound the portion in the second quadrant are $\theta = \dfrac{\pi}{2}$ (which makes sense) and $\theta = \dfrac{3\pi}{4}$. The second value makes $r = 0$ (since the sine and cosine of $\dfrac{3\pi}{4}$ are opposites), which is the intersection at the pole marking the end of the region's presence in the second quadrant.

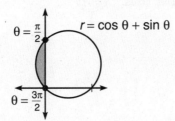

To find the area, use the formula and multiply the squared term out. Mamma will make a return appearance!

$$\frac{1}{2}\int_{\pi/2}^{3\pi/4}\left(\cos\theta+\sin\theta\right)^2 d\theta$$

$$\frac{1}{2}\int_{\pi/2}^{3\pi/4}\left(\cos^2\theta+2\sin\theta\cos\theta+\sin^2\theta\right)d\theta$$

$$\frac{1}{2}\int_{\pi/2}^{3\pi/4}\left(\cos^2\theta+\sin^2\theta+\sin 2\theta\right)d\theta$$

$$\frac{1}{2}\int_{\pi/2}^{3\pi/4}\left(1+\sin 2\theta\right)d\theta$$

That was fantastic—a Mamma substitution *and* a trigonometric substitution. You can integrate this (using *u*-substitution for sin 2θ) as follows:

$$\frac{1}{2}\left(\theta-\frac{1}{2}\cos 2\theta\right)\Big|_{\pi/2}^{3\pi/4}$$

$$\frac{1}{2}\left(\frac{3\pi}{4}-\frac{\pi}{2}-\frac{1}{2}\right)$$

$$\frac{1}{2}\left(\frac{\pi}{4}-\frac{1}{2}\right).$$

It's not a pretty answer, but calculus ain't a beauty contest.

3. The best way to approach this problem is to calculate the area of *one* of the petals and multiply it by 4. This graph is symmetric in just about every possible way, so that makes it a little easier. The petal shaded below is bounded by the radial lines $\theta=\frac{\pi}{4}$ and $\theta=\frac{3\pi}{4}$.

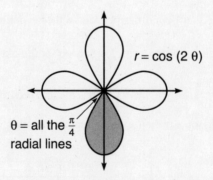

Its area is

$$\frac{1}{2}\int_{\pi/4}^{3\pi/4}\cos^2 2\theta d\theta.$$

In order to integrate this, you'll have to use a power-reduction formula.

$$\frac{1}{2}\int_{\pi/4}^{3\pi/4}\left(\frac{1}{2}+\frac{1}{2}\cos 4\theta\right)d\theta$$

$$\frac{1}{4}\int_{\pi/4}^{3\pi/4}d\theta+\frac{1}{16}\int_{\pi/4}^{3\pi/4}\cos u\,du$$

$$\frac{1}{4}\cdot\frac{\pi}{2}+\frac{1}{16}(0)=\frac{\pi}{8}$$

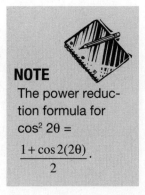

NOTE

The power reduction formula for $\cos^2 2\theta =$

$$\frac{1+\cos 2(2\theta)}{2}.$$

There are 4 petals, so the final answer is $4\cdot\dfrac{\pi}{8}=\dfrac{\pi}{2}$

4. The region in common is constructed, as shown in the diagram below.

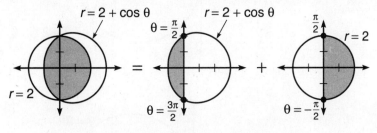

NOTE

In problem 3, $u = 4\theta$.

$$\frac{1}{2}\int_{\pi/2}^{3\pi/2}\left(2+\cos\theta\right)^2 d\theta+\frac{1}{2}\int_{-\pi/2}^{\pi/2}4\,d\theta$$

Use your calculator to evaluate and sum the definite integrals to get the approximate area of 9.352.

5. First, find the intersection points of the two graphs, and *be very careful—there's more here than meets the eye.*

$$3-2\sin\theta-4\cos\theta=0$$

$$\theta=1.299129483=A,\ \theta=5.911351042=B$$

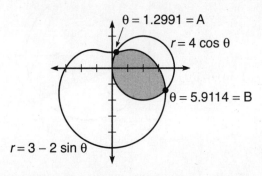

For the sake of not writing decimals until our eyes fall out, we'll use *A* and *B* instead of the gigantic decimal intersection values. What causes trouble here is the circle. It will actually pass through points *A* and *B* *twice*, since it draws its graph completely from $\theta = 0$ to $\theta = \pi$. So, the circle actually hits A when $\theta = A$ *and* when $\theta = A + \pi$. Use your calculator to convince yourself that this is true!

The circle will hit B when $\theta = B$ *and* when $\theta = B - \pi$. Below are the integrals that make up the area (there are 3) and a graph of the portion of the shaded region they represent.

$$\frac{1}{2} \int_0^A (3 - 2 \sin \theta)^2 \, d\theta$$

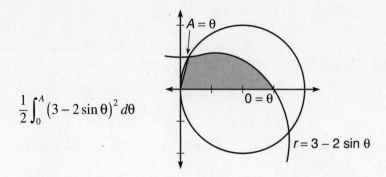

$$\frac{1}{2} \int_A^{B-\pi} (4 \cos \theta)^2 \, d\theta$$

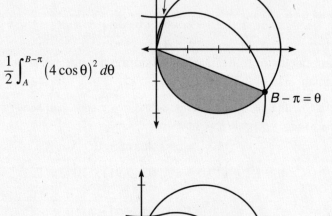

$$\frac{1}{2} \int_B^{2\pi} (3 - 2 \sin \theta)^2 \, d\theta$$

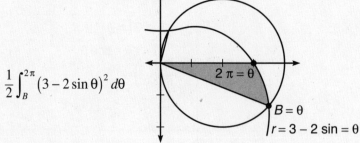

The only tricky integral is the second one. If you try to represent that area with the integral $\int_A^B (4 \cos \theta)^2 \, d\theta$, you will be tracing the entire circle an extra time. Make sure that you take the speed with which the graph draws into account. You need to know the exact values of θ for *that specific graph* that bound the region. When you add the three integrals, you get 2.496741051 + 3.494545773 + 2.116617352 ≈ 8.108.

TECHNOLOGY: USING YOUR CALCULATOR EFFICIENTLY

By now, you know the four major calculator proficiencies you need for the AP test: graphing, finding roots (solving equations), evaluating numeric derivatives, and evaluating definite integrals. However, you may not be using all of the calculator's features. Remember, the AP test is a high-stress situation and is governed by strict time constraints—you want the calculator to do as much as possible for you.

Take, for instance, problem 5 from the last problem set. (Even though it is a BC problem, the skills apply to AB students as well.) You should be using your calculator to serve two major purposes: finding the point of intersection and then evaluating three definite integrals. You already know how to find the solutions to the equation

$$3 - 2\sin\theta - 4\cos\theta = 0,$$

but here's a good tip. Once you find a root, press [Clear] twice to return to the home screen (you can also use [2ⁿᵈ]→[Mode], which is the [Quit] button). If you press [X,T,θ] and [Enter], the calculator gives you the root and shows more decimal places than it did on the graph screen. Now press [Sto→] and then [Alpha]→[Math] (the [A] button).

```
X
        1.299129483
Ans→A
        1.299129483
```

This stores that root as the variable *A* in the calculator's memory; this way, any time you need that number, you can simply use *A* instead of typing it out. Follow the process again, and store the other root to *B*.

```
        1.299129483
Ans→A
        1.299129483
X
        5.911351042
Ans→B
        5.911351042
```

Now, let's shorten the process of constantly typing the equations in as we find definite integrals. If you haven't already, switch to polar mode and graph the two equations: $r_1 = 3 - 2\sin\theta$ and $r_2 = 4\cos\theta$. Look how much easier it is to evaluate the second, troublesome integral from that problem when you use calculator shortcuts:

$$\frac{1}{2}\int_A^{B-\pi}(4\cos\theta)^2\,d\theta.$$

```
.5fnInt(r₂²,θ,A,
B-π)
        3.494545773
```

Instead of typing the limits of integration, just use the variables we defined. Instead of typing $(4\cos\theta)^2$, just use r_2^2. Once you type an equation into the "Y=" (or "r=") screen, you can access those equations by pressing [Vars]→"Y-vars". To get r_2, you then select "Polar…"→"r_2". If you are an AB student, you will use only the "Y=" screen, and all these equations are found under [Vars]→"Y-vars"→"Function…".

Look how much time you save when calculating the final integral if you employ all these shortcuts:

```
.5fnInt(r₁²,θ,0,
A)+.5fnInt(r₂²,θ
,A,B-π)+.5fnInt(
r₁²,θ,B,2π)
        8.107904177
```

These methods help eliminate the errors of mistyping and help you concentrate on the business at hand—passing that stinking AP test.

PRACTICE PROBLEMS

You may use a graphing calculator on problems 4 through 6.

1. If region A is bounded by $y = \frac{3}{x}$ and $y = -x + 4$, find

    (a) the area of A.

(b) the volume generated when *A* is rotated about the *y*-axis *using both the Washer and Shell Methods.*

(c) the volume of the solid with base *A* that has isosceles right triangular cross sections with hypotenuses perpendicular to the *x*-axis.

2. Prove that the volume of a sphere with radius *r* is $\frac{4}{3}\pi r^3$.

3. If region *B* is bounded by $y = -\sqrt{x} + 2$ and $y = 1$, find

(a) the area of *B*.

(b) the volume generated by rotating *B* about the *y*-axis.

(c) the volume generated when *B* is rotated about the line $y = -1$ *using both the Washer and Shell Methods.*

*4. If region *C* is bounded by the polar curve $r = \sin \theta \cos \theta$, find

(a) the area of *C*.

(b) the perimeter of *C*.

*5. Find the perimeter of the region bounded by $x = y^2 - y - 2$ and the *y*-axis.

6. **James' Diabolical Challenge:** A machine part is made from an alloy that costs $130 per cm³. The base of the part is bounded by the area between the functions $y = x$ and $y = x^2$ (*x* is measured in cm). The part can be made using either semicircular or equilateral triangular cross sections (perpendicular to the *x*-axis). Which method is less expensive?

TIP
Remember, you should always use the *entire* decimal when performing calculations; never round them, except when giving a final answer. Variables help you to do so without tedium.

SOLUTIONS

1. (a) The area is the definite integral of *top – bottom* with *x*-boundaries reflecting their intersection.

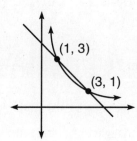

(1, 3)

(3, 1)

The area will be

$$\int_1^3 \left(-x + 4 - 3x^{-1}\right)dx$$

$$\left(-\frac{x^2}{2} + 4x - 3\ln|x|\right)\Bigg|_1^3$$

$$4 - \ln 27.$$

a BC-only question.

(b) *Shell Method*: Because you are rotating about a vertical axis, you use vertical rectangles with the Shell Method. No variables need to be converted—you use x's.

$$2\pi\int_1^3 x\left(-x+4-\frac{3}{x}\right)dx$$

$$2\pi\int_1^3\left(-x^2+4x-3\right)dx$$

$$2\pi\left(-\frac{x^3}{3}+2x^2-3x\right)\Big|_1^3$$

$$2\pi\cdot\frac{4}{3}=\frac{8\pi}{3}$$

Washer Method: You need to convert the equations so that they are in terms of y: $x=4-y$, $x=\frac{3}{y}$. The boundaries of integration are now the y-boundaries of the intersection, which are exactly the same as the x's. The outer radius will be $R(x)=(4-y)$, and the inner radius is $r(x)=\frac{3}{y}$.

$$\pi\int_1^3\left((4-y)^2-\left(\frac{3}{y}\right)^2\right)dy$$

$$\pi\int_1^3\left(16-8y+y^2-9y^{-2}\right)dy$$

$$\pi\left(16y-4y^2+\frac{y^3}{3}+\frac{9}{y}\right)\Big|_1^3$$

$$\pi\left(24-\left(21+\frac{1}{3}\right)\right)=\frac{8\pi}{3}$$

(c) In the problem set for known cross sections, we found that the area of an isosceles right triangle is $\frac{h^2}{4}$, if h is the hypotenuse. The length of each hypotenuse will be $h=-x+4-3x^{-1}$. Therefore, the volume is

$$\frac{1}{4}\int_1^3\left(-x+4+-3x^{-1}\right)^2 dx$$

$$\frac{1}{4}\int_1^3\left(x^2-8x+22-\frac{24}{x}+\frac{9}{x^2}\right)dx$$

$$\frac{1}{4}\left(\frac{x^3}{3} - 4x^2 + 22x - 24\ln|x| - \frac{9}{x}\right)\Big|_1^3$$

$$\frac{1}{4}\left(36 - 24\ln 3 - \frac{28}{3}\right) = \frac{1}{4}\left(\frac{80}{3} - 24\ln 3\right).$$

2. To create a sphere of your very own, you need to rotate a semicircle. A circle centered at the origin has equation $x^2 + y^2 = r^2$. Solving for y gives you a semicircle equation of $y = \sqrt{r^2 - x^2}$. When rotated about the x-axis, this graph produces a sphere.

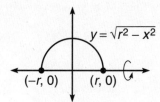

Use the Disk Method to find the volume.

$$\pi\int_{-r}^{r}\left(\sqrt{r^2 - x^2}\right)^2 dx$$

Remember that r is a constant, so treat it as you would any number as you complete the problem.

$$\pi\int_{-r}^{r}\left(r^2 - x^2\right)dx$$

$$\pi\left(r^2 x - \frac{x^3}{3}\right)\Big|_{-r}^{r}$$

Now, plug r and $-r$ into the expression in place of x (just as you have always done with the Fundamental Theorem).

$$\pi\left(\left(r^3 - \frac{r^3}{3}\right) - \left(-r^3 + \frac{r^3}{3}\right)\right)$$

$$\pi \cdot \frac{\left(4r^3\right)}{3}$$

3. (a) The area is *top – bottom* with x-boundary limits.

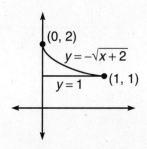

TIP

Even though you don't have to simplify the free-response section of the AP test, it's a good idea to practice simplifying, since most of the multiple-choice answers are in simplest form.

$$\int_0^1 \left(-\sqrt{x} + 2 - 1\right) dx$$

$$\int_0^1 \left(-x^{1/2} + 1\right) dx$$

$$\left(-\frac{2}{3} x^{3/2} + x\right)\Big|_0^1$$

$$-\frac{2}{3} + 1 = \frac{1}{3}$$

(b) There is no hole in the rotational solid, so you can use the Disk Method. To do so, however, you have to put everything in terms of y, since the Disk Method requires horizontal rectangles in this case. (You can use the Shell Method, and you will get the same thing.)

$$y = \sqrt{x} + 2$$
$$y - 2 = \sqrt{x}$$
$$x = (y - 2)^2$$

Now that the equation is in terms of y, use y boundaries and complete the integration to find volume.

$$\pi \int_1^2 \left((y - 2)^2\right)^2 dy$$

$$\pi \int_1^2 (y - 2)^4 dy$$

Set $u = y - 2$, and this is a simple u-substitution problem.

$$\pi \left(\frac{y - 2}{5}\right)\Big|_1^2 = \frac{\pi}{5}$$

(c) *Shell Method*: You must use horizontal rectangles (since they are parallel to $y = -1$), and everything must be in terms of y.

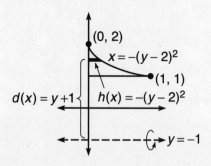

$$2\pi \int_1^2 (y+1)\left[-(y-2)\right]^2 dy$$

$$2\pi \int_1^2 \left(y^3 - 3y^2 + 4\right) dy$$

$$2\pi \left(\frac{y^4}{4} - y^3 + 4y\right)\Big|_1^2$$

$$2\pi \left(4 - \left(\frac{1}{4} - 1 + 4\right)\right)$$

$$2\pi \cdot \frac{3}{4} = \frac{3\pi}{2}$$

Washer Method: The washers will be in terms of x (since horizontal rectangles are perpendicular to the vertical axis of rotation).

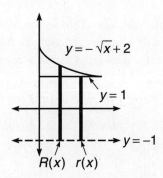

The outer radius is $R(x) = -\sqrt{x} + 3$, and the inner radius is $r(x) = 2$.

$$\pi \int_0^1 \left(\left(-\sqrt{x} + 3\right)^2 - 2^2\right) dx$$

$$\pi \int_0^1 \left(x - 6x^{\frac{1}{2}} + 5\right) dx$$

$$\pi \left(\frac{x^2}{2} - 4x^{\frac{3}{2}} + 5x\right)\Big|_0^1$$

$$\pi \left(\frac{1}{2} - 4 + 5\right) = \frac{3\pi}{2}$$

4. (a) This graph has four distinct petals that are framed by radial lines
   $\theta = 0$, $\theta = \frac{\pi}{2}$, $\theta = \pi$, $\theta = \frac{3\pi}{2}$, and $\theta = 2\pi$.

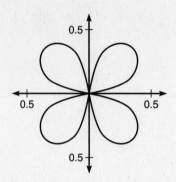

Because of the graph's symmetry, we can find the area of the petal in the first quadrant and multiply it by 4.

$$\frac{1}{2}\int_0^{\pi/2}\left(\cos^2\theta\sin^2\theta\right)d\theta$$

The area of the single petal is .0981747704, so the total area is approximately .393.

> (b) When finding the perimeter of the region, you are technically finding the arc length of its bounding function. To find polar arc length, you must first convert to parametric form by using $x = r\cos\Theta$ and $y = r\sin\theta$:

$$x = \cos^2\theta\sin\theta, \quad y = \sin^2\theta\cos\theta.$$

Now, take the derivative of each with respect to the parameter (which is Θ instead of t in this problem.)

$$\frac{dx}{d\theta} = -2\cos\theta\sin^2\theta + \cos^3\theta$$

$$\frac{dy}{d\theta} = 2\sin\theta\cos^2\theta - \sin^3\theta$$

Now, plug these both into the formula

$$\int_0^{\pi/2}\sqrt{\left(\frac{dx}{d\theta}\right)^2 + \left(\frac{dy}{d\theta}\right)^2}\,d\theta.$$

You're crazy if you don't use your calculator. Type it in *very* carefully. Here's what it should look like:

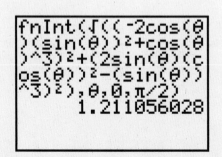

The total perimeter is four times that answer, which is approximately 4.844.

5. Again, you find the perimeter by calculating the arc length of the boundaries.

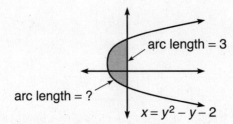

To find the length of the left boundary, you have to use the formula

$$\int_a^b \sqrt{1+\left(\frac{dx}{dy}\right)^2}\ dy.$$

The derivative (with respect to y, since the expression is in terms of y) is $x' = 2y - 1$. Therefore, the arc length is

$$\int_{-1}^2 \sqrt{1+\left(2y-1\right)^2}\ dy$$

$$\int_{-1}^2 \sqrt{4y^2 - 4y + 2}\ dy \approx 5.653$$

therefore, the total perimeter is 5.653 + 3 = 8.653.

6. Begin by drawing the bounded region and a darkened rectangle on it perpendicular to the x-axis.

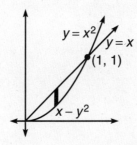

The length of the rectangle is $x - x^2$, and the region has x-boundaries 0 and 1. Use this information to find the cross-sectional volumes separately.

Semicircle cross sections

The area of a semicircle is $\frac{\pi r^2}{2}$, and the dark rectangle would represent a diameter of the semicircle. Therefore, $\frac{x - x^2}{2}$ would be its radius. Now, integrate the formula for the area of a semicircle, substituting in the radius.

$$\int_a^b \frac{\pi r^2}{2}\,dx$$

$$\frac{\pi}{2}\int_0^1 \frac{\left(x-x^2\right)^2}{4}\,dx$$

$$\frac{\pi}{8}\int_0^1 \left(x-x^2\right)^2 dx$$

The volume is .0130899694, so the cost of the part would be approximately $1.70.

Equilateral triangle cross sections

The area of an equilateral triangle is $\frac{\sqrt{3}}{4}s^2$, so the volume of this machine part will be

$$\frac{\sqrt{3}}{4}\int_0^1 \left(x-x^2\right)^2 dx \approx .0144337567.$$

The resultant cost for the part is $1.87. Therefore, the part with the semicircle cross sections is cheaper.

Differential Equations

A *differential equation* is simply an equation that contains a derivative. Your typical goal in a differential equation problem is to find the equation that has the given derivative; in other words, you are trying to find an antiderivative. How is this different from the integrals you have been finding until now? Well, it's not very different at all. You will be using all of your integration techniques to find particular solutions (no $+ C$) using the method of *separation of variables*. However, the vast majority of differential equations in the real world cannot be solved using this method. We will then further examine those solutions using *slope fields* and *Euler's Method*. Finally, we will look at some applications of differential equations in *exponential* and *logarithmic growth*. After we're done, we'll go get an ice cream cone, and I'll buy you that pony you've always wanted.

HANDS-ON ACTIVITY 10.1: SEPARATION OF VARIABLES

Even though the study of differential equations is complex, you are only required to know and understand the easiest of all methods for solving them—separation of variables. The name says it all, and you've already performed all the tasks that are involved in the process. Therefore, this section represents a new topic with nothing substantially new to learn (I love those kinds of topics).

1. What about the equation $\frac{dy}{dx} = -\frac{x}{y}$ makes it a *differential equation*?

_____

_____

2. Your goal will be to find an equation (in the form $y = f(x)$) whose derivative is $-\frac{x}{y}$. Why can't you simply integrate right away? What makes this equation different from the equation $\frac{dy}{dx} = \frac{4}{x^3}$, which you can integrate right away?

_____

_____

3. This topic is very similar to a differentiation topic for the reason you cited in number 2. What is the name of that topic, and why is it similar?

4. Before you can integrate $\frac{dy}{dx} = -\frac{x}{y}$, you must separate the variables (put all the y's on one side and the x's on the other). How can you accomplish this in our equation?

5. Go ahead and separate the variables using the method you named in number 4. Now, you should be able to integrate both sides of the equation separately. Integrate, remembering to include a C for any constant. What geometric shape is the solution to the differential equation?

6. The answer we have is very general (because of the C). What if you knew that the solution curve passed through the point $(0,3)$? Given this information, what would your solution be?

7. Let's try a new differential equation: $\frac{dy}{dx} = \frac{x^2 + x^2y}{3}$. Separate the variables, and integrate both sides separately.

8. Your y expression ends up contained in a natural log. In such cases, it is preferred to solve the equation for y, not $\ln y$. Solve for y.

9. If you knew that the particular solution you were looking for satisfied the condition $y(3) = 5$, what is C, and what is the solution to the differential equation?

_____

_____

_____

SELECTED SOLUTIONS TO HANDS-ON ACTIVITY 10.1

1. It contains a derivative.

2. $\frac{dy}{dx} = -\frac{x}{y}$ contains both x's and y's; until now, all of our integrals have contained just x's.

3. This is similar to implicit differentiation; those expressions contained both x's and y's and thus required a different method.

4. Separation can be accomplished in this problem by cross-multiplying. That may not work for all problems, but separation is usually achieved through very simple methods (see number 7).

5. Cross-multiplying gives you

$$ydy = -xdx$$

$$\int ydy = -\int xdx$$

$$\frac{y^2}{2} = -\frac{x^2}{2} + C.$$

Multiply everything by 2 and move the x term to get

$$x^2 + y^2 = C.$$

Therefore, the solution to the differential equation is a circle. (Often, solutions are written solved for y, but in this case, the answer is more clearly a circle when you leave it in standard form for a circle. Either way, however, the answer is right.)

6. Plug in the $x = 0$ and $y = 3$, since these values (if on the graph) must make the equation true.

$$0^2 + 3^2 = C$$

$$C = 9$$

Now that we know the specific value for C, we can plug it into the solution:

$$x^2 + y^2 = 9.$$

So, this specific solution is a circle of radius 3.

TIP
Even though you could have grouped the – with the y, it's better to leave it (and any constants a problem might have) on the side with the x's; this makes solving for y easier, and most differential equation solutions are in that form.

CAUTION
This is not the same C as in the previous step—it just indicates an arbitrary number. Some textbooks use different constants each time because of this, but that's silly—just remember that C might not ever be the same number.

7. Begin by factoring the x^2 out of the numerator.

$$\frac{dy}{dx} = \frac{x^2(1+y)}{3}$$

Divide both sides by $(1 + y)$, and multiply both sides by dx to separate the variables.

$$\frac{dy}{1+y} = \frac{x^2}{3}dx$$

Notice how the constant stays with the x terms so that you can solve for y more easily. Now, integrate both sides.

$$\ln|y+1| = \frac{x^3}{9} + C$$

8. To solve for y, raise e to the power of both sides to get

$$y+1 = e^{\frac{x^3}{9}+C}.$$

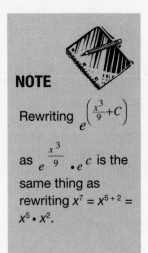

NOTE

Rewriting $e^{\left(\frac{x^3}{9}+C\right)}$ as $e^{\frac{x^3}{9}} \bullet e^c$ is the same thing as rewriting $x^7 = x^{5+2} = x^5 \bullet x^2$.

You can rewrite the right side of the equation as $e^{\frac{x^3}{9}} \bullet e^c$ (using properties of exponents), and e^c is just another constant, which you can then write as C. This gives you

$$y+1 = Ce^{\frac{x^3}{9}}$$

$$y = Ce^{\frac{x^3}{9}} - 1.$$

9. $y(3)=5$ means that plugging a value of 3 into the equation (for x) gives an output of 5; it's similar to saying $f(3) = 5$. If that confuses you, remember that $y(3) = 5$ means the point $(3,5)$ is on the graph. Either way, plug in $x = 3$ and $y = 5$ in order to find C.

$$5 = Ce^{27/9} - 1$$

$$6 = Ce^3$$

$$C = \frac{6}{e^3}$$

Therefore, the solution to this differential equation is

NOTE

When you are given a point and can find C in a differential equation, the resulting solution is called a *particular solution*, since it is only one of many possible solutions if no point were indicated.

$$y = \left(\frac{6}{e^3}\right) e^{\frac{x^3}{9}} - 1 \text{ or}$$

$$y = 6 e^{-3} \cdot e^{\frac{x^3}{9}} - 1 = 6 e^{\left(\frac{x^3}{9} - 3\right)} - 1.$$

PROBLEM SET

Do not use a calculator to solve the following.

1. Find the particular solution to $\dfrac{dy}{dx} = \dfrac{\sqrt{y}}{16 + x^2}$ that satisfies the condition that

    $y(0) = 1$.

2. What function has derivative $f'(x) = e^x \csc y$ and passes through the origin?

3. In some cases, the rate of change of a quantity is proportional to the quantity itself. This is written as

    $$\frac{dy}{dt} = ky,$$

where y is the quantity and k is the *proportionality constant*. What is the general solution to this very important differential equation?

4. A particle moves along the x-axis with acceleration at time t given by $a(t) = 3t$. Find the function describing the particle's position if it travels at a rate of 2 ft/sec when $t = 0$ and is 3 feet to the right of the origin when $t = 1$.

5. If a ball is thrown upward and reaches its maximum height when $t = 2$ seconds, answer the following questions:

    (a) Give the function representing the ball's *velocity*.

    (b) If the ball is 85 feet off the ground when $t = 1.5$ seconds, what is the position function for the ball?

NOTE

Your solution to number 3 is the foundation for the section on exponential growth and decay, where all the quantities in question have the property number 3 describes.

SOLUTIONS

1. Divide both sides by $\sqrt{y}$, and multiply by dx to separate the variables.

    $$\frac{dy}{\sqrt{y}} = \frac{dx}{16 + x^2}$$

    $$\int y^{-1/2}\, dy = \int \frac{dx}{16 + x^2}$$

$$2\sqrt{y} = \frac{1}{4}\arctan\frac{x}{4} + C$$

$$y = \left(\frac{1}{8}\arctan\frac{x}{4} + C\right)^2$$

Now, plug in the given values for x and y, and you find (easily) that $C = 1$. Therefore, the particular solution is

$$y = \left(\frac{1}{8}\arctan\frac{x}{4} + 1\right)^2.$$

2. Begin by rewriting the equation like this:

$$\frac{dy}{dx} = \frac{e^x}{\sin y}.$$

Now, cross-multiplying will separate the variables.

$$\int \sin y \, dy = \int e^x dx$$

$$-\cos y = e^x + C$$

$$y = \arccos(-e^x + C)$$

Now, plug in the point (0,0).

$$\cos(0) = \cos(\arccos(-1 + C))$$

$$-1 + C = 1$$

$$C = 2$$

The final answer is $y = \arccos(-e^x + 2)$.

3. Divide both sides by y, and multiply them by dt to get the necessary separation. Remember that k is a constant.

$$\int\frac{dy}{y} = k\int dt$$

$$\ln|y| = kt + C$$

$$e^{\ln|y|} = e^{kt+C}$$

$$e^{\ln|y|} = e^{kt} \cdot e^c$$

$$y = Ce^{kt}$$

4. (a) The integral of acceleration is velocity, and we know that $v(0) = 2$. So, find the antiderivative of $a(t)$, and plug in the given information.

$$v(t) = \int 3t\, dt$$

$$v(t) = \frac{3}{2}t^2 + C$$

$$v(0) = \frac{3}{2}\cdot 0^2 + C = 2$$

$$C = 2$$

Therefore, the equation for velocity is $v(t) = \frac{3}{2}t^2 + 2$. In order to find position, integrate again, and use the fact that $s(1) = 3$.

$$s(t) = \int\left(\frac{3}{2}t^2 + 2\right)dt$$

$$s(t) = \frac{1}{2}t^3 + 2t + C$$

$$s(1) = \frac{1}{2} + 2 + C = 3$$

$$C = \frac{1}{2}$$

Therefore, the position equation is $s(t) = \frac{1}{2}t^3 + 2t + \frac{1}{2}$.

5. (a) Remember that the equation for projectile position is
    $s(t) = -16t^2 + v_0 t + h_0$, where v_0 is the initial velocity and h_0 is the
    initial height. You are looking for the velocity equation, so take the
    derivative:

$$v(t) = -32t + v_0.$$

The ball reaches its maximum height when $t = 2$. Therefore, the derivative
must equal 0 when $t = 2$. This allows you to find v_0.

$$v(2) = -32(2) + C = 0$$

$$C = 64$$

Therefore, the velocity equation is $v(t) = -32t + 64$.

    (b) To find position, integrate velocity, and use the fact that $s(1.5) = 85$.

$$s(t) = \int(-32t + 64)dt$$

$$s(t) = -16t^2 + 64t + C$$

$$s(1.5) = -36 + 96 + C = 85$$

$$C = 25$$

Therefore, the position equation for the ball is

$$s(t) = -16t^2 + 64t + 25.$$

HANDS-ON ACTIVITY 10.2: SLOPE FIELDS

Slope fields sound like dangerous places to play soccer but are, instead, handy ways to visualize differential equations. Remember back in the section on linear approximations when we discussed the fact that a derivative has values very close to its original function at the point of tangency? When that fact manifests itself all over the coordinate plane, it's truly something to behold. So, you'd better be holding on to something when you undertake this activity.

1. Let's return to the first differential equation from activity 10.1: $\frac{dy}{dx} = -\frac{x}{y}$. What exactly does this equation tell you about the general solution?

    _____

2. If the solution to $\frac{dy}{dx} = -\frac{x}{y}$ contained the point $(0,1)$, what could you determine about the graph of the solution at that point?

    _____

3. What would the tangent line to the solution graph look like at the point $(-2,0)$?

    _____

    _____

4. Calculate the slopes at all of the points indicated on the axes below, and draw a small line segment with the correct slope. The slope at point $(0,1)$ has already been drawn as an example.

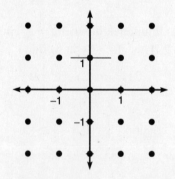

5. The general solution to $\dfrac{dy}{dx} = -\dfrac{x}{y}$ (according to our work in the last chapter) was $x^2 + y^2 = C$. How does the solution relate to the drawing you made in number 4?

_____

_____

6. What is the purpose of a slope field?

_____

_____

7. Draw the particular solution to $\dfrac{dy}{dx} = -\dfrac{x}{y}$ that passes through the point $(2,0)$ on the slope field above.

8. Draw the slope field for $\dfrac{dy}{dx} = 2x^2 y$ on the axes below.

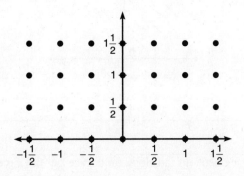

9. Use the slope field to draw an approximate solution graph that contains the point $(-\frac{1}{2}, \frac{1}{2})$.

10. Find the particular solution to $\dfrac{dy}{dx} = 2x^2 y$ that contains $(-\frac{1}{2}, \frac{1}{2})$ using separation of variables.

_____

_____

_____

11. Use your calculator to draw the graph of the particular solution you found in number 10. It should look a lot like the graph you drew in number 9.

SELECTED SOLUTIONS TO HANDS-ON ACTIVITY 10.2

Note: All slope fields in the solution are drawn with a computer and hence with more precision than in the exercises. Your slope fields should look similar, though not as detailed. See the technology section at the end of this chapter for a program to help you draw slope fields with your TI-83.

1. It tells you the slope, $\frac{dy}{dx}$, of the tangent lines to the solution graph at all points (x,y).

2. You would know that the slope of the tangent line to the graph at $(0,1)$ would be $\frac{dy}{dx} = -\frac{0}{1} = 0$. The graph would have a horizontal tangent line there.

3. The slope of the tangent line (also called the *slope of the solution curve* itself) is $\frac{dy}{dx} = \frac{2}{0}$, which is undefined. Therefore, the tangent line there is vertical (since a vertical line has an undefined slope).

4. All you have to do is plug each (x,y) point into $\frac{dy}{dx}$ to get the slope. Draw a small line segment with approximately that slope. It doesn't have to be exact, but a slope of $\frac{1}{2}$ should be *much* shallower than a slope of 2.

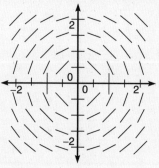

5. The tangent segments in the slope field trace out the circular shape of the solution graph. Remember that a tangent line has values very close to its original graph near the point of tangency. So, if we draw nothing but little tangent lines so small that all the points on the segment are close to the point of tangency, the result looks like the solution graph.

6. A slope field gives you a basic idea of the shape of the solution graph.

7. The particular solution will be a circle centered at the origin with radius 2.

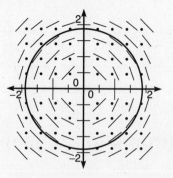

8.

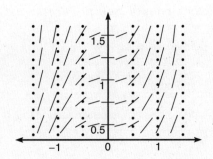

9.

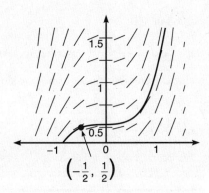

$$\left(-\frac{1}{2}, \frac{1}{2}\right)$$

10. Divide by y and multiply by dx to separate variables.

$$\int \frac{dy}{y} = 2 \int x^2 dx$$

$$\ln|y| = \frac{2x^3}{3} + C$$

$$y = Ce^{\frac{2x^3}{3}}$$

To find C, plug in the point $(-\frac{1}{2}, \frac{1}{2})$.

$$\tfrac{1}{2} = Ce^{-1/12}$$

$$C = \tfrac{1}{2}e^{1/12}$$

The final (unsimplified) solution is $y = \dfrac{e^{1/12}}{2} \cdot e^{\frac{2x^3}{3}}$.

PROBLEM SET

Do not use a calculator on any of these problems.

1. Sketch the slope field of $\dfrac{dy}{dx} = -x^2 + y$.

2. (a) Draw the slope field for $\dfrac{dy}{dx} = \dfrac{y + xy}{x}$.

   (b) Find the solution of the differential equation that passes through the point $(1, 2e)$.

3. (a) Use the slope field of $f'(x) = \dfrac{x}{y}$ to draw an approximate graph of $f(x)$ if $f(-2) = 0$.

   (b) Find $f(x)$ specified in 3(a).

4. Explain how the slope field of $\dfrac{dy}{dx} = 3x^2$ describes its general solution.

5. Which of the following differential equations has the slope field below?

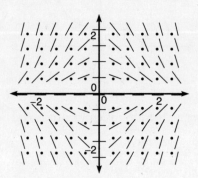

(A) $\dfrac{dy}{dx} = x - y$

(B) $\dfrac{dy}{dx} = \sqrt{y} - \sqrt{x}$

(C) $\dfrac{dy}{dx} = -xy$

(D) $\dfrac{dy}{dx} = \dfrac{2x}{y}$

SOLUTIONS

1. All kinds of weird stuff happens close to the origin. Plot enough points so that you can see what's going on. Remember, all you have to do is plug in *any* points (x,y) into the differential equation $\dfrac{dy}{dx}$; the result is the slope of the line segment you should draw at that point.

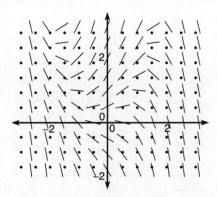

2. (a)

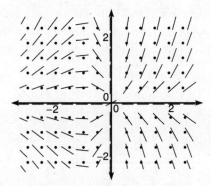

(b) Solve the differential equation using separation of variables. You'll have to factor a y out of the numerator to do so.

$$\frac{dy}{dx} = \frac{y(1+x)}{x} dx$$

$$\int \frac{dy}{y} = \int \frac{1+x}{x} dx$$

$$\int \frac{dy}{y} = \int \left(x^{-1} + 1\right) dx$$

$$\ln|y| = \ln|x| + x + C$$

$$e^{\ln y} = e^{\ln x} \cdot e^x \cdot e^c$$

$$y = Cxe^x$$

Now, plug in the point $(1, 2e)$.

$$2e = C(1)e$$

$$C = 2$$

Therefore, the solution is $y = 2xe^x$. If you graph it using your calculator (although the problem doesn't ask you to do so), you'll see that it fits the slope field perfectly.

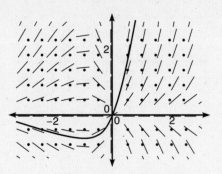

3. (a) This slope field sort of looks like the big-bang theory—everything is exploding out of the origin. The solution just screams, "I am a hyperbola! Love me! Accept me! Tell me that I am handsome!"

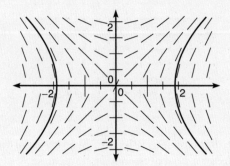

(b) It's the revenge of separation of variables. In fact, this problem is very similar to the equation all of us are growing tired of: $\frac{dy}{dx} = -\frac{x}{y}$. Begin by cross-multiplying.

$$\frac{dy}{dx} = \frac{x}{y}$$

$$\int y\,dy = \int x\,dx$$

$$\frac{y^2}{2} = \frac{x^2}{2} + C$$

$$y^2 = x^2 + C$$

Now, find the C (Caspian).

$$0 = 4 + C$$

$$C = -4$$

All that remains is to plug in everything. Remember that the standard form for a hyberbola always equals 1.

$$y^2 - x^2 = -4$$

$$\frac{x^2}{4} - \frac{y^2}{4} = 1$$

This is the equation of the hyberbola in part 3(a).

4. Elementary integration tells you that if $\frac{dy}{dx} = 3x^2$, then $y = x^3 + C$. The slope field

    for $\frac{dy}{dx} = 3x^2$ outlines the family of curves $y = x^3 + C$.

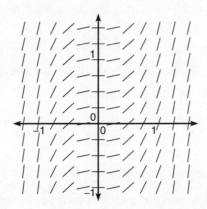

5. If you drew all four slope fields, you wasted valuable time; if this were the AP test, you'd have wasted five precious minutes on a very simple question in disguise. Look at the defining characteristic of the slope field: it has horizontal tangents *all along the x- and y-axes*. In other words, if either the *x*-coordinate

    of the point is 0 *or* the *y*-coordinate of the point is 0, then $\frac{dy}{dx}$ equals 0. That is

    only true for one of the four choices: (C). For example, choice (D) is undefined at points that have $y = 0$; choices (A) and (B) won't have horizontal tangents anywhere except for the origin.

EULER'S METHOD (BC TOPIC ONLY)

If you could solve all differential equations using separation of variables, the world would be a much happier place. However, only a very small portion of differential equations can be solved that way. There are a slew of other methods you'd have to learn to become King (or Queen) of Differential Equations, but luckily you don't have to. Enter Euler, famous Swiss mathematician, and his method of finding *approximate* solutions to nonintegrable differential equations. Even though *Euler's Method* only uses tangent lines and is quite simple in premise, you can use it to find an approximate answer to a differential equation.

NOTE
Euler is pronounced "oiler," not "youler."

Before we dive into Euler's Method, we need to focus on one simple concept. Let's say that a line has slope $\frac{12}{7}$ and contains point $(1,0)$. If you start at the point and go $\frac{1}{2}$ unit to the right, how many units must you go up to stay on the line? What *y*-coordinate completes the point on the following page?

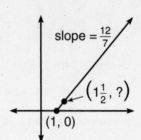

Believe it or not, the answer is quite easy. Take the slope formula that you've used since elementary algebra: $m = \frac{\Delta y}{\Delta x}$. Solve it for Δy.

$$\Delta y = m \cdot \Delta x$$

In our line problem, the slope is $\frac{12}{7}$, and the change in x (Δx) from (1,0) is $\frac{1}{2}$. Plug these values into the new formula.

$$\Delta y = \frac{12}{7} \cdot \frac{1}{2} = \frac{6}{7}$$

This tells you that the point $(\frac{1}{2}, \frac{6}{7})$ is also on the line with slope $\frac{12}{7}$ and point (1,0). If you went $\frac{1}{2}$ unit right of (1,0), you'd have to go $\frac{6}{7}$ unit up to stay on the line. This is a very important part of Euler's Method, although most textbooks do not explain what it means or why it is so important. Now, let's map out the goal of Euler's Method.

A typical differential equation contains a derivative and a point through which the solution graph passes. We will find the slope, m_1, of the tangent line at that starting point. Then, we will travel a fixed distance right or left (Δx) and use the method above to find the corresponding Δy. Basically, we are finding another point on the tangent line we just drew.

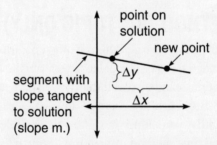

This works because a tangent line has values close to the graph it's tangent to (around the point of tangency). Once we find that new point, we repeat the process and find yet another point until we reach our approximation. The problem will typically tell you how many steps you should take to reach the solution, and (just like integration) the smaller the steps you take, the more accurate your answer. This probably sounds very complicated, but it's quite easy. How many licks does it take to get to the Tootsie-Roll center of Euler's Method? Let's find out.

Example 1: Use Euler's Method to approximate $y(1)$ if $\frac{dy}{dx} = -\frac{x}{y}$ and $y(0) = 3$. Use two steps of length $\Delta x = .5$ when finding the solution.

Solution: This is the same differential equation we have seen numerous times now. We already know that its solution is a circle with radius 3, but let's pretend we don't, since most differential equations using Euler's Method will not be solvable. We will have to repeat the method two times, once when $x = .5$ and once when $x = 1$ (since we are taking two steps of length $\frac{1}{2}$ along the x-axis from $x = 0$ to $x = 1$, as the problem indicates).

What is the slope of the tangent line to the solution curve at (0,3)? Plug the values into $\frac{dy}{dx}$:

$$\frac{dy}{dx} = -\frac{0}{3} = 0.$$

We will move $\frac{1}{2}$ unit to the right along the tangent line (since $\Delta x = \frac{1}{2}$); in order to stay on the line (since it is horizontal), we should move neither up nor down. Our new point will be $(\frac{1}{2},3)$. It's sort of like tightrope walking toward the solution—the tightrope is horizontal, and we edge our way $\frac{1}{2}$ unit to the right along the highwire and stop there.

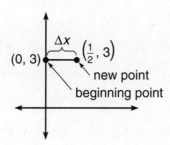

Now, repeat the process with the point $(\frac{1}{2},3)$. The slope of the tangent line here will be

$$\frac{dy}{dx} = -\frac{\frac{1}{2}}{3} = -\frac{1}{6}.$$

Again, we will move $\frac{1}{2}$ unit to the right along this new tangent line. However, how much should we go up or down to stay on the new tangent line? Use the formula from the beginning of the section.

$$\Delta y = m \cdot \Delta x$$

$$\Delta y = -\frac{1}{6} \cdot \frac{1}{2} = -\frac{1}{12}$$

Therefore, the new point is $\left(\frac{1}{2} + \frac{1}{2}, 3 - \frac{1}{12} \right) = \left(1, \frac{35}{12} \right)$, and we're finished.

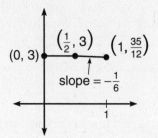

Our approximation for $f(1)$ is $\frac{35}{12}$. Since we already know the actual solution of the differential equation is $x^2 + y^2 = 9$, we can find the actual solution of $\sqrt{8} \approx 2.828427$. Our estimate ($\frac{35}{12}$) is approximately 2.916667, so the approximation has an error of .08824.

Had we used more than two steps (which means a smaller Δx) in Example 1, the answer would have been even more accurate. In the next example, we'll use four steps to travel the same distance. Furthermore, you won't be able to solve Example 2 by separation of variables. You'll be out in the wild, untamed forest with only one weapon—Euler's Method. Make sure you know how to use it!

Example 2: Approximate the value of $y(2)$ for the differential equation $\frac{dy}{dx} = x - y$ using four steps of length $\Delta x = \frac{1}{4}$, given that the point (1,2) is on the solution curve.

Solution: We will have to repeat the process four times. If you want to use decimals instead of fractions, that's okay, but remember not to round off until the very end. However, fractions will give you the exact answer—and don't be shy about your calculator's ability to add, subtract, and multiply fractions faster than you can.

Point (1,2): The slope of the line tangent to the solution is $\frac{dy}{dx} = x - y = 1 - 2 = -1$. The corresponding Δy will be

$$\Delta y = (-1)\left(\frac{1}{4}\right) = -\frac{1}{4}.$$

The new point will be $\left(1 + \frac{1}{4}, 2 - \frac{1}{4} \right) = \left(\frac{5}{4}, \frac{7}{4} \right)$.

Point $\left(\frac{5}{4}, \frac{7}{4} \right)$: $\frac{dy}{dx} = \frac{5}{4} - \frac{7}{4} = -\frac{1}{2}$. The corresponding Δy will be

$$\Delta y = \left(-\frac{1}{2} \right)\left(\frac{1}{4} \right) = -\frac{1}{8}.$$

The new point will be $(\frac{5}{4} + \frac{1}{4}, \frac{7}{4} - \frac{1}{8}) = (\frac{3}{2}, \frac{13}{8})$.

Point $\left(\frac{3}{2}, \frac{13}{8} \right)$: $\frac{dy}{dx} = \frac{3}{2} - \frac{13}{8} = -\frac{1}{8}$. The corresponding Δy will be

$$\Delta y = \left(-\frac{1}{8} \right)\left(\frac{1}{4} \right) = -\frac{1}{32}.$$

The new point will be $(\frac{3}{2} + \frac{1}{4}, \frac{13}{8} - \frac{1}{32}) = (\frac{7}{4}, \frac{51}{32})$.

Point $\left(\frac{7}{4}, \frac{51}{32} \right)$: $\frac{dy}{dx} = \frac{7}{4} - \frac{51}{32} = \frac{5}{32}$. The corresponding Δy will be

$$\Delta y = \left(\frac{5}{32} \right)\left(\frac{1}{4} \right) = \frac{5}{128}.$$

The new point will be $\left(\frac{7}{4} + \frac{1}{4}, \frac{51}{32} + \frac{5}{128} \right) = (2, \frac{209}{128})$. Therefore, your approximation of $y(2) = \frac{209}{128} \approx 1.633$.

There you have it—Euler's Method is quite mechanical, and besides the ugly fractions and/or decimals, it is a very handy way to approximate solutions to differential equations that we, as elementary calculus students, can solve no other way.

PROBLEM SET

You may use a graphing calculator to simplify your arithmetic on all problems, but don't use the Euler's Method calculator program until problem 4.

1. Use Euler's Method to approximate $y(6)$ for $\frac{dy}{dx} = \frac{1}{x}$ given that $y(5) = 2$. Use three steps of size $\Delta x = \frac{1}{3}$.

2. (a) If the points $(1,3)$ and $(4,c)$ are on the solution graph to $\frac{dy}{dx} = \frac{x+y}{x}$, approximate c using Euler's Method and three steps of length $\Delta x = 1$.

    (b) What limitations are evident in this approximation?

3. (a) Use Euler's Method to approximate $y(-\frac{1}{2})$ for $\frac{dy}{dx} = axy$ (where a is a real number) if $y(0) = -2$. Use two steps of length $\Delta x = \frac{1}{4}$.

    (b) What is the error on your approximation if $a = 3$?

4. Approximate $y(0)$ using 10 steps of length $\Delta x = .1$ for $\frac{dy}{dx} = y^2$ given that $y(-1) = 1$. Fill out the chart below; it will help organize your information.

(x, y)	$\frac{dy}{dx}$	$\Delta y = (.1)\frac{dy}{dx}$

5. Give an example of a differential equation for which Euler's Method gives the *exact* value rather than just an approximation.

SOLUTIONS

1. The given point is $(5,2)$ and $\Delta x = \frac{1}{3}$. We'll start there and apply Euler's Method three times.

Point (5,2): $\frac{dy}{dx} = \frac{1}{x} = \frac{1}{5}$. The corresponding Δy is

$$\Delta y = \frac{1}{5} \cdot \frac{1}{3}.$$

The new point will be $\left(5 + \frac{1}{3},\ 2 + \frac{1}{15}\right) = \left(\frac{16}{3},\ \frac{31}{15}\right)$.

Point $\left(\frac{16}{3}, \frac{31}{15}\right)$: $\frac{dy}{dx} = \frac{3}{16}$.

$$\Delta y = \frac{3}{16} \cdot \frac{1}{3} = \frac{1}{16}.$$

The new point will be $\left(\left(\frac{16}{3} + \frac{1}{3},\ \frac{31}{15} + \frac{1}{16}\right) = \left(\frac{17}{3},\ \frac{511}{240}\right)\right.$.

Point $\left(\frac{17}{3}, \frac{511}{240}\right)$: $\frac{dy}{dx} = \frac{3}{17}$.

$$\Delta y = \frac{3}{17} \cdot \frac{1}{3} = \frac{1}{17}.$$

The correct approximation for y is $\frac{511}{240} + \frac{1}{17} \approx 2.188$.

2. (a) This question is just asking you to find $y(4)$ given a starting point of $(1,3)$ and three steps of length 1 to get there.

Point (1,3): $\frac{dy}{dx} = \frac{4}{1} = 4$; $\Delta y = 4 \cdot 1 = 4$. The new point will be $(1 + 1, 3 + 4) = (2,7)$.

Point (2,7): $\frac{dy}{dx} = \frac{9}{2}$; $\Delta y = \frac{9}{2} \cdot 1 = \frac{9}{2}$. The new point will be $\left(2+1, 7+\frac{9}{2}\right)$ $= \left(3, \frac{23}{2}\right)$.

Point $\left(3, \frac{23}{2}\right)$: $\frac{dy}{dx} = \frac{\frac{29}{2}}{3} = \frac{29}{6}$; $\Delta y = 1 \cdot \frac{29}{6}$.

Therefore, $C = \frac{23}{2} + \frac{29}{6} = \frac{49}{3} \approx 16.333$.

(b) The *y*-value is changing pretty dramatically as the *x*-value only changes by 1. With such a huge change in *y*, the approximation can't be very accurate; it turns out to be within about 1 unit of the correct answer, but that is not nearly as accurate of an approximation as you can get with a less steep graph. Of course, you can always increase the number of steps by decreasing Δx; that will always make your approximation more accurate.

NOTE
Check out the slope field for $\frac{dy}{dx} = \frac{x+y}{x}$. It's pretty.

3. (a) This one has only two steps, thank goodness. We will be stepping backward from $x = 0$ to $x = -\frac{1}{2}$, so Δx must be *negative*: $\Delta x = -\frac{1}{4}$.

Point (0,–2): $\frac{dy}{dx} = a(0)(-2) = 0$; $\Delta y = (-\frac{1}{4})(0) = 0$. The new point will be $(0 - \frac{1}{4}, -2 - 0)$.

Point $(-\frac{1}{4},-2)$: $\frac{dy}{dx} = a(-\frac{1}{4})(-2) = \frac{a}{2}$; $\Delta y = -\frac{1}{4} \cdot \frac{a}{2} = -\frac{a}{8}$. The new point will be $(-\frac{1}{2},-2-\frac{a}{8})$.

Therefore, Euler's Method gives an approximation of $-2 - \frac{a}{8}$. The number *a* could be anything, so leave it as *a*.

(b) If $a = 3$, you have the equation $\frac{dy}{dx} = 3xy$; our work above gives an approximation of $y(-\frac{1}{2}) = -2 - \frac{3}{8} = -2.375$. To find out the *actual* value of $y(-\frac{1}{2})$, you have to solve $\frac{dy}{dx}$ using separation of variables:

$$\frac{dy}{dx} = 3xy$$

$$\int \frac{dy}{y} = \int 3x\,dx$$

$$\ln|y| = \frac{3}{2}x^2 + C$$

$$y = Ce^{3x^2/2}.$$

Plug in the given point $(0,-2)$ to find C:

$$-2 = C \cdot 1.$$

The solution equation will be $y = -2e^{3x^2/2}$; its value for $x = -\frac{1}{2}$ (which we predicted to be -2.375) is actually

$$y\left(-\frac{1}{2}\right) = -2e^{3/8} = -2.909982829 .$$

The error in the approximation was about .535.

4. The correct approximation for $y(0)$ is nearly 6.129.

(x, y)	$\frac{dy}{dx}$	$\Delta y = (.1)\frac{dy}{dx}$	
$(-1, 1)$	1	$\Delta y = (.1)1$	
$(-.9, 1.1)$	1.21	$\Delta y = (.1)(1.21)$	
$(-.8, 1.221)$	1.491	$\Delta y = (.1)(1.491)$	
$(-.7, 1.370)$	1.877	$\Delta y = (.1)(1.877)$	
$(-.6, 1.558)$	2.427	$\Delta y = (.1)(2.427)$	
$(-.5, 1.800)$	3.240	$\Delta y = (.1)(3.240)$	
$(-.4, 2.125)$	4.516	$\Delta y = (.1)(4.516)$	
$(-.3, 2.576)$	6.636	$\Delta y = (.1)(6.636)$	$(0, 6.129)$
$(-.2, 3.240)$	10.498	$\Delta y = (.1)(10.498)$	
$(-.1, 4.289)$	18.396	$\Delta y = (.1)(18.396)$	

Note: 3 decimal places are written, but decimals are not rounded in calculations

5. If $\frac{dy}{dx} = c$, where c is any real number, the solution to the differential equation is just a line. Because Euler's Method gets its value from the tangent line (and the tangent line to a linear equation is just the line itself), you will be stepping through the graph of the solution during Euler's Method, and the answer will be exact. Try it, and see for yourself. Pick a value for c and a given point through which the solution line will pass; use Euler's method to approximate something, and then double check it with separation of variables. The values will match.

EXPONENTIAL GROWTH AND DECAY

You have probably alluded to exponential growth in everyday conversation without even realizing it. Perhaps you've said things like, "Ever since I started carrying raw meat in my pockets, the number of times I've been attacked by wild dogs has increased *exponentially*." Exponential growth is sudden, quick, and relentless. Mathematically, exponential growth or decay has one defining characteristic (and this is *key*): the rate of y's growth is directly proportional to y itself. In other words, the bigger y is, the bigger it grows; the smaller y is, the slower it decays.

Mathematically, something exhibiting exponential growth or decay satisfies the differential equation

$$\frac{dy}{dt} = ky,$$

where k is called the *constant of proportionality*. A model ship might be built to a 1:35 scale, which means that any real ship part is 35 times as large as the model. The constant of proportionality in that case is 35. However, k in exponential growth and decay is never so neat and tidy, and it is rarely (if ever) evident from reading a problem. Luckily, it is quite easy to find.

In the first problem set of this chapter (problem 3), you proved that the general solution to $\frac{dy}{dt} = kt$ is $y = Ce^{kt}$. I find the formula easier to remember, however, if you call the constant N instead of C (although that doesn't amount to a hill of beans mathematically). Why is it easier to remember? It sounds like Roseanne pronouncing "naked"—"nekkit."

$$y = Ne^{kt}$$

In this formula, N stands for the original amount of material, k is the proportionality constant, t is time, and y is the amount of N that remains after time t has passed. When approaching exponential growth and decay problems, your first goals should be to find N and k; then, answer whatever question is being posed. Don't be intimidated by these problems—they are very easy.

Example 3: The new theme restaurant in town (Rowdy Rita's Eat and Hurl) is being tested by the health department for cleanliness. Health inspectors find the men's room floor to be a fertile ground for growing bacteria. They have determined that the rate of bacterial growth is proportional to the number of colonies. So, they plant 10 colonies and come back in 15 minutes; when they return, the number of colonies has risen to 35. How many colonies will there be one full hour after they planted the original 10?

Solution: The key phrase in the problem is "*the rate of bacterial growth is proportional to the number of colonies,*" because that means that you can apply exponential growth and decay. They started with 10 colonies, so $N = 10$ (starting amount). *Do not try to figure out what k is in your head—it defies simple calculation.* Instead, we know that there will be 35 colonies after $t = 15$ minutes, so you can set up the equation

$$35 = 10e^{k(15)}.$$

Solve this equation for k. Divide by 10 to begin the process.

$$\frac{7}{2} = e^{15k}$$

$$\ln \frac{7}{2} = 15k$$

$$k = \frac{\ln \frac{7}{2}}{15}$$

$$k = .0835175312$$

Now you have a formula to determine the amount of bacteria for any time *t* minutes after the original planting:

$$y = 10e^{(.0835175312)t}.$$

We want the amount of bacteria growth after 1 hour; since we calculated *k* using minutes, we'll have to express 1 hour as *t* = 60 minutes. Now, find the number of colonies.

$$y = 10e^{(.0835175312)(60)}$$

$$y \approx 1500.625$$

So, almost 1,501 colonies are partying along the surface of the bathroom floor. In one day, the number will grow to 1.7×10^{53} colonies. You may be safer going to the bathroom in the alley behind the restaurant.

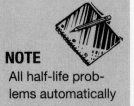

NOTE

All half-life problems automatically satisfy the property $\frac{dy}{dt} = ky$ by their very nature.

TIP

In an exponential *decay* problem such as this, the *k* will be negative.

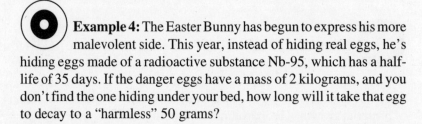

Example 4: The Easter Bunny has begun to express his more malevolent side. This year, instead of hiding real eggs, he's hiding eggs made of a radioactive substance Nb-95, which has a half-life of 35 days. If the danger eggs have a mass of 2 kilograms, and you don't find the one hiding under your bed, how long will it take that egg to decay to a "harmless" 50 grams?

Solution: The egg starts at a mass of 2000 g. A half-life of 35 days means that in 35 days, exactly half of the mass will remain. After 70 days, one fourth of the mass will remain, etc. Therefore, after 35 days, the mass will be 1000. This information will allow us to find *k*.

$$1000 = 2000e^{k(35)}$$

$$\tfrac{1}{2} = e^{35k}$$

$$\ln \tfrac{1}{2} = 35k$$

$$\frac{ln\frac{1}{2}}{35} = k$$

$$k = -.0198042052$$

Now that we know N and k, we want to find t when only 50 grams are left. In this case, t will be in days (since days was the unit of time we used when determining k).

$$50 = 2000e^{-.0198042052t}$$

$$\frac{1}{40} = e^{-.0198042052t}$$

$$\frac{\ln \frac{1}{40}}{-.0198042052} = t$$

$$t \approx 186.267 \text{ days}$$

You should be safe by Thanksgiving. (Nothing wrong with a little premature hair loss and a healthy greenish glow, is there?)

PROBLEM SET

You may use a graphing calculator on all of these problems.

1. If Pu-230 (a particularly stinky radioactive substance) has a half-life of 24,360 years, find an equation that represents the amount of Pu-230 left after time t, if you began with N grams.

2. Most men in the world (except, of course, for me, if my wife is reading this) think that Rebecca Romijn-Stamos is pretty attractive. If left unchecked (and the practice were legal), we can assume the number of her husbands would increase exponentially. As of right now, she has one husband, but if legal restrictions were lifted (and Uncle Jesse couldn't foot the bill), she might have 4 husbands 2 years from now. How many years would it take her to increase her harem to 100 husbands if the number of husbands is proportional to the rate of increase?

3. Assume that the world population's interest in the new boy band, "Hunks o' Love," is growing at a rate proportional to the number of its fans. If the *Hunks* had 2,000 fans one year after they released their first album and 50,000 fans five years after their first album, how many fans did they have the moment the first album was released?

4. Vinny the Talking Dog was an impressive animal for many reasons during his short-lived career. First of all, he was a talking dog, for goodness sakes! However, one of the unfortunate side-effects of this gift was that he increased his size by $\frac{1}{3}$ every two weeks. If he weighed 5 pounds at birth, how many *days* did it take him to reach an enormous 600 pounds (at which point his poor, pitiable, poochie heart puttered out)?

SOLUTIONS

1. Because the rate of decrease is proportional to the amount of substance (the amount decreases by half), we can use exponential growth and decay. In other words, let's get Ne^{kt}. In 24,360 years, N will decrease by half, so we can write

$$\frac{N}{2} = Ne^{k(24360)}.$$

Divide both sides by N, and you get

$$\frac{1}{2} = e^{24360k}$$

$$\frac{\ln\frac{1}{2}}{24360} = k = -.0000284543178.$$

Therefore, the equation $y = Ne^{(-.0000284543178)t}$ will give you the amount of Pu-230 left after time t if you began with N grams.

2. This problem clearly states the proportionality relationship required to use exponential growth and decay. Here, $N = 1$, and $y = 4$ when $t = 2$ years, so you can set up the equation:

$$4 = 1e^{k\cdot2}$$

$$\frac{\ln 4}{2} = k = .6931471806$$

Now that we have k, we need to find t when $y = 100$.

$$100 = 1e^{.6931471806t}$$

$$\ln 100 = .6931471806t$$

$$t \approx 6.644 \text{ years}$$

3. Our job in this problem will be to find N, the original number of fans. We have the following equations based on the given information:

$$2000 = Ne^{k(1)} \text{ and } 50,000 = Ne^{k(5)}.$$

Solve the first equation for N, and you get

$$N = \frac{2000}{e^k}.$$

Plug this value into the other equation, and you can find k.

$$50000 = \frac{2000}{e^k} \cdot e^{5k}$$

$$50000 = 2000e^{5k}e^{-k}$$

$$25 = e^{4k}$$

$$\frac{\ln 25}{4} = k = .8047189562$$

Finally, we can find the value of N by plugging k into $N = \dfrac{2000}{e^k}$.

$$N = \frac{2000}{e^{.8047189562}}$$

$$N \approx 894.427 \text{ original fans}$$

4. Oh, cruel fate. If Vinny weighed 5 pounds at birth, he weighed $5 + \frac{1}{3} \cdot 5 = \frac{20}{3}$ or 6.667 pounds 14 days later. Notice that we will use days rather than weeks as our unit of time, since the final question in the problem asks for days.

$$\frac{20}{3} = 5e^{k(14)}$$

$$\frac{4}{3} = e^{14k}$$

$$\frac{\ln \frac{4}{3}}{14} = k = .0205487195$$

We want to find t when $y = 600$.

$$600 = 5e^{(.0205487195)t}$$

$$\ln 120 = .0205487195t$$

$$t = 232.982 \text{ days}$$

The poor guy almost lived for 8 months. The real tragedy is that even though he could talk, all he wanted to talk about were his misgivings concerning contemporary U.S. foreign policy. His handlers were relieved at his passing. "It was like having to talk to a fuzzy Peter Jennings all the time," they explained.

LOGISTIC GROWTH (BC TOPIC ONLY)

In addition to exponential growth, BC students need to be familiar with *logistic growth*. The major difference between the two is that exponential growth assumes that there are no restrictions on the quantity growing. Although billions upon billions of bacteria colonies can grow in a small area, the same is not true of, let's say, deer. Only so many deer can be supported by a particular ecosystem; this number is called the *carrying capacity* for that species. After the carrying capacity has been reached, there are no longer enough natural resources to support the extended population, and it's almost impossible for deer to get reservations in any deer restaurants (except as the main course). So, once too many deer inhabit an area, nature puts the brakes on deer-population growth.

Logistic growth looks like exponential growth at the beginning, but then changes concavity and levels out towards the carrying capacity (the upper limit for growth).

NOTE

It should be no surprise that the right-hand side of the equation is $\frac{4}{3}$ in the second step, as Vinny's weight is $\frac{4}{3}$ of his original weight every 14 days. In the half-life problems, you may have noticed that this number always turns out to be $\frac{1}{2}$.

NOTE

Logistic growth is also called *restricted growth*, whereas exponential growth is unrestricted.

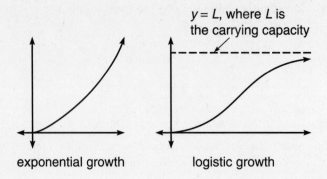

exponential growth   logistic growth

In exponential growth, the rate of change of the quantity is proportional to the quantity itself. In logistic growth, this is also true, but the quantity is *also* proportional to its distance from the upper bound. This translates into the formula

$$\frac{dy}{dt} = ky(L - y),$$

where L is the carrying capacity (upper bound) and k is the constant of proportionality. If you solve this differential equation using a very tricky separation of variables, you get the general solution:

$$y = \frac{L}{1 + ce^{-Lkt}}.$$

It's not worth actually solving, but both of the formulas above need to be memorized.

Example 5: A highly contagious "pinkeye" (scientific name: *Conjunctivitus itchlikecrazius*) is ravaging the local elementary school. The population of the school is 900 (including students and staff), and the rate of infection is proportional both to the number infected and the number of students whose eyes are pus-free. If seventy-five people were infected on December 15 and 250 have contracted pinkeye by December 20, how many people will have gotten the gift of crusty eyes by Christmas Day?

Solution: Because of the proportionality statements in the problem, logistic growth is the approach we should take. The upper limit for the disease will be $L = 900$; it is impossible for more than 900 people to be infected since the school only contains 900 people. This gives us the equation

$$y = \frac{900}{1 + ce^{-900kt}}.$$

We will interpret $t = 0$ as December 15, since that is the earliest information given. Therefore, we know that $y(0) = 75$. Plug that information into the equation.

$$75 = \frac{900}{1 + ce^0}$$

$$75 + 75c = 900$$

$$c = \frac{825}{75} = 11$$

Five days later, 250 people have contracted pinkeye, so plug that information (and the c we just found) to find k:

$$250 = \frac{900}{1 + 11e^{-900k \cdot 5}}$$

$$250(1 + 11e^{-4500k}) = 900$$

$$11e^{-4500k} = \frac{13}{5}$$

$$-4500k = -1.442383838$$

$$k = .0003205297.$$

Finally, we have the equation $y = \dfrac{900}{1 + 11e^{-900(.0003205297)t}}$. We want to find the number of infections on December 25, so $t = 10$.

$$y = \frac{900}{1 + 11e^{-900(.0003205297)(10)}}$$

$$y \approx 557.432$$

So, almost 558 students have contracted pinkeye in time to open presents.

Logistic growth can model a lot of real-world behavior; in fact, some retailers have harnessed a key feature of logistic growth. Every year, there seems to be a new toy on the market that everyone wants (Tickle Me Elmo, Furby, Cabbage Patch Kid), and fad toys like these follow a logistic sales pattern. Some stores compute the logistic growth curve (just as we did in Example 4) and stop purchasing the fad toys when the concavity of the curve changes (which means that the sales are still increasing but now at a decreasing rate). This practice helps cut down on piles of unsold product and thus decreases unnecessary spending by the store.

PROBLEM SET

You may use a graphing calculator on these problems.

1. The big toy for the holiday season this year is the Tickle Me Ben Stein Doll. When you gently squeeze his tummy, Ben will giggle and talk to you about his work in the Nixon administration. The kids love it! Research shows that the rate of sales is proportional to the number sold and the number in the target audience who haven't yet purchased it. Assume that the target audience number is 4 million. If 950,000 have been sold by 12 a.m. on October 1, and 3.5 million have been sold by 12 a.m. on December 1, on what day did the company sell to exactly half of its target audience?

2. The carrying capacity for deer in a particular small town is 2,200, and the rate of increase in their numbers is proportional to both the number, n, of deer and $2,200 - n$. If there were 1,000 deer one month ago and 1,150 deer now, how many months will it take the deer to number 2,100?

3. Assume that the rate of fans being seated in Oriole Park at Camden Yards (home to the Baltimore Orioles) is proportional both to the number of fans already seated and the number of empty seats; the park has a capacity of 48,262 fans. One hour before game time, only 10 percent of the seats are filled. At game time, 85 percent of the seats are filled. Assuming that no one leaves early, what percentage of the seats are filled 2 hours into the game?

4. For what value of y is $\dfrac{dy}{dt}$ increasing the fastest in a logistic growth curve? Justify your answer mathematically.

SOLUTIONS

1. In this problem, $L = 4,000,000$. The first objective is to find c using the fact that $y(0) = 950,000$. (For our purposes, $t = 0$ is October 1.)

$$950,000 = \frac{4,000,000}{1 + ce^{-4,000,000 \cdot k \cdot 0}}$$

$$950,000 = \frac{4,000,000}{1 + c}$$

$$950,000(1 + c) = 4,000,000$$

$$c = \frac{61}{19}$$

If $t = 0$ equates to the beginning of October 1, then $t = 1$ is October 2, $t = 30$ is October 31, and $t = 61$ is December 1. We know that $y(61) = 3,500,000$.

$$3,500,000 = \frac{4,000,000}{1 + \frac{61}{19}e^{-4,000,000 \cdot k \cdot 61}}$$

$$\frac{61}{19}e^{-4,000,000 \cdot k \cdot 61} = \frac{1}{7}$$

$$e^{-244,000,000\,k} = \frac{19}{427}$$

$$k = \frac{\ln\frac{19}{427}}{-244,000,000} = .0000000128$$

Now that we have c and k, we want to find t when $y = 2,000,000$ (half of the target audience).

$$2,000,000 = \frac{4,000,000}{1 + \frac{61}{19}e^{-4,000,000(.0000000128)t}}$$

$$1 + \frac{61}{19}e^{-.0512t} = 2$$

$$e^{-.0512t} = \frac{19}{61}$$

$$t = \frac{\ln\frac{19}{61}}{-.0512} \approx 22.782$$

Therefore, the sales figure was reached after $t = 22$ but before $t = 23$; this translates to October 23.

2. $L = 2,200$; $y(0)$ will translate to a month ago and $y(0) = 1,000$; $y(1) = 1,150$. Use all of this information the same way you did in number 1 to find c and k. First, find c.

$$1000 = \frac{2200}{1 + ce^{-2200k \cdot 0}}$$

$$1000(1 + c) = 2200$$

$$c = \frac{6}{5}$$

Now find k.

$$1150 = \frac{2200}{1 + \frac{6}{5}e^{-2200k \cdot 1}}$$

$$1 + \frac{6}{5}e^{-2200k} = \frac{44}{23}$$

$$e^{-2200k} = \frac{35}{46}$$

$$k = \frac{\ln\frac{35}{46}}{-2200} = .0001242242$$

Our final goal in this problem is to calculate t when $y = 2,100$.

$$2100 = \frac{2200}{1+\frac{6}{5}e^{-2200(.0001242242)t}}$$

$$1+\frac{6}{5}e^{-2200(.0001242242)t} = \frac{22}{21}$$

$$e^{-2200(.0001242242)t} = \frac{5}{126}$$

$$t = \frac{\ln\frac{5}{126}}{-.273293335} = 11.807$$

Since $t = 1$ translates to this month, $t = 11.807$ translates to 10.807 months from now.

3. $L = 48,262$; $y(0)$ is one hour before game time, and $y(0) = .10L = 4826.2$; $y(1) = .85L = 41,022.7$. Begin by finding c.

$$4826.2 = \frac{48,262}{1+ce^{-48,262k\cdot0}}$$

$$4826.2 = \frac{48,262}{1+c}$$

$$c = 9$$

Now find k using $y(1)$.

$$41,022.7 = \frac{48262}{1+9e^{-48,262k\cdot1}}$$

$$1 + 9e^{-48,262k} = 1.176470588$$

$$e^{-48,262k} = .0196078431$$

$$k = .0000814684$$

The problem asks us to find y when $t = 3$.

$$y = \frac{48,262}{1+9e^{-48,262(.0000814684)3}}$$

$$y = 48,258.7258$$

Two hours into the game, $\frac{48,258.7258}{48,262} = 99.993$percent of the seats are filled.

4. In this problem, you are trying to maximize $\frac{dy}{dt}$. Treat this like any optimization problem—start by finding its derivative and setting it equal to 0. (Remember, the derivative of $\frac{dy}{dt}$ is written $\frac{d^2y}{dt^2}$.) You'll have to use the Product Rule. Because a logistic relationship is evident, we use its general form.

$$\frac{dy}{dt} = ky(L - y)$$

$$\frac{d^2y}{dt^2} = k(L - y)\frac{dy}{dt} + ky\left(-\frac{dy}{dt}\right) = 0$$

$$k \cdot \frac{dy}{dt}(L - y - y) = 0$$

Since $\frac{dy}{dt}$ is not zero, we can divide both sides by $\left(k \cdot \frac{dy}{dy}\right)$ and say that

$$L - 2y = 0$$
$$y = \frac{L}{2}$$

Therefore, the rate of change is the greatest halfway to the upper boundary for the logistic growth. This makes sense geometrically, because the slope of the tangent lines will increase to that point and then become less steep afterward.

TECHNOLOGY: A DIFFERENTIAL EQUATIONS CALCULATOR PROGRAM

BC students may be a little disappointed in the BC topics so far. All of them seem to be very computational in nature. Arc lengths, slope fields, etc. are all basically formulas you have to memorize. Even worse, the major BC topic in differential equations (Euler's Method) requires mindless, repetitive work. Luckily, the calculator can save the day. By the way, a warm welcome to AB students, who get to subject themselves to *slope fields* for the first time on the 2004 exam. Until now, slope fields were a BC-only topic, but it seems that you guys are just lucky I guess, and you get to learn them now, too! Don't ever say that the College Board doesn't have a warm heart.

Thanks to Greg Hoerst, a former 5-er on the BC test and a student of mine, your calculator can help drag you through the doldrums of slope fields and Euler's Method. Some students were so appreciative of Greg's program, they set up small shrines in their basement in honor of him, and have voted for him as a write-in candidate for every presidential election ever since.

At the end of this section, you'll find a relatively short program that will take the drudgery out of Euler's and slope fields. All you have to do is type in some basic information, and the calculator takes care of the rest. Does this mean you shouldn't know the formulas and techniques for the two methods it simplifies? No, no, no! Essentially half of the AP test is now done without a calculator, so make sure you don't grow too dependent on the calculator. If you'd rather not type the program in, and you (or your teacher) have a TI Graphlink that hooks into your computer, you can download the program from my Web site on the students page: www.geocities.com/impactedcolon/students.html.

NOTE

You may use this program as long as you type it in exactly as listed, including the author's name and copyright date. Do not sell this program to anyone—it is meant to be free.

Example 6: Let's look back at the differential equation $\frac{dy}{dx} = 2x^2y$; you graphed its slope field back in Hands-on Activity 10.2. Of course graphing it with the calculator will be significantly easier. Before we run the program, make sure to set the window you want. In the exercises, we used an x interval of $[-1.5,1.5]$ and a y interval of $[0,1.5]$. Push the [Window] button, and enter these values into the appropriate spots.

```
WINDOW
 Xmin=-1.5
 Xmax=1.5
 Xscl=1
 Ymin=0
 Ymax=1.5
 Yscl=1
 Xres=1
```

Now, run the program by pressing [Prgm], pressing [Enter] to select the program, and then pressing [Enter] again; you should see the title screen. Press [Enter] to go to the menu, and select "Slope Field."

The program now asks you to type the numerator and denominator of $\frac{dy}{dx}$ separately. In this example, the denominator is simply 1. Use the [Alpha] button to get X and Y. The program then asks you what the length of the line segments should be. Because this window setting is so detailed, you should use a small number; I used length = .25. On larger windows, you will need to use a larger length, such as 1.

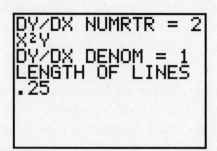

```
DY/DX NUMRTR = 2
X²Y
DY/DX DENOM = 1
LENGTH OF LINES
.25
```

Now, wait as the calculator draws the slope field for you. Once it's finished, press [Enter] to return to the menu screen.

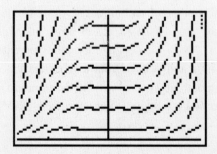

The only drawback to this program is that you can't tell the calculator exactly where to draw the line segments, but in the grand scheme of things, that matters little. If you have a PC, go to the Web site http://www.graphmatica.com and download the program *Graphmatica*. It draws incredible slope fields and easily draws solution graphs to a slope field when you click the initial point (x,y).

Example 7: Use Euler's Method to approximate $y(6)$ for $\frac{dy}{dx} = \frac{1}{x}$ given that $y(5) = 2$; use three steps of size $\Delta x = \frac{1}{3}$.

Solution: This is one of the problems following the section on Euler's method. Let's solve it with the calculator. Run the program, and select "Eulers" from the menu. Enter all the information requested by the program. The "Known X" and "Known Y" prompts correspond to the point (x,y) that is on the solution to the differential equation; in this case, that point is (5,2). Before it begins its calculations, the program asks "See Steps? (Y=1). If you want to see each step of Euler's method, press [1] and then press [Enter]. If you just want the solution, press any other number and press [Enter]. Press [1] so we can see all of the steps.

```
DY/DX= 1/X
KNOWN X= 5
KNOWN Y= 2
INCREMENT= 1/3
FIND AT X= 6
SEE STEPS? (Y=1)
1
```

The calculator begins with the given point and takes a step toward the final approximation each time you press [Enter].

```
            5.666666667
Y=
            2.129166667
X=
                      6
Y=
            2.187990196
```

The answer is 2.188, which matches the answer we got the hard way.

Differential Equation Tools, By Greg Hoerst

```
ClrHome
Disp "DIFF EQ TOOLS"
Disp "BY GREG HOERST"
Disp "(C) 2000"
Pause
Lbl M
ClrHome
Menu("DIFF EQ TOOLS","SLOPE
  FIELD",1,"EULERS",2,"QUIT",3)
Lbl 1
0→A
Input "DY/DX NUMRTR = ",Str1
String▶Equ(Str1,Y1
DelVar Str1
Input "DY/DX DENOM = ",Str1
String▶Equ(Str1,Y2
Input "LENGTH OF LINES ",N
FnOff
For(X,Xmin+N/2,Xmax-N/2,N)
For(A,Ymin+N/2,Ymax-N/2,N)
A→Y
If Y2(X)≠0
Then
Y1(X)/Y2(X)→S
√((N²/(S²+1)))/2→D
Line(X-D,Y-D*S,X+D,Y+D*S)
Else
Line(X,Y-S,X,Y+S)
End
End
End
Pause
Goto E
Lbl 2
Input "DY/DX= ",Str1
String▶Equ(Str1,Y1)
Input "KNOWN X= ",B
```

```
Input "KNOWN Y= ",Y
Input "INCREMENT= ",D
Input "FIND AT X= ",F
Input "SEE STEPS? (Y=1)",A
For(X,B,F,D)
If A=1
Then
Disp "X= ",X
Disp "Y= ",Y
Pause
End
If (A≠1) and (X=F)
Then
Disp "X= ",X
Disp "Y= ",Y
Pause
End
Y+D*Y1(X)→Y
End
Goto E
Lbl E
DelVar Str1
DelVar Y1
DelVar Y
DelVar D
DelVar S
DelVar N
DelVar X
DelVar A
DelVar Y2
DelVar B
DelVar C
DelVar F
ClrDraw
Goto M
Lbl 3
```

PRACTICE PROBLEMS

You may use a graphing calculator on problems 5 through 8.

1. If $f''(x) = 3x + 1$, $f'(1) = 2$, and $f(2) = 3$, find $f(x)$.

2. Which conic section is the solution to the differential equation $\dfrac{dy}{dx} = \dfrac{18 - 4x}{9y}$?

   Justify your answer mathematically.

3. What is the position function of a particle moving along the x-axis with velocity $v(t) = \dfrac{3}{2}t^3 + 4t + 3$ if the particle passes through the origin when $t = 2$?

4. If $\dfrac{dy}{dx} = x^2 - y^2$,

   (a) draw the slope field for $\dfrac{dy}{dx}$ at all indicated points on the axes below.

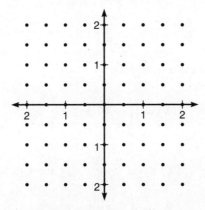

   *(b) use Euler's Method to approximate $y(2)$ using 4 steps if $y(0) = 1$.

5. A human zygote consists of 1 cell at conception, and the number of cells grows to 8 by the end of one week. Assuming that the rate of cell increase is proportional to the number of cells, how many weeks will it take the baby-in-process to amass 1,000 cells?

6.* Sarah likes mollies above all other tropical fish. Her fish must really like each other, because they are reproducing like crazy. The rate of increase of the fish is proportional to both the population and the number of additional fish the tank could support. Her tank has a carrying capacity of 50 mollies. If she bought 10 fish to start the tank two months ago (none of which died) and has 25 fish now, how many fish will she have in one month?

7. Newton's Law of Cooling states that an object cools down at a rate proportional to the difference between its temperature and the temperature of the ambient (surrounding) air. If my coffee was 100°F ten minutes ago, the temperature of my room is a constant 75°F, and the coffee is only 90°F now, what will the temperature of my coffee be in 15 minutes?

8. **James' Diabolical Challenge:** The population of a species of jellyfish in a small harbor, appropriately called *Sting Harbor*, is directly proportional to $450 - Q(t)$, where Q is the population (in thousands) and t is the time (in years).

CAUTION
Problems 7 and 8 represent neither straightforward exponential nor logistic growth. You have to begin with the differential equation expressed in the problem.

*a BC-only question

At $t = 0$ (1990) the population was 100,000, and in 1992, the population was 300,000.

(a) What was the population in 1993?

(b) In what year did the population reach 400,000?

SOLUTIONS

1. First, find $f'(x)$ by integrating $f''(x)$:

$$f'(x) = \int (3x+1)dx$$

$$f'(x) = \tfrac{3}{2}x^2 + x + C.$$

Since you know that $f'(1) = 2$, use that information to find C.

$$f'(1) = \tfrac{3}{2} + 1 + C = 2$$

$$C = -\tfrac{1}{2}$$

Now, repeat the process to find $f(x)$.

$$f(x) = \int \left(\frac{3}{2}x^2 + x - \frac{1}{2} \right) dx$$

$$f(x) = \frac{1}{2}x^3 + \frac{x^2}{2} - \frac{1}{2}x + C$$

$$f(2) = 4 + 2 - 1 + C = 3$$

$$C = -2$$

Therefore, $f(x) = \dfrac{1}{2}x^3 + \dfrac{x^2}{2} - \dfrac{1}{2}x - 2$.

2. Solve this differential equation by separating the variables.

$$\int 9y\,dy = \int (18 - 4x)dx$$

$$\frac{9}{2}y^2 = 18x - 2x^2 + C$$

$$2x^2 - 18x + \frac{9}{2}y^2 = C$$

$$4x^2 - 36x + 9y^2 = C$$

This is an ellipse, since the squared terms have unlike coefficients with the same sign. There's no need to put it in standard form.

3. We are looking for $s(t)$ knowing that $s(t) = \int v(t)dt$ and $s(2) = 0$.

$$s(t) = \int \left(\frac{3}{2}t^3 + 4t + 3 \right) dt$$

$$s(t) = \frac{3}{8}t^4 + 2t^2 + 3t + C$$

$$s(2) = 6 + 8 + 6 + C = 0$$

$$C = -20$$

Therefore, the position equation is $s(t) = \frac{3}{8}t^4 + 2t^2 + 3t - 20$.

4. (a)

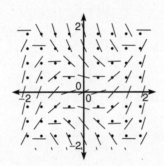

(b) Start at point (0,1), and take 4 steps of size $\Delta x = \frac{1}{2}$.

Point (0,1): $\frac{dy}{dx} = 0 - 1 = -1$; $\Delta y = \frac{1}{2} \cdot (-1) = -\frac{1}{2}$. The new point will be
$(0 + \frac{1}{2}, 1 - \frac{1}{2})$.

Point $(\frac{1}{2}, \frac{1}{2})$: $\frac{dy}{dx} = \frac{1}{4} - \frac{1}{4} = 0$; $\Delta y = 0 \cdot \frac{1}{2} = 0$. The new point will be
$(\frac{1}{2} + \frac{1}{2}, \frac{1}{2} + 0)$.

Point $(1, \frac{1}{2})$: $\frac{dy}{dx} = 1 - \frac{1}{4} = \frac{3}{4}$; $\Delta y = \frac{3}{4} \cdot \frac{1}{2} = \frac{3}{8}$. The new point will be
$(1 + \frac{1}{2}, \frac{1}{2} + \frac{3}{8})$.

Point $(\frac{3}{2}, \frac{7}{8})$: $\frac{dy}{dx} = \frac{9}{4} - \frac{49}{64} = \frac{95}{64}$; $\Delta y = \frac{95}{64} \cdot \frac{1}{2} = \frac{95}{128}$. The new point will be
$\left(\frac{3}{2} + \frac{1}{2}, \frac{7}{8} + \frac{95}{128} \right)$.

The approximation given by Euler's method is $y(2) = \frac{207}{128} \approx 1.617$.

5. Because of the proportional relationship described, it's time to use Ne^{kt}.
    Clearly, the initial value (N) is 1; find k if t is measured in weeks.

$$8 = 1e^{k(1)}$$

$$k = \ln 8$$

You can write k as a decimal, but for once, k is not messy in its exact form, so you can leave it for now. Time to find t when $y = 1000$:

$$1000 = 1e^{(\ln 8)t}$$

$$t \approx 3.322 \text{ weeks.}$$

That baby's growing like a weed.

6. This situation calls for a logistic growth model, and we know that $y(0) = 10$ (if t equals two months ago):

$$10 = \frac{50}{1 + ce^{-50k \cdot 0}}$$

$$10 = \frac{50}{1 + c}$$

$$10c = 40$$

$$c = 4.$$

Now, use c and the fact that $y(2) = 25$ to find k:

$$25 = \frac{50}{1 + 4e^{-100k}}$$

$$4e^{-100k} = 1$$

$$k = \frac{\ln \frac{1}{4}}{-100} = .0138629436.$$

The problem asks us to find y (number of fish) when $t = 3$ (one month from now):

$$y = \frac{50}{1 + 4e^{-50(.0138629436)(3)}}$$

$$y = 33.333.$$

Sarah will have 33 fish and some fish "on the way."

7. If the rate of change of the temperature is proportional to the difference between the temperature and the ambient air, then this translates to the differential equation

$$\frac{dy}{dt} = k(R - y),$$

where y is temperature of the cooling object, R is the room temperature, and k is the constant of proportionality. You cannot use $y = Ne^{kt}$ or logistic growth to model the temperature, as the rate of growth is not proportional to the temperature. However, it is not difficult to translate the given information into the above differential equation. Since R is 75, let's plug it in and then solve by separation of variables.

$$\frac{dy}{dt} = k(75 - y)$$

$$\int \frac{dy}{75 - y} = \int k \, dt$$

$$\ln |75 - y| = kt + C$$

$$75 - y = Ce^{kt}$$

$$y = 75 - Ce^{kt}$$

Use the fact that $y(0) = 100$ (temperature 10 minutes ago) to find C:

$$100 = 75 - Ce^0$$

$$C = -25.$$

Now, use C and the fact that $y(10) = 90$ to find k:

$$90 = 75 - (-25e^{10k})$$

$$15 = 25e^{10k}$$

$$\frac{\ln \frac{3}{5}}{10} = k = -.0510825624.$$

The temperature of my coffee 15 minutes later ($t = 25$) will be

$$y = 75 - (-25e^{25(-.0510825624)})$$

$$y = 81.971°F.$$

8. Much like number 7, this problem translates into the differential equation

$$\frac{dQ}{dt} = k(450 - Q).$$

If you solve it using separation of variables, you get

$$Q = 450 - Ce^{kt}.$$

Now, use the fact that $Q(0) = 100$ to find that $C = 350$. Then, use the fact that $Q(2) = 300$ to find k:

$$300 = 450 - 350e^{2k}$$

$$\frac{\ln \frac{3}{7}}{2} = k = -.4236489302.$$

Now we can solve the individual parts of the question.

(a) Find Q if $t = 3$:

$$Q = 450 - 350e^{-.4236489302 \cdot 3}$$

$$Q \approx 351.802.$$

CAUTION
Remember that Q is in thousands, so don't write unnecessary zeros.

There should be almost 352,000 jellyfish in Sting Harbor.

(b) We want to find t when $Q = 400$:

$$400 = 450 - 350e^{-.4236489302 \cdot t}$$

$$\frac{\ln \frac{1}{7}}{-.4236489302} = t \approx 4.593.$$

The population will reach 400,000 during mid-1994.

Sequences and Series (BC Topics Only)

As a BC calculus student, you have come a long way since your first limits and derivatives. There may have been a time when you feared the Chain Rule, but that time is long since past. You are now a member of the elite Calculus Club. (First rule of Calculus Club: Don't talk about Calculus Club.) And your membership is complete with sequences and series.

Once you get an idea of what sequences and series are, we will focus primarily on infinite series. For a couple of sections, you'll use various tests to determine the convergence of infinite series. Once that is complete, we'll discuss power series and use Taylor and Maclaurin series to approximate the values of functions. Sound good? Your Jedi training is almost complete…the force is strong with this one.

INTRODUCTION TO SEQUENCES AND SERIES, *N*TH TERM DIVERGENCE TEST

A sequence is basically a list of numbers based on some defining rule. Nearly every calculus book begins with the same example, and it's so darn fine that I will bow to peer pressure and use it as well. Consider the sequence $\left\{ \left(\frac{1}{2} \right)^n \right\}$.

The number n will take on all integer values beginning at 1, so the resulting sequence of numbers will be

$$\left\{ \left(\frac{1}{2} \right)^n \right\} = \left(\frac{1}{2} \right)^1, \left(\frac{1}{2} \right)^2, \left(\frac{1}{2} \right)^3, \dots$$

$$\left\{ \left(\frac{1}{2} \right)^n \right\} = \frac{1}{2}, \frac{1}{4}, \frac{1}{8}, \frac{1}{16}, \frac{1}{32}, \frac{1}{64}, \dots$$

In some cases, for the sake of ease, we will let n begin with 0 instead of 1—but you'll know exactly when to do that, so don't get stressed out or confused.

NOTE
The Force is with you always, and it equals mass times acceleration.

NOTE
The Technology section at the end of this chapter explains how to graph sequences and series on your TI-83. You should probably read that before you go any further.

Our singular goal in sequences is to determine whether or not they *converge*. In other words, is the sequence heading in some direction—toward some limiting number? You can graph the sequence above to see that it is headed toward a limit of 0.

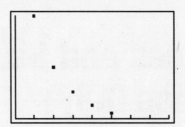

Each term of the sequence is half as large as the term before, and the sequence approaches a limit of 0 very quickly. Mathematically, we write

$$\lim_{n \to \infty}\left(\frac{1}{2}\right)^n = 0.$$

Because this sequence has a limit as *n* approaches infinity, the sequence is said to *converge*; if no limit existed, the sequence would be described as *divergent*.

Example 1: Determine whether or not $\left\{\frac{3n^3 - 2n - 7}{4 - 5n^2 + 6n^3}\right\}$ is a convergent or divergent sequence.

Solution: The sequence will converge if its limit at infinity exists and will diverge if the limit does not exist.

$$\lim_{n \to \infty} \frac{3n^3 - 2n - 7}{4 - 5n^2 + 6n^3}$$

This is a rational function with equal degrees in the numerator and denominator; therefore, the limit is the ratio of the leading coefficients: $\frac{3}{6} = \frac{1}{2}$.

Therefore, the sequence converges to $\frac{1}{2}$. This is further evidenced by the graph of the sequence below:

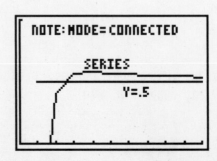

A *series* is similar to a sequence, but in a series, you add all the terms together. Series are written using sigma notation and look like this:

$$\sum_{n=1}^{\infty} \left(\frac{1}{2}\right)^n = \frac{1}{2} + \frac{1}{4} + \frac{1}{8} + \frac{1}{16} + \frac{1}{32} + \frac{1}{64} + \dots$$

The notation is read "the sum from 1 to infinity of $\left(\frac{1}{2}\right)^n$" and gets its value from the sum of all the terms in the sequence. However, how can you tell if a series with an infinite number of terms has a finite sum? At first glance, it seems impossible—how can you add infinitely many numbers together to get a real sum? Consider the diagram below, which should help you

visualize the infinite series $\sum_{n=1}^{\infty} \left(\frac{1}{2}\right)^n$:

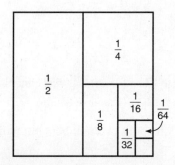

If the large box represents one unit and you continuously divide the box into halves, the sum of all the pieces will eventually (if you add forver and ever)

equal the entire box. Thus, $\sum_{n=1}^{\infty} \left(\frac{1}{2}\right)^n = 1$.

You won't be able to draw pictures like this for the majority of series, but in the next section, you'll learn a much easier way to find the above sum. Mathematically, you need to know that a series gets its value from the *sequence of its partial sums* (SOPS). A *partial sum* is the sum of a piece of the series, rather than the entire thing; it is written S_n, where n is the number

of terms being summed. Let's use good old $\sum_{n=1}^{\infty} \left(\frac{1}{2}\right)^n$ as an example:

$$S_1 = \frac{1}{2}$$

$$S_2 = \frac{1}{2} + \frac{1}{4} = .75$$

$$S_3 = \frac{1}{2} + \frac{1}{4} + \frac{1}{8} = .875.$$

Therefore, the SOPS is .5, .75, .875, …. This is the important thing to remember: *If the SOPS converges, then the infinite series converges, and its sum is equal to the limit at which the SOPS converged*. For the most part, you will use the fantastic formula alluded to earlier in order to find sums. However, if all else fails, you can use the SOPS to find the sum of a series.

 Example 2: Find the sum of the series $\sum_{n=1}^{\infty}\left(\dfrac{1}{n}-\dfrac{1}{n+2}\right)$.

Solution: Construct the sequence of partial sums to gain some insight on this series:

$$S_1 = 1 - \frac{1}{3}$$

$$S_2 = 1 - \frac{1}{3} + \frac{1}{2} - \frac{1}{4}$$

$$S_3 = 1 - \frac{1}{3} + \frac{1}{2} - \frac{1}{4} + \frac{1}{3} - \frac{1}{5}.$$

See what happened there? The $-\frac{1}{3}$ and $\frac{1}{3}$ will cancel out. In S_4, the $-\frac{1}{4}$ and $\frac{1}{4}$ will cancel out. In fact, each partial sum will cancel out another term all the way to infinity, and the only two numbers left will be $1 + \frac{1}{2}$. Therefore, the sum of the series is $\frac{3}{2}$. You can use your calculator to calculate S_{500} to verify that the SOPS is indeed approaching $\frac{3}{2}$:

```
sum(seq((1/N)-(1
/(N+2)),N,1,500,
1))
          1.49601196
```

It makes sense that each term in a sequence needs to get smaller if the series is going to converge. You are adding numbers for an infinite amount of time; if you are not eventually adding 0 in this infinite loop, your sum will grow and grow and never approach a limiting value. We showed that the sequence of the terms that make up $\sum_{n=1}^{\infty}\left(\frac{1}{2}\right)^n$ have a limit of 0 early in this section, and that infinite series has a sum. However, if the limit, as n approaches infinity, of the sequence that forms an infinite series does not equal 0, then that infinite series *cannot* converge. This is called the *nth Term Divergence Test,* and it is the easiest way to immediately tell if a series is going to diverge.

***n*th Term Divergence Test:** If $\lim\limits_{n\to\infty} a_n \neq 0$, then $\sum\limits_{n=1}^{\infty} a_n$ is a divergent series.

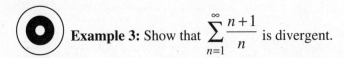 **Example 3:** Show that $\sum\limits_{n=1}^{\infty} \dfrac{n+1}{n}$ is divergent.

Solution: Because $\lim\limits_{n\to\infty} \frac{n+1}{n} = 1$ (and this limit must equal 0 for the series to be convergent), the series diverges by the *n*th Term Divergence Test. That's all there is to it. If you think about it, since the limit at infinity is 1, as *n* approaches infinity, you'd be adding $1 + 1 + 1 + 1 + 1 + 1$ forever, and that clearly approaches no limiting or maximum value.

Be careful! Just because $\lim\limits_{n\to\infty} a_n = 0$, that does *not* mean that the series will converge! For example, the *harmonic series* $\sum\limits_{n=1}^{\infty} \frac{1}{n}$ is divergent even though $\lim\limits_{n\to\infty} \frac{1}{n} = 0$. *If the limit at infinity is 0, you can conclude nothing from the nth Term Divergence Test.* It can only be used to show that a series diverges (if its limit at infinity does not equal 1); it can *never* be used to show that a series converges.

PROBLEM SET

You may use your graphing calculator on problem 4 only.

1. Determine if the sequence $\left\{ \dfrac{\ln x}{x^2} \right\}$ converges.

2. Find the *n*th term of each sequence (in other words, find the pattern evidenced by the sequence), and use it to determine whether or not the sequence converges.

    (a) $2, \dfrac{3}{4}, \dfrac{4}{9}, \dfrac{5}{16}, \dfrac{6}{25}, \ldots$

    (b) $1, \dfrac{1}{2}, \dfrac{1}{6}, \dfrac{1}{24}, \dfrac{1}{120}, \ldots$

CAUTION

The sequence $\{a_n\}$

converges if

$\lim\limits_{n\to\infty} a_n = 0$;

however, the

series $\sum\limits_{n=1}^{\infty} a_n$ does

not necessarily

converge if

$\lim\limits_{n\to\infty} a_n = 0$.

3. Use the *n*th Term Divergence Test to determine whether or not the following *series* converge:

(a) $\displaystyle\sum_{n=1}^{\infty} \frac{1 + 3n^2 + n^3}{4n^3 - 5n + 2}$

(b) $\displaystyle\sum_{n=1}^{\infty} \frac{1}{x^2}$

4. (a) What is the sum of $\displaystyle\sum_{n=1}^{\infty} \left(\frac{1}{n+1} - \frac{1}{n+3} \right)$?

(b) Calculate S_{500} to verify that the SOPS is bounded by the sum you found.

SOLUTIONS

1. The sequence converges if $\lim\limits_{n\to\infty} \dfrac{\ln x}{x^2} = 0$. You'll have to use L'Hôpital's Rule:

$$\lim_{n\to\infty} \frac{\frac{1}{x}}{2x}$$

$$\lim_{n\to\infty} \frac{1}{2x^2} = 0.$$

The sequence converges.

2. (a) This is the sequence $\displaystyle\sum_{n=1}^{\infty} \left\{ \frac{n+1}{n^2} \right\}$. Use L'Hôpital's to show that the sequence converges.

$$\lim_{n\to\infty} \frac{1}{2n} = 0$$

(b) This is the sequence $\displaystyle\sum_{n=1}^{\infty} \frac{1}{n!}$. How can we tell that it converges? It is

very clear that $\lim\limits_{n\to\infty} \dfrac{1}{n!} = 0$; in fact, this sequence converges to 0

significantly faster than $\left\{ \frac{1}{x} \right\}$, since the former's denominator will grow larger much faster than the latter's.

3. (a) $\lim\limits_{n\to\infty} \dfrac{1+3n^2+n^3}{4n^3-5n+2} = \dfrac{1}{4}$, since the expression is rational and the degrees of the numerator and denominator are equal. Because the limit does not equal 0, this series diverges by the *n*th Term Divergence Test.

(b) $\lim\limits_{n\to\infty} \dfrac{1}{x^2} = 0$, but this does not necessarily mean that the series is convergent. You will find out that it does converge very soon (in the Integral Test subsection), but you can never conclude that any series converges using the *n*th Term Divergence Test; it can only be used to prove divergence. Therefore, you can draw no conclusion.

4. (a) This is a telescoping series; if you write out the fourth partial sum (S_4), you can see what terms will cancel out in the long run and which ones will remain:

$$S_4 = \frac{1}{2} - \frac{1}{4} + \frac{1}{3} - \frac{1}{5} + \frac{1}{4} - \frac{1}{6} + \frac{1}{5} - \frac{1}{7}$$

The only terms that will remain as *n* approaches infinity are $\frac{1}{2} + \frac{1}{3}$, so the infinite sum is $\frac{5}{6} \approx .833333$.

(b) Use your calculator to find S_{500}. Of course, this is not the limit at infinity, but it will give us an idea of where SOPS is heading at that point.

```
sum(seq(1/(N+1)-
1/(N+3),N,1,500,
1))
          .8293532299
```

S_{500} is closing in on .8333, but it has all of infinity to get there—so what's the rush?

CONVERGENCE TESTS FOR INFINITE SERIES

Your major focus in infinite series will be in determining whether or not the series converge; in some rare instances, you will also be able to provide the sum at which the series converge. Each of these tests works best for certain kinds of series, so make sure to learn the series characteristics that indicate which tests to use as well as how to apply the tests themselves. So, don't just learn one or two and expect to be able to apply them in all situations; you need to know them all. I know, no rest for the wicked …

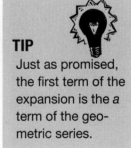

TIP
Just as promised, the first term of the expansion is the *a* term of the geometric series.

GEOMETRIC SERIES

Geometric series have the easiest test for convergence of them all, but first you need to know what a geometric series is. Every geometric series has the form

$$\sum_{n=0}^{\infty} ar^n$$

where a is a constant that can be factored out of each term and r is the *ratio*, the identical portion of each term raised to an increasing power as the terms increase. The major difference between this series and most other infinite series is that geometric series begin with $n = 0$ and not $n = 1$. However, that makes the a term very easy to find; the a term will be the first term in the geometric series when expanded, since it is multiplied by r^0, which is just 1.

Let's expand the geometric series $\displaystyle\sum_{n=0}^{\infty} 2\left(\frac{1}{3}\right)^n$ to get a feel for these puppies:

$$2 + 2\left(\frac{1}{3}\right) + 2\left(\frac{1}{3}\right)^2 + 2\left(\frac{1}{3}\right)^3 + 2\left(\frac{1}{3}\right)^4 + \dots$$

$$2 + \frac{2}{3} + \frac{2}{9} + \frac{2}{27} + \frac{2}{81} + \dots$$

Not only will you be able to tell if this series actually adds up to something (converges), but we can actually find out *exactly* what the sum of the series is.

Geometric Series: The geometric series $\displaystyle\sum_{n=0}^{\infty} ar^n$ will diverge for $|r| \ge 1$. It will only converge if $0 < |r| < 1$. If the series does converge, it will converge to the sum $\frac{a}{1-r}$.

Therefore, the series $\displaystyle\sum_{n=0}^{\infty} 2\left(\frac{1}{3}\right)^n$ converges since the ratio, $\frac{1}{3}$, is between 0 and 1. Furthermore, the sum of the series will be

$$\frac{a}{1-r} = \frac{2}{\frac{2}{3}} = 2 \cdot \frac{3}{2} = 3.$$

 Example 4: Determine whether or not the following series converge; if they do, find the sum of the series.

(a) $3 + \dfrac{15}{4} + \dfrac{75}{16} + \dfrac{375}{64} + \dots$

The first term is a, so $a = 3$. You also know that each term of the series contains a, so factor it out to determine r:

$$3\left(1 + \frac{5}{4} + \frac{25}{16} + \frac{125}{64} + \dots\right).$$

So, the $n = 0$ term is 1, the $n = 1$ term is $\frac{25}{16}$, and the $n = 2$ term is $\frac{125}{64}$. The numerators are powers of 5, and the denominators are powers of 4. This is just the series $\sum\limits_{n=0}^{\infty} 3\left(\frac{5}{4}\right)^n$. However, $\frac{5}{4} \geq 1$, so the series diverges.

(b) $\dfrac{1}{2} + \dfrac{1}{4} + \dfrac{1}{8} + \dfrac{1}{16} + \dfrac{1}{32} + \dfrac{1}{64} + \ldots$

Recognize this problem? We have already decided that the answer must be 1, so let's finally prove it. First of all, $a = \frac{1}{2}$, so factor that out of everything:

$$\frac{1}{2}\left(1 + \frac{1}{2} + \frac{1}{4} + \frac{1}{8} + \frac{1}{16} + \frac{1}{32} + \ldots\right).$$

The ratio is definitely $\left(\frac{1}{2}\right)^n$, making the series $\sum\limits_{n=0}^{\infty} \left(\frac{1}{2}\right)\left(\frac{1}{2}\right)^n$. Therefore, the sum will be

$$\frac{\frac{1}{2}}{1 - \frac{1}{2}} = \frac{\frac{1}{2}}{\frac{1}{2}} = 1,$$

just as we thought it would be.

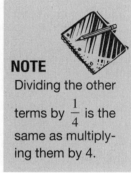

NOTE
Dividing the other terms by $\frac{1}{4}$ is the same as multiplying them by 4.

(c) $\dfrac{1}{4} + \dfrac{1}{6} + \dfrac{1}{9} + \dfrac{2}{27} + \dfrac{4}{81} + \ldots$

Again, since a is the first term, $\frac{1}{4}$, divide it out of all the other terms.

$$\frac{1}{4}\left(1 + \frac{4}{6} + \frac{4}{9} + \frac{8}{27} + \frac{16}{81} + \ldots\right)$$

The $\frac{4}{6}$ term can be reduced to $\frac{2}{3}$, and the series in parentheses clearly has ratio $\frac{2}{3}$. Thus, the series converges (since $\frac{2}{3}$ is between 0 and 1), and it has sum

$$\frac{\frac{1}{4}}{1 - \frac{2}{3}} = \frac{\frac{1}{4}}{\frac{1}{3}} = \frac{1}{12}.$$

THE INTEGRAL TEST AND P-SERIES

To help you remember the necessary conditions that must exist in order to apply the integral test, memorize the following phrase: **I**n **p**rison, **d**ogs **c**urse. The mneumonic device reinforces that in order to apply the **I**ntegral Test, the function must be **p**ositive, **d**ecreasing, and **c**ontinuous. In essence, the Integral Test shows that an infinite series and an integral with an infinite upper bound have a lot in common: either they will both converge or both diverge.

The Integral Test: If $a_n = f(n)$ is a positive, decreasing, continuous function, then $\sum_{n=1}^{\infty} a_n$ and $\int_1^{\infty} f(n)dn$ either both converge or both diverge.

As an example, consider $\sum_{n=1}^{\infty} \frac{1}{n}$, the harmonic series. We already know that the sequence $\left\{\frac{1}{n}\right\}$ converges, but does the infinite series converge? According to the Integral Test, it will if the integral $\int_1^{\infty} \frac{1}{n} \, dn$ does. Therefore, try to evaluate this improper integral:

$$\lim_{b \to \infty} \int_1^b \frac{1}{n} dn$$

$$\lim_{b \to \infty} \ln|n| \Big|_1^b$$

$$\lim_{b \to \infty} \ln b.$$

As x approaches infinity, $\ln x$ will approach a height of infinity; the graph has no horizontal asymptote or limiting value and will increase forever (although more slowly than many other graphs). Therefore, the integral diverges, and the infinite series must diverge, too.

The justification for the Integral Test lies in geometry. Remember that an integral is really the sum of an infinite number of rectangle areas; the above series was an infinite sum of numbers. If a function creates rectangles whose areas grow and result in divergence, then that same function will create outputs whose sums grow too much and also create divergence.

This leads us back to a brief discussion we began in the Improper Integrals section of Chapter 8. At that time, we discussed the set of improper integrals of the form $\int_1^{\infty} \frac{1}{x^p} \, dx$. We said that this integral will diverge for any $p \leq 1$ but will converge for $p > 1$. We can now apply this fact to infinite series.

P-series: A p-series is a series of the form $\sum_{n=1}^{\infty} \frac{1}{n^p}$, where p is a positive number. A p-series will converge if $p > 1$ but will diverge if $p \leq 1$.

In the case of the harmonic series $\frac{1}{n}$, $p = 1$, so it will diverge; you don't have to apply the Integral Test as we did above, but if you do, it will become very clear exactly why the harmonic series diverges.

Example 5: Determine whether or not the following series converge, and explain how you arrived at your answer:

(a) $\displaystyle\sum_{n=1}^{\infty} n^{-\frac{2}{3}}$

If you rewrite the series as $\displaystyle\sum_{n=1}^{\infty} \frac{1}{n^{2/3}}$, it is clearly a p-series with $p = \frac{2}{3}$. Since $\frac{2}{3} \leq 1$, the series will diverge.

(b) $\displaystyle\sum_{n=1}^{\infty} \frac{3n}{2n^2 + 3}$

According to the Integral Test, the convergence of this series is correlated with the convergence of the related improper integral. Therefore, try to integrate:

$$\int_1^{\infty} \frac{3n}{2n^2 + 3} \, dn$$

$$\lim_{b \to \infty} 3 \int_1^{b} \frac{n}{2n^2 + 3} \, dn.$$

You should use u-substitution, with $u = 2n^2 + 3$. Don't forget to replace 1 and b with u boundaries.

$$\lim_{b \to \infty} 3 \left(\frac{1}{4} \ln(u) \right) \Big|_5^{2b^2 + 3}$$

This integral definitely grows infinitely large as x approaches infinity, so both it and the series will diverge.

THE COMPARISON TEST

This test is useful when the series at hand looks similar to a series for which you already know—or can easily determine—the convergence. Now that you know how to apply the integral, geometric, and p-series tests, you can use them in conjunction with the Comparison Test in order to determine the convergence of series that are almost (but not quite) geometric or p-series or that are nearly (but not quite) integrable. As was the case with the Integral Test, all the terms of the series involved must be positive to apply the Comparison Test.

The Comparison Test: If Σa_n and Σb_n are both positive series and every term of Σa_n is less than or equal to the corresponding Σb_n term,

(1) if Σb_n converges, then Σa_n converges.

(2) if Σa_n diverges, then Σb_n diverges.

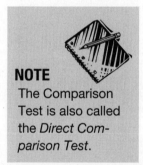

NOTE
The Comparison Test is also called the *Direct Comparison Test*.

CAUTION
If $a_n \leq b_n$, and b_n is divergent or a_n is convergent, you can draw no conclusions about the other series.

Consider the following diagram of the first few terms of each sequence. Notice that $a_n < b_n$ for all n. The figure also contains the functions $a(x)$ and $b(x)$ from which the terms of the series get their value.

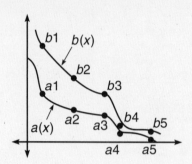

NOTE

The trickiest part of the Comparison Test is deciding what to compare the given series to. Most of the time, you pick a series that is close to, but simpler than, the given series.

If $\sum\limits_{n=1}^{\infty} b_n$ converges, then according to the Integral Test, $\int_1^{\infty} b(x)dx$ is finite. In other words, there is a number that represents the area trapped between $b(x)$ and the x-axis. Since the area beneath $a(x)$ *must* be less than the area beneath $b(x)$ (since all the function values for a are less than or equal to b's values), then the area beneath $a(x)$ must also be finite. How can an area less than a finite area be infinite? Similarly, if $\sum\limits_{n=1}^{\infty} a(x)$ diverges, then the area $\int_1^{\infty} a(x)dx$ must be infinite. If b has larger values than this divergent integral, then b_n must be divergent, too.

This test works just like the restriction sign at kiddie rides at theme parks: "You cannot be taller than this to ride this ride." I am too tall to ride kids' rides, but Papa Smurf can ride any of them. If an individual comes along who is taller than I am, then he will *not* be able to ride either (the divergence part of the rule). However, if someone shorter than Papa Smurf strolls up, then he will be able to ride (the convergence part of the rule).

CAUTION

The comparison series in Example 6(b) is still geometric, even though it does not begin with $n = 0$. However, the sum of $\sum\limits_{n=1}^{\infty} \frac{3^n}{7^n}$ will *not* be $\frac{a}{1-r}$ since it begins with $n = 1$ and not $n = 0$.

 Example 6: Use the Comparison Test to determine the convergence of the following series:

(a) $\sum\limits_{n=1}^{\infty} \frac{e^n}{n+3}$

Compare this to the series $\sum\limits_{n=1}^{\infty} \frac{1}{n+3}$. Without a doubt, the numerator e^n is much larger than the other numerator, 1. Therefore, each term of the series $\sum\limits_{n=1}^{\infty} \frac{e^n}{n+3}$ will be larger than the comparsion series, $\sum\limits_{n=1}^{\infty} \frac{1}{n+3}$. However, using the Integral Test, it is easy to see that $\sum\limits_{n=1}^{\infty} \frac{1}{n+3}$ is divergent. Because

$\sum\limits_{n=1}^{\infty} \dfrac{e^n}{n+3}$ is larger than a divergent series, it must also be divergent by the Comparison Test.

(b) $\sum\limits_{n=1}^{\infty} \dfrac{3^n}{7^n+1}$

Compare this to the series $\sum\limits_{n=1}^{\infty} \left(\dfrac{3}{7}\right)^n$, which is a convergent geometric series (since $r = \dfrac{3}{7}$, which is less than 1). Since each of the denominators of $\sum\limits_{n=1}^{\infty} \dfrac{3^n}{7^n+1}$ are greater than the corresponding denominators in the geometric series $\sum\limits_{n=1}^{\infty} \left(\dfrac{3}{7}\right)^n$, that will result in a smaller overall value for each term (in the same way that $\dfrac{1}{6} < \dfrac{1}{5}$, even though $6 > 5$). Therefore, $\sum\limits_{n=1}^{\infty} \dfrac{3^n}{7^n+1}$ is smaller than a convergent series, so it must be convergent, too.

THE LIMIT COMPARISON TEST

This test operates in a slightly different way from previous tests but has many of the same characteristics as the Comparison Test. For one thing, you'll need to invent a comparison series, and both your original and the new series must contain positive terms. In practice, the Limit Comparison Test allows you to take ugly series and compare them to very simple *p*-series to determine their convergence. There is even good news, pessimistic mathematician: It's very easy to pick a comparison series with this test. When you've stopped cheering, you may continue reading.

The Limit Comparison Test: If Σa_n and Σb_n are positive series, and

$$\lim_{n\to\infty} \frac{a_n}{b_n} = N$$

(where *N* is a positive *number*), then both series either converge or diverge. If the limit does not equal a finite number, you can draw no conclusion from this test.

Translation: To determine whether or not a series Σa_n converges, you will first invent a comparison series Σb_n. Then, find the limit at infinity of the quotient $\dfrac{a_n}{b_n}$. If that limit exists (in other words, if the limit is a finite number), then either both series will converge or both will diverge. Because of this, you should choose a Σb_n that is obviously convergent or divergent to cut down on your work.

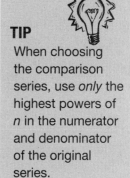

TIP
When choosing the comparison series, use *only* the highest powers of *n* in the numerator and denominator of the original series.

Just like the Comparison Test, the convergence or divergence of the series, a_n, is dependent upon the convergence of the series to which it is compared, b_n. Although a bit strange, the Limit Comparison Test is more straightforward than its predecessor and is, in practice, easier to use.

Example 7: Use the Limit Comparison Test to determine whether or not the following series converge:

(a) $\displaystyle\sum_{n=1}^{\infty} \frac{3n+6}{1-5n+7n^2}$

As the tip above explains, use the highest powers of n in the numerator and denominator to create a comparison series of $\frac{n}{n^2}$ or $\frac{1}{n}$.

$$\lim_{n\to\infty} \frac{\frac{3n+6}{1-5n+7n^2}}{\frac{1}{n}}$$

Multiply the numerator and denominator of this mega fraction by $\frac{n}{1}$ to get

$$\lim_{n\to\infty} \frac{3n^2+6n}{1-5n+7n^2}.$$

Because this is a rational function with equal degrees in the numerator and denominator, the limit is the ratio of the leading coefficients: $\frac{3}{7}$. Therefore, both series will either converge or diverge; so which is it? We already know that $\sum \frac{1}{n}$ diverges, so both series diverge. If you forgot that it was a harmonic series, you still know that $\sum \frac{1}{n}$ is a divergent p-series (since $p = 1$).

(b) $\displaystyle\sum_{n=1}^{\infty} \frac{n+5}{3n \cdot 4^n}$

CAUTION
A limit of $\frac{1}{3}$ does *not* mean that either series has a sum of $\frac{1}{3}$. All that's important is that $\frac{1}{3}$ is a finite number (not ∞).

This one is trickier. You're supposed to use the highest powers of n only, so what do you do with the 4^n? Answer: include it also—with an n power, it has to be important. Therefore, the comparison series is $\frac{n}{n \cdot 4^n}$ or $\frac{1}{4^n}$.

$$\lim_{n\to\infty} \frac{\frac{n+5}{3n \cdot 4^n}}{\frac{1}{4^n}}$$

Multiply the top and bottom of that giant fraction by 4^n to get

$$\lim_{n\to\infty} \frac{n+5}{3n} = \frac{1}{3}$$

We got the number we needed—now, to determine if the series converge or not. The comparison series can be rewritten as $\displaystyle\sum_{n=1}^{\infty}\left(\frac{1}{4}\right)^{n}$, which is a convergent geometric series (since $r = \frac{1}{4} < 1$). Therefore, both of the series must be convergent.

THE RATIO TEST

Yet again, this test works only for series with positive terms. (The next series test will finally address series with positive and negative terms.) The Ratio Test works best for series that contain things that grow extremely large as n increases, like powers of n or factorials involving n. In essence, you take the limit at infinity of the ratio of a generic series term a_n and the next consecutive generic term a_{n+1}; you can determine the convergence of the series based on what happens after you find the limit.

The Ratio Test: If Σa_n is a series whose terms are positive, and

$$\lim_{n\to\infty}\frac{a_{n+1}}{a_n} = N$$

(where N is a real number), then

(1) Σa_n converges if $N < 1$.

(2) Σa_n diverges if $N > 1$.

(3) If $N = 1$, we don't know diddly squat: the series could converge or diverge, but to determine which, we'll have to use a different test; the Ratio Test doesn't help.

 Example 8: Use the Ratio Test to determine whether or not the following series converge:

(a) $\displaystyle\sum_{n=1}^{\infty}\frac{n^3}{n!}$

Your best bet is the Ratio Test because of the factorial in the series. In this case, $a_n = \frac{n^3}{n!}$; to get a_{n+1}, substitute $(n + 1)$ for n:

$$a_{n+1} = \frac{(n+1)^3}{(n+1)!}.$$

Now you can set up the limit that is the heart of the Ratio Test:

$$\lim_{n\to\infty}\frac{a_{n+1}}{a_n}$$

$$\lim_{n\to\infty}\frac{\frac{(n+1)^3}{(n+1)!}}{\frac{n^3}{n!}}.$$

To simplify the fraction, multiply the numerator and denominator by the reciprocal of the denominator. (This is always the second step.)

$$\lim_{n \to \infty} \frac{(n+1)^3}{(n+1)!} \cdot \frac{n!}{n^3}$$

$$\lim_{n \to \infty} \frac{(n+1)^3 \cdot n!}{(n+1) \cdot n! \cdot n^3}$$

$$\lim_{n \to \infty} \frac{(n+1)^2}{n^3}$$

$$\lim_{n \to \infty} \frac{n^2 + 2n + 1}{n^3} = 0$$

The limit at infinity is 0; since $0 < 1$, the series converges according to the Ratio Test.

(b) $\sum_{n=1}^{\infty} \frac{2}{n^2}$

This doesn't have anything that grows so large as to make the Ratio Test our first choice, but we have to follow directions.

$$\lim_{n \to \infty} \frac{\frac{2}{(n+1)^2}}{\frac{2}{n^2}}$$

$$\lim_{n \to \infty} \frac{n^2}{n^2 + 2n + 1} = 1$$

Since the limit equals 1, no conclusion can be drawn using the Ratio Test. So, let's try something else even though the directions don't tell us to—we're rebels! Notice that $\frac{2}{n^2}$ is almost a p-series, so the better bet would have been the Limit Comparison Test. A good comparison series is $\frac{1}{n^2}$:

$$\lim_{n \to \infty} \frac{\frac{2}{n^2}}{\frac{1}{n^2}} = 2$$

Because 2 is a finite number (and $\frac{1}{n^2}$ is a convergent p-series) both series converge according to the Limit Comparison Test.

THE ALTERNATING SERIES TEST

Until now, you have determined convergence for lots of different series. Most of those tests required the terms of the series to be positive. Finally (exhale here), the Alternating Series Test (inhale nervously here) allows you to consider series with negative terms. However, these series must alternate back and forth between positive and negative terms. For example, $\sum_{n=1}^{\infty}\frac{(-1)^{n+1}}{n}$ is an alternating series:

$$\sum_{n=1}^{\infty}\frac{-1^{n+1}}{n}=1-\frac{1}{2}+\frac{1}{3}-\frac{1}{4}+\frac{1}{5}-\frac{1}{6}+\dots$$

The series has a positive term, then a negative term, then a positive term, etc. In contrast, the series

$$1-\frac{1}{2}-\frac{1}{3}+\frac{1}{4}-\frac{1}{5}-\frac{1}{6}+\dots$$

is not an alternating series. Not only do the signs have to alternate, but they also have to alternate every other term.

The Alternating Series Test: If Σa_n is an alternating series; each term, a_{n+1}, is smaller than the preceding term a_n; and $\lim_{n\to\infty} a_n = 0$, then Σa_n converges.

Translation: In order for an alternating series to converge, (1) the terms *must* decrease as *n* increases (in fact, each term *has* to be smaller than the term before it), and (2) the series must pass the *n*th Term Divergence Test—the *n*th term *must* have a limit of 0. If both conditions are satisfied, the alternating series converges.

 Example 9: Determine whether or not the following series converge using the Alternating Series Test:

(a) $\sum_{n=1}^{\infty}\frac{(-1)^{n+1}}{n}$

It's the alternating series form of a harmonic series. You may want to answer "divergent" as a reflex. Let's see what happens. If we write out a few terms

$$1-\frac{1}{2}+\frac{1}{3}-\frac{1}{4}+\frac{1}{5}-\frac{1}{6}+\dots,$$

it's clear that this is an alternating series and that each term is smaller than the previous one. Don't worry about the positive and negative signs. Clearly, $\frac{1}{3}>\frac{1}{4}>\frac{1}{5}$, etc. That satisfies the first requirement of the test. The second requirement involves the following limit:

$$\lim_{n\to\infty}\frac{1}{n}=0.$$

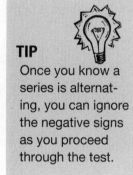

The limit is very easy to find, and we again ignore the possibility of a negative sign when evaluating the limit. The test takes care of the negative sign without our having to worry about it. Because both conditions are satisfied, this series converges.

(b) $\displaystyle\sum_{n=1}^{\infty}\frac{\left(-1^{n}\right)\bullet(n+3)}{2n}$

To begin, let's write out a few terms of the series:

$$-2+\frac{5}{4}-1+\frac{7}{8}-\frac{4}{5}+\frac{3}{4}+\dots$$

The series is definitely alternating, and (ignoring the signs) each term is less than the term preceeding it. However,

$$\lim_{n\to\infty}\frac{n+3}{2n}=\frac{1}{2}\neq 0\,;$$

therefore, this series fails the second condition of the Alternating Series Test. In fact, this series outright *fails* the nth Term Divergence Test and therefore diverges. Even though it's an alternating series, it (like all series) *must pass* the nth Term Divergence Test in order to converge.

In some series, the alternating series test will not apply, even though we want it to so badly. Let's take, for example, $\displaystyle\sum_{n=1}^{\infty}(-1)^{n}\bullet\frac{4^{n}}{n!}$. The series definitely alternates; that's not the problem. If we write a few terms out, we get

$$-4+8-\frac{32}{3}+\frac{32}{3}-\frac{128}{15}+\dots$$

The terms (at least in this small sample) definitely don't seem to be getting uniformly smaller. They will later, but at first, things are pretty weird. Furthermore, let's try to find the required limit for the Alternating Series Test:

$$\lim_{n\to\infty}\frac{4^{n}}{n!}$$

This is the indeterminate form $\frac{\infty}{\infty}$, so our natural instinct would be to use L'Hôpital's Rule. How do we find the derivative of $(n!)$? Things are looking ugly—the Alternating Series Test isn't cutting the mustard. In cases such as these, we will look for *absolute convergence*. It will allow us to ignore the pesky negative signs and *still* determine if the alternating series converges.

Absolute Convergence: If $\displaystyle\sum\left|a_{n}\right|$ converges, then $\displaystyle\sum a_{n}$ is said to *converge absolutely*, and the original series, $\displaystyle\sum a_{n}$, automatically converges. If $\displaystyle\sum\left|a_{n}\right|$ diverges but $\displaystyle\sum a_{n}$ converges, the series is said to converge *conditionally*.

Example 10: Does $\sum_{n=1}^{\infty}(-1)^n \cdot \dfrac{4^n}{n!}$ converge?

Solution: Because the Alternating Series Test doesn't work, as we showed above, we will test to see if

$$\sum_{n=1}^{\infty}\left|(-1)^n \cdot \frac{4^n}{n!}\right| = \sum_{n=1}^{\infty}\frac{4^n}{n!}$$

converges instead. Because of the large quantities involved, the Ratio Test is your best bet.

$$\lim_{n \to \infty} \frac{\dfrac{4^{n+1}}{(n+1)!}}{\dfrac{4^n}{n!}}$$

$$\lim_{n \to \infty} \frac{4}{n+1} = 0$$

Because $0 < 1$, this series converges by the Ratio Test. Here's the great part: because the absolute value of the series converged, then the series converges

absolutely, and $\sum_{n=1}^{\infty}(-1)^n \cdot \dfrac{4^n}{n!}$ converges automatically! Handy, eh?

You may wonder why absolute convergence works. In essence, it says to ignore all negative terms in a series and make them positive. If that all-positive series *still* converges to a single number, then allowing some of those terms to be negative will not cause that sum to get any larger and possibly diverge. In fact, the inclusion of negative terms will make that sum smaller.

The final important characteristic of alternating series is the error-bound that they can report. Like most other series tests (excluding geometric), you cannot find the sum of the infinite series. You still can't with Alternating Series (Booooo!), but you *can* tell approximately how close your partial sum approximation is. Here's the important factoid to remember: *the error inherent in an alternating series' partial sum S_n is less than the absolute value of the next term, a_{n+1}*. This may sound tricky, but it's really easy!

Example 11: Find the interval in which the actual sum of $\sum_{n=1}^{\infty}\dfrac{(-1)^{n+1}}{n^2}$ is contained if S_5 is used to approximate it.

Solution: S_5 is the partial sum that includes only the first five terms of the series. So, this problem is approximating the sum of the series by adding only those terms together:

$$S_5 = 1 - \frac{1}{4} + \frac{1}{9} - \frac{1}{16} + \frac{1}{25} \approx .838611 \, .$$

Although that's not the sum of the infinite series, we will soon be able to tell just how close of an approximation it is. The maximum possible error will be the absolute value of the next term, $|a_6| = \left| -\frac{1}{36} \right| \approx .027777$. The actual infinite sum for the series will fall within .027777 of .838611. Therefore, the actual sum is somewhere in the interval

$$(.838611 - .027777, .838611 + .027777)$$

$$(.810833, .866388).$$

CONCLUSION

After learning all of these tests for series convergence, most students think that it will be a miracle if they can simply remember them all on a test—I agree. However, it is a miracle that helps me remember all the tests. Whenever you face an ugly, squirming, gross series problem, remember Moses parting the Red Sea. Not only is it relaxing, it also forms a mneumonic phrase to help you remember all the series you've learned:

PARTING C

(Okay, I had to be a little creative with "C," but cut me some slack.) Each letter represents one of the tests you just learned. As you attempt each of the problems in the Problem Set, use the Moses phrase to help decide which test to use:

P   *P*-series: Is the series in the form $\frac{1}{n^p}$?

A   Alternating series: Does the series alternate? If it does, are the terms getting smaller, and is the *n*th term 0?

R   Ratio Test: Does the series contain things that grow very large as *n* increases?

T   Telescoping series: Will all but a couple of the terms in the series cancel out?

I   Integral Test: Can you easily integrate the expression that defines the series?

N   *N*th Term Divergence Test: Is the *n*th term something other than 0?

G   Geometric series: Is the series of the form $\sum\limits_{n=0}^{\infty} ar^n$?

C   Comparison Tests: Is the series *almost* another kind of series (e.g., *p*-series or geometric series)? Which would be better to use: the Comparison or the Limit Comparison Test?

Just think of it: Moses was fleeing from the Egyptians and captivity. You've probably never had to part anything larger than your hair. Unless you are a member of the 80s band Poison, the tasks aren't even comparable.

PROBLEM SET

Do not use a graphing calculator on any of these problems.

For 1 through 12, determine whether or not the series converge using the appropriate convergence test (there may be more than one applicable test). *If possible*, give the sum of the series.

1. $\displaystyle\sum_{n=0}^{\infty}\left(\frac{2}{7}\right)^{n}$

2. $\displaystyle\sum_{n=1}^{\infty}\frac{4}{n^{3}}$

3. $\displaystyle\sum_{n=1}^{\infty}\frac{n^{2}}{5^{n}}$

4. $\displaystyle\sum_{n=1}^{\infty}\frac{1}{\sqrt[3]{n^{5}+5}}$

5. $\displaystyle\sum_{n=1}^{\infty}\frac{n^{n}}{n!}$

6. $\dfrac{1}{5}+\dfrac{1}{6}+\dfrac{1}{7}+\dfrac{1}{8}+\dfrac{1}{9}+\ldots$

7. $2+\dfrac{1}{2}+\dfrac{1}{8}+\dfrac{1}{32}+\ldots$

8. $\displaystyle\sum_{n=1}^{\infty}\frac{5n^{2}-6n+3}{n^{3}-7n+8}$

9. $\displaystyle\sum_{n=1}^{\infty}\frac{\cos n\pi}{\sqrt{n}}$

10. $\displaystyle\sum_{n=1}^{\infty}\frac{3^{n}+4}{2^{n}}$

11. $\displaystyle\sum_{n=1}^{\infty}\frac{8n^{3}-6n^{5}}{12n^{4}+9n^{5}}$

12. $\displaystyle\sum_{n=1}^{\infty}\sqrt{\frac{3n+1}{n^{5}+2}}$

13. Determine if the series $\displaystyle\sum_{n=1}^{\infty} \frac{(-1)^n}{\sqrt[5]{3n+4}}$ converges absolutely, converges conditionally, or diverges.

SOLUTIONS

1. This is a geometric series with $a = 1$ and $r = \frac{2}{7}$. Because $0 < \frac{2}{7} < 1$, the series will converge, and it will converge to

$$\frac{a}{1-r} = \frac{1}{1-\frac{2}{7}} = \frac{1}{\frac{5}{7}} = \frac{7}{5}.$$

2. This looks like a p-series with $p = 3$. However, the 4 makes it a little different. Because the 4 will be multiplied by every term in the series, we can rewrite the series as

$$4\sum_{n=1}^{\infty} \frac{1}{n^3}.$$

(You can pull out the constant, just like you did with definite integrals.) Therefore, the 4 does not affect the series at all. Since $p > 1$, the series converges.

3. This series is a good candidate for the Ratio Test because of 5_n; this quantity will grow large quickly as n increases.

$$\lim_{n \to \infty} \frac{\dfrac{(n+1)^2}{5^{n+1}}}{\dfrac{n^2}{5^n}}$$

$$\lim_{n \to \infty} \frac{(n+1)^2}{5^{n+1}} \cdot \frac{5^n}{n^2}$$

$$\lim_{n \to \infty} \frac{n^2+2n+1}{5n^2} = \frac{1}{5}$$

Because $\frac{1}{5} < 1$, this series converges according to the Ratio Test. (Remember, $\frac{1}{5}$ just tells you that the series converges—it does not mean that the sum of the series is $\frac{1}{5}$.)

4. Compare this to the series $\displaystyle\sum_{n=1}^{\infty} \frac{1}{n^{5/3}}$. The original series has denominators that are slightly greater than each corresponding denominator in the second series. A larger denominator means a *smaller* value. Therefore, each term of

$$\sum_{n=1}^{\infty} \frac{1}{\sqrt[3]{n^5}+5}$$ will be smaller than the corresponding term of $\displaystyle\sum_{n=1}^{\infty} \frac{1}{n^{\frac{5}{3}}}$. Notice

that $\displaystyle\sum_{n=1}^{\infty} \frac{1}{n^{\frac{5}{3}}}$ is a simple *p*-series, and since $p = \frac{5}{3} > 1$, it converges. Because

$\displaystyle\sum_{n=1}^{\infty} \frac{1}{\sqrt[3]{n^5 + 5}}$ is smaller than a convergent series, it must converge by the Comparison Test, too.

5. The Ratio Test is our best bet, as nn and $n!$ will grow large quickly as n increases.

$$\lim_{n \to \infty} \frac{\dfrac{(n+1)^{n+1}}{(n+1)!}}{\dfrac{n^n}{n!}}$$

$$\lim_{n \to \infty} \frac{(n+1)^{n+1} \cdot n!}{(n+1)! \cdot n^n}$$

$$\lim_{n \to \infty} \frac{(n+1)^{n+1}}{(n+1) \cdot n^n}$$

You can cancel an $(n + 1)$ term out of the top and bottom, leaving you with

$$\lim_{n \to \infty} \frac{(n+1)^n}{n^n}.$$

Here's the tricky part: If you rewrite this limit, you get

$$\lim_{n \to \infty} \left(\frac{n+1}{n}\right)^n$$

$$\lim_{n \to \infty} \left(1 + \frac{1}{n}\right)^n = e.$$

Remember that limit you memorized a long time ago? There it is again. Since $e > 1$ (e is approximately 2.718), the series diverges.

6. This is actually the series $\displaystyle\sum_{n=1}^{\infty} \frac{1}{n+4}$. Use the Integral Test to see if it converges.

$$\int_1^\infty \frac{dn}{n+4}$$

$$\lim_{b \to \infty} \int_1^b \frac{dn}{n+4}$$

$$\lim_{b \to \infty} \ln|n+4| \Big|_1^b$$

$$\lim_{b \to \infty} \left[\ln(b+4) - \ln 5\right] = \infty$$

Therefore, both the integral and the series diverge.

7. This is a geometric series with $a = 2$. Factor 2 out of all the terms in the series to get the following ratio:

$$2\left(1 + \frac{1}{4} + \frac{1}{16} + \frac{1}{64} + \cdots\right)$$

The denominators are consecutive powers of 4, so the ratio is $r = \frac{1}{4}$, and the

series is $\sum_{n=0}^{\infty} 2\left(\frac{1}{4}\right)^n$. The series converges since $\frac{1}{4}$ is between 0 and 1, and

it has sum

$$\frac{a}{1-r} = \frac{2}{\frac{3}{4}} = 2 \cdot \frac{4}{3} = \frac{8}{3}.$$

8. This is a Limit Comparison problem. Take the highest powers of the numerator

and denominator to create the comparison series $\frac{n^2}{n^3} = \frac{1}{n}$.

$$\lim_{n \to \infty} \frac{\frac{5n^2 - 6n + 3}{n^3 - 7n + 8}}{\frac{1}{n}}$$

$$\lim_{n \to \infty} \frac{5n^3 - 6n^2 + 3n}{n^3 - 7n + 8} = \frac{5}{1} = 5$$

Since $\frac{1}{n}$ is a divergent p-series, and the above limit exists, both series diverge by the Limit Comparison Test.

9. Although this series looks funky, it's just an alternating series. The numerators will be $\cos \pi$, $\cos 2\pi$, $\cos 3\pi$, $\cos 4\pi$, etc., which are just the numbers 1, –1, 1, –1, etc. Each successive term will definitely get smaller (since the denominators will grow steadily). Thererfore, all that's left in the Alternating Series Test is the limit:

$$\lim_{n \to \infty} \frac{1}{n^{1/2}}.$$

The limit does equal zero (consider the graph). Therefore, this alternating series diverges by the nth Term Divergence Test.

10. Each term of this series is greater than each corresponding term of the

$\sum_{n=1}^{\infty} \left(\frac{3}{2}\right)^n$, which is a divergent geometric series. Because $\sum_{n=1}^{\infty} \frac{3^n + 4}{2^n}$ is greater

than a divergent series, then it must diverge by the Comparison Test, too.

11. Although this problem may look complicated, it is actually quite easy. This series diverges by the nth Term Divergence Test. Notice that

$$\lim_{n\to\infty}\frac{8n^3-6n^5}{12n^4+9n^5}=-\frac{6}{9}=-\frac{2}{3}.$$

Because the limit at infinity of the *n*th term is not 0, this series cannot converge (you will eventually be adding $-\frac{2}{3}$ forever).

12. If you rewrite the series as

$$\sum_{n=1}^{\infty}\frac{\sqrt{3n+1}}{\sqrt{n^5+2}},$$

you can use the Limit Comparison Test. Take the highest powers of *n* in the numerator and denominator to create the comparison series $\frac{n^{1/2}}{n^{5/2}}=\frac{1}{n^{4/2}}$.

$$\lim_{n\to\infty}\frac{\dfrac{\sqrt{3n+1}}{\sqrt{n^5+2}}}{\dfrac{1}{\sqrt{n^4}}}$$

(If you leave $\sqrt{n^4}$ unsimplified, it's more obvious to multiply the radicals together.)

$$\lim_{n\to\infty}\frac{\sqrt{3n^5+n^4}}{\sqrt{n^5+2}}=\frac{\sqrt{3}}{\sqrt{1}}=\sqrt{3}$$

Because $\displaystyle\sum_{n=1}^{\infty}\frac{1}{\sqrt{n^4}}=\frac{1}{n^2}$ is a convergent *p*-series, both series must converge.

13. First of all, test the absolute convergence by ignoring the $(-1)^n$. This gives the series $\displaystyle\sum_{n=1}^{\infty}\frac{1}{\sqrt[5]{3n+4}}$. This is similar to the *p*-series $\frac{1}{n^{1/5}}$, so use the Limit Comparison Test:

$$\lim_{n\to\infty}\frac{\dfrac{1}{\sqrt[5]{3n+4}}}{\dfrac{1}{n^{1/5}}}$$

$$\lim_{n\to\infty}\frac{\sqrt[5]{n}}{\sqrt[5]{3n+4}}.$$

Because this is a rational function with equal powers in both parts of the fraction, the limit is $\sqrt[5]{\frac{1}{3}}$. Because that is a finite number, both series diverge according to the Limit Comparison Test. The conclusion? This series *does not* converge absolutely. Next, we need to determine whether or not the series converges conditionally by leaving the $(-1)^n$ in place.

The series $\displaystyle\sum_{n=1}^{\infty} \frac{(-1)^n}{\sqrt[5]{3n+4}}$ is definitely alternating, and, because the denominators will grow steadily, the terms will lessen in value as n increases. Now, to test the nth term (the last hurdle in the Alternating Series Test):

$$\lim_{n\to\infty} \frac{1}{\sqrt[5]{3n+4}} = 0.$$

According to the Alternating Series Test, this series converges but only *conditionally,* since the absolute value of the series diverged.

POWER SERIES

Thus far, you have dealt exclusively with series of the form Σa_n; all the series have been runs of constants. The final two sections of this chapter deal with series that contain variables, and the AP test has more questions on these topics than all the other sequences and series topics combined. Power series, the first of the essential topics, are series of the form $\displaystyle\sum_{n=0}^{\infty} a_n x^n$ or $\displaystyle\sum_{n=0}^{\infty} a_n(x-c)^n$; the first is said to be centered about $x=0$, while the second form is centered about $x=c$. Strangely enough, most of our information concerning power series does not come from mathematicians, but rather from Shirley McClaine, famous new-age celebrity. In her most recent book *I Have the Power*, she claims that, among her many reincarnations, three of them were spent as power series. "It wasn't an exciting life," she says in the book, "but I always felt centered swimming in the c."

Your major goal with power series will be (surprise, surprise) to determine where they converge. Notice that we are not trying to determine *if* they converge, but rather *where* they converge. All power series will converge at the x-value at which they are centered ($x=c$). To test this, plug $x=c$ into the generic power series centered at c:

$$\sum_{n=0}^{\infty} a_n(x-c)^n = a_n + a_n(c-c)^1 + a_n(c-c)^2 + \ldots = a_n.$$

Will power series converge at other points? Possibly. One of three things will happen:

(1) The power series will only converge when $x=c$.

(2) The power series will converge at c and some distance around c (called the *radius of convergence*, or ROC). For example, if a series centered about 3 ($c=3$) has a radius of convergence of 4, then the series will converge when x is between -1 and 7.

(3) The power series will converge for all x (in this case, the radius of convergence is ∞).

To find the ROC for a power series, you'll use the Ratio Test, since power series by definition contain powers of n, which grow large quickly as n increases. In fact, *you will always test for absolute convergence when finding the ROC* to simplify matters. Once you find the ROC, you will sometimes be asked whether or not the series converges at the endpoints of that interval. For example, our series centered about 3 with ROC = 4 have an *interval of* convergence (IOC) of $[-1,7),(-1,7],[-1,7]$, or $(-1,7)$, depending upon which endpoints cause the series to converge.

 Example 12: Find the radius of convergence for the following power series:

(a) $\displaystyle\sum_{n=0}^{\infty} \frac{(-1)^{n-1} \cdot x^n}{n^2 + 5n + 4}$

Use the Ratio Test to see if the series converges absolutely:

$$\lim_{n \to \infty} \left| \frac{\dfrac{x^{n+1}}{(n+1)^2 + 5(n+1) + 4}}{\dfrac{x^n}{n^2 + 5n + 4}} \right|$$

$$\lim_{n \to \infty} \left| \frac{\left(n^2 + 5n + 4\right)x}{n^2 + 7n + 10} \right|.$$

As n approaches infinity, $\dfrac{n^2 + 5n + 4}{n^2 + 7n + 10}$ approaches 1.

$$\lim_{n \to \infty} \left| 1 \cdot x \right|$$

Remember, the Ratio Test only guarantees convergence if that limit is less than one. Since the limit is x, the series will only converge if $|x| < 1$ or $-1 < x < 1$. The series is centered at $c = 0$ and has ROC = 1.

(b) $\displaystyle\sum_{n=0}^{\infty} \frac{3x^n}{(n+1)!}$

This is another power series centered at $c = 0$. To see where it converges, use the Ratio Test to see if the series converges absolutely:

$$\lim_{n \to \infty} \left| \frac{\dfrac{3x^{n+1}}{(n+2)!}}{\dfrac{3x^n}{(n+1)!}} \right|$$

$$\lim_{n \to \infty} \left| \frac{x}{n+2} \right|.$$

Remember that x is some number you'll plug in later, so even though it's technically a variable, you can treat it like a number. Whatever number it is, it's irrelevant in this problem. The denominator will grow infinitely large, so

$$\lim_{n \to \infty} \left| \frac{x}{n+2} \right| = 0.$$

The Ratio Test only guarantees convergence when the limit is less than 1. In this case, the limit is 0, which is *always* less than one, regardless of what x is. Therefore, this series converges for all x, and the radius of convergence is ∞.

 Example 13: On what interval does the series

$$\sum_{n=0}^{\infty} \frac{(-1)^n (x+2)^n}{n \cdot 2^n} \text{ converge?}$$

Solution: For a change, this power series is centered at $c = -2$, since $(x - (-2)) = (x - c)$. This question asks for the IOC; we still need to find the ROC first:

$$\lim_{n \to \infty} \left| \frac{\dfrac{(x+2)^{n+1}}{(n+1) \cdot 2^{n+1}}}{\dfrac{(x+2)^n}{n \cdot 2^n}} \right|$$

$$\lim_{n \to \infty} \left| \frac{n(x+2)}{2(n+1)} \right|.$$

As n approaches infinity, $\frac{n}{2(n+1)}$ approaches $\frac{1}{2}$, so the limit is $\left| \frac{x+2}{2} \right|$. Remember that this must be less than one in order for the series to converge:

$$\left| \frac{x+2}{2} \right| < 1$$

$$|x+2| < 2.$$

The radius of convergence is 2. The series is centered at $x = -2$, so the series converges on $(-2 - 2, -2 + 2) = (-4, 0)$.

We're not done yet! We need to see if the series converges at the endpoints of the interval, $x = -4$ and $x = 0$. Let's plug each separately into the original series to see what happens.

$x = -4$:

$$\sum_{n=0}^{\infty} \frac{(-1)^n (-4+2)^n}{n \cdot 2^n}$$

$$\sum_{n=0}^{\infty} \frac{(-1)^n (-2)^n}{n \cdot 2^n}$$

Regardless of what n is, the numerator will be positive in this series. In essence, you will have the series

$$\sum_{n=0}^{\infty} \frac{2^n}{n \cdot 2^n}.$$

If you cancel out the 2^n, you have

$$\sum_{n=0}^{\infty} \frac{1}{n},$$

which is the divergent harmonic series. Therefore, the series does *not* converge for $x = -4$.

$x = 0$:

$$\sum_{n=0}^{\infty} \frac{(-1)^n \cdot 2^n}{n \cdot 2^n}$$

$$\sum_{n=0}^{\infty} \frac{(-1)^n}{n}.$$

This is the alternating harmonic series that satisfies all the conditions of the Alternating Series Test and thus converges. Therefore, the series *does* converge when $x = 0$.

The IOC for $\displaystyle\sum_{n=0}^{\infty} \frac{(-1)^n (x+2)^n}{n \cdot 2^n}$ is $(-4,0]$.

Because power series contain x's, they can also be used to define functions:

$$\sum_{n=0}^{\infty} a_n x^n = f(x).$$

As long as the series converges on the interval (c,d), it will return a sum for any value between c and d; that sum will be the function value for the given input. Functions defined as power series have two important characteristics:

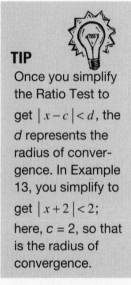

TIP
Once you simplify the Ratio Test to get $|x-c| < d$, the d represents the radius of convergence. In Example 13, you simplify to get $|x+2| < 2$; here, $c = 2$, so that is the radius of convergence.

(1) Derivatives or integrals of functions defined by power series have the *same* radius of convergence as the original function. However, the endpoints may act differently, so you'll have to check them again. Thus, although the *radius* of convergence will be equal, the *interval* of convergence may not be.

(2) To find the derivative or integral of a function defined by a power series, simply differentiate or integrate the given power series *just as you have done previously.* All of the techniques you have used to find derivatives and integrals of regular functions still apply for power series functions.

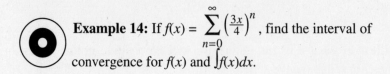

Example 14: If $f(x) = \sum_{n=0}^{\infty} \left(\frac{3x}{4}\right)^n$, find the interval of convergence for $f(x)$ and $\int f(x)dx$.

Solution: This power series (centered at $x = 0$) is actually just a geometric series with $a = 1$ and $r = \frac{3x}{4}$. Geometric series only converge when $|r| < 1$:

$$\left|\frac{3x}{4}\right| < 1$$

$$|x| < \frac{4}{3}.$$

The radius of convergence for this series is $\frac{4}{3}$, so the series converges on $\left(-\frac{4}{3}, \frac{4}{3}\right)$. To find the interval of convergence, plug in the endpoints:

$x = -\frac{4}{3}$:

$$\sum_{n=0}^{\infty} (-1)^n.$$

This is a divergent alternating series.

$x = \frac{4}{3}$:

$$\sum_{n=0}^{\infty} 1^n.$$

This diverges by the *n*th Term Divergence Test.

Therefore, the IOC for $f(x) = \left(-\frac{4}{3}, \frac{4}{3}\right)$. The next part of the question asks you to find the IOC for $\int f(x)dx$, so begin by integrating the series with u-substitution:

$$\int \left(\frac{3x}{4}\right)^n dx$$

$$u = \frac{3x}{4}; du = \frac{3}{4}dx; \frac{4}{3}du = dx$$

$$\frac{4}{3}\int u^n du$$

$$\frac{4}{3}\cdot\frac{u^{n+1}}{n+1}$$

$$\int f(x)dx = \sum_{n=0}^{\infty} \frac{4}{3(n+1)}\left(\frac{3x}{4}\right)^{n+1}.$$

The integral will have the same ROC, but we need to recheck the endpoints to see if the IOC changes:

$x = -\frac{4}{3}$:

$$\sum_{n=0}^{\infty} \frac{4}{3(n+1)}(-1)^{n+1}.$$

This is a convergent alternating series.

$x = \frac{4}{3}$:

$$\sum_{n=0}^{\infty} \frac{4}{3(n+1)}(-1)^{n+1}$$

$$\sum_{n=0}^{\infty} \frac{4}{3n+3}.$$

You can compare this with the series $\frac{1}{n}$ and apply the Limit Comparison Test:

$$\lim_{n\to\infty} \frac{\frac{4}{3n+3}}{\frac{1}{n}} = \frac{4}{3}.$$

Therefore, both series diverge.

The interval of convergence for $\int f(x)dx = \left[-\frac{4}{3}, \frac{4}{3}\right)$.

PROBLEM SET

Do not use a graphing calculator for these problems.

1. Find the radius of convergence for each of the following:

   (a) $\displaystyle\sum_{n=0}^{\infty} \frac{(-1)^n \cdot x^{2n}}{n!}$

   (b) $\displaystyle\sum_{n=1}^{\infty} \frac{n!(x+3)^n}{n^2}$

2. The series $\displaystyle\sum_{n=0}^{\infty} \frac{n^2 x^n}{4^n}$ converges on what interval?

3. If $f(x) = (x+1) + \dfrac{(x+1)^2}{2} + \dfrac{(x+1)^3}{3} + \ldots$, give the interval of convergence for $f'(x)$.

SOLUTIONS

1. (a) Apply the Ratio Test to see if the series converges absolutely:

$$\lim_{n\to\infty}\left|\frac{\dfrac{x^{2(n+1)}}{(n+1)!}}{\dfrac{x^{2n}}{n!}}\right|$$

$$\lim_{n\to\infty}\left|\frac{x^2}{n+1}\right| = 0.$$

Remember, the n approaches infinity while the x stays the same; therefore, the limit is 0, regardless of x. Because $0 < 1$ (still), this series converges for all x, and the radius of convergence is ∞. The interval of convergence is $(-\infty, \infty)$, if you're curious.

(b) Apply the Ratio Test. Once you simplify the complex fraction, you get

$$\lim_{n\to\infty}\left|\frac{(n+1)!(x+3)^{n+1}}{(n+1)^2} \cdot \frac{n^2}{n!(x+3)^n}\right|$$

$$\lim_{n\to\infty}\left|\frac{n^2(n+1)(x+3)}{(n+1)^2}\right|$$

$$\lim_{n\to\infty}\left|\frac{(x+3)(n^3+n^2)}{n^2+2n+1}\right| = \infty.$$

Because the degree of n in the numerator is higher, this rational expression will become infinitely large. Therefore, this limit is *never* less than one and can never converge, according to the Ratio Test. Remember that all power series converge—at least for the x value at which they are centered. Therefore, the series converges only for $x = -3$, and the radius of convergence is 0.

2. Begin by finding the radius of convergence:

$$\lim_{n \to \infty} \left| \frac{\frac{(n+1)^2 \cdot x^{n+1}}{4^{n+1}}}{\frac{n^2 x^n}{4^n}} \right|$$

$$\lim_{n \to \infty} \left| \frac{x(n^2 + 2n + 1)}{4n^2} \right| = \left| x \cdot \frac{1}{4} \right| = \left| \frac{x}{4} \right|.$$

According to the Ratio Test, the series converges if

$$\left| \frac{x}{4} \right| < 1$$

$$|x| < 4.$$

So, the radius of convergence is 4 for this series centered at 0. The series must converge on $(0 - 4, 0 + 4) = (-4, 4)$. Now you have to see whether or not the series converges at the endpoints by plugging them in for x:

$x = -4$:

$$\sum_{n=0}^{\infty} \frac{n^2(-4)^n}{4^n}$$

$$\sum_{n=0}^{\infty} n^2 \cdot (-1)^n$$

This diverges by the nth Term Divergence Test.

$x = 4$:

$$\sum_{n=0}^{\infty} \frac{n^2 4^n}{4^n}$$

$$\sum_{n=0}^{\infty} n^2.$$

This diverges by the nth Term Divergence Test, too, and both endpoints have failed us.

Therefore, the series $\displaystyle\sum_{n=0}^{\infty} \frac{n^2 x^n}{4^n}$ converges on $(-4, 4)$.

3.   The function and its derivative will both have the same radius of convergence, so you can find the ROC of the original function first.

$$\lim_{n \to \infty} \left| \frac{\frac{(x+1)^{n+1}}{n+1}}{\frac{(x+1)^n}{n}} \right|$$

$$\lim_{n \to \infty} \left| \frac{n(x+1)}{n+1} \right|$$

As n approaches infinity, $\frac{n}{n+1}$ approaches 1.

$$\lim_{n \to \infty} |x+1| = |x+1|$$

According to the Ratio Test, the series converges if $|x+1| < 1$, so the radius of convergence (as always) is that number on the right side once we've solved for $|x-c|$; in this case, the ROC is 1. Since the power series is centered at $c = -1$, we know (so far) that the series converges on the interval $(-1 - 1, -1 + 1) = (-2,0)$.

Now, it's time to take the derivative of the series and test the endpoints $x = -2$ and $x = 0$. Use the Power Rule to take the derivative:

$$\sum_{n=1}^{\infty} \frac{n(x+1)^{n-1}}{n}$$

$$\sum_{n=1}^{\infty} (x+1)^{n-1}.$$

Both endpoints make the series diverge by the nth Term Divergence Test if you plug them in, so the interval of convergence is $(-2,0)$.

NOTE
Just like power series, Taylor series are stressed very heavily on the AP test.

TAYLOR AND MACLAURIN SERIES

At the end of the power series section, we saw that a function can be defined using a power series. *Taylor series* are specific forms of the power series that are used to approximate function values. For example, you know $\cos \pi$ and $\cos \frac{\pi}{2}$ by heart, but if asked to evaluate $\cos \frac{1}{2}$, you'd probably be stumped.

(Arccos $\frac{1}{2}$ is very easy; that is $\frac{\pi}{3}$, but $\cos \frac{1}{2}$ is rough.) We can create a very simple Taylor series that will approximate $\cos \frac{1}{2}$ very nicely. We will create a power series centered around a very easily obtained value of cosine that is also close to $\frac{1}{2}$. The best choice is $c = 0$, since 0 is close to $\frac{1}{2}$ and $\cos 0$ is easy to evaluate; $\cos 0 = 1$.

Taylor series for $f(x)$ centered about $x = c$: $\displaystyle\sum_{n=0}^{\infty} \frac{f^{(n)}(c)(x-c)^n}{n!}$.

$$f(x) = f(c) + f'(c)(x-c) + \frac{f''(c)(x-c)^2}{2!} + \frac{f'''(c)(x-c)^3}{3!} + \ldots$$

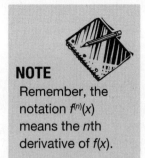

Most of the time, you will not use an infinite series to approximate function values. Instead, you will use only a finite number of the series' terms. In these cases, the Taylor series is often called a *Taylor polynomial* of degree n (where n is the highest power of the resulting polynomial). A *Maclaurin series* is the specific case of a Taylor series that is centered at $c = 0$, resulting

in the simpler-looking series $\displaystyle\sum_{n=0}^{\infty} \frac{f^{(n)}(0)x^n}{n!}$.

$$f(x) = f(0) + f'(0)x + \frac{f''(0)x^2}{2!} + \frac{f'''(0)x^3}{3!} + \ldots + \frac{f^{(n)}(0)x^n}{n!} + \ldots$$

Example 15: Use a fourth-degree Taylor polynomial of order (degree) 4 centered at 0 to approximate $\cos\left(\frac{1}{2}\right)$.

Solution: Since this Taylor series is centered at $c = 0$, it is actually a Maclaurin series. We will have to use the Maclaurin series expansion up to $n = 4$, since the requested degree is 4. In order to find the series, we will have to find $f(0), f'(0), f''(0), f'''(0)$, and $f^{(4)}(x)$:

$$f(x) = \cos x; f(0) = 1$$

$$f'(x) = -\sin x; f'(0) = 0$$

$$f''(x) = -\cos x; f''(0) = -1$$

$$f'''(x) = \sin x; f'''(0) = 0$$

$$f^{(4)}(x) = \cos x; f(0) = 1.$$

To get the Maclaurin polynomial, plug these into the Maclaurin formula and stop when $n = 4$:

$$\cos x \approx f(0) + f'(0)x + \frac{f''(0)x^2}{2!} + \frac{f'''(0)x^3}{3!} + \frac{f^{(4)}(0)x^4}{4!}$$

$$\cos x \approx 1 + 0 \cdot x - \frac{x^2}{2!} + 0 \cdot \frac{x^3}{3!} + 1 \cdot \frac{x^4}{4!}$$

$$\cos x \approx 1 - \frac{x^2}{2!} + \frac{x^4}{4!}.$$

The resulting polynomial will give you the approximate value of cos x. To find the approximate value of cos $\frac{1}{2}$, plug $\frac{1}{2}$ in for x:

$$\cos\frac{1}{2} \approx 1 - \frac{\left(\frac{1}{2}\right)^2}{2!} + \frac{\left(\frac{1}{2}\right)^4}{4!}$$

$$\cos\frac{1}{2} \approx 1 - \frac{1}{8} + \frac{1}{384} \approx .8776041667.$$

The actual value for cos $\frac{1}{2}$ (according to the calculator) is .8775825619, so the approximation wasn't too shabby at all. Just like Riemann sums, the accuracy of your prediction will increase as you increase the number of terms in your Taylor polynomial; in other words, the greater the n, the more accurate the result. Below, the graph of $y = \cos x$ is compared with the graphs of three Maclaurin polynomials for cos x centered about 0.

Maclaurin polynomial, $n = 0$

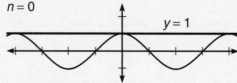

Maclaurin polynomial, $n = 2$

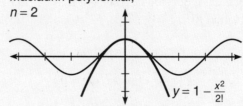

Maclaurin polynomial, $n = 4$

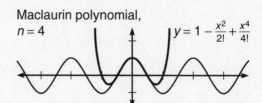

A couple of things are clear from the graphs. First of all, the greater the degree of the polynomial, the closer its graph is to the graph of cos x. However, none of the approximations are very good for approximating values far away from $x = 0$. If you need to approximate other values, you will have to use a Taylor polynomial centered about a different value. For example, to estimate cos (3.2), you might use a Taylor series centered about $c = \pi$, since π is close to 3.2.

Although it wasn't too difficult to come up with the Maclaurin polynomial for cos x, you shouldn't have to construct it like that on the AP test. It is one of four functions for which you should have the Maclaurin expansions memorized—doing so will save you much-needed time.

Maclaurin Series to Memorize

$$\cos x = 1 - \frac{x^2}{2!} + \frac{x^4}{4!} - \frac{x^6}{6!} + \dots$$

$$\sin x = x - \frac{x^3}{3!} + \frac{x^5}{5!} - \frac{x^7}{7!} + \dots$$

$$e^x = 1 + x + \frac{x^2}{2!} + \frac{x^3}{3!} + \frac{x^4}{4!} + \dots$$

$$\frac{1}{1-x} = 1 + x + x^2 + x^3 + x^4 + \dots$$

Not all of the series on the AP test will be based on these four functions, but most of them will. Any other series can be constructed using the method of Example 15.

 Example 16: Derive the Maclaurin series for sin x from the Maclaurin series for cos x.

Solution: We know that $\int \cos x\, dx = \sin x$. Taylor series act just like their parent functions; if you integrate each term of the cos x Maclaurin series, you will end up with the Maclaurin series for sin x. This is not only useful for impressing your friends, however.

$$\int \cos x\, dx = \int \left(1 - \frac{x^2}{2!} + \frac{x^4}{4!} + \dots \right) dx$$

$$\sin x = x - \frac{x^3}{3 \cdot 2!} + \frac{x^5}{5 \cdot 4!} + \dots$$

$$\sin x = x - \frac{x^3}{3!} + \frac{x^5}{5!} + \dots$$

The resulting series is exactly the one for sin x that you are to memorize.

 Example 17: Determine a power series for $\sin x^2$.

CAUTION Each of the Maclaurin series listed will converge on $(-\infty, \infty)$, except for $\frac{1}{1-x}$. That series has an interval of convergence of $(-1,1)$ and will not work well for x's outside that interval.

CAUTION Even though the sin x series is written "$x - \frac{x^3}{3!} + \frac{x^5}{5!} + \dots$," the next term in the series, $-\frac{x^7}{7!}$, is still negative. The series still alternates; it's just common notation to write a "+" at the end of an infinite series, regardless of the sign of the next term.

Solution: Besides acting like their parent functions, Taylor series are also handy because they are very flexible. We already know the Maclaurin series for sin x (and Taylor and Maclaurin series are just special power series anyway). To find the series for x^2, just plug x^2 in for x. That's all there is to it.

$$\sin x = x - \frac{x^3}{3!} + \frac{x^5}{5!} - \frac{x^7}{7!} + \dots$$

$$\sin x^2 = x^2 - \frac{\left(x^2\right)^3}{3!} + \frac{\left(x^2\right)^5}{5!} - \frac{\left(x^2\right)^7}{7!} + \dots$$

$$\sin x^2 = x^2 - \frac{x^6}{3!} + \frac{x^{10}}{5!} - \frac{x^{14}}{7!} + \dots$$

If the question had asked you to find a series for sin $(2x + 3)$, all you would do is plug $(2x + 3)$ in for x. Simple!

Like alternating series, there is a way to tell how accurately your Taylor polynomial approximates the actual function value: you use something called the *Lagrange remainder* or *Lagrange error bound.* It is the trickiest part of Taylor series.

Lagrange Remainder: If you use a Taylor polynomial of degree n centered about c to approximate the value x, then the actual function value falls within the error bound

$$R_n(x) = \frac{f^{(n+1)}(z)(x-c)^{n+1}}{(n+1)!},$$

where z is some number between x and c.

Translation: Similar to alternating series, the error bound is given by the *next* term in the series, $n + 1$. The only tricky part is that you evaluate $f^{(n+1)}$, the $(n + 1)$th derivative, at z, not c. What the heck is z, you ask? It is the number that makes $f^{(n+1)}(z)$ as large as it can be. This error bound is supposed to tell you how far off you are from the real number, so we want to assume the worst. We want the error bound to represent the largest possible error. In practice, picking z is relatively easy—really, you'll see.

Example 18: Approximate cos $(.1)$ using a fourth-degree Maclaurin polynomial, and find the associated Lagrange remainder.

Solution: We already know the fourth-degree Maclaurin polynomial for cosine, so plug .1 in for x to get the approximation:

$$\cos x = 1 - \frac{x^2}{2!} + \frac{x^4}{4!}$$

$$\cos .1 = 1 - \frac{.1^2}{2!} + \frac{.1^4}{4!} \approx .99500416667.$$

The associated Lagrange remainder for $n = 4$ (denoted $R_4(x)$) is

$$R_4(x) = \frac{f^{(5)}(z)(x-c)^5}{5!}.$$

The fifth derivative of $\cos x$ is $-\sin x$, so $f^{(5)}(z) = -\sin z$. Now, plug in $x = .1$ and $c = 0$ to get

$$R_4(.1) = \frac{(-\sin z)(.1)^5}{5!}.$$

We need $-\sin z$ to be as large as it can possibly be. The largest value of $-\sin x$ is 1, since $-\sin x$, like $\sin x$, has a range of $[-1,1]$. By assuming $-\sin z$ is the largest possible value, we are creating the largest possible error; so, plug in 1 for $-\sin z$. The actual remainder will be less than this largest possible value.

$$R_4(.1) < \frac{1 \cdot .1^5}{5!} = \frac{.1^5}{5!} = .0000000833$$

Therefore, our approximation of .99500416667 is off by no more than .0000000833. In fact, it is only off by .0000000014.

PROBLEM SET

You may use a graphing calculator on problems 3 and 4.

1. (a) Verify that the Maclaurin expansion for

$$f(x) = e^x \text{ is } 1 + x + \frac{x^2}{2!} + \frac{x^3}{3!} + \dots$$

   (b) Show that the Maclaurin series for e^x (like the function it represents) is its own derivative.

2. (a) Create a Maclaurin series for $g(x) = \cos(e^x)$.

   (b) Use a sixth-degree Mauclaurin series to approximate $\cos e^2$.

   (c) Explain why the approximation in 2(b) is so horrid.

3. Estimate the value of $\sqrt{1.3}$ using the third-degree Taylor polynomial for $y = \sqrt{x}$ centered about $x = 1$.

4. Let $P(x) = 4 - (x-2) + 3(x-2)^2 - 5(x-2)^3$ be a Taylor polynomial of degree 3 for $f(x)$ centered about 2.

TIP
$f^{(n+1)}(z)$ will often have a value of 1 on AP problems, making the Lagrange remainder simply the value of the next term in the series, as it turned out to be in Example 18.

(a) What is $f''(2)$?

(b) Use a second-degree Taylor polynomial to approximate $f'(2.1)$.

SOLUTIONS

1. (a) The Maclaurin series is $\displaystyle\sum_{n=0}^{\infty} \frac{f^{(n)}(0)x^n}{n!}$. To write out the expansion, we'll need $f(0), f'(0), f''(0)$, etc.

$$f(x) = e^x; f(0) = e^0 = 1$$

$$f'(x) = e^x; f'(0) = e^0 = 1$$

In fact, each derivative of ex is ex, and each derivative's resulting value at $x = 0$ will be 1. Therefore, the series is

$$f(0) + f'(0)x + \frac{f''(0)x^2}{2!} + \frac{f'''(0)x^3}{3!} + \dots$$

$$1 + 1 \cdot x + \frac{1 \cdot x^2}{2!} + \frac{1 \cdot x^3}{3!} + \dots,$$

which is the expansion we all know and love.

(b) Find the derivative of each term of the series separately:

$$\frac{d}{dx}\left(e^x\right) = \frac{d}{dx}\left(1 + x + \frac{x^2}{2!} + \frac{x^3}{3!} + \frac{x^4}{4!} + \dots\right)$$

$$\frac{d}{dx}\left(e^x\right) = 0 + 1 + \frac{2x}{2!} + \frac{3x^2}{3!} + \frac{4x^3}{4!} + \dots$$

$$\frac{d}{dx}\left(e^x\right) = 1 + x + \frac{x^2}{2!} + \frac{x^3}{3!} + \dots$$

2. (a) You already know the Maclaurin series for $\cos x$, so just plug ex in for each x:

$$\cos\left(e^x\right) = 1 - \frac{\left(e^x\right)^2}{2!} + \frac{\left(e^x\right)^4}{4!} + \dots$$

$$\cos\left(e^x\right) = 1 - \frac{e^{2x}}{2!} + \frac{e^{4x}}{4!} + \dots$$

(b) Based on your work for 2(a), the sixth-degree Maclaurin polynomial for $\cos x$ is

$$\cos\left(e^x\right) \approx 1 - \frac{e^{2x}}{2!} + \frac{e^{4x}}{4!} - \frac{e^{6x}}{6!}.$$

To approximate $\cos e^2$, plug 2 in for x:

$$\cos e^2 \approx 1 - \frac{e^4}{2!} + \frac{e^8}{4!} - \frac{e^{12}}{6!} \approx -128.1408.$$

Whoa, how can cosine have a value lower than −1? Yikes!

(c) Remember that a Taylor series is only accurate around the value at which it is centered. Because this is a Maclaurin series, it's centered at $c = 0$. Using this approximation to evaluate $\cos (e^2)$ is irresponsible, since $e^2 \approx 7.389$, which is nowhere close to $c = 0$. As you can see from the graph of $y = \cos (e^x)$ and the Maclaurin series in 2(b), the graphs are nowhere close to each other as you get farther away from $x = 0$.

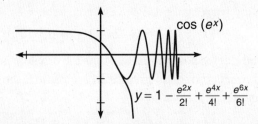

3. To find the third-degree Taylor polynomial for $f(x)$ centered at $c = 1$, we'll need the value of the first three derivatives of f evaluated at 1; these are required by the formula.

$$f(x) = x^{1/2}; f(1) = 1$$

$$f'(x) = \frac{1}{2}x^{-1/2}; f'(1) = \frac{1}{2}$$

$$f''(x) = -\frac{1}{4}x^{-3/2}; f''(1) = -\frac{1}{4}$$

$$f'''(x) = \frac{3}{8}x^{-5/2}; f'''(1) = \frac{3}{8}$$

Therefore, the Taylor polynomial is

$$\sqrt{x} \approx f(1) + f'(1)(x-1) + \frac{f''(1)(x-1)^2}{2!} + \frac{f'''(1)(x-1)^3}{3!}$$

$$\sqrt{x} \approx 1 + \frac{x-1}{2} - \frac{(x-1)^2}{4 \cdot 2!} + \frac{3(x-1)^3}{8 \cdot 3!}$$

$$\sqrt{x} \approx 1 + \frac{x-1}{2} - \frac{(x-1)^2}{8} + \frac{(x-1)^3}{16}.$$

Finally, plug in $x = 1.3$ to get the approximation of 1.1404375, which is relatively close to the actual value of 1.140175.

4. (a) We know that the squared term in any Taylor polynomial is given by $\frac{f''(c)(x-c)^2}{2!}$. In this problem, that term should be $\frac{f''(2)(x-2)^2}{2!}$. In the actual expansion, the squared term is $3(x-2)^2$. Therefore,

$$\frac{f''(2)(x-2)^2}{2!} = 3(x-2)^2$$

$$\frac{f''(2)}{2!} = 3$$

$$f''(2) = 6.$$

(b) This question asks you to approximate the value of the *derivative* of f. Since a Taylor series acts like its parent function, you can approximate $f'(x)$ by taking the derivative of each term:

$$P'(x) = -1 + 6(x-2) - 15(x-2)^2$$

This is the second-degree polynomial to which the problem is alluding. You can use it to find your approximation since 2.1 is close to 2, the value at which the series is centered:

$$f'(2.1) \approx -1 + 6(.1) - 15(.1)^2 \approx -.55.$$

Isn't that bizarre? We don't even know the function that P approximates, but we can still approximate its derivative.

TECHNOLOGY: VIEWING AND CALCULATING SEQUENCES AND SERIES WITH A GRAPHING CALCULATOR

Your calculator can serve three major functions to assist you in this chapter: it can graph sequences, graph series, and calculate partial sums. Be warned ahead of time: The calculator commands to accomplish these tasks are not as friendly as the commands used to evaluate a definite integral. It may take a bit of practice before these techniques become second nature.

In the first problem set way back in the beginning of this chapter, you had to determine whether or not the series $\sum_{n=1}^{\infty} \frac{1+3n^2+n^3}{4n^3-5n+2}$ converged. Hopefully, this problem is much easier now; the series clearly diverges because of the nth Term Divergence Test, since

$$\lim_{n\to\infty} \frac{1+3n^2+n^3}{4n^3-5n+2} = \frac{1}{4}.$$

In order to show that the *sequence* approaches $\frac{1}{4}$ as n approaches infinity, you'll have to change the [Mode] to "seq"uence. This changes the [Y=]

screen to the "$u(n)$" screen; type the sequence in for $u(n)$. (The $[x,t,\theta,n]$ button will now display an "n.")

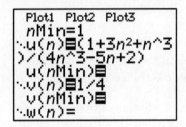

If you choose friendly [Window] settings and press [Graph], the sequence clearly levels off. If you enter a second sequence of $\frac{1}{4}$ for $v(n)$, it's easier to see that the sequence does indeed begin to level off at $\frac{1}{4}$.

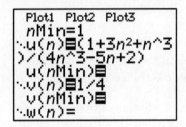 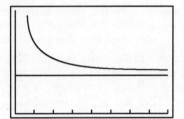

Remember that although the sequence converges to $\frac{1}{4}$, the series will diverge by the nth Term Divergence Test. In order to visualize this, we can graph that series. To do so, however, you'll need to make some minor setting changes on your calculator. First of all, go to the [Mode] screen and select "Par"ametric mode and "Dot" rather than connected. Now, go to the [Y=] screen and set $X_{1T} = T$. The Y_{1T} is the tricky part:

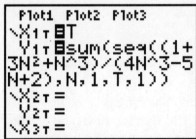

To get the "sum" command, you need to press [2nd]→[Stat]→"Math" →"sum",(and the "seq" command comes from [2nd]→[Stat]→"Ops"→ "seq".(The "N" used is the letter N, which is the result of pressing [Alpha]→[Log]. The syntax for a sequence is

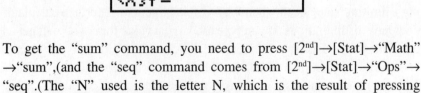

$$\text{seq}(sequence\ of\ n,n,whatever\ n=\ in\ sigma,\ T,1).$$

Choose nice [Window] settings for the graph (it might take a couple of tries to pick good settings), and you get a good picture of the series:

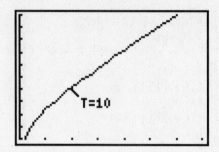

This series is definitely divergent—it approaches no limiting height. Furthermore, at approximately $n = 10$ (on the graph $t = 10$), the terms progress almost in a linear fashion. Can you guess what the slope of that line is? I'll spare you the suspense: it's $\frac{1}{4}$, since the sequence tells us we will be adding approximately $\frac{1}{4}$ to each term forever.

If you aren't all that impressed with the series-graphing capability of the calculator (we really did have to force it, didn't we?), then you'll probably be equally unimpressed with the calculator's ability to calculate partial sums. The process is very similar to the series graph. Although the series $\sum_{n=1}^{\infty} \frac{1 + 3n^2 + n^3}{4n^3 - 5n + 2}$ diverges, you can still find the sum of its first n terms. Let's use the calculator to find S_{200}. You do not have to use the [Y=] screen, as this is simply a command you can type out on the regular screen:

Therefore, $S_{200} \approx 58.869$. You can easily tell that this series diverges if you calculate $S_{300} \approx 84.173$. Clearly, the series is not "leveling out" and approaching a limiting value even when the n is this large. However, the calculator does have its limitations. If you type too large a value for n, you will get an error message. Texas Instruments technicians explain that this occurs because the calculator is "tired of adding so many dang numbers, for Pete's sake...let it do something interesting for a change."

PRACTICE PROBLEMS

You may use a graphing calculator on problems 6 and 7 only.

1. Determine whether or not the series $\sum_{n=1}^{\infty}\left(1+\frac{1}{n}\right)^{-n}$ converges, and justify your answer.

2. Find the interval of convergence for the power series:

$$1-\frac{x}{3}+\frac{x^2}{2\cdot 9}-\frac{x^3}{6\cdot 27}+\ldots$$

3. If $P(x) = 2 + 3(x + 1) - 6(x + 1)^2$ is a Taylor polynomial for $f(x)$, write a third-degree Taylor polynomial for $m(x) = \int_{-1}^{x} P(t)\,dt$.

4. Does the series $\sum_{n=0}^{\infty} \frac{(-1)^{n+1}}{(3n+2)!}$ converge absolutely, converge conditionally, or diverge?

5. Find the value of a so that the radius of convergence for the series

$$\sum_{n=1}^{\infty} \frac{\left(a^2\right)^n (x-1)^n}{3n^2} \text{ is } \frac{1}{4} \text{ and } a > 0.$$

6. If $g(x) = e^{\sin x}$, use a fourth-degree Maclaurin polynomial to approximate $g'(.3)$.

7. Find the sum of each of the following series. If you cannot find the *exact* sum, find it accurate to four decimal places:

   (a) $3 - \frac{3}{2} + \frac{3}{4} - \frac{3}{8} + \ldots$

   (b) $\sum_{n=1}^{\infty} (-1)^n \cdot e^{-n^3}$

   (c) $\sum_{n=1}^{\infty} \left(\frac{1}{n} - \frac{1}{n+3}\right)$

8. **James' Diabolical Challenge:** Prove the convergence or divergence of the series $\sum_{n=1}^{\infty} \frac{3}{\pi n^3}$ using *four* different convergence tests.

SOLUTIONS

1. The nth term of the series is $\frac{1}{\left(1+\frac{1}{n}\right)^n}$; as n approaches infinity, you get $\frac{1}{e}$, which is not 0. Therefore, this series fails the nth Term Divergence Test and therefore converges.

2. This is the series $\sum_{n=0}^{\infty} \frac{(-1)^n x^n}{n! \cdot 3^n}$. To find the interval of convergence, use the Ratio Test to see where the series converges absolutely:

$$\lim_{n \to \infty} \left| \frac{x^{n+1}}{(n+1)! \cdot 3^{n+1}} \cdot \frac{n! 3^n}{x^n} \right|$$

$$\lim_{n \to \infty} \left| \frac{x}{3(n+1)} \right|.$$

Regardless of x, this limit will be 0 as the denominator will grow infinitely large. Therefore, this limit is always less than one, so this series always converges. The interval of convergence will be $(-\infty, \infty)$. No need to test endpoints, since an interval can't be closed at an unbounded (infinite) endpoint.

3. To find $m(x)$, integrate:

$$\int_{-1}^{x} \left(2 + 3(t+1) - 6(t+1)^2 \right) dt$$

$$2t + \frac{3}{2}(t+1)^2 - \frac{6}{3}(t+1)^3 \Big|_{-1}^{x}$$

Apply Fundamental Theorem Part One:

$$2x + \frac{3}{2}(x+1)^2 - 2(x+1)^3 - \left(2(-1) + 0 \right)$$

All of the polynomial terms will cancel out when you plug in $x = -1$, except for $2(-1)$. You can write your final answer by simplifying:

$$2 + 2x + \frac{3}{2}(x+1)^2 - 2(x+1)^3$$

or by factoring out a 2.

$$2(x+1) + \frac{3}{2}(x+1)^2 - 2(x+1)^3$$

Either of those answers are acceptable.

4. First, ignore the fact that it is an alternating series to see if the convergence is absolute. The Ratio Test serves the best to test for convergence because of the factorial.

$$\lim_{n \to \infty} \frac{\dfrac{1}{\left(3(n+1) + 2 \right)!}}{\dfrac{1}{(3n+2)!}}$$

$$\lim_{n \to \infty} \frac{(3n+2)!}{(3n+5)!}$$

$$\lim_{n \to \infty} \frac{1}{(3n+5)(3n+4)(3n+3)}$$

As n approaches infinity, this value will grow extremely small. Each of the binomials in the denominator will be huge, and their product will be even larger. One divided by a high number is 0. Since $0 < 1$, this series converges absolutely. Remember, this means that the original alternating series converges automatically.

5. Find the radius of convergence as usual, using the Ratio Test:

$$\lim_{n \to \infty} \left| \frac{\dfrac{\left(a^2\right)^{n+1}(x-1)^{n+1}}{3(n+1)^2}}{\dfrac{\left(a^2\right)^{n}(x-1)^{n}}{3n^2}} \right|$$

$$\lim_{n \to \infty} \left| \frac{a^2(x-1)n^2}{(n+1)^2} \right|$$

As n approaches infinity, the limit of $\frac{n}{(n+1)^2}$ is 1, making the overall limit

$$\left| a^2(x-1) \right|$$

This limit must be less than 1 to make the series converge.

$$\left| a^2(x-1) \right| < 1$$

$$\left| x-1 \right| < \frac{1}{a^2}$$

We also know that the radius of convergence is $\frac{1}{4}$; we had a shortcut that said whenever we solve for $\left| x-c \right|$, the number on the other side of the inequality is the radius of convergence. Thus,

$$\frac{1}{a^2} = \frac{1}{4}$$

$$a^2 = 4$$

$$a = 2.$$

a cannot equal -2, since the problem specified that $a > 0$.

6. We already know the Maclaurin series for e^x, so plug $\sin x$ in for x to get the Maclaurin polynomial for $e^{\sin x}$:

$$e^x = 1 + x + \frac{x^2}{2!} + \frac{x^3}{3!} + \frac{x^4}{4!} + \dots$$

$$g(x) = e^{\sin x} = 1 + \sin x + \frac{\sin^2 x}{2!} + \frac{\sin^3 x}{3!} + \frac{\sin^4 x}{4!} + \dots$$

If you read the question carefully, you see that it asks you to find an approximation for $g'(x)$, the derivative of the series above. However, the series above has one too few terms written. Since the equation above is of degree 4, its derivative will have degree 3, so add another term when you take the derivative.

$$g'(x) \approx \cos x + \frac{2\sin x \cos x}{2!} + \frac{3\sin^2 x \cos x}{3!} + \frac{4\sin^3 x \cos x}{4!} + \frac{5\sin^4 x \cos x}{5!}$$

Use this ugly monster to approximate $g'(.3)$ by plugging .3 in for x. The resulting approximation is 1.2837864. The actual value for $g'(x) = (\cos x)(e^{\sin x}) = 1.2838053$.

7. (a) This is the geometric series $\displaystyle\sum_{n=0}^{\infty} 3\left(-\frac{1}{2}\right)^n$. The infinite sum is given by

$$S = \frac{a}{1-r} = \frac{3}{1-\left(-\frac{1}{2}\right)}$$

$$S = \frac{3}{\frac{3}{2}} = 2.$$

(b) This is a convergent alternating series, according to the Alternating Series Test. However, you cannot usually find the sum of such a series (with the exception of part (a) above). Instead, remember that the remainder (or error bound) in an alternating series is the absolute value of the first omitted term. Therefore, we should calculate the values of a few terms in the alternating series until one of the term's values ensures accuracy to four decimal places.

$$|a_2| = \frac{1}{e^{2^3}} = \frac{1}{e^8} = .000335463$$

The term a_2 is the remainder for S_1, the sum of the $n = 1$ term. The above means that S_1 is accurate to three decimal places, but the fourth decimal place could be off by as much as 3, so try a_3:

$$|a_3| = \frac{1}{e^{3^3}} = \frac{1}{e^{27}} = 1.880 \times 10^{-12}.$$

This one definitely cinches it. The sum S_2 will be accurate to 11 decimal places. All that remains is to find S_2:

$$S_2 = a_1 + a_2 = -\frac{1}{e} + \frac{1}{e^8} \approx -.3675.$$

If you are dubious, use your calculator to find S_{500}, and you'll find that we were quite accurate, even after only two terms.

(c) This is a telescoping series, and you can find the exact sum by expanding the series to determine which terms will cancel out as n approaches infinity:

$$\sum_{n=1}^{\infty}\left(\frac{1}{n} - \frac{1}{n+3}\right) = \left(1 - \frac{1}{4}\right) + \left(\frac{1}{2} - \frac{1}{5}\right) + \left(\frac{1}{3} - \frac{1}{6}\right) + \left(\frac{1}{4} - \frac{1}{7}\right) + \cdots$$

Every number from $\frac{1}{4}$ lower will be canceled out by its opposite as the sum gets longer and longer. Therefore, the exact sum is composed of the only numbers that do not get canceled: $1 + \frac{1}{2} + \frac{1}{3} = \frac{11}{6} \approx 1.833333$.

8. Okay, so four tests is a little overkill; although it's true that only one is necessary, this problem wouldn't be diabolical otherwise, would it?

P-series Test: If you factor the $\frac{3}{\pi}$ out of the series, you get $\frac{3}{\pi}\sum_{n=1}^{\infty}\frac{1}{n^3}$. $\frac{1}{n^3}$ is a p-series with $p = 3$, so the series converges. The fact that you multiply the sum by $\frac{3}{\pi}$ when you're finished does not change the fact that the sum is a finite number.

Integral Test: The series will converge if

$$\lim_{b \to \infty}\left(\frac{3}{\pi}\int_1^b x^{-3}\,dx\right)$$

exists.

$$\lim_{b \to \infty}\left(-\frac{3}{2\pi}\left(\frac{1}{x^2}\right)\Big|_1^b\right)$$

$$\lim_{b \to \infty}\left(-\frac{3}{2\pi}\left(\frac{1}{b^2} - 1\right)\right)$$

$$-\frac{3}{2\pi}(-1) = \frac{3}{2\pi}$$

Because the limit is a finite number, both the integral and the series converge.

Limit Comparison Test: Compare the series to $\frac{1}{n^3}$.

$$\lim_{n \to \infty} \frac{\frac{3}{\pi \, n^3}}{\frac{1}{n^3}} = \frac{3}{\pi}$$

Because the limit is a finite number, and $\frac{1}{n^3}$ is a convergent p-series, then both series must converge.

Comparison Test: Each term in the series $\sum_{n=1}^{\infty} \frac{3}{\pi \, n^3}$ is less than the corresponding term in the series $\sum_{n=1}^{\infty} \frac{1}{n^3}$. You know this because $\frac{3}{\pi} < 1$. When you multiply $\frac{1}{n^3}$ by a value less than one, the result is smaller than $\frac{1}{n^3}$. Notice that $\frac{1}{n^3}$ is a convergent p-series. Because the terms of $\sum_{n=1}^{\infty} \frac{3}{\pi \, n^3}$ are less than a convergent series, then it must also converge by the Comparison Test.

PART

3

Practice AP Exams

PREVIEW

Answer Sheet

FIRST MODEL AP CALCULUS AB EXAM

Section I Part A

1. Ⓐ Ⓑ Ⓒ Ⓓ Ⓔ 7. Ⓐ Ⓑ Ⓒ Ⓓ Ⓔ 13. Ⓐ Ⓑ Ⓒ Ⓓ Ⓔ 19. Ⓐ Ⓑ Ⓒ Ⓓ Ⓔ 25. Ⓐ Ⓑ Ⓒ Ⓓ Ⓔ

2. Ⓐ Ⓑ Ⓒ Ⓓ Ⓔ 8. Ⓐ Ⓑ Ⓒ Ⓓ Ⓔ 14. Ⓐ Ⓑ Ⓒ Ⓓ Ⓔ 20. Ⓐ Ⓑ Ⓒ Ⓓ Ⓔ 26. Ⓐ Ⓑ Ⓒ Ⓓ Ⓔ

3. Ⓐ Ⓑ Ⓒ Ⓓ Ⓔ 9. Ⓐ Ⓑ Ⓒ Ⓓ Ⓔ 15. Ⓐ Ⓑ Ⓒ Ⓓ Ⓔ 21. Ⓐ Ⓑ Ⓒ Ⓓ Ⓔ 27. Ⓐ Ⓑ Ⓒ Ⓓ Ⓔ

4. Ⓐ Ⓑ Ⓒ Ⓓ Ⓔ 10. Ⓐ Ⓑ Ⓒ Ⓓ Ⓔ 16. Ⓐ Ⓑ Ⓒ Ⓓ Ⓔ 22. Ⓐ Ⓑ Ⓒ Ⓓ Ⓔ 28. Ⓐ Ⓑ Ⓒ Ⓓ Ⓔ

5. Ⓐ Ⓑ Ⓒ Ⓓ Ⓔ 11. Ⓐ Ⓑ Ⓒ Ⓓ Ⓔ 17. Ⓐ Ⓑ Ⓒ Ⓓ Ⓔ 23. Ⓐ Ⓑ Ⓒ Ⓓ Ⓔ

6. Ⓐ Ⓑ Ⓒ Ⓓ Ⓔ 12. Ⓐ Ⓑ Ⓒ Ⓓ Ⓔ 18. Ⓐ Ⓑ Ⓒ Ⓓ Ⓔ 24. Ⓐ Ⓑ Ⓒ Ⓓ Ⓔ

Section I Part B

29. Ⓐ Ⓑ Ⓒ Ⓓ Ⓔ 32. Ⓐ Ⓑ Ⓒ Ⓓ Ⓔ 35. Ⓐ Ⓑ Ⓒ Ⓓ Ⓔ 38. Ⓐ Ⓑ Ⓒ Ⓓ Ⓔ 42. Ⓐ Ⓑ Ⓒ Ⓓ Ⓔ

30. Ⓐ Ⓑ Ⓒ Ⓓ Ⓔ 33. Ⓐ Ⓑ Ⓒ Ⓓ Ⓔ 36. Ⓐ Ⓑ Ⓒ Ⓓ Ⓔ 39. Ⓐ Ⓑ Ⓒ Ⓓ Ⓔ 43. Ⓐ Ⓑ Ⓒ Ⓓ Ⓔ

31. Ⓐ Ⓑ Ⓒ Ⓓ Ⓔ 34. Ⓐ Ⓑ Ⓒ Ⓓ Ⓔ 37. Ⓐ Ⓑ Ⓒ Ⓓ Ⓔ 40. Ⓐ Ⓑ Ⓒ Ⓓ Ⓔ 44. Ⓐ Ⓑ Ⓒ Ⓓ Ⓔ

41. Ⓐ Ⓑ Ⓒ Ⓓ Ⓔ 45. Ⓐ Ⓑ Ⓒ Ⓓ Ⓔ

First Model AP Calculus AB Exam

SECTION I, PART A

Time – 55 Minutes

Number of Questions - 28

A CALCULATOR MAY NOT BE USED ON THIS PART OF THE EXAMINATION.

Directions: Solve each of the following problems, using the available space for scratchwork. After examining the form of the choices, decide which is the best of the choices given and fill in the corresponding oval on the answer sheet. No credit will be given for anything written in the test book. Do not spend too much time on any one problem.

In this test: Unless otherwise specified, the domain of a function f is assumed to be the set of all real numbers x for which $f(x)$ is a real number.

1. $\int_0^1 e^{2x} dx =$

    (A) $e^2 - 1$

    (B) e^2

    (C) $\dfrac{e^2}{2}$

    (D) $\dfrac{e^2 - 1}{2}$

    (E) $2e^2 - 2$

2. If $f(x) = \tan(e^{\sin x})$, then $f'(x) =$

(A) $-e^{\sin x} \cos x \sec^2\left(e^{\sin x}\right)$

(B) $e^{\sin x} \cos x \sec^2\left(e^{\sin x}\right)$

(C) $-e^{\sin x} \sec\left(e^{\sin x}\right)\tan\left(e^{\sin x}\right)$

(D) $e^{\sin x} \sec^2\left(e^{\sin x}\right)$

(E) $e^{\sin x} \sec\left(e^{\sin x}\right)\tan\left(e^{\sin x}\right)$

3. If $F(x) = \int_{2}^{x^2} t^2\,dt$, then $F(2) =$

(A) $\dfrac{64}{3}$

(B) 64

(C) $\dfrac{16}{3}$

(D) 16

(E) $\dfrac{56}{3}$

4. If $f(x) = \tan^2 x + \sin x$, then $f'\left(\dfrac{\pi}{4}\right) =$

(A) $\dfrac{4+\sqrt{2}}{2}$

(B) $\dfrac{2+\sqrt{2}}{2}$

(C) $\dfrac{8+\sqrt{2}}{2}$

(D) $\dfrac{8-\sqrt{2}}{2}$

(E) $\dfrac{4-\sqrt{2}}{2}$.

5. At which of the following points is the graph of $f(x) = x^4 - 2x^3 - 2x^2 - 7$ decreasing and concave down?

(A) $(1, -10)$

(B) $(2, -15)$

(C) $(3,2)$

(D) $(-1, -6)$

(E) $(-2, 17)$

6. Which of the following are antiderivatives of $f(x) = \cos^3 x \sin x$?

I. $F(x) = \dfrac{-\cos^4 x}{4}$

II. $F(x) = \dfrac{\sin^2 x}{2} - \dfrac{\sin^4 x}{4}$

III. $F(x) = \dfrac{1 - \cos^4 x}{4}$

(A) I only

(B) II only

(C) III only

(D) I and III only

(E) I, II, and III

7. $\left(\dfrac{d}{dx}\right)\left(e^{\sin 2x}\right) =$

(A) $-\cos 2x e^{\sin 2x}$

(B) $\cos 2x e^{\sin 2x}$

(C) $2e^{\sin 2x}$

(D) $2\cos 2x e^{\sin 2x}$

(E) $-2\cos 2x e^{\sin 2x}$

Questions 8 and 9 refer to the graph below of the velocity of a moving object as a function of time.

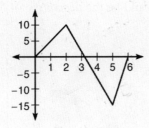

8. At what time has the object reached its maximum speed?

(A) 0

(B) 2

(C) 3

(D) 5

(E) 6

GO ON TO THE NEXT PAGE

9. Over what interval does the object have the greatest acceleration?

(A) [0,2]

(B) [2,3]

(C) [2,4]

(D) [3,5]

(E) [5,6]

10. An equation of the line tangent to $y = \sin x + 2\cos x$ at $\left(\dfrac{\pi}{2}, 1\right)$ is

(A) $2x - y = \pi - 1$

(B) $2x + y = \pi + 1$

(C) $2x - 2y = 2 - \pi$

(D) $4x + 2y = 2 - \pi$

(E) None of the above

11. The graph of the function f is given below.

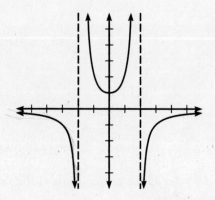

Which of these graphs could be the derivative of f?

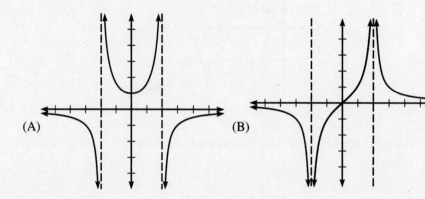

(A)   (B)

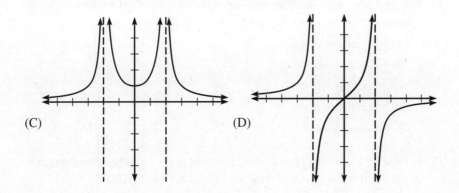

(C)

(D)

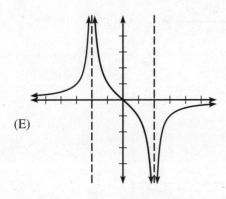

(E)

12. The function f is given by $f(x) = x^4 - 8x^3 + 24x^2 - 32x + 15$. All of these statements are true EXCEPT

   (A) 1 and 3 are zeros of f.

   (B) $f'(2) = 0$.

   (C) $f''(2) = 0$.

   (D) $(2, -1)$ is a point of inflection of f.

   (E) $(2, -1)$ is a local minimum of f.

13. The function f is given by $f(x) = 3e^{\sin x}$. f is decreasing over which interval?

   (A) $[0, \pi]$

   (B) $\left[-\dfrac{\pi}{2}, \dfrac{\pi}{2} \right]$

   (C) $\left[\dfrac{\pi}{2}, \dfrac{3\pi}{2} \right]$

   (D) $\left[\dfrac{3\pi}{2}, \dfrac{5\pi}{2} \right]$

   (E) $[-\infty, \infty]$

GO ON TO THE NEXT PAGE

14. Let f and g be twice differentiable functions such that $f'(x) \geq 0$ for all x in the domain of f. If $h(x) = f(g'(x))$ and $h'(3) = -2$, then at $x = 3$

(A) h is concave down.

(B) g is decreasing.

(C) f is concave down.

(D) g is concave down.

(E) f is decreasing.

15. In the diagram below, f has a vertical tangent at $x = 1$ and horizontal tangents at $x = 2$ and at $x = 5$. All of these statements are true EXCEPT

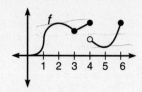

(A) $\displaystyle\lim_{x \to 3^+} f(x) = \lim_{x \to 3^-} f(x)$

(B) $\displaystyle\lim_{x \to 5} f(x) = f(5)$

(C) $\displaystyle\lim_{h \to 0} \frac{f(2+h) - f(2)}{h} = 0$

(D) $\displaystyle\lim_{h \to 0^-} \frac{f(4+h) - f(4)}{h} = \lim_{h \to 0^+} \frac{f(4+h) - f(4)}{h}$

(E) $\displaystyle\lim_{x \to 2.5} f(x) > \lim_{h \to 0} \frac{f(2.5+h) - f(2.5)}{h}$

16. What is the area of the region bounded by the curves $y = x^3 + 1$ and $y = -x^2$ from $x = 0$ to $x = 2$?

(A) $-\dfrac{26}{3}$

(B) $-\dfrac{10}{3}$

(C) $\dfrac{10}{3}$

(D) $\dfrac{20}{3}$

(E) $\dfrac{26}{3}$

17. Determine $\dfrac{dy}{dx}$ for the curve defined by $x^3 + y^3 = 3xy$.

    (A) $\dfrac{x^2}{y^2 - x}$

    (B) $\dfrac{x^2}{x - y^2}$

    (C) $\dfrac{y - x^2}{y^2 - x}$

    (D) $\dfrac{1 - x}{y - 1}$

    (E) $\dfrac{x^2 - y}{y^2 - x}$

18. $\displaystyle\int_0^{\pi/4} \sin 2x \, dx =$

    (A) -1

    (B) $-\dfrac{1}{2}$

    (C) 0

    (D) $\dfrac{1}{2}$

    (E) 1

19. The graph of $f(x) = (x - 4)^3(3x - 1)^3$ has a local minimum at $x =$

    (A) -4

    (B) $-\dfrac{1}{3}$

    (C) $\dfrac{1}{3}$

    (D) $\dfrac{13}{6}$

    (E) 4

GO ON TO THE NEXT PAGE

20. What is the average value of $y = \sin 2x$ over $\left[\frac{\pi}{4}, \frac{\pi}{3}\right]$?

   (A) $-\dfrac{6}{\pi}$

   (B) $-\dfrac{1}{6\pi}$

   (C) $\dfrac{3}{\pi}$

   (D) 3π

   (E) $\dfrac{6}{\pi}$

21. $\lim\limits_{x \to \infty} \dfrac{5x^2 + 7x - 3}{2 + 3x - 11x^2} =$

   (A) $-\dfrac{3}{2}$

   (B) $-\dfrac{5}{11}$

   (C) 0

   (D) $\dfrac{7}{3}$

   (E) It is nonexistent.

22. The graph of $f(x) = \dfrac{1-x}{x^2-1}$ is concave down over which interval(s)?

   (A) $(-\infty, -1)$
   (B) $(-1, \infty)$
   (C) $(-1, 1) \bigcup (1, \infty)$
   (D) $(-\infty, 1)$
   (E) $(-\infty, \infty)$

23.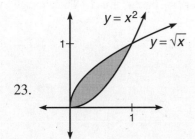

The area of the shaded region in the diagram above is equivalent to

(A) $\int_0^1 \left(x^2 - \sqrt{x}\right) dx$

(B) $\pi \int_0^1 \left(x^4 - x\right) dx$

(C) $\int_0^1 \left(\sqrt{x} - x^2\right) dx$

(D) $2\pi \int_0^1 \left(x\left(\sqrt{x} - x^2\right)\right) dx$

(E) $\pi \int_0^1 \left(\sqrt{x} - x^2\right)^2 dx$

24. $\displaystyle \lim_{h \to 0} \frac{\tan 2\left(\frac{\pi}{8} + h\right) - \tan \frac{\pi}{4}}{h} =$

(A) $\dfrac{3}{2}$

(B) 2

(C) $2\sqrt{2}$

(D) 4

(E) $4\sqrt{2}$

25. $\displaystyle \int_1^{e^2} \left(\frac{\ln^2 x}{x}\right) dx =$

(A) $\dfrac{7}{3e^2}$

(B) $\dfrac{4}{e^2}$

(C) 2

(D) $\dfrac{7}{3}$

(E) $\dfrac{8}{3}$

26. A particle's position is given by $s(t) = \sin t + 2\cos t + \frac{t}{\pi} + 2$. The average velocity of the particle over $[0, 2\pi]$ is

   (A) $-\dfrac{\pi+1}{\pi}$

   (B) $-\dfrac{1}{3}$

   (C) 0

   (D) $\dfrac{1}{\pi}$

   (E) $\dfrac{\pi+1}{\pi}$

27. If $f(x) = \begin{cases} e^x, & x < \ln 2 \\ 2, & x \geq \ln 2 \end{cases}$

   then $\displaystyle\lim_{x \to \ln 2} f(x) =$

   (A) $\dfrac{1}{2}$

   (B) $\ln 2$

   (C) 2

   (D) e^2

   (E) It is nonexistent.

28.

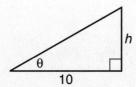

In the triangle shown above, θ is increasing at a constant rate of $\frac{15}{26}$ radians per minute. At what rate is the area of the triangle increasing, in square units per minute, when h is 24 units?

(A) $\dfrac{338}{5}$

(B) 39

(C) $\dfrac{195}{4}$

(D) 182

(E) 195

END OF PART A SECTION I

STOP

END OF PART A SECTION 1. IF YOU HAVE ANY TIME LEFT, GO OVER YOUR WORK IN THIS SECTION ONLY. DO NOT WORK IN ANY OTHER SECTION OF THE TEST.

SECTION I, PART B

Time – 50 Minutes

Number of Questions - 17

A GRAPHING CALCULATOR IS REQUIRED FOR SOME QUES-TIONS ON THIS PART OF THE EXAMINATION.

Directions: Solve each of the following problems, using the available space for scratchwork. After examining the form of the choices, decide which is the best of the choices given and fill in the corresponding oval on the answer sheet. No credit will be given for anything written in the test book. Do not spend too much time on any one problem.

In this test: (1) The exact numerical value of the correct answer does not always appear among the choices given. When this happens, select from among the choices the number that best approximates the exact numerical value. (2) Unless otherwise specified, the domain of a function f is assumed to be the set of all real numbers x for which $f(x)$ is a real number.

29. If $f(x) = \dfrac{e^{3x}}{\sin x^2}$ then $f'(x) =$

(A) $e^{3x} \dfrac{3\sin x^2 - 2x\cos x^2}{\sin^2 x^2}$

(B) $\dfrac{3e^{3x}}{2x\cos x^2}$

(C) $e^{3x} \dfrac{2x\cos x^2 - 3\sin x^2}{\sin^2 x^2}$

(D) $e^{3x} \dfrac{3\sin x^2 + 2x\cos x^2}{\sin^2 x^2}$

(E) $-\dfrac{3e^{3x}}{2x\cos x^2}$

30. Which of the following is an equation for a line tangent to the graph of $f(x) = e^{2x}$ when $f'(x) = 10$?

  (A) $y = 10x - 8.05$

  (B) $y = x - 8.05$

  (C) $y = x - 3.05$

  (D) $y = 10x - 11.5$

  (E) $y = 10x - 3.05$

31.

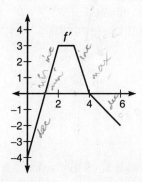

The graph of the derivative of f is shown above. Which of the following statements is true?

  (A) $f(0) < f(6) < f(2) < f(4)$

  (B) $f(6) < f(0) < f(2) < f(4)$

  (C) $f(0) < f(2) < f(4) < f(6)$

  (D) $f(2) < f(0) < f(6) < f(4)$

  (E) $f(0) < f(2) < f(6) < f(4)$

32. Let f be a function such that $\lim\limits_{h \to 0} \dfrac{f(5+h) - f(5)}{h} = 3$. Which of the following must be true?

  I. $f(5) = 3$

  II. $f'(5) = 3$

  III. f is continuous and differentiable at $x = 5$.

  (A) I only

  (B) II only

  (C) III only

  (D) I and II only

  (E) II and III only

GO ON TO THE NEXT PAGE

33. The function f whose derivative is given by $f'(x) = 5x^3 - 15x + 7$ has a local maximum at $x =$

    (A) – 1.930

    (B) – 1.000

    (C) 0.511

    (D) 1.000

    (E) 1.419

34. Car A is travelling south at 40 mph toward Millville, and Car B is traveling west at 30 mph toward Millville. If both cars began traveling 100 miles outside of Millville at the same time, then at what rate, in mph, is the distance between them decreasing after 90 minutes?

    (A) 35.00

    (B) 47.79

    (C) 50.00

    (D) 55.14

    (E) 68.01

35. Let $f(x) = \dfrac{|x^2 - 1|}{x - 1}$. Which of these statements is true?

    I. f is continuous at $x = -1$.

    II. f is differentiable at $x = 1$.

    III. f has a local maximum at $x = -1$.

    (A) I only

    (B) II only

    (C) III only

    (D) I and III only

    (E) II and III only

36. If $y = 3x - 7$ and $x \geq 0$, what is the minimum product of x^2y?

    (A) – 5.646

    (B) 0

    (C) 1.555

    (D) 2.813

    (E) 3.841

37. What is the area of the region bounded by $y = \sin x$, $y = \frac{1}{4}x - 1$, and the y-axis?

    (A) 0.772

    (B) 2.815

    (C) 3.926

    (D) 5.552

    (E) 34.882

38. A region R located in the first quadrant is bounded by the x-axis, $y = \sin x$, and $y = \left(\frac{1}{2}\right)x$. Determine the volume of the solid formed when R is rotated about the y-axis.

    (A) 1.130

    (B) 2.724

    (C) 3.265

    (D) 16.875

    (E) 17.117

39. Let f be the function given by $f(x) = \dfrac{3x^3}{e^x}$. For what value of x is the slope of the line tangent to f equal to -1.024?

    (A) -9.004

    (B) -4.732

    (C) 1.029

    (D) 1.277

    (E) 4.797

40.

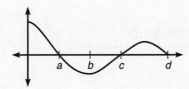

    The graph of f is shown above. If $g(x) = \displaystyle\int_a^x f(t)\, dt$, for what value of x does $g(x)$ have a relative minimum?

    (A) a

    (B) b

    (C) c

    (D) d

    (E) It cannot be determined from the information given.

41. The graph of the function $y = x^5 - x^2 + \sin x$ changes concavity at $x =$

    (A) 0.324

    (B) 0.499

    (C) 0.506

    (D) 0.611

    (E) 0.704

GO ON TO THE NEXT PAGE

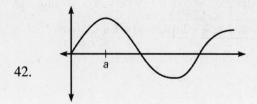

42.

Let $f(x) = \int_0^x h(t)\, dt$, where h is the graph shown above. Which of the following could be the graph of f?

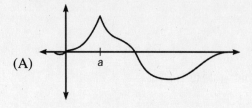

(A)

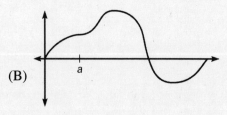

(B)

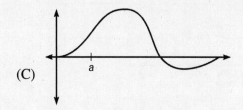

(C)

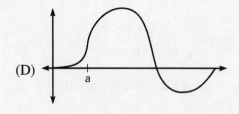

(D)

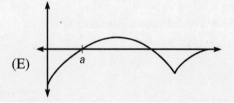

(E)

43.

x	0	1	2	3	4	5	6
f(x)	1	2	4	1	3	2	5

A table of values for a continuous function f is shown above. If three equal subintervals are used for [0,6], which of the following is equivalent to a right-hand Riemann Sum approximation for $\int_0^6 f(x)\,dx$?

(A) 14

(B) 17

(C) 20

(D) 24

(E) 27

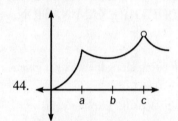

44.

The graph of a function f is shown above. Which of these statements about f is false?

(A) f is continuous but not differentiable at $x = a$.

(B) $\lim_{x \to a^-} \left(\frac{f(x+a)-f(x)}{a} \right) \neq \lim_{x \to a^+} \left(\frac{f(x+a)-f(x)}{a} \right)$.

(C) $f(a)$ is defined, but $f(c)$ is not.

(D) $f'(b) = 0$.

(E) $\lim_{x \to c^-} \left(f(x) \right) \neq \lim_{x \to c^+} \left(f(x) \right)$.

45. Let f be defined as follows: $f(x) = \begin{cases} -x^2, & x \leq 0 \\ \sqrt{x}, & x > 0 \end{cases}$

Let $g(x) = \int_{-2}^{x} f(t)\,dt$. For what value of $x \neq -2$ would $g(x) = 0$?

(A) 0

(B) $\sqrt{2}$

(C) 2

(D) $2\sqrt[3]{2}$

(E) $2\sqrt{2}$

END OF SECTION I

STOP

END OF PART B SECTION 1. IF YOU HAVE ANY TIME LEFT, GO OVER YOUR WORK IN THIS SECTION ONLY. DO NOT WORK IN ANY OTHER SECTION OF THE TEST.

SECTION II PART A

Time – 45 Minutes

Number of Questions – 3

SHOW ALL YOUR WORK. It is important to show your setups for these problems because partial credit will be awarded. If you use decimal approximations, they should be accurate to three decimal places.

A GRAPHING CALCULATOR IS REQUIRED FOR SOME PROBLEMS OR PARTS OF PROBLEMS ON THIS PART OF THE EXAMINATION.

1. At time t, $0 \leq t \leq 10$, the velocity of a particle moving along the x-axis is given by the following equation: $v(t) = 1 - 4\sin(2t) - 7\cos t$.

    A) Is the particle moving left or right at $t = 5$ seconds? Explain your reasoning.

    B) What is the average velocity of the particle from $t = 0$ to $t = 10$?

    C) What is the average acceleration of the particle from $t = 0$ to $t = 10$?

    D) Given that $p(t)$ is the position of the particle at time t and $p(0) = 5$, find $p(2)$.

2. Let R be the region bound by $y = 2x^2 - 8x + 11$ and $y = x^2 - 4x + 10$.

    A) Sketch the region on the axes provided.

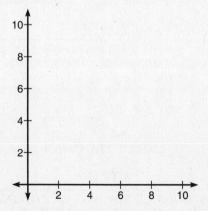

B) Determine the area of R.

C) Determine the volume when R is rotated about the y-axis.

D) The line $x = k$ divides the region R into two regions such that when these two regions are rotated about the y-axis, they generate solids with equal volume. Find the value of k.

3. Water is leaking out of a conical reservoir at a rate proportional to the amount of water in the reservoir; that is, $\dfrac{dy}{dx} = ky$ where y is the amount of water left for any time t. Initially, there were 100 gallons in the reservoir, and after 10 hours, there were 70 gallons.

    A) Write an expression for $A(t)$—the amount of water in the reservoir for any time t.

    B) How much water would have leaked out after 5 hours?

    C) What is the average amount of water in the reservoir during the first 20 hours?

    D) After how many hours of leaking will the amount from part C be in the reservoir?

END OF SECTION II PART A

STOP

END OF SECTION II PART A. IF YOU HAVE ANY TIME LEFT, GO OVER YOUR WORK IN THIS SECTION ONLY. DO NOT WORK IN ANY OTHER SECTION OF THE TEST.

SECTION II, PART B

Time – 45 Minutes

Number of Questions – 3

A CALCULATOR IS NOT PERMITTED ON THIS PART OF THE EXAMINATION.

4. Consider the differential equation $\dfrac{dy}{dx} = \dfrac{x+1}{y}$. All of the following questions refer to this differential equation.

   A) Draw the slope field for $\dfrac{dy}{dx}$ at the indicated points on the below coordinate axis.

   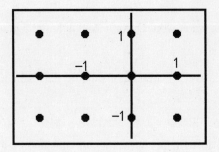

   B) The solution to this differential equation can be classified as which conic section? Justify your answer mathematically.

   C) Find the specific solution to the differential equation, given that it contains the point (2,4).

5. Consider the curve defined by $y^4 = y^2 - x^2$.

   A) Verify that $\dfrac{dy}{dx} = \dfrac{x}{y - 2y^3}$.

   B) Write the equation for any horizontal tangents of the curve.

   C) Write the equation for any vertical tangents of the curve.

   D) At what ordered pair (x,y) is the line $4x\sqrt{3} - 4y = 1$ tangent to the curve $y^4 = y^2 - x^2$?

6. The graph of a function f consists of two quarter circles and two line segments, as shown below. Let g be the function given by $g(x) = \int_3^x f(x)\,dx$.

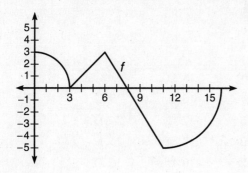

A) Find $g(0)$ and $g(8)$.

B) What is the maximum value of g on $[3,16]$?

C) Write the equation for the line tangent to g at $(11, g(11))$.

D) Find the x-coordinate of any points of inflection of g on $[0,16]$.

END OF SECTION II PART B

STOP

END OF SECTION II PART B. IF YOU HAVE ANY TIME LEFT, GO OVER YOUR WORK IN THIS SECTION ONLY. DO NOT WORK IN ANY OTHER SECTION OF THE TEST.

Answer Key

SECTION 1

PART A

1. D	8. D	15. D	22. C
2. B	9. E	16. E	23. C
3. E	10. B	17. C	24. D
4. C	11. B	18. D	25. E
5. A	12. D	19. D	26. D
6. E	13. C	20. C	27. C
7. D	14. D	21. B	28. E

PART B

29. A	34. B	39. E	44. E
30. E	35. D	40. C	45. D
31. D	36. A	41. B	
32. E	37. C	42. C	
33. C	38. E	43. D	

SOLUTIONS FOR SECTION I

PART A

1. **The correct answer is (D).** Solve this integral using u-substitution. Let $u = 2x$, so $du = 2\ dx$. $\int_0^1 e^{2x} dx$ becomes $\frac{1}{2}\int_0^2 e^u du$, which yields $\frac{1}{2}\left(e^2 - 1\right)$.

2. **The correct answer is (B).** This is a rather complicated Chain Rule application. The derivative of $\tan u$ is $\sec^2 u\ du$, but we mustn't forget to also take the derivative of $e^{\sin x}$. Since $\frac{d}{dx}\left(e^{\sin x}\right) = \cos x\ e^{\sin x}$,
    $\frac{d}{dx}\tan\left(e^{\sin x}\right) = \sec^2\left(e^{\sin x}\right)\cos x\ e^{\sin x}$.

3. **The correct answer is (E).** Be careful here. Although it resembles a Fundamental Theorem Part Two problem, it is not. The problem asks for $F(2)$, not
    $F'(2)$! So, $F(2) = \int_2^4 t^2\ dt = \frac{56}{3}$.

4. **The correct answer is (C).** Straight-forward evaluation of a derivative at a point problem: $f'(x) = 2\tan x \sec^2 x + \cos x$. So, $f'\left(\frac{\pi}{4}\right) = 4 + \frac{\sqrt{2}}{2}$, which is
    $\frac{8 + \sqrt{2}}{2}$.

5. **The correct answer is (A).** We need both the first and second derivatives to be negative for this function to be decreasing and concave down. $f'(x) = 4x^3 - 6x^2 - 4x$ and $f''(x) = 12x^2 - 12x - 4$. By using the wiggle graph below,

    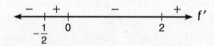

    we can easily see that choices (B) and (C) can be eliminated, so we must check out the values of $f''(-2)$, $f''(-1)$, and $f''(1)$. $f''(1) = -4 < 0$.

6. **The correct answer is (E).** Here, we should take the derivative of each I, II, and III and see what we get.

    $$\frac{d}{dx}\left(\frac{-\cos^4 x}{4}\right) = \frac{-4\cos^3 x}{4}(-\sin x) = \cos^3 x \sin x.$$

    $$\frac{d}{dx}\left(\frac{\sin^2 x}{2} - \frac{\sin^4 x}{4}\right) = \sin x \cos x - \frac{4\sin^3 x \cos x}{4}$$

$$= \sin x\cos x - \sin^3 x\cos x = \sin x\cos x(1-\sin^2 x)\sin x\cos x(\cos^2 x) = \cos^3 x\sin x$$

$$\frac{d}{dx}\left(\frac{1-\cos^4 x}{4}\right) = \frac{-4\cos^3 x(-\sin x)}{4} = \cos^3 x\sin x.$$

7. **The correct answer is (D).** Another lengthy Chain Rule—just don't forget to take the derivative of $2x$, the argument of the argument. Since the derivative of e^u is $e^u\,du$ and the derivative of $\sin u$ is $\cos u\,du$, then
$$\frac{d}{dx}e^{\sin 2x} = 2\cos 2x e^{\sin 2x}.$$

8. **The correct answer is (D).** Remember, speed is the absolute value of velocity. Since $|-15| > |10|$, the maximum speed is reached at $t = 5$ seconds.

9. **The correct answer is (E).** Acceleration can be thought of as the absolute value or slope of velocity. The slope of the velocity curve is steepest on [5,6].

10. **The correct answer is (B).** In order to write the equation for a line, we need its slope and a point on that line. We already have the point, $\left(\frac{\pi}{2},1\right)$, so the big problem is determining its slope, which is the derivative of the curve when $x = \frac{\pi}{2}$. $y' = \cos x - 2\sin x$, so $y'\left(\frac{\pi}{2}\right) = -2$. Using point slope form, the equation for the line could be written as $y - 1 = -2\left(x - \frac{\pi}{2}\right)$. Since this is not a choice, we must change this to standard form, $2x + y = \pi + 1$.

11. **The correct answer is (B).** Since the f is decreasing over $(-\infty,-1)$, its derivative, f', must be negative over this same interval. This eliminates choices (C), (D), and (E). Examining the interval $(-1,0)$, the graph of f is decreasing here; thus, the graph of f' must be negative. Only choice (B) meets this requirement.

12. **The correct answer is (D).** By examining the second-derivative wiggle graph below, we can see that the second derivative is positive at before and after 2. Therefore there is no point of inflection at $x = 2$.

13. **The correct answer is (C).** Pretty simple problem—we determine the derivative, set it equal to zero, and use a wiggle graph.
$$f'(x) = 3e^{\sin x}\cos x = 0$$

Since $3e^{\sin x}$ will never be 0, we set $\cos x = 0$ and solve.

$\cos x = 0$ when $x = \frac{\pi}{2} + n\pi$ for any integer n. By examining the wiggle graph below, we can see that the derivative is negative over $[\frac{\pi}{2}, \frac{3\pi}{2}]$.

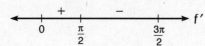

14. **The correct answer is (D).** This one seems tricky, but it actually works out quite quickly. If $h(x) = f(g'(x))$, then by using the Chain Rule, $h'(x) = f'(g'(x))g''(x)$. The problem tells us that $f'(x)$ will always be positive. Since $h'(3)$ somehow becomes negative, $g''(3)$ must be negative. Therefore, g must be concave down at $x = 3$.

15. **The correct answer is (D).** Do we understand the definitions of continuity and differentiability?

    (A)—Does the limit as x approaches 3 exist? Yes.

    (B)—Is f continuous at $x = 5$? Yes.

    (C)—Does $f'(2) = 0$? Yes. (It has a horizontal tangent line.)

    (D)—Does $f'(4)$ exist? No.

    (E)—Is $f(2.5) > f'(2.5)$? Yes. ($f(2.5) > 0$, and since f is decreasing at $x = 2.5$, $f'(2.5) < 0$.)

16. **The correct answer is (E).** We should always sketch the region.

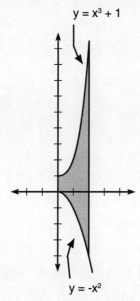

As we can see, $y = x^3 + 1$ is above $y = -x^2$ over the entire interval. So,

$$A = \int_0^2 \left(x^3 + 1 - \left(-x^2 \right) \right) dx$$

$$= \frac{x^4}{4} + x + \frac{x^3}{3} \Big|_0^2$$

$$= 4 + 2 + \frac{8}{3} = \frac{26}{3}$$

17. **The correct answer is (C).** Some implicit differentiation: Remember, everything here is differentiated with respect to x. Don't forget the Product Rule for $3xy$.

$$\left(3x^2 + 3y^2 \right) \frac{dy}{dx} = 3x \frac{dy}{dx} + 3y$$

$$\left(3y^2 - 3x \right) \frac{dy}{dx} = 3y - 3x^2$$

$$\frac{dy}{dx} = \frac{3y - 3x^2}{3y^2 - 3x}$$

$$\frac{dy}{dx} = \frac{y - x^2}{y^2 - x}$$

18. **The correct answer is (D).** This is a very simple u-substitution integral.

Let $u = 2x$, so $du = 2\ dx$.

It follows that $\int_0^{\pi/4} \sin 2x\ dx$ becomes

$$\frac{1}{2} \int_0^{\pi/2} \sin u\ du = -\frac{1}{2} \cos u \Big|_0^{\pi/2}$$

$$= -\frac{1}{2}(0-1) = \frac{1}{2}.$$

19. **The correct answer is (D).** Again, we will rely on the magic of the wiggle graph to supply us with the solution to this differentiation problem.

$$f'(x) = 3(x-4)(3x-1)^2[3(x-4) + (3x-1)]$$

$$= 3(x-4)^2(3x-1)^2(6x-13)$$

By setting $f'(x) = 0$ and solving, we quickly discover that the zeros of the derivative are $\frac{1}{3}$, $\frac{13}{6}$, and 4. By examining the wiggle graph below, we can see that the only value where the derivative changes from negative to positive is at $x = \frac{13}{6}$.

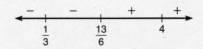

20. **The correct answer is (C).** Whenever the problem asks for the "average value of the function," we should immediately think of the Mean Value Theorem for Integration. We are looking for $f(c)$ in this formula: $f(c) = \frac{1}{b-a} \int_a^b f(x)\, dx$.
Applying the MVT for Integration here yields the following equation:

$$
\begin{aligned}
f(c) &= \frac{1}{\frac{\pi}{3} - \frac{\pi}{4}} \, 4 \int_{\pi/4}^{\pi/3} \sin 2x\, dx \\
&= \frac{1}{\frac{\pi}{12}} \cdot \frac{1}{2} \int_{\pi/2}^{2\pi/3} \sin u\, du \\
&= \frac{6}{\pi}(-\cos u)\Big|_{\pi/2}^{2\pi/3} \\
&= -\frac{6}{\pi}\left(-\frac{1}{2} - 0\right) \\
&= \frac{3}{\pi}.
\end{aligned}
$$

21. **The correct answer is (B).** Remember, limits at infinity are like horizontal asymptotes. If the top degree is greater than the bottom degree, the limit does not exist. If the bottom degree is greater than the top degree, the limit is zero. If, as in this case, the degrees are equal, then the limit is the ratio of the leading coefficient of the numerator over that of the denominator. Here, that ratio is $-\frac{5}{11}$.

22. **The correct answer is (C).** Sketch the curve. We may simplify the expression first by using cancellation. However, we must remember that by canceling out a term involving x, we are removing a discontinuity. So, the graph of the original function will have a point discontinuity.

$$
f(x) = \frac{1-x}{x^2 - 1} = \frac{1-x}{(x-1)(x+1)}
$$

$$
= -\frac{1}{x+1} \text{ with a point discontinuity at } x = 1.
$$

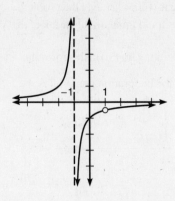

By examining the graph above, we can see that it is concave down to the right of the asymptote. These are actually two intervals, because the function is not continuous at $x = 1$. They are $(-1,1)$ and $(1,\infty)$.

23. **The correct answer is (C).** The two curves, $y = \sqrt{x}$ and $y = x^2$ intersect at $x = 0$ and $x = 1$. So, these will be our limits of integration. For x values between 0 and 1, $\sqrt{x} > x^2$. This being the case, when we determine the area of the region, we should subtract $\sqrt{x} - x^2$. Therefore, $A = \int_0^1 \left(\sqrt{x} - x^2 \right) dx$.

24. **The correct answer is (D).** This is just the derivative of $f(x) = \tan 2x$ evaluated at $x = \frac{\pi}{8}$. $f'(x) = 2\sec^2 2x$, so $f'\left(\frac{\pi}{8}\right) = 2\sec^2 \frac{\pi}{4} = 4$.

25. **The correct answer is (E).** A little, tricky u-substitution integral.

$$\text{Let } u = \ln x, \text{ then } du = \frac{dx}{x}.$$

It follows that $$\int_1^{e^2} \left(\frac{\ln^2 x}{x} \right) dx = \int_0^2 u^2\, du$$

$$= \frac{u^3}{3}\Big|_0^2 = \frac{8}{3} - 0 = \frac{8}{3}$$

26. **The correct answer is (D).** Since the question asks for average velocity and we are given the position equation, we should determine the slope of the secant line:

$$\text{Average velocity } = \frac{s(2\pi) - s(0)}{2\pi - 0}$$

$$= \frac{0 + 2 + 2 + 2 - (2 + 2)}{2\pi}$$

$$= \frac{2}{2\pi} = \frac{1}{\pi}$$

27. **The correct answer is (C).**

    Since $\lim\limits_{x \to (\ln 2)^-} f(x) = \lim\limits_{x \to (\ln 2)^+} f(x) = e^{\ln 2} = 2$, then $\lim\limits_{x \to \ln 2} f(x)$ exists and

    is equal to 2 as well.

28. **The correct answer is (E).** This is a rather challenging related-rates problem.

    We are looking for $\frac{dA}{dt}$ when $h = 24$. First, we need a primary equation. This
    will be the formula for the area of a triangle:

    $$A = \frac{1}{2}bh.$$

    Since the base is a constant, 10, this becomes

    $$A = \frac{1}{2}h(10) = 5h.$$

    Differentiating with respect to t yields

    $$\frac{dA}{dt} = 5\frac{dh}{dt}.$$

    How do we determine $\frac{dh}{dt}$? We must find a relationship other than $A = \frac{1}{2}bh$
    involving h. How about using the tangent equation?

    $$\tan\theta = \frac{h}{10} \text{ or } h = 10\tan\theta$$

    Differentiating this with respect to t yields

    $$\frac{dh}{dt} = 10\sec^2\theta\frac{d\theta}{dt}.$$

    Since we know $\frac{d\theta}{dt} = \frac{15}{26}$, this equation becomes

    $$\frac{dh}{dt} = 10\sec^2\theta\frac{15}{26}.$$

    The last unknown to identify is $10\sec^2\theta$. We can use the Pythagorean
    Theorem to help here. Since $h = 24$ and $b = 10$, the hypotenuse must be 26.
    So,

    $$10\sec^2\theta = 10\left(\frac{26}{10}\right)^2.$$

Plugging this expression in for $\frac{dh}{dt}$ gives us

$$\frac{dA}{dt} = (5)(10)\left(\frac{26}{10}\right)^2\left(\frac{15}{26}\right)$$

$$= 50\left(\frac{13}{5}\right)\left(\frac{3}{2}\right) = 195 \ .$$

PART B

29. **The correct answer is (A).** We must use the Quotient Rule to evaluate the following derivative:

$$f'(x) = \frac{3e^{3x}\sin x^2 - 2xe^{3x}\cos x^2}{\sin^2 x^2}$$

$$= e^{3x}\left(\frac{3\sin x^2 - 2x\cos x^2}{\sin^2 x^2}\right)$$

30. **The correct answer is (E).** To write the equation for a line, we need the slope of the line and a point on the line. We already have its slope, since $f'(x) = 10$; the slope of the tangent line is 10 as well. To find the point on the line, we must set the derivative of f equal to 10 and solve for x; then, substitute this x value into f to determine the corresponding y value.

$$f'(x) = 2e^{2x} = 10$$

$$e^{2x} = 5$$

$$2x = \ln 5$$

$$x = \frac{\ln 5}{2}$$

Now, we substitute this value into f and the result is

$$f\left(\frac{\ln 5}{2}\right) = e^{\ln 5} = 5 \ .$$

Our problem has now been reduced to determining the equation for a line that passes through $\left(\frac{\ln 5}{2}, 5\right)$ and has slope 10. Point-slope form of this equation is

$$y - 5 = 10\left(x - \frac{\ln 5}{2}\right).$$

Converting to slope-intercept form and using the calculator to evaluate the value of $\frac{\ln 5}{2}$, we get

$$y = 10x - 3.047.$$

31. **The correct answer is (D).** This is an area accumulation problem. The function decreases from $x = 0$ to $x = 1$ by an amount equivalent to the area between the graph of f' and the x-axis, which is -2 units squared. This was determined by finding the area of the triangle. The function then increases from $x = 1$ to $x = 2$ by $\frac{3}{2}$ units squared. From $x = 2$ to $x = 4$, it increases $3 + \frac{3}{2}$ or $\frac{9}{2}$ units squared. From $x = 4$ to $x = 6$, the function decreases 2 units squared. Putting all of this together, we can see that the function's value is greatest at $x = 4$, followed by at $x = 6$, then at $x = 0$, and least at $x = 2$.

32. **The correct answer is (E).** This is the limit of the difference quotient that is the definition of the derivative. All this means is that $f'(5) = 3$. For I, does the function's value necessarily equal the derivative's value? No, so I is out. Since the derivative exists at $x = 5$, the function is differentiable there. Remember that differentiability implies continuity, so the function must be continuous at $x = 5$ as well.

33. **The correct answer is (C).** Use the calculator to graph the derivative given. Where the graph changes from positive to negative will be the local maximum. This occurs at $x = 0.511$.

34. **The correct answer is (B).** Related rates—oh boy! First, lets draw the following diagram:

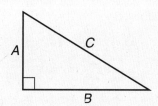

We are looking for $\frac{dC}{dt}$. We know that $\frac{dA}{dt} = -40$ and $\frac{dB}{dt} = -30$. Now, we need an equation to relate A, B, and C. Since this is a right triangle, we can certainly use the Pythagorean Theorem:

$$A^2 + B^2 = C^2$$

Differentiating with respect to t, which we do in every related-rates problem, yields

$$2A\frac{dA}{dt} + 2B\frac{dB}{dt} = 2C\frac{dC}{dt}$$

Solving this equation for $\frac{dC}{dt}$ gives us

$$\frac{A\left(\frac{dA}{dt} + B\frac{dB}{dt}\right)}{C} = \frac{dC}{dt}.$$

What are the values of A, B, and C? To answer this, we use the facts that the cars each started 100 miles from Millville and have been travelling for 90 minutes or $\frac{3}{2}$ hours. Car A has travelled 60 miles, so $A = 40$. Car B has travelled 45 miles, so $B = 55$. By the Pythagorean Thereom, $C = \sqrt{40^2 + 55^2}$. Substituting these values into the equation yields

$$\frac{dc}{dt} = \frac{40(-40) + 55(-30)}{\sqrt{40^2 + 55^2}} = -47.79$$

35. **The correct answer is (D).** This problem is best answered using the graphing calculator. If we examine the graph, we can see that it is continuous at $x = -1$. The graph has a jump discontinuity at $x = 1$; it is not differentiable there. Since the graph is increasing before and decreasing after $x = -1$, there is a local maximum at $x = -1$.

36. **The correct answer is (A).** This is an optimization problem. Let's first express the product $x^2 y$ only as a function of x:

$$p(x) = x^2(3x - 7) = 3x^3 - 7x^2$$

Next, we differentiate and set the derivative equal to zero and solve for x to determine our critical values:

$$p'(x) = 9x^2 - 14x = 0$$

$$x(9x - 14) = 0$$

$$x = 0 \text{ or } x = \frac{14}{9}.$$

By examining the wiggle graph below, we can see that the function's minimum occurs at $x = \frac{14}{9}$.

The problem asks for the minimum product. So, we must substitute $\frac{14}{9}$ back into $p(x)$, and we get -5.656.

37. **The correct answer is (C).** Use your calculator to determine where these two curves intersect. This intersection point will give us a limit of integration. Since the graph of $y = \sin x$ is above the graph of $y = \dfrac{1}{4}x - 1$, we integrate to find the area:

$$A = \int_0^{3.314}\left(\sin x - \left(\frac{1}{4}x - 1\right)\right)dx.$$

Our calculator will then do all the work and give us

$$A = 3.926.$$

38. **The correct answer is (E).** Be careful here—make sure you have the right region, as shown in the diagram below:

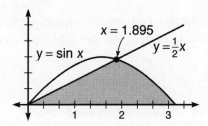

The best method to use, since we are rotating about the y-axis, would be shells. However, we will need to break it up into two regions.

$$V = 2\pi\left(\int_0^{1.895}\left(x\left(\frac{1}{2}x\right)\right)dx + \int_{1.895}^{\pi}(x(\sin x))dx\right)$$

Using our calculator,

$$V = 17.117.$$

39. **The correct answer is (E).** Set the derivative of f equal to -1.024, and solve. By the Quotient Rule, we have

$$f'(x) = \frac{e^x 9x^2 - e^x 3x^3}{e^{2x}} = -1.024.$$

Graphing and determining the intercept yields $x = 4.797$.

40. **The correct answer is (C).** Since g is an antiderivative of f, then f is the derivative of g. The graph of the derivative of g changes from negative to positive at $x = c$, so g has a minimum there.

41. **The correct answer is (B).** Use your calculator for this one. First, determine the second derivative of f. Graph it, and find the x-intercept.

$$f'(x) = 5x^4 - 2x + \cos x$$

$$f''(x) = 20x^3 - 2 - \sin x$$

Our calculator shows us that the x-intercept of this second derivative is 0.4985.

42. **The correct answer is (C).** We have the graph of the derivative, but we are looking for the graph of the function. Since the derivative's graph is continuous, there are no discontinuities, cusps, or vertical tangents on the function's graph. This eliminates choices (A), (D), and (E). Since the derivative is positive and increasing from $x = 0$ to $x = a$, the function must be increasing and concave up over this same interval. Between the two choices remaining, only (C) meets this requirement.

43. **The correct answer is (D).** This problem requires a little drawing. We should plot the seven points, draw the rectangles, find the area of each one, and add them up.

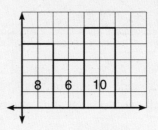

$$8 + 6 + 10 = 24.$$

44. **The correct answer is (E).** Let's examine each statement: Since there is an obvious cusp at $x = a$, (A) is a true statement. (B) says that the left-hand derivative does not equal the right-hand derivative at $x = a$, which means that there must be a cusp there—which there is. So, this is true also. The hole at $x = c$ would indicate that $f(c)$ is undefined, so (C) is true. (D) is true because there is a horizontal tangent at $x = b$. Since the discontinuity at $x = c$ is removable, the

$\lim\limits_{x \to c} f(x)$ exists, $\lim\limits_{x \to c^-} \left(f(x) \right) \neq \lim\limits_{x \to c^+} \left(f(x) \right)$ is false.

45. **The correct answer is (D).** In this problem, the question is for what value of k will the area under $\int_0^k \left(\sqrt{x} \; dx \right)$ equal $\int_0^2 \left(x^2 \; dx \right)$? Setting these two integrals equal to each other and solving for x yields

$$\int_0^k \left(\sqrt{x} \; dx \right) = \int_0^2 x^2$$

$$\frac{2}{3} x^{3/2} \Big|_0^k = \frac{x}{x^3} \Big|_0^2$$

$$\frac{2}{3} k^{3/2} = \frac{8}{3}$$

$$k^{3/2} = 4$$

$$k = 4^{3/2} = \sqrt[3]{16} = 2\sqrt[3]{2}.$$

SOLUTIONS FOR SECTION II

PART A

1.   A) At $t = 5$, $v(t) = 1.1904 > 0$, which indicates that the velocity is positive. Therefore, the particle is moving to the right.

B) Since the velocity function is given, this is an application of the Mean Value Theorem for Integration, which says that

$$\text{Average value} = \frac{1}{b-a} \int_a^b f(x) \; dx.$$

Applying that formula to this problem leads to

$$f(c) = \frac{1}{10-0} \int_0^{10} \left(1 - 4\sin 2x - \cos 7x \right) dx$$

$$= 1.262$$

C) Remember that acceleration is the derivative or slope of the velocity curve. So, we want the average slope of the velocity from $t = 0$ to $t = 10$. In other words, we want the slope of the secant line from $t = 0$ to $t = 10$.

$$m(\sec) = \frac{v(10) - v(0)}{10 - 0}$$

$$= \frac{3.2217 - -6}{10} = 0.922$$

D) Here, we need to determine the position equation, $p(t)$. In order to determine the position equation, we should find an antiderivative of the velocity equation, $v(t)$.

$$p(t) = \int \left(1 - 4\sin 2x - 7\cos x \right) dx$$

$$= x + 2\cos 2x - 7\sin x + C$$

To determine the correct value of C, the constant of integration, we should use the condition given to us, $p(0) = 5$:

$$p(0) = 5 = 0 + 2\cos 0 - 7\sin 0 + C$$

$$5 = 2 + C$$

$$3 = C$$

Substituting this value back into the position equation gives us

$$p(t) = x + 2\cos 2x - 7\sin x + 3.$$

Now, to answer the question:

$$p(2) = 2 + 2\cos 4 - 7\sin 2 + 3$$

$$= -2.672$$

2.　A)

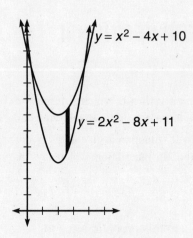

$y = x^2 - 4x + 10$

$y = 2x^2 - 8x + 11$

B) We use the definite integral to determine the area of this region. We must first use the calculator to determine the points of intersection of the graphs. The x-coordinates will give us our limits of intersection.

$$A = \int_{.2679}^{3.7321} \left(x^2 - 4x + 10 - \left(2x^2 - 8x + 11 \right) \right) dx$$

$$= \int_{.2679}^{3.7321} \left(-x^2 + 4x - 1 \right) dx$$

$$= 6.928$$

C) To determine the volume when the region R is rotated about the y-axis, we are going to use the shell method:

$$V = 2\pi \int_{a}^{b} \left(d(x)h(x) \right) dx.$$

$$V_y = 2\pi \int_{0.2679}^{3.7321} \left(x\left(-x^2 + 4x - 1 \right) \right) dx$$

$$= 87.06236948$$

D) Insert k for the upper limit of integration, set the volume expression equal to half of the solution for part C, and solve for k.

$$2\pi \int_{0.2679}^{k} \left(x - \left(-x^2 + 4x - 1\right)\right) dx = 43.531$$

$$2\pi \left(-\frac{x^4}{4} + \frac{4x^3}{3} - \frac{x^2}{2}\right)\Big|_{0.2679}^{k} = 43.531$$

$$-\frac{k^4}{4} + \frac{4k^3}{3} - \frac{k^2}{2} - (-0.0115) = 6.928$$

$$-\frac{k^4}{4} + \frac{4k^3}{3} - \frac{k^2}{2} - 6.91667 = 0$$

$$k = 2.350$$

k was determined by graphing the function and determining the x-intercept.

3.   A) The equation $\frac{dy}{dx} = ky$ should indicate that we are dealing with this exponential growth or decay model:

$$A(t) = Ne^{kt}.$$

Since we are given that the initial amount of water was 100 gallons, we know that $N = 100$. We can determine k by substituting values for N, t, and $A(t)$ and solving for k, as such:

$$A(10) = 70 = 100e^{10k}$$

$$0.7 = e^{10k}$$

$$\ln 0.7 = 10^k$$

$$k = \frac{\ln\ 0.7}{10}.$$

Now that we have detemined both constants, N and k, we can write our expression $A(t)$:

$$A(t) = 100e^{\frac{\ln 0.7}{10}t}$$

B) This problem is kind of tricky. To determine the amount of water that had leaked out after 5 hours, we should find how much is still in the reservoir, which is $A(5)$, and subtract that from the initial amount, 100 gallons.

$$A(5) = 100e^{5\frac{\ln 0.7}{10}}$$

$$= 100e^{\frac{\ln 0.7}{2}}$$

$$= 83.667 \text{ gallons}.$$

This represents the amount still in the reservoir. The amount of leakage would be

$$100 - 83.667 = 16.333 \text{ gallons.}$$

C) By applying the Mean Value Theorem for Integration, we get

$$\text{Average amount} = \frac{1}{20} \int_0^{20} \left(100 e^{\frac{\ln 0.7}{10} t} \right) dt$$

$$= 71.494 \text{ gallons}$$

D) We know the amount, $A(t)$, is equal to 71.494. We set Equation (2) equal to this and solve for t.

$$71.494 = 100 e^{\frac{\ln 0.7}{10} t}$$

$$\frac{71.494}{100} = e^{\left(\frac{\ln 0.7}{10} \right) t}$$

$$\ln 0.71494 = \left(\frac{\ln 0.7}{10} \right) t$$

$$t = 9.408 \text{ hours}$$

PART B

4.   A) To draw a slope field, you plug each of the indicated coordinates into $\frac{dy}{dx}$ for x and y. If the resulting slope is undefined, you can indicate that with a small vertical line (since the slope of a vertical line is undefined). Your answer should look something like this

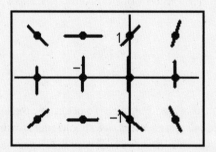

B) The slope field's shape suggests a hyperbola centered at $(-1,0)$, but the question requires us to justify our answer mathematically. The easiest way to justify the answer is by actually solving the differential equation by separation. (Remember, in the following steps, C represents any constant. Therefore, multiplying a constant by any number or adding any number to a constant results in another constant. For the sake of ease, we just continue to use the same symbol for that constant—C.) Begin by cross-multiplying to separate x's and y's

$$y\,dy = (x+1)\,dx$$

Now, integrate both sides of the equation and multiply by 2 to make it prettier.

$$\frac{y^2}{2} = \frac{x^2}{2} + x + C$$
$$y^2 = x^2 + 2x + C$$

To put this into standard form for a conic, complete the square for x.

$$y^2 + 1 = x^2 + 2x + 1 + C$$
$$y^2 + 1 = (x+1)^2 + C$$

If you subtract y^2, you get the equation of a hyperbola. This is all the justification you need to get the question completely correct.

$$(x+1)^2 - y^2 = 1 - C$$

C) If the hyperbola contains point (2,4), substitute these values into your answer for part (B), and you get the corresponding value for the constant C.

$$(2+1)^2 - (-4)^2 = 1 - C$$
$$9 - 16 = 1 - C$$
$$C = 8$$

Therefore, the exact solution to that particular differential equation is

$$(x+1)^2 - y^2 = -7$$

If you want to put it into standard form, you can

$$\frac{y^2}{7} - \frac{(x+1)^2}{7} = 1$$

5.   A) Here, we must use implicit differentiation, because x and y are not separated for us. We will differentiate both sides of the equation with respect to x, group the terms with a $\frac{dy}{dx}$, and solve for $\frac{dy}{dx}$.

$$4y^3 \frac{dy}{dx} = 2y\frac{dy}{dx} - 2x$$
$$4y^3 \frac{dy}{dx} - 2y\frac{dy}{dx} = 2x$$
$$\left(4y^3 - 2y\right)\frac{dy}{dx} = -2x$$
$$\frac{dy}{dx} = -\frac{2x}{4y^3 - 2y}$$
$$= -\frac{x}{2y^3 - y}$$
$$= \frac{x}{y - 2y^3}$$

B) In order to have a horizontal tangent, the derivative must equal zero. Let's set $\frac{dy}{dx}$ equal to zero and solve for x. Remember, in order for a rational expression like our $\frac{dy}{dx}$ to equal zero, the numerator must equal zero. So, we'll just set the numerator of our expression for $\frac{dy}{dx}$ equal to zero, as such:

$$2x = 0, \text{ so } x = 0.$$

Now, we will determine the corresponding y-value(s):

$$y^4 = y^2 - 0$$

$$y^4 - y^2 = 0$$

$$y^2(y^2 - 1) = 0$$

$$y = 0, y = -1, \text{ or } y = 1.$$

Upon closer examination, $y \neq 0$ because the function does not exist when $y = 0$; that is, the denominator will equal zero there. So, our two points when the tangent line is horizontal are $(0, -1)$ and $(0,1)$. This leads us to the equations for the horizontal tangents:

$$y = -1 \text{ and } y = 1.$$

C) Remember that the slope of a vertical line is undefined. That means that we look for points where the derivative is undefined (i.e., where the denominator of the derivative is equal to zero). This time we will set the $y - 2y^3$ equal to zero and solve for y.

$$y - 2y^3 = 0$$

$$y\left(1 - 2y^2\right) = 0$$

$$y = 0, \ y = -\frac{\sqrt{2}}{2}, \text{ or } y = \frac{\sqrt{2}}{2}$$

The corresponding x-values are $x = 0, x = -\frac{1}{2}$, and $x = \frac{1}{2}$. Since vertical lines have equations of the form $x = a$, the x-values are all we need to write the equations for the vertical tangent lines:

$$x = 0, x = -\tfrac{1}{2}, \text{ and } x = \tfrac{1}{2}.$$

D) In order for the line and the curve to be tangent, two things must be true: They must intersect and have equal slopes. For the slopes to be equal, we should set their derivatives equal. Let's determine $\frac{dy}{dx}$ for the line:

$$4x\sqrt{3} - 4y = 1$$

$$4\sqrt{3} - 4\frac{dy}{dx} = 0$$

$$\frac{dy}{dx} = \sqrt{3}.$$

Next, we will solve the equation of the line for x in terms of y:

$$4x\sqrt{3} - 4y = 1$$

$$x = \frac{1+4y}{4\sqrt{3}}.$$

When the derivatives are equal, they are equal to $\sqrt{3}$. So, now we will set the expression for the derivative of the curve equal to $\sqrt{3}$.

$$\sqrt{3} = \frac{x}{y - 2y^3}$$

We already have $x = \frac{1+4y}{4\sqrt{3}}$.

$$\sqrt{3} = \frac{1+4y}{\left(4\sqrt{3}\right)\left(y - 2y^3\right)}$$

$$12\left(y - 2y^3\right) = 1+4y$$

$$24y^3 - 8y + 1 = 0$$

How do we solve this without our calculator? Let's try synthetic division with $\frac{1}{2}$. Why $\frac{1}{2}$? Remember the Rational Root Theorem? It told us to try factors of the constant term over factors of the leading coefficient. Since $\frac{1}{2}$ worked in our synthetic division, $\frac{1}{2}$ is a solution. We will now find the x-coordinate of the point of tangency by substituting this y-value:

$$x = \frac{1 + 4 \cdot \frac{1}{2}}{4\sqrt{3}}$$

$$= \frac{3}{4\sqrt{3}}$$

$$= \frac{\sqrt{3}}{4}.$$

We have both our x- and y-coordinates now, so we can determine that our line and curve are tangent at $\left(\dfrac{\sqrt{3}}{4}, \dfrac{1}{2} \right)$.

6. A) To determine $g(0)$, we have to find the area under the curve from $x = 3$ to $x = 0$. This is the opposite of the area under the curve from $x = 0$ to $x = 3$. Since the area of a circle of radius 3 is 9π, then the area of this quarter circle must be $\dfrac{9\pi}{4}$. So, $g(0) = -\dfrac{9\pi}{4}$.

$g(8)$ would be the area of the triangle with base 5 and height 3. So, $g(8) = \dfrac{15}{2}$.

B) The only interval for which g is increasing is $[3,8]$. So, the maximum value of g is at $x = 8$. The maximum value of g is $\dfrac{15}{2}$.

C) We need a point and a slope. Since $f(11) = -5$, then $g'(11) = -5$. Hence, the slope of the tangent line is -5. To determine the y-coordinate of the point on the line, we must determine $g(11)$. By area accumulation, we can see that $g(11) = 0$. This leads us to the following equation for the tangent line:

$$y = -5(x - 11)$$

or

$$y = -5x + 55.$$

D) Points of inflection occur only when the derivative of the derivative is equal to zero. Since the graph of f, the derivative of g, does not have any horizontal tangents, then f', or g'', will never be zero. Thus, g has no points of inflection.

Answer Sheet

SECOND MODEL AP CALCULUS AB EXAM

Section I Part A

1. Ⓐ Ⓑ Ⓒ Ⓓ Ⓔ   7. Ⓐ Ⓑ Ⓒ Ⓓ Ⓔ   13. Ⓐ Ⓑ Ⓒ Ⓓ Ⓔ   19. Ⓐ Ⓑ Ⓒ Ⓓ Ⓔ   25. Ⓐ Ⓑ Ⓒ Ⓓ Ⓔ

2. Ⓐ Ⓑ Ⓒ Ⓓ Ⓔ   8. Ⓐ Ⓑ Ⓒ Ⓓ Ⓔ   14. Ⓐ Ⓑ Ⓒ Ⓓ Ⓔ   20. Ⓐ Ⓑ Ⓒ Ⓓ Ⓔ   26. Ⓐ Ⓑ Ⓒ Ⓓ Ⓔ

3. Ⓐ Ⓑ Ⓒ Ⓓ Ⓔ   9. Ⓐ Ⓑ Ⓒ Ⓓ Ⓔ   15. Ⓐ Ⓑ Ⓒ Ⓓ Ⓔ   21. Ⓐ Ⓑ Ⓒ Ⓓ Ⓔ   27. Ⓐ Ⓑ Ⓒ Ⓓ Ⓔ

4. Ⓐ Ⓑ Ⓒ Ⓓ Ⓔ   10. Ⓐ Ⓑ Ⓒ Ⓓ Ⓔ   16. Ⓐ Ⓑ Ⓒ Ⓓ Ⓔ   22. Ⓐ Ⓑ Ⓒ Ⓓ Ⓔ   28. Ⓐ Ⓑ Ⓒ Ⓓ Ⓔ

5. Ⓐ Ⓑ Ⓒ Ⓓ Ⓔ   11. Ⓐ Ⓑ Ⓒ Ⓓ Ⓔ   17. Ⓐ Ⓑ Ⓒ Ⓓ Ⓔ   23. Ⓐ Ⓑ Ⓒ Ⓓ Ⓔ

6. Ⓐ Ⓑ Ⓒ Ⓓ Ⓔ   12. Ⓐ Ⓑ Ⓒ Ⓓ Ⓔ   18. Ⓐ Ⓑ Ⓒ Ⓓ Ⓔ   24. Ⓐ Ⓑ Ⓒ Ⓓ Ⓔ

Section I Part B

29. Ⓐ Ⓑ Ⓒ Ⓓ Ⓔ   32. Ⓐ Ⓑ Ⓒ Ⓓ Ⓔ   35. Ⓐ Ⓑ Ⓒ Ⓓ Ⓔ   38. Ⓐ Ⓑ Ⓒ Ⓓ Ⓔ   42. Ⓐ Ⓑ Ⓒ Ⓓ Ⓔ

30. Ⓐ Ⓑ Ⓒ Ⓓ Ⓔ   33. Ⓐ Ⓑ Ⓒ Ⓓ Ⓔ   36. Ⓐ Ⓑ Ⓒ Ⓓ Ⓔ   39. Ⓐ Ⓑ Ⓒ Ⓓ Ⓔ   43. Ⓐ Ⓑ Ⓒ Ⓓ Ⓔ

31. Ⓐ Ⓑ Ⓒ Ⓓ Ⓔ   34. Ⓐ Ⓑ Ⓒ Ⓓ Ⓔ   37. Ⓐ Ⓑ Ⓒ Ⓓ Ⓔ   40. Ⓐ Ⓑ Ⓒ Ⓓ Ⓔ   44. Ⓐ Ⓑ Ⓒ Ⓓ Ⓔ

41. Ⓐ Ⓑ Ⓒ Ⓓ Ⓔ   45. Ⓐ Ⓑ Ⓒ Ⓓ Ⓔ

Second Model AP Calculus AB Exam

SECTION 1, PART A

Time – 55 Minutes

Number of Questions – 28

A CALCULATOR MAY NOT BE USED ON THIS PART OF THE EXAMINATION.

Directions: Solve each of the following problems, using the available space for scratchwork. After examining the form of the choices, decide which is the best of the choices given and fill in the corresponding oval on the answer sheet. No credit will be given for anything written in the test book. Do not spend too much time on any one problem.

In this test: Unless otherwise specified, the domain of a function f is assumed to be the set of all real numbers x for which $f(x)$ is a real number.

1. What is the instantaneous rate of change for $f(x) = \dfrac{x^3 + 3x^2 + 3x + 1}{x+1}$ at $x = 2$?

   (A) -27

   (B) -6

   (C) 6

   (D) 9

   (E) 27

2.

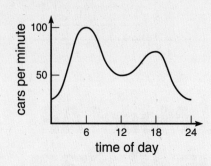

The rate at which cars cross a bridge in cars per minute is given by the graph shown above. A good approximation for the total number of cars that crossed the bridge by 12:00 noon is

(A) 50

(B) 825

(C) 1,200

(D) 45,000

(E) 49,500

3. $\int_1^5 \left(\dfrac{3x}{x^3} \right) dx =$

(A) $-\dfrac{18}{5}$

(B) $-\dfrac{72}{25}$

(C) $\dfrac{124}{125}$

(D) $\dfrac{126}{125}$

(E) $\dfrac{12}{5}$

4.

x	0	1	2
f(x)	3	4	9

The function *f* is continuous on the closed interval [0,2] and has values as defined by the table above. Which of the following statements must be true?

(A) *f* must be increasing on [0,2].

(B) *f* must be concave up on (0,2).

(C) $f'\left(\dfrac{3}{2}\right) > f'\left(\dfrac{1}{2}\right)$.

(D) The average rate of increase of *f* over [0,2] is 3.

(E) *f* has no points of inflection on [0,2].

5. $\displaystyle\int_0^{\pi/3}\left(\dfrac{\tan x e^{\sec x}}{\cos x}\right)dx =$

(A) e^2

(B) $e^2 - 1$

(C) $\sqrt{e}$

(D) $\sqrt{e} - e$

(E) $e^2 - e$

6. What is the slope of the curve defined by $3x^2 + 2xy + 6y^2 - 3x - 8y = 0$ at the point (1,1)?

(A) $-\dfrac{5}{6}$

(B) $-\dfrac{1}{2}$

(C) 0

(D) $\dfrac{1}{2}$

(E) It is undefined.

7. $\displaystyle\int_0^1\left(\sqrt{x}\right)\left(x^2 + 3x - 8\right)dx =$

(A) -4

(B) $\dfrac{-404}{105}$

(C) $\dfrac{-8}{5}$

(D) $\dfrac{-28}{105}$

(E) 1

GO ON TO THE NEXT PAGE

8. The radius of a sphere is increasing at a rate of 2 inches per minute. At what rate (in cubic inches per minute) is the volume increasing when the surface area of the sphere is 9π square inches?

(A) 2

(B) 2π

(C) 9π

(D) 18

(E) 18π

9.

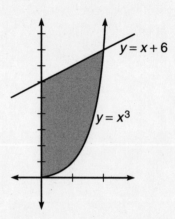

The area of the shaded region in the diagram shown above is equivalent to

(A) $\int_0^8 \left(x + 6 - x^3\right)dx$

(B) $\int_0^8 \left(x^3 - x - 6\right)dx$

(C) $\int_0^2 \left(x + 6 - x^3\right)dx$

(D) $\int_0^2 \left(x^3 - x - 6\right)dx$

(E) $\int_0^2 \left(x + 6 + x^3\right)dx$

10. What is the average rate of change of $f(x) = x^3 - 3x^2 + x - 1$ over $[-1,4]$?

(A) $\dfrac{13}{5}$

(B) 3

(C) 5

(D) 10

(E) 25

11. If the graph of the second derivative of some function, f, is a line of slope 6, then f could be which type of function?

    (A) constant

    (B) linear

    (C) quadratic

    (D) cubic

    (E) quartic

12. Let f be defined as

$$P(x) = \begin{cases} \sqrt{-x}, & x < 0 \\ x^2, & x \geq 0 \end{cases}$$

What is the average value of f over $[-4, 4]$?

    (A) $\dfrac{2}{3}$

    (B) $\dfrac{8}{3}$

    (C) $\dfrac{10}{3}$

    (D) $\dfrac{16}{3}$

    (E) $\dfrac{80}{3}$

13.

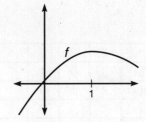

    f is a twice differentiable function with a horizontal tangent line at $x = 1$, as shown in the diagram above. Which of these statements must be true?

    (A) $f'(1) < f(1) < f''(1)$

    (B) $f(1) < f''(1) < f'(1)$

    (C) $f(1) < f'(1) < f''(1)$

    (D) $f''(1) < f(1) < f'(1)$

    (E) $f''(1) < f'(1) < f(1)$

GO ON TO THE NEXT PAGE

14. Let f be a continuous function on $[-4, 12]$. If $f(-4) = -2$ and $f(12) = 6$, then the Mean Value Theorem guarantees that

(A) $f(4) = 2$.

(B) $f'(4) = \dfrac{1}{2}$.

(C) $f'(c) = \dfrac{1}{2}$ for at least one c between -4 and 12.

(D) $f(c) = 0$ for at least one c between -4 and 12.

(E) $f(4) = 0$.

15. $\left(\dfrac{d}{dx}\right)\displaystyle\int_{3}^{2x^2}\left(e^t\right)dt =$

(A) e^{2x^2}

(B) $4xe^{2x^2}$

(C) $e^{2x^2} - e^3$

(D) $4xe^{2x^2} - e^3$

(E) e^x

16. Let $f(x) = e^x$. If the rate of change of f at $x = c$ is e^3 times its rate of change at $x = 2$, then $c =$

(A) 1

(B) 2

(C) 3

(D) 4

(E) 5

17.

x	1	2	3	4
$f(x)$	3	0	1	2
$g(x)$	2	3	4	1
$f'(x)$	1	1	0	2
$g'(x)$	2	2	1	1

Let f, g, and their derivatives be defined by the table above. If $h(x) = f(g(x))$, then for what value, c, is $h(c) = h'(c)$?

(A) 1

(B) 2

(C) 3

(D) 4

(E) None of the above

18. Let f be a differentiable function over $[0,10]$ such that $f(0) = 0$ and $f(10) = 3$. If there are exactly two solutions to $f(x) = 4$ over $(0,10)$, then which of these statements must be true?

(A) $f'(c) = 0$ for some c on $(0,10)$.

(B) f has a local maximum at $x = 5$.

(C) $f''(c) = 0$ for some c on $(0,10)$.

(D) 0 is the absolute minimum of f.

(E) f is strictly monotonic.

19. The normal line to the curve $y = \sqrt{8 - x^2}$ at the point $(2,2)$ has slope

(A) -2

(B) $-\dfrac{1}{2}$

(C) $\dfrac{1}{2}$

(D) 1

(E) 2

20. What are all the values for k such that $\displaystyle\int_{-2}^{k} x^3 dx = 0$?

(A) 0

(B) 2

(C) -2 and 2

(D) -2, 0, and 2

(E) 0 and 2

21. If the rate of change of y is directly proportional to y, then it's possible that

(A) $y = 3te^{2/3}$

(B) $y = 5e^{1.5t}$

(C) $y = \frac{3}{2}t^2$

(D) $y = \ln\left(\frac{3}{2}t\right)$

(E) $y = t^{3/2}$

GO ON TO THE NEXT PAGE

22. The graph of $y = 3x^3 - 2x^2 + 6x - 2$ is decreasing for which interval(s)?

    (A) $\left(-\infty, \dfrac{2}{9}\right)$

    (B) $\left(\dfrac{2}{9}, \infty\right)$

    (C) $\left[0, \dfrac{2}{9}\right]$

    (D) $(-\infty, \infty)$

    (E) None of the above

23. Determine the value for c on $[2,5]$ that satisfies the Mean Value Theorem for
    $$f(x) = \frac{x^2 - 3}{x - 1}.$$

    (A) -1

    (B) 2

    (C) 3

    (D) 4

    (E) 5

24. Below is the slope field graph of some differential equation $\dfrac{dy}{dx} = f'(x)$.
    (Note: Each dot on the axes marks one unit.)

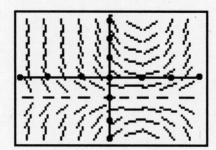

    (A) $x^2(y + 1)$

    (B) $xy - x - y - 1$

    (C) $xy + y$

    (D) $\dfrac{x-1}{y+1}$

    (E) $xy + 3xy - 1$

25.

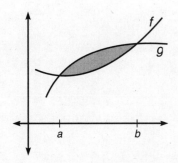

The area of the shaded region in the diagram on the previous page is

(A) $\int_a^b \left(f(x) - g(x) \right) dx$

(B) $\int_b^a \left(f(x) - g(x) \right) dx$

(C) $\int_a^b \left(g(x) + f(x) \right) dx$

(D) $\int_b^a \left(g(x) - f(x) \right) dx$

(E) $\int_b^a \left(g(x) + f(x) \right) dx$

26. The function f is continuous on the closed interval $[0,2]$. It is given that $f(0) = -1$ and $f(2) = 2$. If $f'(x) > 0$ for all x on $[0,2]$ and $f''(x) < 0$ for all x on $(0,2)$, then $f(1)$ could be

(A) 0

(B) $\dfrac{1}{2}$

(C) 1

(D) 2

(E) $\dfrac{5}{2}$

27. The water level in a cylindrical barrel is falling at a rate of one inch per minute. If the radius of the barrel is ten inches, what is the rate that water is leaving the barrel (in cubic inches per minute) when the volume is 500π cubic inches?

(A) 1

(B) π

(C) $100\,\pi$

(D) $200\,\pi$

(E) $500\,\pi$

GO ON TO THE NEXT PAGE

28. If $f(x) = \arctan(x^2)$, then $f'\left(\sqrt{3}\right) =$

(A) $\dfrac{1}{5}$

(B) $\dfrac{1}{4}$

(C) $\dfrac{\sqrt{3}}{4}$

(D) $\dfrac{\sqrt{3}}{5}$

(E) $\dfrac{2\sqrt{3}}{5}$

END OF SECTION I PART A

STOP

END OF SECTION I PART A. IF YOU HAVE ANY TIME LEFT, GO OVER YOUR WORK IN THIS SECTION ONLY. DO NOT WORK IN ANY OTHER SECTION OF THE TEST.

SECTION I, PART B

Time – 50 Minutes

Number of Questions – 17

A GRAPHING CALCULATOR IS REQUIRED FOR SOME QUES-
TIONS ON THIS PART OF THE EXAMINATION.

Directions: Solve each of the following problems, using the available
space for scratchwork. After examining the form of the choices, decide
which is the best of the choices given and fill in the corresponding oval
on the answer sheet. No credit will be given for anything written in the
test book. Do not spend too much time on any one problem.

In this test: (1) The exact numerical value of the correct answer does
not always appear among the choices given. When this happens, select
from among the choices the number that best approximates the exact
numerical value. (2) Unless otherwise specified, the domain of a
function f is assumed to be the set of all real numbers x for which $f(x)$
is a real number.

29. A particle starts at the origin and moves along the x-axis with decreasing positive
velocity. Which of these could be the graph of the distance, $s(t)$, of the particle
from the origin at time t?

   (A)   (B)

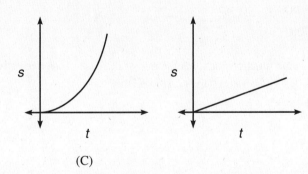

   (C)

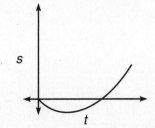

GO ON TO THE NEXT PAGE

(D)

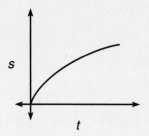

(E)

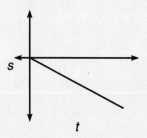

30. Let f be the function given by $f(x) = 3 \ln 2x$, and let g be the function given by $g(x) = x^3 + 2x$. At what value of x do the graphs of f and g have parallel tangent lines?

(A) – 0.782

(B) – 0.301

(C) 0.521

(D) 0.782

(E) 1.000

31. Let f be some function such that the rate of increase of the derivative of f is 2 for all x. If $f'(2) = 4$ and $f(1) = 2$, find $f(3)$.

(A) 3

(B) 6

(C) 7

(D) 9

(E) 10

32. $\displaystyle \lim_{x \to a} \frac{x-a}{x^3 - a^3} =$

    (A) $\dfrac{1}{a^2}$

    (B) $\dfrac{1}{3a^2}$

    (C) $\dfrac{1}{4a^2}$

    (D) 0

    (E) It is nonexistent.

33.

x	3	6	9	12
f(x)	3	2	4	5

Let f be a continuous function with values as represented in the table above. Approximate $\displaystyle \int_{3}^{12} f(x)\,dx$ using a right-hand Riemann Sum with three subintervals of equal length.

    (A) 14

    (B) 27

    (C) 33

    (D) 42

    (E) 48

34.

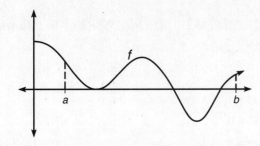

The graph of f', the derivative of f, is shown above. Which of the following describes all relative extrema of f on (a,b)?

    (A) One relative maximum and one relative minimum

    (B) Two relative maximums and one relative minimum

    (C) One relative maximums and no relative minimums

    (D) No relative maximums and two relative minimums

    (E) One relative maximum and two relative minimums

GO ON TO THE NEXT PAGE

35. Let $f(x) = \int_0^{2x} \left(e^t\right) dt$. What value on [0,4] satisfies the Mean Value Theorem for f?

    (A) 2.960

    (B) 2.971

    (C) 3.307

    (D) 3.653

    (E) 4.000

36. The position for a particle moving on the x-axis is given by $s(t) = -t^3 + 2t^2 + \dfrac{1}{2}$. At what time, t, on [0,3] is the particle's instaneous velocity equal to its average velocity over [0,3]?

    (A) 0.535

    (B) 1.387

    (C) 1.821

    (D) 1.869

    (E) 2.333

37. Let f be defined as

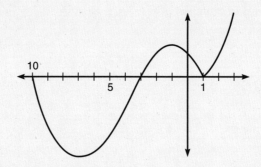

    and g be defined as $g(x) = \int_{-10}^{x} f(x) dx$. Which of the following statements about f and g is *false*?

    (A) $g'(-3) = 0$.

    (B) g has a local minimum at $x = -3$.

    (C) $g(-10) = 0$.

    (D) $f'(1)$ does not exist.

    (E) g has a local maximum at $x = 1$.

38. Let $f(x) = x^2 + 3$. Using the Trapezoidal Rule, with $n = 5$, approximate $\int_0^3 f(x)\, dx$.

    (A) 11.34
    (B) 17.82
    (C) 18.00
    (D) 18.18
    (E) 22.68

39. Population y grows according to the equation $\dfrac{dy}{dt} = ky$, where k is a constant and t is measured in years. If the population triples every five years, then $k =$

    (A) 0.110
    (B) 0.139
    (C) 0.220
    (D) 0.300
    (E) 1.099

40. The circumference of a circle is increasing at a rate of $\dfrac{2\pi}{5}$ inches per minute. When the circumference is 10π inches, how fast is the area of the circle increasing in square inches per minute?

    (A) $\dfrac{1}{5}$

    (B) $\dfrac{\pi}{5}$

    (C) 2

    (D) 2π

    (E) 25π

41.

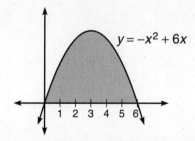

$y = -x^2 + 6x$

The base of a solid is the region in the first quadrant bounded by the x-axis and the parabola $y = -x^2 + 6x$, as shown in the figure above. If cross sections perpendicular to the x-axis are equilateral triangles, what is the volume of the solid?

    (A) 15.588
    (B) 62.354
    (C) 112.237
    (D) 129.600
    (E) 259.200

GO ON TO THE NEXT PAGE

42. Let f be the function given by $f(x) = x^2 + 4x - 8$. The tangent line to the graph at $x = 2$ is used to approximate values of f. For what value(s) of x is the tangent line approximation twice that of f?

   I. $-\sqrt{2}$

   II. 1

   III. $\sqrt{2}$

   (A) I only
   (B) II only
   (C) III only
   (D) I and II only
   (E) I and III only

43. The first derivative of a function, f, is given by $f'(x) = \dfrac{e^{-x}}{x^2} - \sin x$. How many critical values does f have on the open interval $(0,10)$?

   (A) One
   (B) Two
   (C) Three
   (D) Four
   (E) Five

44. $\dfrac{d}{dx}\left(\displaystyle\int_{2x}^{3} f'(t)\,dt \right) =$

   (A) $f'(3)$
   (B) $2f(-2x)$
   (C) $-2f(2x)$
   (D) $2f'(2x)$
   (E) $-2f'(2x)$

45. Let f be defined as

$$f(x) = \begin{cases} \sqrt[3]{x} + k, \, x < 1 \\ \ln(x), \, x \geq 1 \end{cases}$$

for a constant, k. For what value of k will $\lim_{x \to 1^-} f(x) = \lim_{x \to 1^+} f(x)$?

    (A) – 2

    (B) – 1

    (C) 0

    (D) 1

    (E) None of the above

END OF SECTION I

STOP

END OF SECTION I PART A. IF YOU HAVE ANY TIME LEFT, GO OVER YOUR WORK IN THIS SECTION ONLY. DO NOT WORK IN ANY OTHER SECTION OF THE TEST.

SECTION II, Part A

Time – 45 Minutes

Number of Questions – 3

SHOW ALL YOUR WORK. It is important to show your setups for these problems because partial credit will be awarded. If you use decimal approximations, they should be accurate to three decimal places.

A GRAPHING CALCULATOR IS REQUIRED FOR SOME PROBLEMS OR PARTS OF PROBLEMS ON THIS PART OF THE EXAMINATION.

1. Examine the function, f, defined as $f(x) = \frac{x+\sqrt{x}}{3}$ for $0 \le x \le 10$.

    A. Use a Riemann Sum with five equal subintervals evaluated at the midpoint to approximate the area under f from $x = 0$ to $x = 10$.

    B. Again using five equal subintervals, use the Trapezoidal Rule to approximate the area under f from $x = 0$ to $x = 10$.

    C. Using your result from part B, approximate the average value of the function, f, from $x = 0$ to $x = 10$.

    D. Determine the actual average value of the function, f, from $x = 0$ to $x = 10$.

2. A man is observing a horserace. He is standing at some point, O, 100 feet from the track. The line of sight from the observer to some point P located on the track forms a 30° angle with the track, as shown in the diagram below. Horse H is galloping at a constant rate of 45 feet per second.

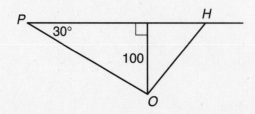

A. At what rate is the distance from the horse to the observer changing 4 seconds after the horse passes point P?

B. At what rate is the area of the triangle formed by P, H, and O changing 4 seconds after the horse passes point P?

C. At the instant the horse gallops past him, the observer begins running at a constant rate of 10 feet per second on a line perpendicular to and toward the track. At what rate is the distance between the observer and the horse changing when the observer is 50 feet from the track?

3. Let $v(t)$ be the velocity, in feet per second, of a race car at time t seconds, $t \geq 0$. At time $t = 0$, while traveling at 197.28 feet per second, the driver applies the brakes such that the car's velocity satisfies the differential equation $\frac{dv}{dt} = -\frac{11}{25}t - 7$.

A. Find an expression for v in terms of t where t is measured in seconds.

B. How far does the car travel before coming to a stop?

C. Write an equation for the tangent line to the velocity curve at $t = 9$ seconds.

D. Find the car's average velocity from $t = 0$ until it stops.

END OF SECTION II PART A

STOP

END OF SECTION I PART A. IF YOU HAVE ANY TIME LEFT, GO OVER YOUR WORK IN THIS SECTION ONLY. DO NOT WORK IN ANY OTHER SECTION OF THE TEST.

SECTION II, Part B

Time – 45 Minutes

Number of Questions – 3

A CALCULATOR IS NOT PERMITTED ON THIS PART OF THE EXAMINATION.

4. Let R be defined as the region in the first quadrant bounded by the curves $y = x^2$ and $y = 8 - x^2$.

   A. Sketch and label the region on the axes provided.

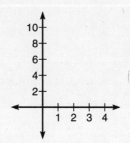

   B. Determine the area of R.

   C. Determine the volume of the solid formed when R is rotated about the x-axis.

   D. Determine the volume of the solid whose base is R and whose cross sections perpendicular to the x-axis are semicircles.

5. The graph below represents the derivative, f', of some function f.

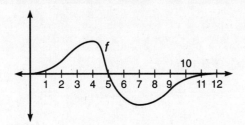

A. At what value of x does f achieve a local maximum? Explain your reasoning.

B. Put these values in order from least to greatest: $f(4), f(5)$, and $f(7)$. Explain your reasoning.

C. Does f have any points of inflection? If so, what are they? Explain your reasoning.

6. Examine the curve defined by $2e^{xy} - y = 0$.

A. Verify $\dfrac{dy}{dx} = \dfrac{2ye^{xy}}{1 - 2xe^{xy}}$.

B. Find $\dfrac{dy}{dx}$ for the family of curves $be^{xy} - y = 0$.

C. Determine the y-intercept(s) of $be^{xy} - y = 0$.

D. Write the equation for the tangent line at the y-intercept.

END OF SECTION II

STOP

END OF SECTION I PART A. IF YOU HAVE ANY TIME LEFT, GO OVER YOUR WORK IN THIS SECTION ONLY. DO NOT WORK IN ANY OTHER SECTION OF THE TEST.

Answer Key

SECTION I

PART A

1. C	8. E	15. B	22. E
2. E	9. C	16. E	23. C
3. E	10. C	17. C	24. B
4. D	11. D	18. A	25. B
5. E	12. C	19. D	26. C
6. A	13. E	20. C	27. C
7. B	14. C	21. B	28. D

PART B

29. D	34. A	39.C	44.E
30. D	35. A	40.D	45.B
31. E	36. D	41.C	
32. B	37. E	42.E	
33. C	38.D	43.D	

SOLUTIONS FOR SECTION I

PART A

1. **The correct answer is (C).** Instantaneous rates of change always imply differentiation. To quickly determine this derivative, it is helpful to recognize $x^3 + 3x^2 + 3x + 1$ as $(x + 1)^3$. We then simplify the orginal function to

$$f(x) = \frac{(x+1)^3}{x+1} = (x+1)^2$$

Now, use the Power and Chain Rules:

$$f'(x) = 2(x + 1).$$

To find the instanteous rate of change when $x = 2$,

$$f'(2) = 2(2 + 1) = 6.$$

2. **The correct answer is (E).** To approximate the actual number of cars crossing the bridge, approximate the area under this graph. One way to do this is to divide the interval from $t = 0$ to $t = 12$ into 2 equal subintervals. Both of these regions resemble trapezoids. The area of the left one is

$$A = \frac{1}{2}h(b_1 + b_2) = \frac{1}{2}(6)(100 + 25) = 375.$$

The area of the right trapezoid would be

$$A = \frac{1}{2}(6)(100 + 50) = 450.$$

So, the area under the curve would be approximately 825. But we must be careful here. The rate is in cars per minute. Since the x-axis is in hours, we must convert the rate to cars per hour. To do this, we multiply our 825 by 60 (minutes per hour) and get 49,500 cars.

3. **The correct answer is (E).** Find the definite integral:

$$\int_1^5 \frac{3x}{x^3} dx = \int_1^5 \frac{3}{x^2} dx$$

$$= -\frac{3}{x}\Big|_1^5 = -\frac{3}{5} + 3$$

$$= \frac{12}{5}.$$

GO ON TO THE NEXT PAGE

4. **The correct answer is (D).** There is not enough information to determine whether or not choices (A), (B), (C), or (E) are true. Look at choice (D):

$$\text{Average rate of change} = \frac{f(2) - f(0)}{2 - 0}$$

$$= \frac{9 - 3}{2} = \frac{6}{2} = 3.$$

5. **The correct answer is (E).** We must recognize that $\frac{1}{\cos x} = \sec x$. This lets us rewrite the integral as

$$\int_0^{\pi/3} \left(\sec x \tan x e^{\sec x} \right) dx.$$

Next, we can evaluate this integral using u-substitution. If we let $u = \sec x$ and $du = \sec x \tan x \, dx$, we get

$$\int_1^2 e^u \, du = e^u \Big|_1^2 = e^2 - e.$$

6. **The correct answer is (A).** We need the derivative of the curve. Since x and y are not separated for us, we must use implicit differentiation. Differentiate everything with respect to x.

$$3x^2 + 2xy + 6y^2 - 3x - 8y = 0$$

$$6x + 2x\frac{dy}{dx} + 2y + 12y\frac{dy}{dx} - 3 - 8\frac{dy}{dx} = 0$$

Now, group all terms with $\frac{dy}{dx}$, and solve for $\frac{dy}{dx}$.

$$\frac{dy}{dx}(2x + 12y - 8) = 3 - 6x - 2y$$

$$\frac{dy}{dx} = \frac{3 - 6x - 2y}{2x + 12y - 8}$$

Finally, we just substitute our point $(1,1)$ into our expression for $\frac{dy}{dx}$ and get

$$\frac{dy}{dx}(1,1) = \frac{3 - 6 - 2}{2 + 12 - 8} = -\frac{5}{6}.$$

7. **The correct answer is (B).** Before we try to integrate anything here, distribute that $\sqrt{x}$ and change the notation to that of rational exponents. After these two steps, we get

$$\int_0^1 \left(x^{5/2} + 3x^{3/2} - 8x^{1/2} \right) dx.$$

Integrating leads to

$$\left(\frac{2}{7}x^{7/2} + \frac{6}{5}x^{5/2} - \frac{16}{3}x^{3/2}\right)\Big|_0^1 = \left(\frac{2}{7} + \frac{6}{5} - \frac{16}{3}\right) - 0 = -\frac{404}{105}$$

8. **The correct answer is (E).** This is a related-rates problem. We are given $\dfrac{dr}{dt}$, the rate at which the radius is increasing, and need to find $\dfrac{dV}{dt}$, the rate at which the volume is increasing when A, the surface area, is 9π. Our primary equation is the volume equation for a sphere:

$$V = \frac{4}{3}\pi r^3$$

As in all such problems, we differentiate with respect to t.

$$\frac{dV}{dt} = 4\pi r^2 \frac{dr}{dt}$$

Knowledge of basic formulas is useful here. $4\pi r^2$ is merely the surface area formula for a sphere. We were given that the surface area is equal to 9π and that $\frac{dr}{dt} = 2$. Substituting these values yields

$$\frac{dV}{dt} = 9\pi(2) = 18\pi.$$

9. **The correct answer is (C).** To find the area of a region bounded by two curves, we should apply the following formula:

$$\int_a^b \left(f(x) - g(x)\right)dx,$$

where a and b are the endpoints of the interval. $f(x)$ represents the top curve, while $g(x)$ represents the bottom curve. Finding a and b is simple enough. Since the region begins at the y-axis, $a = 0$. To find b, we will determine what value satisfies the following equation:

$$x^3 = x + 6.$$

By inspection, we can see that $x = 2$. Now, the area of the region would be given by

$$\int_0^2 \left(x + 6 - x^3\right)dx.$$

10. **The correct answer is (C).** Whenever the average rate of change is requested, we just need to compute the slope of the secant line.

$$m = \frac{f(b) - f(a)}{b - a}$$
$$= \frac{f(4) - f(-1)}{4 - (-1)}$$
$$= \frac{19 - (-6)}{5} = 5$$

11. **The correct answer is (D).** Since we know that the second derivative is a line of slope 6, we can say that

$$f''(x) = 6x + C_1$$

That implies that $f'(x) = 3x^2 + C_1 x + C_2,$

which in turn implies that $f(x) = x^3 + C_3 x^2 + C_2 x + C_4.$

That is a cubic function.

12. **The correct answer is (C).** The MVT for Integrals says that the average value of a function over a given interval is the area under the curve divided by the length of the interval. So, the average value, $f(c)$, of f over $[-4,4]$ could be found like this:

$$f(c) = \left(\frac{1}{8}\right)\left(\int_{-4}^{0}\left(\sqrt{-x}\right)dx + \int_{0}^{4}x^2 dx\right),$$

which is equivalent to

$$f(c) = \frac{1}{8}\left(\int_{0}^{4}\left(\sqrt{x}\right)dx + \int_{0}^{4}x^2 dx\right)$$

$$= \frac{1}{8}\left(\frac{2}{3}x^{3/2}\Big|_{0}^{4} + \frac{x^3}{3}\Big|_{0}^{4}\right)$$

$$= \frac{1}{8}\left(\frac{16}{3} + \frac{64}{3}\right) = \frac{10}{3}.$$

13. **The correct answer is (E).** Since f has a horizontal tangent at $x = 1$, we know that $f'(1) = 0$. By reading the graph, we can see that $f(1) > 0$. Since the graph is concave down at $x = 1$, $f''(1) < 0$. Hence, $f''(1) < f'(1) < f(1)$.

14. **The correct answer is (C).** The MVT states that at some point c on the interval $[a,b], f'(c) = \dfrac{f(b) - f(a)}{b - a}$. Since $\dfrac{f(12) - f(-4)}{12 - (-4)} = \dfrac{1}{2}$, then at some point c on $[a,b], f'(c) = \dfrac{1}{2}$.

15. **The correct answer is (B).** This is an application of the Fundamental Theorem of Calculus, Part Two, which states

$$\frac{d}{dx}\left(\int_a^x f(t)\,dt\right) = f(x).$$

We must remember that when the upper limit of integration is some function of x, such as $2x^2$, we must multiply $f(x)$ by the derivative of that function with respect to x. Hence,

$$\frac{d}{dx}\left(\int_3^{2x^2} e^t\,dt\right) = e^{2x^2}\cdot 4x$$

$$= 4xe^{2x^2}.$$

16. **The correct answer is (E).** Since we know that rate of change implies derivative, from the information in the problem, we can write

$$f'(c) = e^3 f'(2).$$

We are also told that $f(x) = e^x$, so $f'(x) = e^x$. So, the above equation becomes

$$e^c = e^3 \bullet e^2 = e^5$$

So, $c = 5$.

17. **The correct answer is (C).** This problem is testing if we can apply the Chain Rule to functions defined by a table. If $h(x) = f(g(x))$, then $h'(x) = f'(g(x))g'(x)$. This is why, when $x = 3$,

$$h(3) = f(g(x)) = f(4) = 2$$
$$h'(3) = f'(g(3))g'(3) = f'(4)g'(3) = 2 \bullet 1 = 2.$$
$$\text{So, } h(3) = h'(3).$$

18. **The correct answer is (A).** Since there are exactly two points on $(0,10)$ where f has a value of 4, the graph of f must cross the line $x = 4$ twice: Once on the way up and once on the way down. The fact that f is differentiable over the interval insures no cusps or discontinuities. Since the curve turns around somewhere on $(0,10)$, there must be at least one horizontal tangent on $(0,10)$. Horizontal tangents are places where the derivative is equal to zero.

19. **The correct answer is (D).** Normal lines are perpendicular to tangent lines. The slopes of two perpendicular lines are opposite reciprocals of each other. So, this problem needs us to determine the slope of the curve at $x = 2$, and then determine the opposite reciprocal of that slope. Since slope of the curve is determined by the value of its derivative,

$$\text{tangent slope} = y' = \frac{1}{2}\left(8 - x^2\right)^{-1/2}(-2x) = \frac{-x}{\sqrt{8 - x^2}}$$

$$= 2 = \frac{-2}{\sqrt{8-4}} = \frac{-2}{2} = -1.$$

The slope of the normal line is the opposite reciprocal

$$\frac{-1}{1} = 1.$$

20. **The correct answer is (C).** Evaluate the definite integral and apply the Fundamental Theorem, Part One:

$$\frac{x^4}{4}\Big|_{-2}^{k} = 0$$

$$\frac{k^4}{4} - \frac{2^4}{4} = 0$$

$$k^4 - 16 = 0$$

$$k = \pm 2$$

21. **The correct answer is (B).** The rate of change of y being directly proportional to y is the same statement as

$$y' = ky,$$

which we know leads to

$$y = Ne^{kt}.$$

22. **The correct answer is (E).** The question to answer here is when, if ever, is the derivative of $y = 3x^3 - 2x^2 + 6x - 2$ negative? We should try to determine the derivative, find any critical values, and examine a wiggle graph.

$$y' = 9x^2 - 4x + 6 = 0$$

This is an unfactorable trinomial. Since we are not permitted to use our calculators, we'd better use the quadratic formula. So,

$$x = \frac{4 \pm \sqrt{16 - 216}}{18}.$$

Aha! The radicand, $16 - 216$, is less than zero, which would indicate that the equation has no real solutions, which would imply that the derivative of $y = 3x^3 - 2x^2 + 6x - 2$ is never zero. Since it is a polynomial function, then it must be continuous; hence, $y = 3x^3 - 2x^2 + 6x - 2$ is strictly monotonic. Now, we should determine the value of the derivative at one x value to determine if the derivative is always positive or always negative. Using the equation, let's determine the value of the derivative when $x = 0$:

$$y'(0) = 0 - 0 + 6 = 6 > 0.$$

Since the derivative is always positive, the function $y = 3x^3 - 2x^2 + 6x - 2$ is never decreasing.

23. **The correct answer is (C).** Remember, the MVT guarantees that for some c on [2,5],

$$f'(c) = \frac{f(5) - f(2)}{(5-2)}$$

$$= \frac{\frac{11}{2} - 1}{3} = \frac{3}{2}.$$

Now, we should determine the derivative of $f(x) = \frac{x^2 - 3}{x - 1}$, set it equal to $\frac{3}{2}$, and solve for x.

$$f'(x) = 2x(x-1) - \frac{x^2 - 3}{(x-1)^2} = \frac{3}{2}$$

$$\frac{x^2 - 2x + 3}{(x-1)^2} = \frac{3}{2}$$

$$2(x^2 - 2x + 3) = 3(x-1)^2$$

$$2x^2 - 4x + 6 = 3x^2 - 6x + 3$$

$$x^2 - 2x - 3 = 0$$

$$(x-3)(x+1) = 0$$

$$x = 3 \text{ or } x = -1$$

Since -1 is not on [2,5], we throw that value out and the value on [2,5] that satisfies the MVT is 3.

24. **The correct answer is (B).** The slope field shows us that $f(x)$ will have a derivative of zero when $y = -1$ and when $x = 1$ (since the slopes are horizontal there). The easiest possible differential equation with such characteristics is $f'(x) = (x-1)(y+1)$, since plugging in -1 for y or 1 for x makes the slope 0. If you factor choice (B) by grouping, that is exactly what you get

$$xy - x - y - 1$$
$$x(y-1) - 1(y+1)$$
$$(x-1)(y+1)$$

25. **The correct answer is (B).** This problem is not as clear as it may initially appear. After examining the diagram, we should look for

$$\int_a^b \big(g(x) - f(x)\big)dx.$$

However, this is not an answer choice. One of the choices must be equivalent to ours. Remember that when you switch the limits of integration, you get the opposite value. Switching our limits gives us

$$-\int_b^a \big(g(x) - f(x)\big)dx.$$

This is still not a choice. What if we treat that negative sign as if it were the constant – 1 and distribute it through the integral? We get

$$\int_b^a \big(f(x) - g(x)\big)dx.$$ Eureka!

26. **The correct answer is (C).** The derivative being positive over [0,2] implies that the function is increasing over this interval. The second derivative being negative means that the function is concave down. The curve must look something like the curve drawn below:

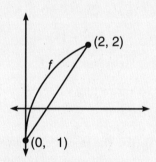

Notice that every y-coordinate over the interval (0,2) is less than 2, so $f(1) < 2$. Notice that the entire curve is above the secant line from $(0, -1)$ to $(2,2)$. (This is true due to the concavity of the curve.) Since the secant line segment passes through $\left(1, \dfrac{1}{2}\right)$, $f(1) > \dfrac{1}{2}$. Therefore, $f(1)$ could be 1.

27. **The correct answer is (C).** This is another related-rates problem. We know that $\dfrac{dh}{dt} = -1$, where h represents the water level in the barrel. We are looking for $\dfrac{dV}{dt}$, with V representing the volume of the barrel. Our primary equation is the formula for volume of a cylinder:

$$V = \pi r^2 h.$$

Since it is given that the radius, r, is a constant of 10, we can substitute this into the equation and get

$$V = 100\pi h.$$

Now, as in any related-rates problem, we should differentiate with respect to t:

$$\frac{dV}{dt} = 100\pi \frac{dh}{dt}.$$

Substituting $\frac{dh}{dt} = -1$ into the equation yields

$$\frac{dV}{dt} = -100\pi.$$

Therefore, the water is leaving the barrel at 100π in^3/min.

Notice that the information that the volume was 500π cubic inches was unnecessary.

28. **The correct answer is (D).** All we need here is the derivative of $f(x) = \arctan u$.

$$f'(x) = \frac{u'}{1 + u^2}$$

Since our function is $u = x^2$, the equation becomes

$$f'(x) = \frac{2x}{1 + x^4}.$$

Now, we substitute $x = \sqrt{3}$ and get

$$f'\left(\sqrt{3}\right) = \frac{2\sqrt{3}}{9 + 1} = \frac{2\sqrt{3}}{10} = \frac{\sqrt{3}}{5}.$$

PART B

29. **The correct answer is (D).** The particle's positive velocity indicates that the position function's graph is increasing. The decreasing velocity indicates that the position function's graph should be concave down.

30. **The correct answer is (D).** In order for these two functions to have parallel tangent lines, their derivatives must be equal. So, we should find the derivatives of both functions, set them equal to each other, and solve for x. Since $f(x) = 3\ln(2x)$, $f'(x) = \frac{3}{x}$. The derivative of $g(x) = x^3 + 2x$ is $g'(x) = 3x^2 + 2$. Now, we will set these two expressions equal and use our calculator to solve for x:

$$\frac{3}{x} = 3x^2 + 2$$

$$\frac{3}{x} - 3x^2 - 2 = 0$$

$$x = 0.782.$$

31. **The correct answer is (E).** The rate of increase of the derivative is the second derivative. So,

$$f''(x) = 2.$$

To find an expression for the first derivative, we can find an antiderivative:

$$f'(x) = 2x + C_1$$

In order to determine C_1, we can use the information given to us that $f'(2) = 4$. We will substitute this and solve for C_1:

$$f'(2) = 4 = 4 + C_1$$

$$C_1 = 0.$$

Substituting this value into the second equation yields

$$f'(x) = 2x.$$

Now, we will determine $f(x)$ and find an antiderivative of the last equation:

$$f(x) = x^2 + C_2.$$

To solve for C_2, we can use the fact that $f(1) = 2$, so

$$f(1) = 2 = 1 + C_2;$$

$$C_2 = 1.$$

Then, $f(x) = x^2 + 1$.

Finally, we can determine $f(3)$:

$$f(3) = 9 + 1 = 10.$$

32. **The correct answer is (B).** We must remember how to factor the difference of perfect cubes:

$$a^3 - b^3 = (a - b)(a^2 + ab + b^2).$$

Using this formula, we can simplify the limit:

$$\lim_{x \to a} \frac{x - a}{x^3 - a^3} = \lim_{x \to a} \frac{1}{x^2 + ax + a^2},$$

which we can evaluate by substitution:

$$= \frac{1}{a^2 + a^2 + a^2} = \frac{1}{3a^2}.$$

33. **The correct answer is (C).** The best way to attack this problem would be to plot the 4 points given and sketch the 3 rectangles, as shown in the diagram below:

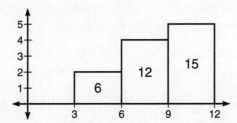

Notice that the heights of the rectangles are determined by the y-value corresponding to the right endpoint of the subintervals. Next, we determine the area of each rectangle and then add them up:

$$\int_3^{12} f(x)\, dx \approx 6 + 12 + 15 = 33.$$

34. **The correct answer is (A).** Since the graph of the derivative of f crosses the x-axis twice, there will be two relative extrema. There will be one maximum because the derivative changes from positive to negative once. There will also be one minimum since the derivative changes from negative to positive once as well.

35. **The correct answer is (A).** This is a rather complicated application of the MVT. We will also have to use the Fundamental Theorem of Calculus Part Two. First, let's determine the value of $f'(c)$:

$$f'(c) = \frac{f(b) - f(a)}{b - a}$$

$$= \frac{f(4) - f(0)}{4} = \frac{\left(e^8 - e^0\right) - 0}{4}$$

$$= \frac{e^8 - 1}{4}$$

Note: $f(4) = \int_0^8 e^t dt = e^t \Big|_0^8 = e^8 - e^0$

Now, we will determine the derivative of $f(x) = \int_0^{2x} e^t dt$:

$$f'(x) = 2e^{2x}.$$

Next, we will set our value for $f'(c)$ equal to our expression for $f'(x)$ and use our calculator to solve for x:

$$2e^{2x} = \frac{e^8 - 1}{4}$$

$$x = 2.960.$$

36. **The correct answer is (D).** This problem asks where is the slope of the tangent line, which is the instantaneous velocity, equal to the slope of the secant line, which is the average velocity, over $[0,3]$.

$$m_{sec} = \frac{s(3) - s(0)}{3} = \frac{\left(-27 + 18 + \frac{1}{2}\right) - \frac{1}{2}}{3} = -3$$

To find the slope of the tangent line, find the derivative of the curve:

$$m_{tan} = -3t^2 + 4t.$$

To determine where the two slopes are the same we will set m_{sec} equal to m_{tan} and solve for x using our calculator:

$$-3t^2 + 4t = -3$$

$$-3t^2 + 4t + 3 = 0$$

$$t = 1.869.$$

37. **The correct answer is (E).** For g to have a local maximum at $x = 1$, the derivative of g, which is f, must change from positive to negative at $x = 1$. It does not.

38. **The correct answer is (D).** This is a tedious example of the Trapezoidal Rule. Since we have 5 subintervals and the interval is 3 units long, we will be dealing with some messy numbers. Anyway, we still have to remember the Trapezoidal Rule:

$$\int_a^b f(x)dx \approx \left(\frac{b-a}{2n}\right)\left(f(a) + 2f(x_1) + 2f(x_2) + \ldots 2f(x_{n-1}) + f(b)\right)$$

Applying it to this function, we get

$$\int_0^3 (x^2 + 3)dx \approx \frac{3}{10}\left(3 + 2(3.36) + 2(4.44) + 2(6.24) + 2(8.76) + 12\right)$$

$$= 0.3 \cdot (3 + 6.72 + 8.88 + 12.48 + 17.52 + 12)$$

$$= .3 \cdot 60.6 = 18.18.$$

39. **The correct answer is (C).** Since $\frac{dy}{dx} = ky$, we can immediately say that we are dealing with an exponential function of the following form:

$$y = Ne^{kt}.$$

We know that after five years, the population will be three times what it was initially. If we substitute $3C$ for y and 5 for t and solve for k, we get

$$3C = Ne^{5k}$$

$$3 = e^{5k}$$

$$\ln 3 = 5k$$

$$k = \frac{\ln 3}{5} \approx 0.220.$$

40. **The correct answer is (D).** In this related-rates problem, we are going to need $\frac{dr}{dt}$. To quickly find $\frac{dr}{dt}$, let's use the formula for the circumference, differentiate with respect to t, and solve for $\frac{dr}{dt}$:

$$C = 2\pi r$$

$$\frac{dC}{dt} = 2\pi \frac{dr}{dt}$$

$$\frac{dr}{dt} = \frac{\frac{dC}{dt}}{2\pi}.$$

We are given that $\frac{dC}{dt} = \frac{2\pi}{5}$. We can now substitue this value into the equation to get a value for $\frac{dr}{dt}$:

$$\frac{dr}{dt} = \frac{\frac{2\pi}{5}}{2\pi} = \frac{1}{5}.$$

The question is asking us about the rate at which the area is increasing, $\frac{dA}{dt}$, when the circumference is 10π inches. We will take the formula for the area of a circle and differentiate with respect to t:

$$A = \pi r^2$$

$$\frac{dA}{dt} = 2\pi r \frac{dr}{dt}.$$

Notice that we have the expression $2\pi r$. This is just the circumference that we know to be 10π. We can substitute this value and the value $\frac{1}{5}$ for $\frac{dr}{dt}$ to determine $\frac{dA}{dt}$:

$$\frac{dA}{dt} = 10\pi \cdot \frac{1}{5} = 2\pi.$$

41. **The correct answer is (C).** Remember that the formula for the volume of a solid with known cross sections is

$$V = \int_a^b A(x)\, dx,$$

where $A(x)$ represents the area of the cross sections. In this problem, we are dealing with cross sections that are equilateral triangles. The formula for the area of an equilateral triangle is

$$A = \frac{\sqrt{3}}{4} s^2$$

where s is the length of one side. As we can see from the diagram, the interval is from $x = 0$ to $x = 6$. Therefore, the volume of this solid is

$$V = \frac{\sqrt{3}}{4} \cdot \int_0^6 \left(-x^2 + 6x\right)^2 dx$$

Our calculator will now do the rest and get

$$V = 112.237.$$

42. **The correct answer is (E).** First, we must determine an equation for the tangent line to this curve at $x = 2$. We need a point on the line and the slope of the line. First the point: Since $f(2) = 4$, $(2,4)$ is on the line. Now the slope:

$$f'(x) = 2x + 4$$

$$f'(2) = 4 + 4 = 8.$$

The equation for the tangent line is

$$y - 4 = 8(x - 2)$$

or

$$y = 8x - 12.$$

Now, let's examine our choices.

   I. The value of the function at $x = -\sqrt{2}$ is $-4\sqrt{2} - 6$, and the tangent line approximation is $-8\sqrt{2} - 12$, which is twice the value of the function. So, I checks out.

   II. The function value at $x = 1$ is -3, while the tangent line approximation is -4.

   III. The function value at $x = \sqrt{2}$ is $4\sqrt{2} - 6$, and the tangent line approximation is $8\sqrt{2} - 12$. So, III applies too.

43. **The correct answer is (D).** In order to determine the number of critical values of the function, we can count the zeros of the derivative. This would require us to graph the derivative on the calculator and count how many times it crosses the x–axis. It crosses four times.

44. **The correct answer is (E).** This is a tricky Fundamental Theorem of Calculus Part Two problem. First, we should rewrite it as such:

$$\frac{d}{dx}\left(\int_{2x}^{3} f'(t)\,dt\right) = -\frac{d}{dx}\left(\int_{3}^{2x} f'(t)\,dt\right).$$

Once we've rewritten the problem like this, it's not so difficult:

$$-f'(2x)\cdot\frac{d}{dx}(2x)$$
$$= -f'(2x)\cdot 2$$
$$= -2f'(2x)$$

45. **The correct answer is (B).** In order for the left-hand and right-hand limits to be equal, the function must be continuous. So, we need to find the value of k for which this equation is true:

$$\sqrt[3]{1} + k = \ln 1.$$

This is relatively simple to solve:

$$1 + k = 0$$
$$k = -1.$$

SOLUTIONS FOR SECTION II

PART A

1.   A. The five subintervals would each be of length 2 and would be [0,2], [2,4], [4,6], [6,8], and [8,10]. The midpoints of these subintervals would be 1, 3, 5, 7, and 9, respectively. The Riemann Sum that we are looking for is just the sum of five rectangles, each of width 2 and height $f(m_i)$, where m_i is the midpoint of the ith subinterval. So,

$$A = 2(f(1) + f(3) + f(5) + f(7) + f(9))$$

$$= 2(0.667 + 1.577 + 2.412 + 3.215 + 4)$$

$$= 2(11.871)$$

$$= 23.743.$$

  B. Recall the Trapezoidal Rule:

$$\int_a^b f(x)dx \approx \frac{b-a}{2n}\Big(f(a)+2f(x_1)+2f(x_2)+\ldots+2f(x_{n-1})+f(b)\Big).$$

$$A \approx \frac{10}{10}\Big(f(0)+2f(2)+2f(4)+2f(6)+2f(8)+f(10)\Big)$$

$$= (0 + 2(1.138) + 2(2) + 2(2.817) = 2(3.610) + 4.387)$$

$$= 23.516.$$

  C. The average value, $f(c)$, of the function is the area under the curve divided by the length of the interval. So, we can approximate $f(c)$ like this:

$$f(c) \approx \frac{23.516}{10} = 2.352.$$

  D. Here, we should determine the exact area under the curve and divide it by the length of the interval:

$$f(c) = \frac{1}{10}\int_0^{10}\left(\frac{x+\sqrt{x}}{3}\right)dx = \frac{1}{10}\cdot(23.694) = 2.369.$$

2.   A. We'll start by labeling a fourth point, Q, as the point on the track directly in front of observer O. We will also define some variables: x will be the distance from the horse H to the point Q, y will be the distance from the observer O to the point Q, and z will be the distance

between the horse H and the observer O. All of this is shown in the diagram below.

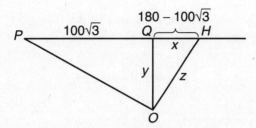

How long is the distance from P to Q? We can use the 30-60-90 Triangle Theorem to determine that it is $100\sqrt{3}$ or 173.2051 feet. Since the horse is running at 45 feet per second, he has run 180 feet after 4 seconds. So, $x = 180 - 173.2051 = 6.79492$.

Now we can use the Pythagorean Theorem to write our primary equation:

$$x^2 + 100^2 = z^2.$$

Substituting $x = 6.79492$ into the equation and solving for z gives us

$$6.79492^2 + 100^2 = z^2$$

$$10046.171 = z^2$$

$$z = 100.23059.$$

The question asked us for the rate that the distance from the horse to the observer is increasing after four seconds. In other words, what is $\frac{dz}{dt}$ when $t = 4$? To answer this, let's differentiate with respect to t and solve for $\frac{dz}{dt}$.

$$x^2 + 100^2 = z^2$$

$$2x\frac{dx}{dt} = 2z\frac{dz}{dt}$$

$$\frac{dz}{dt} = \frac{x\frac{dx}{dt}}{z}$$

$$= \frac{6.79492 \cdot 45}{100.23059}$$

$$= 3.051$$

So, when $t = 4$, the distance from the horse to the observer is increasing at 3.051 feet per second.

B. This is not a difficult problem. The area of a triangle is

$$A = \frac{1}{2} \cdot \text{base} \cdot \text{height} \cdot$$

The height of this triangle is a constant, 100 feet, so

$$A = 50 \cdot \text{base}.$$

To determine the rate at which the area of the triangle is changing, lets differentiate with respect to t:

$$\frac{dA}{dt} = 50\frac{d}{dt}(\text{base}).$$

What is $\frac{d}{dt}(\text{base})$? That's the rate at which the base is changing, which is merely the speed of the horse, 45 feet per second. Now we have

$$\frac{dA}{dt} = 50 \cdot 45 = 2250.$$

So, the area of the triangle formed by P, H, and O is increasing at a constant rate of 2250 feet^2 per second.

C. We will use x, y, and z to represent the same distances as in part A. It can be easily determined that the horse has galloped 225 feet in the same amount of time that the man ran 50 feet. So, $y = 50$, $x = 225$, and z can be determined as such:

$$x^2 + y^2 = z^2$$

$$50^2 + 225^2 = z^2$$

$$z = 230.489.$$

To determine $\frac{dz}{dt}$, we should differentiate with respect to t:

$$2x\frac{dx}{dt} + 2y\frac{dy}{dt} = 2z\frac{dz}{dt}.$$

Solving for $\frac{dz}{dt}$ gives us

$$\frac{dz}{dt} = \frac{x\frac{dx}{dt} + y\frac{dy}{dt}}{z}.$$

Now, we will substitute the following values into the equation: $x = 225$, $y = 50$, $z = 230.489$, $\frac{dx}{dt} = 45$, and $\frac{dy}{dt} = -10$. $\frac{dy}{dt}$ is negative because y is getting shorter.

$$\frac{dz}{dt} = \frac{50 \cdot -10 + 225 \cdot 45}{230.489} = 41.759$$

The distance from the horse to the observer is increasing at 41.759 feet per second.

3. This problem involves solving the separable differential equation $\frac{dv}{dt} = -\frac{11}{25}t - 7$. First, we should separate the v's and t's:

$$dv = \left(-\frac{11}{25}t - 7\right)dt.$$

Integrate both sides:

$$v = -\frac{11}{50}t^2 - 7t + C.$$

To determine the value of C, we use the initial condition given to us in the problem. Since $v(0) = 197.28$, then

$$v(0) = 197.28 = -\frac{11}{50}\cdot 0^2 - 7\cdot 0 + C$$

and $C = 197.28$. Now, we have our expression for v in terms of t:

$$v(t) = -\frac{11}{50}t^2 - 7t + 197.28.$$

B. This is a two-part question. First, we should determine how much time it takes the car to stop, and then we should integrate the velocity curve using that value. In order for the car to stop, $v(t) = 0$.

$$-\frac{11}{50}t^2 - 7t + 197.28 = 0$$

$$t = 18.$$

It takes the car 18 seconds to come to a stop. Now, to determine how far the car travels in those 18 seconds, we should find the area under the velocity curve from $t = 0$ to $t = 18$:

$$\int_0^{18}\left(-\frac{11}{50}t^2 - 7t + 197.28\right)dt$$

$$= 1989.36.$$

The car travels 1989.36 feet while slowing down.

C. To write the equation for a line, we need a point on the line and the slope of the line. To determine the y coordinate of the point, we will evaluate v at $t = 9$:

$$v(9) = -\frac{11}{50} \cdot 9^2 - 7 \cdot 9 + 197.28$$

$$= 116.46.$$

This tells us that (9116.46) is on our tangent line. Now, to determine the slope by evaluating the derivative (which we already know from the problem itself $\frac{dV}{dt} = -\frac{11}{25}t - 7$) at $t = 9$:

$$f'(9) = -\frac{11}{25} \cdot 9 - 7 = -10.96$$

The tangent line passes through $(9, 116.46)$ and has a slope of -10.96. We can now write its equation using point-slope form

$$v - 116.46 = -10.96(t - 9)$$

Or in slope intercept form:

$$v = -10.96t + 215.1.$$

D. This problem calls for the average value formula applied to the velocity equation you found in part (A).

$$\frac{1}{18 - 0} \int_0^{18} v(t)\,dt$$

You already know the value of the integral from your work in part (B).

$$\frac{1}{18}(1989.36)$$

$$= 110.52 \text{ ft/sec.}$$

END OF SECTION II PART A

PART B

4.   A.

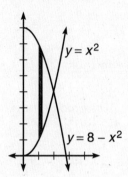

B. The two curves intersect at (2,4). The area of the region R can be determined using the following definite integral:

$$A = \int_0^2 \left(8 - x^2 - x^2\right)dx$$

$$= \int_0^2 \left(8 - 2x^2\right)dx$$

$$= \left(8x - \frac{2x^3}{3}\right)\Big|_0^2$$

$$= 16 - \frac{16}{3} = \frac{32}{3}$$

C. The volume is easiest to determine using the washer method. The outer radius, $R(x)$, will be $8 - x^2$, and the inner radius, $r(x)$, is x^2:

$$V = \pi \int_0^2 \left[\left(8 - x^2\right) - \left(x^2\right)^2\right]dx$$

$$= \pi \int_0^2 \left(64 - 16x + x^4 - x^4\right)dx$$

$$= \pi \left(64x - \frac{16}{3}x^3\right)\Big|_0^2$$

$$= \pi \left(128 - \frac{128}{3}\right)$$

$$= \frac{256\pi}{3}$$

D. The volume of a solid with known cross sections can be determined like this:

$$V = \int_a^b A(x)dx.$$

The cross sections are semicircles whose area formula is $A = \frac{1}{2}\pi r^2$. Now, we need an expression in terms of x for the radius of one of these semicircles. Because the height of R is the diameter of a semicircle, the radius would be

$$r(x) = \frac{1}{2}\left(8 - 2x^2\right) = 4 - x^2$$

This leads to the area of a semicircle:

$$A(x) = \frac{1}{2}\pi(4 - x^2)^2$$

$$= \frac{1}{2}\pi(16 - 8x^2 + x^4),$$

which gives us the volume of the solid:

$$V = \frac{1}{2}\pi\int_0^2 \left(16 - 8x^2 + x^4\right)dx$$

$$= \frac{1}{2}\pi\left(16x - \frac{8x^3}{3} + \frac{x^5}{5}\right)\Big|_0^2$$

$$= \frac{\pi}{2}\left(32 - \frac{64}{3} + \frac{32}{5}\right) = \frac{128\pi}{15}.$$

5. 　A. The local maximum occurs at $x = 5$ because the derivative changes from positive to negative there. This means that the function changes from increasing to decreasing there as well.

B. $f(7) < f(4) < f(5)$

Since the function increases over [4,5] and decreases over [5,7], f(5) is the greatest of the three. To determine which is greater, f(4) or f(7), we examine the accumulated area over [4,7]. Since this area is negative, the function has a net decrease over [4,7]. Thus, f(4) > f(7).

C. f has two points of inflection: one at $x = 4$ and one at $x = 7$. Points of inflection are places where the graph changes concavity. The graph changes concavity whenever the derivative changes from increasing to decreasing or from decreasing to increasing. The derivative changes from increasing to decreasing at $x = 4$ and from decreasing to increasing at $x = 7$.

6.   A. Since the x and y are not separated, we should differentiate implicitly.

$$2e^{xy} - y = 0$$

$$2e^{xy}\left(x\frac{dy}{dx} + y\right) - \frac{dy}{dx} = 0$$

$$2xe^{xy}\frac{dy}{dx} + 2ye^{xy} - \frac{dy}{dx} = 0$$

$$\frac{dy}{dx}\left(2xe^{xy} - 1\right) = -2ye^{xy}$$

$$\frac{dy}{dx} = -\frac{2ye^{xy}}{2xe^{xy} - 1} = \frac{2ye^{xy}}{1 - 2xe^{xy}}$$

B. Differentiating implicitly again yields

$$\frac{dy}{dx} = \frac{bye^{xy}}{1 - bxe^{xy}}.$$

C. To determine the y-intercept, we let $x = 0$ and solve for y:

$$be^{xy} - y = 0$$

$$be^0 = y$$

$$y = b.$$

So, the y-intercept is $(0,b)$.

D. We just need the slope when $x = 0$ and $y = b$. We will substitute these values into our expression for $\frac{dy}{dx}$:

$$\frac{dy}{dx} = \frac{b \cdot be^0}{1 - 0} = b^2$$

Now, we write the equation for a line with y-intercept b and slope b^2:

$$y = b^2x + b.$$

Answer Sheet

FIRST MODEL AP CALCULUS BC EXAM

Section I Part A

1. Ⓐ Ⓑ Ⓒ Ⓓ Ⓔ   7. Ⓐ Ⓑ Ⓒ Ⓓ Ⓔ   13. Ⓐ Ⓑ Ⓒ Ⓓ Ⓔ   19. Ⓐ Ⓑ Ⓒ Ⓓ Ⓔ   25. Ⓐ Ⓑ Ⓒ Ⓓ Ⓔ

2. Ⓐ Ⓑ Ⓒ Ⓓ Ⓔ   8. Ⓐ Ⓑ Ⓒ Ⓓ Ⓔ   14. Ⓐ Ⓑ Ⓒ Ⓓ Ⓔ   20. Ⓐ Ⓑ Ⓒ Ⓓ Ⓔ   26. Ⓐ Ⓑ Ⓒ Ⓓ Ⓔ

3. Ⓐ Ⓑ Ⓒ Ⓓ Ⓔ   9. Ⓐ Ⓑ Ⓒ Ⓓ Ⓔ   15. Ⓐ Ⓑ Ⓒ Ⓓ Ⓔ   21. Ⓐ Ⓑ Ⓒ Ⓓ Ⓔ   27. Ⓐ Ⓑ Ⓒ Ⓓ Ⓔ

4. Ⓐ Ⓑ Ⓒ Ⓓ Ⓔ   10. Ⓐ Ⓑ Ⓒ Ⓓ Ⓔ   16. Ⓐ Ⓑ Ⓒ Ⓓ Ⓔ   22. Ⓐ Ⓑ Ⓒ Ⓓ Ⓔ   28. Ⓐ Ⓑ Ⓒ Ⓓ Ⓔ

5. Ⓐ Ⓑ Ⓒ Ⓓ Ⓔ   11. Ⓐ Ⓑ Ⓒ Ⓓ Ⓔ   17. Ⓐ Ⓑ Ⓒ Ⓓ Ⓔ   23. Ⓐ Ⓑ Ⓒ Ⓓ Ⓔ

6. Ⓐ Ⓑ Ⓒ Ⓓ Ⓔ   12. Ⓐ Ⓑ Ⓒ Ⓓ Ⓔ   18. Ⓐ Ⓑ Ⓒ Ⓓ Ⓔ   24. Ⓐ Ⓑ Ⓒ Ⓓ Ⓔ

Section I Part B

29. Ⓐ Ⓑ Ⓒ Ⓓ Ⓔ   32. Ⓐ Ⓑ Ⓒ Ⓓ Ⓔ   35. Ⓐ Ⓑ Ⓒ Ⓓ Ⓔ   38. Ⓐ Ⓑ Ⓒ Ⓓ Ⓔ   42. Ⓐ Ⓑ Ⓒ Ⓓ Ⓔ

30. Ⓐ Ⓑ Ⓒ Ⓓ Ⓔ   33. Ⓐ Ⓑ Ⓒ Ⓓ Ⓔ   36. Ⓐ Ⓑ Ⓒ Ⓓ Ⓔ   39. Ⓐ Ⓑ Ⓒ Ⓓ Ⓔ   43. Ⓐ Ⓑ Ⓒ Ⓓ Ⓔ

31. Ⓐ Ⓑ Ⓒ Ⓓ Ⓔ   34. Ⓐ Ⓑ Ⓒ Ⓓ Ⓔ   37. Ⓐ Ⓑ Ⓒ Ⓓ Ⓔ   40. Ⓐ Ⓑ Ⓒ Ⓓ Ⓔ   44. Ⓐ Ⓑ Ⓒ Ⓓ Ⓔ

41. Ⓐ Ⓑ Ⓒ Ⓓ Ⓔ   45. Ⓐ Ⓑ Ⓒ Ⓓ Ⓔ

First Model AP Calculus BC Exam

SECTION 1, PART A

Time – 55 Minutes

Number of Questions – 28

A CALCULATOR MAY NOT BE USED ON THIS PART OF THE EXAMINATION.

> **Directions:** Solve each of the following problems, using the available space for scratchwork. After examining the form of the choices, decide which is the best of the choices given and fill in the corresponding oval on the answer sheet. No credit will be given for anything written in the test book. Do not spend too much time on any one problem.
>
> **In this test:** Unless otherwise specified, the domain of a function f is assumed to be the set of all real numbers x for which $f(x)$ is a real number.

1. The function f is given by $f(x) = 3x^4 - 2x^3 + 7x - 2$. On which of the following intervals is f' decreasing?

   (A) $(-\infty, \infty)$

   (B) $(-\infty, 0)$

   (C) $\left(\dfrac{1}{3}, \infty\right)$

   (D) $\left(0, \dfrac{1}{3}\right)$

   (E) $\left(-\dfrac{1}{3}, 0\right)$

2. What is the area under the curve described by the parametric equations $x = \sin t$ and $y = \cos^2 t$ for $0 \le t \le \frac{\pi}{2}$?

(A) $\dfrac{1}{3}$

(B) $\dfrac{1}{2}$

(C) $\dfrac{2}{3}$

(D) 1

(E) $\dfrac{4}{3}$

3. The function f is given by $f(x) = 8x^3 + 36x^2 + 54x + 27$. All of these statements are true EXCEPT

(A) $-\dfrac{3}{2}$ is a zero of f.

(B) $-\dfrac{3}{2}$ is a point of inflection of f.

(C) $-\dfrac{3}{2}$ is a local extremum of f.

(D) $-\dfrac{3}{2}$ is a zero of the derivative of f.

(E) f is strictly monotonic.

4. $\displaystyle\int x \ln x \, dx =$

(A) $\dfrac{x^2 \ln x}{2} + \dfrac{x^2}{4} + C$

(B) $\dfrac{x^2}{4}(2 \ln x - 1) + C$

(C) $\dfrac{x}{2}(x \ln x - 2 + C)$

(D) $x \ln x - \dfrac{x^2}{4} + C$

(E) $\dfrac{(\ln x)}{x} - \dfrac{x^2}{4} + C$

5. Let $h(x) = \ln|g(x)|$. If g is decreasing for all x in its domain, then

    (A) h is strictly increasing.

    (B) h is strictly decreasing.

    (C) h has no relative extrema.

    (D) both (B) and (C).

    (E) none of the above.

Questions 6, 7, and 8 refer to the diagram and information below.

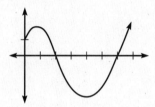

The function f is defined on [0,7]. The graph of its derivative, f', is shown above.

6. The point (2,5) is on the graph of $y = f(x)$. An equation of the line tangent to the graph of f at (2,5) is

    (A) $y = 2$

    (B) $y = 5$

    (C) $y = 0$

    (D) $y = 2x + 5$

    (E) $y = 2x - 5$

7. How many points of inflection does the graph $y = f(x)$ have over [0,7]?

    (A) 0

    (B) 1

    (C) 2

    (D) 3

    (E) 4

8. At what value of x does the absolute maximum value of f occur?

    (A) 1

    (B) 2

    (C) 4

    (D) 6

    (E) 7

GO ON TO THE NEXT PAGE

9. $\int_1^e \left(\frac{x^2 + 4}{x} \right) dx =$

    (A) $\dfrac{e^2 + 9}{2}$

    (B) $\dfrac{e^2 - 9}{2}$

    (C) $\dfrac{e^2 + 7}{2}$

    (D) $\dfrac{e^2 + 8}{2}$

    (E) $\dfrac{e^2 - 4}{2}$

10. The function f given by $f(x) = 3x^5 - 4x^3 - 3x$ is increasing and concave up over which of these intervals?

    (A) $\left(-\infty, -\sqrt{\dfrac{2}{5}} \right)$

    (B) $\left(-\sqrt{\dfrac{2}{5}}, 0 \right)$

    (C) $(-1, 1)$

    (D) $\left(\sqrt{\dfrac{2}{5}}, \infty \right)$

    (E) $(1, \infty)$

11. If $y = 2xy - x^2 + 3$, then when $x = 1$, $\dfrac{dy}{dx} =$

    (A) -6

    (B) -2

    (C) $-\dfrac{2}{3}$

    (D) 2

    (E) 6

12. The length of the curve described by the parametric equations $x = 2t^3$ and $y = t^3$ where $0 \le t \le 1$ is

    (A) $\dfrac{5}{7}$

    (B) $\dfrac{\sqrt{5}}{2}$

    (C) $\dfrac{3}{2}$

    (D) $\sqrt{5}$

    (E) 3

13. What is the average value of $f(x) = 3\sin^2 x - \cos^2 x$ over $\left[0, \frac{\pi}{2}\right]$?

    (A) 0

    (B) 1

    (C) $\sqrt{2}$

    (D) $\sqrt{3}$

    (E) $\frac{\pi}{2}$

14. Let f be defined as

$$f(x) = \begin{cases} \sqrt[3]{x} + kx, & x < 1 \\ \ln x, & x \ge 1 \end{cases}$$

for some constant k. For what value of k will f be differentiable over its whole domain?

    (A) -2

    (B) -1

    (C) $\dfrac{2}{3}$

    (D) 1

    (E) None of the above

GO ON TO THE NEXT PAGE

15. What is the approximation of the value of e^3 obtained by using a fourth-degree Taylor polynomial about $x = 0$ for e^x?

    (A) $1+3+\dfrac{9}{2}+\dfrac{9}{2}+\dfrac{27}{8}$

    (B) $1+3+9+\dfrac{27}{8}$

    (C) $1+3+\dfrac{27}{8}$

    (D) $3-\dfrac{9}{2}+\dfrac{9}{2}-\dfrac{27}{4}$

    (E) $3+9+\dfrac{27}{8}$

16. $\displaystyle\int 6x^3 e^{3x}\,dx =$

    (A) $e^{3x}(9x^3 - 9x^2 + 6x - 2) + C$

    (B) $e^{3x}\left(2x^3 - 2x^2 - \dfrac{4}{3}x + \dfrac{4}{9}\right) + C$

    (C) $\dfrac{2}{9}e^{3x}\left(2x^3 - 2x^2 + \dfrac{4}{3}x - \dfrac{4}{9}\right) + C$

    (D) $\dfrac{2}{9}e^{3x}\left(9x^3 - 9x^2 - 6x - 2\right) + C$

    (E) $\dfrac{2}{9}e^{3x}\left(9x^3 - 9x^2 + 6x - 2\right) + C$

17. If $f(x) = \sec x$, then $f'(x)$ has how many zeros over the closed interval $[0, 2\pi]$?

    (A) 0
    (B) 1
    (C) 2
    (D) 3
    (E) 4

18. Consider the region in the first quadrant bounded by $y = x^2$ over $[0,3]$. Let L_3 represent the Riemann approximation of the area of this region using left endpoints and three rectangles, R_3 represent the Riemann approximation using right endpoints and three rectangles, M_3 represent the Riemann approximation using midpoints and three rectangles, and T_3 represent the Trapezoidal approximation with three trapezoids. Which of these statements is true?

(A) $R_3 < T_3 < \int_0^3 x^2\,dx < M_3 < L_3$

(B) $L_3 < M_3 < T_3 < R_3 < \int_0^3 x^2\,dx$

(C) $M_3 < L_3 < \int_0^3 x^2\,dx < T_3 < R_3$

(D) $L_3 < M_3 < \int_0^3 x^2\,dx < R_3 < T_3$

(E) $L_3 < M_3 < \int_0^3 x^2\,dx < T_3 < R_3$

19. Which of the following series converge?

I. $\sum_{n=1}^{\infty}\left(\dfrac{2^n}{n+1}\right)$

II. $\sum_{n=1}^{\infty}\dfrac{3}{n}$

III. $\sum_{n=1}^{\infty}\left(\dfrac{\cos 2n\pi}{n^2}\right)$

(A) I only
(B) II only
(C) III only
(D) I and II only
(E) I and III only

20. The area of the region inside the polar curve $r = 4\sin\theta$ but outside the polar curve $r = 2\sqrt{2}$ is given by

(A) $2\int_{\pi/4}^{3\pi/4}\left(4\sin^2\theta - 1\right)d\theta$

(B) $\dfrac{1}{2}\int_{\pi/4}^{3\pi/4}\left(4\sin\theta - 2\sqrt{2}\right)^2 d\theta$

(C) $\dfrac{1}{2}\int_{\pi/4}^{3\pi/4}\left(4\sin\theta - 2\sqrt{2}\right)d\theta$

(D) $\dfrac{1}{2}\int_{\pi/4}^{3\pi/4}\left(16\sin^2\theta - 8\right)d\theta$

(E) $\dfrac{1}{2}\int_{\pi/4}^{3\pi/4}\left(4\sin^2\theta - 1\right)d\theta$

GO ON TO THE NEXT PAGE

21. When $x = 16$, the rate at which $x^{3/4}$ is increasing is k times the rate at which $\sqrt{x}$ is increasing. What is the value of k?

   (A) $\dfrac{1}{8}$

   (B) $\dfrac{3}{8}$

   (C) 2

   (D) 3

   (E) 8

22. The length of the path described by the parametric equations $x = 2\cos 2t$ and $y = \sin^2 t$ for $0 \le t \le \pi$ is given by

   (A) $\displaystyle\int_0^\pi \sqrt{4\cos^2 2t + \sin^4 t}\; dt$

   (B) $\displaystyle\int_0^\pi \sqrt{2\sin t \cos t - 4\sin 2t}\; dt$

   (C) $\displaystyle\int_0^\pi \sqrt{4\sin^2 t \cos^2 t - 16\sin^2 2t}\; dt$

   (D) $\displaystyle\int_0^\pi \sqrt{4\sin^2 2t + 4\sin^2 t \cos^2 t}\; dt$

   (E) $\displaystyle\int_0^\pi \sqrt{16\sin^2 2t + 4\sin^2 t \cos^2 t}\; dt$

23. Determine the interval of convergence for the series $\displaystyle\sum_{n=0}^{\infty} \left(\frac{(3x-2)^{n+2}}{n^{5/2}} \right)$.

   (A) $-\dfrac{1}{3} \le x \le \dfrac{1}{3}$

   (B) $-\dfrac{1}{3} < x < 1$

   (C) $-\dfrac{1}{3} \le x \le 1$

   (D) $\dfrac{1}{3} \le x \le 1$

   (E) $-\dfrac{1}{3} \le x \le -1$

24. $f(x) = \dfrac{(3x+4)(2x-1)}{(2x-3)(2x+1)}$ has a horizontal asymptote at $x =$

    (A) $\dfrac{3}{2}$

    (B) $\dfrac{3}{2}$ and $-\dfrac{1}{2}$

    (C) 0

    (D) $-\dfrac{3}{4}$ and $\dfrac{1}{2}$

    (E) None of the above

25.

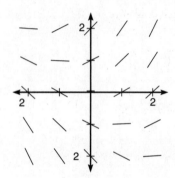

Shown above is the slopefield for which of the following differential equations?

    (A) $\dfrac{dy}{dx} = 1 + x$

    (B) $\dfrac{dy}{dx} = x - y$

    (C) $\dfrac{dy}{dx} = \dfrac{x+y}{2}$

    (D) $\dfrac{dy}{dx} = y - x$

    (E) $\dfrac{dy}{dx} = y + 1$

26. $\displaystyle\int_{2}^{\infty} \dfrac{x^2}{e^x}\,dx =$

    (A) $\dfrac{5}{e}$

    (B) $10e^2$

    (C) $\dfrac{10}{e^2}$

    (D) 2

    (E) $5e$

GO ON TO THE NEXT PAGE

27. The population $P(t)$ of a species satisfies the logistic differential equation $\frac{dP}{dt} = \frac{2}{3}P\left(5 - \frac{P}{100}\right)$. What is $\lim\limits_{t \to \infty} P(t)$?

   (A) 100

   (B) 200

   (C) 300

   (D) 400

   (E) 500

28. If $\sum\limits_{n=0}^{\infty} a_n(x - c)^n$ is a Taylor series that converges to $f(x)$ for every real x, then $f''(c) =$

   (A) 0

   (B) $n(n-1)a_n$

   (C) $\sum\limits_{n=0}^{\infty} na_n(x - c)^{n-1}$

   (D) $\sum\limits_{n=0}^{\infty} a_n$

   (E) $\sum\limits_{n=0}^{\infty} n(n-1)a_n(x - c)^{n-2}$

END OF PART A

STOP

END OF SECTION 1 PART A. IF YOU HAVE ANY TIME LEFT, GO OVER YOUR WORK IN THIS SECTION ONLY. DO NOT WORK IN ANY OTHER SECTION OF THE TEST.

PART B

Time – 50 Minutes

Number of Questions – 17

A GRAPHING CALCULATOR IS REQUIRED FOR SOME QUESTIONS ON THIS PART OF THE EXAMINATION.

Directions: Solve each of the following problems, using the available space for scratchwork. After examining the form of the choices, decide which is the best of the choices given and fill in the corresponding oval on the answer sheet. No credit will be given for anything written in the test book. Do not spend too much time on any one problem.

In this test: (1) The exact numerical value of the correct answer does not always appear among the choices given. When this happens, select from among the choices the number that best approximates the exact numerical value. (2) Unless otherwise specified, the domain of a function f is assumed to be the set of all real numbers x for which $f(x)$ is a real number.

29. The graph of the function represented by the Taylor series, centered at $x = 1$, $1 - (x - 1) + (x - 1)^2 - (x - 1)^3 + \ldots = (-1)^n(x - 1)^n$ intersects the graph of $y = e^x$ at $x =$

    (A) – 9.425

    (B) 0.567

    (C) 0.703

    (D) 0.773

    (E) 1.763

30. If f is a vector-valued function defined by $f(t) = <\cos^2 t, \ln t>$, then $f''(t) =$

    (A) $\left\langle -2\cos t \sin t, \dfrac{1}{t} \right\rangle$

    (B) $\left\langle 2\cos t, \dfrac{1}{t} \right\rangle$

    (C) $\left\langle 2\cos t \ \sin t, \dfrac{1}{t} \right\rangle$

    (D) $\left\langle -2\cos^2 t + 2\sin^2 t, -\dfrac{1}{t^2} \right\rangle$

    (E) $\left\langle -2, -\dfrac{1}{t^2} \right\rangle$

GO ON TO THE NEXT PAGE

31. The diagonal of a square is increasing at a constant rate of $\sqrt{2}$ centimeters per second. In terms of the perimeter, P, what is the rate of change of the area of the square in square centimeters per second?

   (A) $\dfrac{\sqrt{2}}{4}P$

   (B) $\dfrac{4}{\sqrt{2}}P$

   (C) $2P$

   (D) P

   (E) $\dfrac{P}{2}$

32. If f is continuous over the set of real numbers and f is defined as $f(x) = \dfrac{x^2 - 3x + 2}{x - 2}$ for all $x \neq 2$, then $f(2) =$

   (A) -2

   (B) -1

   (C) 0

   (D) 1

   (E) 2

33. If $0 \leq k \leq 2$ and the area between the curves $y = x^2 + 4$ and $y = x^3$ from $x = 0$ to $x = k$ is 5, then $k =$

   (A) 1.239

   (B) 1.142

   (C) 1.029

   (D) 0.941

   (E) 0.876

34. Determine $\dfrac{dy}{dx}$ for the curve defined by $x\sin y = 1$.

   (A) $-\dfrac{\tan y}{x}$

   (B) $\dfrac{\tan y}{x}$

   (C) $\dfrac{\sec y - \tan y}{x}$

   (D) $\dfrac{\sec y}{x}$

   (E) $-\dfrac{\sec y}{x}$

35. If $f(x) = h(x) + g(x)$ for $0 \le x \le 10$, then $\int_0^{10} (f(x) - 2h(x) + 3)dx =$

   (A) $2\int_0^{10} (g(x) - h(x) + 3)dx =$

   (B) $g(10) - h(10) + 30$

   (C) $g(10) - h(10) + 30 - g(0) - h(0)$

   (D) $\int_0^{10} (g(x) - h(x))dx + 30$

   (E) $\int_0^{10} (g(x) - 2h(x))dx + 30$

36. Use a fifth-degree Taylor polynomial centered at $x = 0$ to estimate e^2.

   (A) 7.000

   (B) 7.267

   (C) 7.356

   (D) 7.389

   (E) 7.667

37. What are all the values of x for which the series $\sum_{n=1}^{\infty} \left(\frac{(x+2)^n}{(n\sqrt{n}3^n)} \right)$ converges?

   (A) $-3 < x < 3$

   (B) $-3 \le x \le 3$

   (C) $-5 < x < 1$

   (D) $-5 \le x \le 1$

   (E) $-5 \le x < 1$

38. Let $f(x) = |x^2 - 4|$. Let R be the region bounded by f, the x-axis, and the vertical lines $x = -3$ and $x = 3$. Let T_6 represent the approximation of the area of R using the Trapezoidal Rule with $n = 6$. The quotient $\dfrac{T_6}{\int_{-3}^{3} f(x)dx} =$

   (A) 0.334

   (B) 0.978

   (C) 1.022

   (D) 1.304

   (E) 4.666

39. Let R be the region bounded by $y = 3 - x^2$, $y = x^3 + 1$, and $x = 0$. If R is rotated about the x-axis, the volume of the solid formed could be determined by

GO ON TO THE NEXT PAGE

(A) $\pi \int_0^1 \left(\left(x^3 + 1 \right)^2 - \left(3 - x^2 \right)^2 \right) dx$

(B) $-\pi \int_1^0 \left(\left(x^3 + 1 \right)^2 - \left(3 - x^2 \right)^2 \right) dx$

(C) $2\pi \int_0^1 \left(x \left(-x^3 - x^2 + 2 \right) \right) dx$

(D) $\pi \int_1^0 \left(\left(x^3 + 1 \right)^2 - \left(3 - x^2 \right)^2 \right) dx$

(E) $2\pi \int_0^1 \left(x \left(x^3 + x^2 - 2 \right) \right) dx$

40. Let f be defined as

$$f(x) = \begin{cases} -x^2, & x \le 0 \\ \sqrt{x}, & x > 0 \end{cases}$$

and g be defined as $g(x) = \int_{-4}^x f(t)dt$ for $-4 \le t \le 4$. Which of these is an equation for the tangent line to g at $x = 2$?

(A) $4x + 3y = 4\sqrt{2} + 72$

(B) $3x\sqrt{2} - 3y = -64 - 2\sqrt{2}$

(C) $3x\sqrt{2} - 3y = 64 - 2\sqrt{2}$

(D) $3x\sqrt{2} - 3y = 64 + 2\sqrt{2}$

(E) $4x + 3y = 4\sqrt{2} - 56$

41.

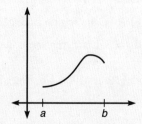

Let $g(x) = \int_a^x f(t)dt\, t$, where $a \le x \le b$. The figure above shows the graph of g on $[a,b]$. Which of the following could be the graph of f on $[a,b]$?

(A)

(B)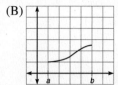

(C)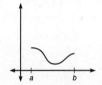

(D)

(E)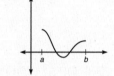

42. The sum of the infinite geometric series $\frac{4}{5} + \frac{8}{35} + \frac{16}{245} + \frac{32}{1715} + \ldots$ is

   (A) 0.622

   (B) 0.893

   (C) 1.120

   (D) 1.429

   (E) 2.800

43. Let f be a strictly monotonic differentiable function on the closed interval [5,10] such that $f(5) = 6$ and $f(10) = 26$. Which of the following must be true for the function f on the interval [5,10]?

   I. The average rate of change of f is 4.

   II. The absolute maximum value of f is 26.

   III. $f'(8) > 0$.

   (A) I only

   (B) II only

   (C) III only

   (D) I and II only

   (E) I, II, and III

44. Let $F(x)$ be an antiderivative of $f(x) = e^{2x}$. If $F(0) = 2.5$, then $F(5) =$

   (A) 150.413

   (B) 11013.233

   (C) 11015.233

   (D) 22026.466

   (E) 22028.466

GO ON TO THE NEXT PAGE ▶

45. The base of a solid is the region in the first quadrant bounded by $y = -x^2 + 3$. The cross sections perpendicular to the x–axis are squares. Find the volume of the solid.

    (A) 3.464

    (B) 8.314

    (C) 8.321

    (D) 16.628

    (E) 21.600

END OF SECTION I

STOP

END OF SECTION 1. IF YOU HAVE ANY TIME LEFT, GO OVER YOUR WORK IN THIS SECTION ONLY. DO NOT WORK IN ANY OTHER SECTION OF THE TEST.

SECTION II, PART A

Time – 45 Minutes

Number of Questions – 3

A GRAPHING CALCULATOR IS REQUIRED FOR SOME PROB-
LEMS OR PARTS OF PROBLEMS ON THIS PART OF THE
EXAMINATION.

SHOW ALL YOUR WORK. It is important to show your setups for
these problems because partial credit will be awarded. If you use
decimal approximations, they should be accurate to three decimal
places.

1. Let f be a function that has derivatives of all orders for all real numbers. Assume
$f(1) = 3, f'(1) = -1, f''(1) = 4$, and $f'''(1) = -2$.

    A. Write the third-degree Taylor polynomial for f about $x = 1$, and use it
    to approximate $f(1.1)$.

    B. Write the second-degree Taylor polynomial for f' about $x = 1$, and use
    it to approximate $f'(1.1)$.

    C. Write the fourth-degree Taylor polynomial for $g(x) = \int_1^x f(t)dt$.

    D. Can $f(2)$ be determined from the information given? Justify your
    answer.

2. Consider the differential equation $\dfrac{dy}{dx} = \dfrac{3x^2 + 2x}{e^y}$.

    A. Find a solution $y = f(x)$ to the differential equation that satisfies
    $f(0) = 2$.

    B. What is the domain of f?

    C. For what value(s) of x does f have a point of inflection?

GO ON TO THE NEXT PAGE

3. Let R be the region enclosed by the graphs of $y = -x^2 + 3$ and $y = \tan^{-1}x$.

A. Determine the area of R.

B. Write an expression involving one or more integrals that gives the length of the boundary of R. Do not evaluate.

C. The base of a solid is the region R. The cross sections perpendicular to the x–axis are semicircles. Write an expression involving one or more integrals that gives the volume of the solid. Do not evaluate.

END OF PART A

STOP

END OF SECTION II PART A. IF YOU HAVE ANY TIME LEFT, GO OVER YOUR WORK IN THIS SECTION ONLY. DO NOT WORK IN ANY OTHER SECTION OF THE TEST.

PART B

Time – 45 Minutes

Number of Questions – 3

A CALCULATOR IS NOT PERMITTED ON THIS PART OF THE EXAMINATION.

4.

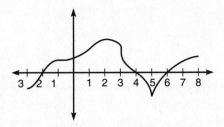

The figure above shows the graph of f', the derivative of some function f, for $-3 \le x \le 8$. The graph of f' has horizontal tangent lines at $x = -1$ and $x = 2$, a vertical tangent line at $x = 3$, and a cusp at $x = 5$.

    A. Find all values of x for which f attains a relative minimum on $(-3,8)$. Explain.

    B. Find all values of x for which f attains a relative maximum on $(-3,8)$. Explain.

    C. For what value of x, $-3 \le x \le 8$, does f attain its absolute minimum? Explain.

    D. For what value(s) of x, for $-3 < x < 8$, does $f''(x)$ not exist?

GO ON TO THE NEXT PAGE

5. Consider the differential equation $\frac{dy}{dx} = x(y-2)$.

    A. On the axes provided below, sketch a slopefield for the given differential equation at the nine points indicated.

    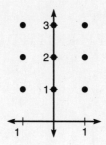

    B. Let $y = f(x)$ be a particular solution to the given differential equation with the initial condition $f(0) = 3$. Use Euler's Method starting at $x = 0$ with a step size of 0.2 to approximate $f(0.4)$. Show the work that leads to your answer.

    C. Find the particular solution $y = f(x)$ to the differential equation with the initial condition $f(0) = 3$.

6. A moving particle has position $(x(t), y(t))$ at time t. The position of the particle at time $t = 1$ is $(7,0)$, and the velocity vector at any time $t > 0$ is given by $\left\langle 3 - \frac{3}{t^2}, 4 + \frac{2}{t^2} \right\rangle$.

    A. Find the position of the particle at $t = 3$.

    B. Will the line tangent to the path of the particle at $(x(t), y(t))$ ever have a slope of zero? If so, when? If not, why not?

    ## END OF SECTION II

    ## STOP

    END OF SECTION II. IF YOU HAVE ANY TIME LEFT, GO OVER YOUR WORK IN THIS SECTION ONLY. DO NOT WORK IN ANY OTHER SECTION OF THE TEST.

Answer Key

PART A

1. D	7. C	13. B	19. C	25. C
2. C	8. B	14. E	20. D	26. C
3. C	9. C	15. A	21. D	27. E
4. B	10. E	16. E	22. E	28. A
5. C	11. E	17. D	23. D	
6. B	12. D	18. E	24. A	

PART B

29. B	33. A	37. D	41. A	45. B
30. D	34. A	38. B	42. C	
31. E	35. D	39. D	43. E	
32. D	36. B	40. D	44. C	

SOLUTIONS FOR SECTION I

PART A

1. **The correct answer is (D).** To determine where the derivative of a function is increasing, we should find the zeros of the second derivative and examine a wiggle graph.

$$f(x) = 3x^4 - 2x^3 + 7x - 2$$

$$f'(x) = 12x^3 - 6x^2 + 7$$

$$f''(x) = 36x^2 - 12x = 0$$

The second derivative is equal to zero when $x = 0$ and when $x = \frac{1}{3}$. By examining the wiggle graph below, we can determine the interval on which the derivative is decreasing.

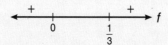

The second derivative is negative over $\left(0, \frac{1}{3}\right)$.

2. **The correct answer is (C).** We can easily convert these parametric equations into the following Cartesian equation: $y = 1 - x^2$. So, the area under the curve would be given by

$$A = \int_0^1 \left(1 - x^2\right) dx$$

$$= \frac{2}{3}.$$

3. **The correct answer is (C).** Although $-\frac{3}{2}$ is a zero of the derivative, the derivative does not change signs there.

4. **The correct answer is (B).** Use integration by parts. Letting $u = \ln x$ and $dv = x\, dx$ yields

$$\int x \ln x\, dx = \frac{x^2}{2} \ln x - \int \frac{x^2}{(2x)} dx$$

$$= \frac{x^2}{2} \ln x - \frac{x^2}{4} + C$$

$$= \frac{x^2}{4}\left(2 \ln x - 1\right) + C.$$

5. **The correct answer is (C).** Let's begin by examining $h'(x)$:

$$h'(x) = \frac{g'(x)}{g(x)}.$$

In order for h to have any relative extrema, its derivative, h', would have to equal zero at some point. Since g is always decreasing, g' is never zero and since g' is the numerator of h', h' is never zero. Therefore, h has no relative extrema.

6. **The correct answer is (B).** By reading the graph, we learn that $f'(2) = 0$. Using point-slope form, we get an equation of the tangent at (2,5) to be

$$y - 5 = 0(x - 2),$$

which becomes

$$y = 5.$$

7. **The correct answer is (C).** Points of inflection correspond with horizontal tangents of the derivative. Since there are two such tangents, there are two points of inflection.

8. **The correct answer is (B).** The maximum accumulated area under the graph of f' occurs at $x = 2$.

9. **The correct answer is (C).** Since the degree of the numerator is greater than the degree of the denominator, we should first divide and integrate the quotient.

$$\int_1^e \frac{x^2 + 4\,dx}{x} = \int_1^e \left(x + \frac{4}{x}\right)dx$$

$$= \frac{e^2}{2} + 4 - \frac{1}{2} = \frac{e^2 + 7}{2}$$

10. **The correct answer is (E).** This question is asking for an interval where both the first and second derivatives are positive.

$$f(x) = 3x^5 - 4x^3 - 3x$$

$$f'(x) = 15x^4 - 12x^2 - 3 = 0$$

$$(15x^2 + 3)(x^2 - 1) = 0$$

$$x = \pm 1$$

$$f''(x) = 60x^3 - 24x = 12x(5x^2 - 2) = 0$$

$$x = \pm\sqrt{\frac{2}{5}}, 0$$

By examining both of the wiggle graphs above, we can see that the curve increases and is concave up from 1 to infinity.

11. **The correct answer is (E).** First, let's determine the value of y when $x = 1$.

$$y = 2xy - x^2 + 3$$

$$\text{let } x = 1$$

$$y = 2y - 1 + 3$$

$$y = -2$$

Now, we differentiate the equation with respect to x:

$$\frac{dy}{dx} = 2x\frac{dy}{dx} + 2y - 2x$$

$$\frac{dy}{dx} = \frac{2y - 2x}{1 - 2x}$$

$$\text{let } x = 1 \text{ and } y = -2$$

$$\frac{dy}{dx} = 6.$$

12. **The correct answer is (D).** The length of a curve defined parametrically is given by

$$l = \int_a^b \sqrt{\left(\frac{dx}{dt}\right)^2 + \left(\frac{dy}{dt}\right)^2}\, dt$$

Applying the formula above gives us

$$l = \int_0^1 \sqrt{36t^4 + 9t^4}\, dt$$

$$= 3\sqrt{5}\int_0^1 t^2\, dt$$

$$= \sqrt{5}.$$

13. **The correct answer is (B).** Since we are asked for the average value, we use the MVT for integrals.

$$f(c) = \frac{2}{\pi}\int_0^{\pi/2}\left(3\sin^2 x - \cos^2 x\right)dx$$

We need to use power reducing formulas.

$$3\sin^2 x - \cos^2 x$$

$$3\left(\frac{1 - \cos 2x}{2}\right) - \left(\frac{1 + \cos 2x}{2}\right)$$

$$\frac{3 - 3\cos 2x - 1 - \cos 2x}{2}$$

$$\frac{2 - 4\cos 2x}{2}$$

$$1 - 2\cos 2x$$

Now, integrate to get the answer.

$$\frac{2}{\pi}\int_0^{\pi/2}(1-2\cos 2x)dx$$

$$\frac{2}{\pi}\cdot\frac{\pi}{2}=1$$

14. **The correct answer is (E).** For what value of k will the left- and right-hand derivatives be equal? If $k=\frac{2}{3}$, then the derivatives will be the same; however, the function is then discontinuous because the left- and right-hand limits are different.

15. **The correct answer is (A).** This is a Taylor or Maclaurin series that you should commit to memory.

$$e^x = 1 + x + \frac{x^2}{2!} + \frac{x^3}{3!} + \frac{x^4}{4!}$$

Substituting 3 for x yields

$$e^x = 1 + 3 + \frac{9}{2} + \frac{27}{6} + \frac{81}{24}$$

$$= 1 + 3 + \frac{9}{2} + \frac{9}{2} + \frac{27}{8}.$$

16. **The correct answer is (E).** This is a very involved integration-by-parts problem. Use a chart

u	dv	+/−1
$6x^3$	e^{3x}	+1
$18x^2$	$\frac{e^{3x}}{3}$	−1
$36x$	$\frac{e^{3x}}{9}$	+1
36	$\frac{e^{3x}}{27}$	−1
0	$\frac{e^{3x}}{81}$	+1
		−1

$$= 2x^3 e^{3x} - 2x^2 e^{3x} + \frac{4xe^{3x}}{3} - \frac{4}{9}e^{3x} + C$$

$$= \frac{2}{9}e^{3x}\left(9x^3 - 9x^2 + 6x - 2\right) + C$$

17. **The correct answer is (D).** Since $f(x) = \sec x$, $f'(x) = \sec x \tan x$. $\sec x$ is never zero, and $\tan x = 0$ when $x = 0$, $x = \pi$, or when $x = 2\pi$. So, the answer is 3.

18. **The correct answer is (E).** To determine R_3, L_3, and M_3, we need to be able to sum the areas of the rectangles. $R_3 = 14$, $L_3 = 5$, and $M_3 = \frac{35}{4}$. To determine T_3, we need to find the area of a triangle and two trapezoids. $T_3 = \frac{19}{2}$. Using the Fundamental Theorem, $\int_0^3 x^2 \, dx = 9$.

19. **The correct answer is (C).** Applying the ratio test to the first series,

$$\lim_{n \to \infty} \frac{2^{n+1}}{n+2} \cdot \frac{n+1}{2^n}$$

$$= \lim_{n \to \infty} 2 \frac{(n+1)}{(n+2)} = 2$$

$$= 2 > 1, \text{ so I is divergent.}$$

Applying the comparison test to the second series and comparing it to the harmonic series helps us conclude that II is divergent as well.

The third series is really just $\sum_{n=1}^{\infty} \frac{1}{n^2}$, which is a p – series with $p > 1$, so it is convergent.

20. **The correct answer is (D).** The two curves intersect at $\theta = \frac{\pi}{4}$ and $\theta = \frac{3\pi}{4}$. So, the area would be given by

$$A = \frac{1}{2} \int_{\pi/4}^{3\pi/4} \left(16 \sin^2 \theta - 8 \right) d\theta.$$

21. **The correct answer is (D).** We must set the two derivatives equal to each other and solve for k.

$$\frac{3}{4} x^{-1/4} = \frac{k}{\left(2\sqrt{x} \right)}$$

$$k = 3.$$

22. **The correct answer is (E).** We apply the following formula:

$$l = \int_a^b \sqrt{\left(\frac{dx}{dt} \right)^2 + \left(\frac{dy}{dt} \right)^2} \, dt$$

$x' = -4\sin 2t$ and $y' = 2\sin t \cos t$. So,

$$l = \int_0^\pi \sqrt{16 \sin^2 2t + 4 \sin^2 t \cos^2 t} \, dt.$$

23. **The correct answer is (D).** First, we'll take the limit of the ratio test:

$$\lim_{n \to \infty} \left| \frac{\left((3x-2)^{n+3} \right)}{\left((n+1)^{\frac{5}{2}} \right)} \cdot \frac{\left(n^{\frac{5}{2}} \right)}{\left((3x-2)^{n+2} \right)} \right| = |3x - 2|$$

$$|3x-2|<1$$
$$\frac{1}{3}<x<1.$$

In order to test the endpoints, we substitute each endpoint into the original series and test for convergence. By letting $x=\frac{1}{3}$, we get

$$\sum_{n=0}^{\infty}\frac{(-1)^{n+2}}{n^{5/2}}, \text{ which converges.}$$

If we let $x=1$, we get

$$\sum_{n=0}^{\infty}\frac{1}{n^{5/2}}, \text{ which converges as well.}$$

So, the interval of convergence is $\frac{1}{3}\le x\le 1$.

24. **The correct answer is (A).** Horizontal asymptotes are determined by finding the limit at infinity. If we multiply the binomials we can see that the ratio of the leading coefficients is $\frac{3}{2}$.

25. **The correct answer is (C).** Notice that all of the slopes on the line $y=-x$ are zero (horizontal). Any point on this line would make $\frac{x+y}{2}$ be zero, since x and y are opposites.

26. **The correct answer is (C).** This is an improper integral and a tricky integration-by-parts problem. First, we'll deal with the improper integral by taking the limit of a definite integral:

$$\int_2^{\infty}\frac{x^2}{e^x}dx=\lim_{p\to\infty}\int_2^p\frac{x^2}{e^x}dx.$$

We now have to use integration by parts on $\int\frac{x^2}{e^x}dx$. We'll choose $u=x^2$ and $dv=e^{-x}dx$ and get

$$\int\frac{x^2}{e^x}dx=\frac{-x^2}{e^x}+\int 2xe^{-x}dx.$$

Now, we'll let $u=2x$ and $dv=e^{-x}dx$ and get

$$=\frac{-x^2}{e^x}-\frac{2x}{e^x}+2\int e^{-x}dx$$
$$=\frac{-x^2}{e^x}-\frac{2x}{e^x}-\frac{2}{e^x}.$$

Now, we have to evaluate the integral using the limits of integration and take the limit as p goes to infinity, so

$$\lim_{p \to \infty} \int_2^p \frac{x^2}{e^x}dx = \lim_{p \to \infty}\left(\frac{-x^2}{e^x} - \frac{2x}{e^x} - \frac{2}{e^x} \right)\Big|_2^p$$

$$= \lim_{p \to \infty}\left(\frac{-p^2}{e^p} - \frac{2p}{e^p} - \frac{2}{e^p} \right) - \left(\frac{-4}{e^2} - \frac{4}{e^2} - \frac{2}{e^2} \right)$$

$$= \frac{4}{e^2} + \frac{4}{e^2} + \frac{2}{e^2}$$

$$= \frac{10}{e^2}.$$

27. **The correct answer is (E).** If we factor out a $\frac{1}{100}$ from this expression, we get

$$\frac{dP}{dt} = \frac{2}{300}P(500 - P).$$

This indicates that the maximum population, P, would be 500; anything greater and the growth rate would be negative.

28. **The correct answer is (A).** $f''(x) = (n-1)na_n(x-c)^{n-2}$. So, $f''(c) = (n-1)na_n(c-c)^{n-2} = 0$.

PART B

29. **The correct answer is (B).** This is the Taylor series for $y = \frac{1}{x}$. We can use our calculator to determine that these two graphs intersect at $x = 0.567$.

30. **The correct answer is (D).** This is a second derivative problem.

$$f'(t) = \left\langle -\cos t \sin t, \frac{1}{t} \right\rangle$$

$$f''(t) = \left\langle -2\cos^2 t + 2\sin^2 t, \frac{-1}{t^2} \right\rangle$$

31. **The correct answer is (E).** The formula for the area of a square is $A = \frac{x^2}{2}$, where x is the length of the diagonal. If we differentiate this formula with respect to t, we get

$$\frac{dA}{dt} = x\frac{dx}{dt}.$$

Since we know that $\frac{dx}{dt} = \sqrt{2}$,

$$\frac{dA}{dt} = x\sqrt{2}.$$

Now, we have to express x, the diagonal, in terms of P, the perimeter.

$x = s\sqrt{2}$, where s is the length of a side.

So,

$$P = \frac{4x}{\sqrt{2}}$$

and

$$x = \frac{P\sqrt{2}}{4}.$$

Substituting gives us

$$\frac{dA}{dt} = \frac{P}{2}.$$

32. **The correct answer is (D).** We need

$$\lim_{x \to 2}\left(\frac{x^2 - 3x + 2}{(x-2)}\right).$$

If we factor and cancel, we get

$$\lim_{x \to 2}(x - 1) = 1.$$

33. **The correct answer is (A).** For this problem, we have to solve an equation for a limit of integration. This is the equation we must solve:

$$\int_0^k \left(x^2 + 4 - x^3\right)dx = 5.$$

If we integrate and apply the Fundamental Theorem, we get

$$\frac{k^3}{3} + 4k - \frac{k^4}{4} - 5 = 0.$$

We can use our calculator to determine that $k = 1.239$.

34. **The correct answer is (A).** We must differentiate implicitly with respect to x.

$$x \sin y = 1$$

$$x \cos y \frac{dy}{dx} + \sin y = 0$$

$$\frac{dy}{dx} = \frac{-\sin y}{x \cos y}$$

$$= \frac{-\tan y}{x}$$

35. **The correct answer is (D).** This problem involves simple substitution and the properties of the definite integral.

$$\int_0^{10} (f(x) - 2h(x) + 3)dx = \int_0^{10} (h(x) + g(x) - 2h(x) + 3)dx$$

$$= \int_0^{10} (g(x) - h(x))dx + \int 3dx$$

$$= \int_0^{10} (g(x) - h(x))dx + (3x)\Big|_0^{10}$$

$$= \int_0^{10} (g(x) - h(x))dx + 30$$

36. **The correct answer is (B).** The fifth-degree Taylor polynomial for e^x centered at $x = 0$ is

$$f(x) = 1 + x + \frac{x^2}{2} + \frac{x^3}{6} + \frac{x^4}{24} + \frac{x^5}{120}.$$

So,

$$f(2) = 1 + 2 + 2 + \frac{4}{3} + \frac{2}{3} + \frac{4}{15}$$

$$= 7.267.$$

37. **The correct answer is (D).** We first want to take the limit of the ratio test.

$$\lim_{n \to \infty} \left| \frac{\left((x+2)^{n+1}\right)}{\left((n+1)\sqrt{n+1}\ 3^{n+1}\right)} \right| \cdot \left| \frac{\left(n\sqrt{n}3^n\right)}{\left((x+2)^n\right)} \right|$$

$$= \lim_{n \to \infty} \left| \frac{(x+2)n\sqrt{n}}{(n+1)\sqrt{n+1}\ 3} \right| = \left| \frac{x+2}{3} \right|$$

$$\left| \frac{x+2}{3} \right| < 1$$

So,

$$-5 < x < 1.$$

If we test the endpoints, we'll find that the series converges at both of them, so the radius of convergence is

$$-5 \leq x \leq 1.$$

38. **The correct answer is (B).** First, figure the trapezoidal approximation using $n = 6$:

$$T_6 = \frac{1}{2}(5 + 2(0) + 2(3) + 2(4) + 2(3) + 2(0) + 5)$$
$$= 15.$$

Now, we can use our calculator to divide $\dfrac{15}{\int_{-3}^{3}\left| x^2 - 4 \right| dx}$. This comes out to

0.978.

39. **The correct answer is (D).** If we examine the figure, we'll see that $y = 3 - x^2$ is the top curve.

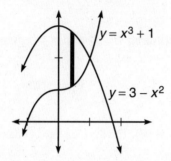

Since the two curves intersect at (1,2), the limits of integration are 0 and 1. Using the Washer Method, the volume would be

$$V = \pi \int_0^1 \left(\left(3 - x^2\right)^2 - \left(x^3 + 1\right)^2 \right) dx.$$

Since this is not a choice, we should switch the limits of integration and factor out a negative and get

$$V = -\pi \int_1^0 \left(\left(x^3 + 1\right)^2 - \left(3 - x^2\right)^2 \right) dx.$$

40. **The correct answer is (D).** In order to determine the tangent line, we need two things: a point and the slope. To find the slope, let's find $g'(2)$. This is a simple application of the Second Fundamental Theorem:

$$g'(2) = \sqrt{2}.$$

To find a point on the tangent line, we need to evaluate $g(2)$:

$$g(2) = \int_{-4}^{0} \left(-x^2 \right) dx + \int_0^2 \sqrt{x}\, dx$$
$$= \frac{4\sqrt{2} - 64}{3}.$$

So, we write the equation for the line through $\left(2, \frac{4\sqrt{2}-64}{3} \right)$ with a slope of $\sqrt{2}$.

$$y - \frac{4\sqrt{2} - 64}{3} = \sqrt{2}(x - 2)$$

This can be transformed into

$$3x\sqrt{2} - 3y = 64 + 2\sqrt{2}.$$

41. **The correct answer is (A).** We are looking for the graph of the derivative of the given graph. Since g has only one horizontal tangent, we can expect its derivative to have only one zero.

42. **The correct answer is (C).** The formula for the sum of an infinite geometric series is

$$S = \frac{a}{1-r}.$$

Substituting $a = \frac{4}{5}$ and $r = \frac{2}{7}$ gives us

$$S = \frac{\frac{4}{5}}{1-\frac{2}{7}} = \frac{28}{25} = 1.120.$$

43. **The correct answer is (E).** The average rate of change is just the slope of the secant, which is

$$m = \frac{26-6}{10-5} = 4.$$

Since it is strictly monotonic and $f(10) > f(5)$, then f is increasing over the interval $[5,10]$ and the absolute maximum must occur at $x = 10$. The absolute maximum is 26. Since 8 is on the interval $[5,10]$ and f is increasing over this interval, $f'(8) > 0$.

44. **The correct answer is (C).** We are going to find the antiderivative of $f(x) = e^{2x}$.

$$F(x) = \int e^{2x} dx$$
$$F(x) = \frac{1}{2}e^{2x} + C$$

Since we are given the initial condition that $F(0) = 2.5$,

$$2.5 = \frac{1}{2}e^0 + C$$
$$C = 2.$$

Substituting this gives us

$$F(x) = \frac{1}{2}e^{2x} + 2.$$

Now, using our calculator, we can determine $F(5)$ to be 11015.233.

45. **The correct answer is (B).** To find the volume of a solid with known cross sections, we integrate the area of these cross sections. So, the volume would be given by

$$V = \int_0^{\sqrt{3}} \left(-x^2 + 3\right)^2 dx$$
$$= 8.314.$$

SOLUTIONS FOR SECTION II

PART A

1. A. The formula for a Taylor series expansion is

$$\sum_{n=0}^{\infty} \frac{f^{(n)}(a)}{n}(x-a)^n = f(a) + f'(a)(x-a) + \left(\frac{f''(a)}{2!}\right)(x-a)^2 + \ldots + \left(\frac{f^n(a)}{n!}\right)(x-a)^n + \ldots$$

We are given the values of the function and the first three derivatives when $x = 1$. We can just plug these into the formula and get

$$f(x) \approx 3 + (-1)(x-1) + \frac{4(x-1)^2}{2} + \frac{(-2)(x-1)^3}{6}$$
$$= 3 - (x-1) + 2(x-1)^2 - \frac{(x-1)^3}{3}.$$

Now, we use this polynomial to find $f(1.1) \approx 2.920$.

B. This is the derivative of the polynomial in part A.
$$f'(x) \approx -1 + 4(x-1) - (x-1)^2$$
$$f'(1.1) \approx -0.61$$

C.
$$\int_1^x f(t)\, dt = \left(3t - \frac{(t-1)^2}{2} + \frac{2(t-1)^3}{3} - \frac{(t-1)^4}{12}\right)\Big|_1^x$$
$$= 3x - \frac{(x-1)^2}{2} + \frac{2(x-1)^3}{3} - \frac{(x-1)^4}{12} - 3$$

D. Can $f(2)$ be determined from the information given? Justify your answer.

No, we only have information about $f(1)$. We can only *approximate* values other than that.

$$\frac{dy}{dx} = \frac{3x^2 + 2x}{e^y}$$

2. A.
$$\int e^y dy = \int (3x^2 + 2x) dx$$
$$e^y = x^3 + x^2 + C$$

Since $f(0) = 2$, we substitute 0 for x and 2 for y:

$$e^2 = C.$$

Substituting back, we get

$$e^y = x^3 + x^2 + e^2.$$

Now, we solve for y by taking the natural log of both sides:

$$y = \ln(x^3 + x^2 + e^2).$$

B. Remember, the domain of a natural log function is the set of all numbers for which the argument is positive. So, using the calculator, we can determine that $x^3 + x^2 + e^2$ is positive for all $x > -2.344$.

C. Where does the second derivative change signs?

$$y = \ln\left(x^3 + x^2 + e^2\right)$$

$$y' = \frac{\left(3x^2 + 2x\right)}{\left(x^3 + x^2 + e^2\right)}$$

$$y'' = \frac{\left(x^3 + x^2 + e^2\right)(6x + 2) - \left(3x^2 + 2x\right)\left(3x^2 + 2x\right)}{\left(x^3 + x^2 + e^2\right)^2}$$

$$= \frac{-3x^4 - 4x^3 - 2x^2 + 6e^2 + 2e^2}{\left(x^3 + x^2 + e^2\right)^2}$$

We are really concerned about where the numerator is zero, so we'll set it equal to zero and use our calculator to solve for x.

$$-3x^4 - 4x^3 - 2x^2 + 6e^2x + 2e^2 = 0$$

The graph of $y = -3x^4 - 4x^3 - 2x^2 + 6e^2x + 2e^2$ crosses the x-axis in two places: $x = -0.331$ and $x = 2.128$. So, this function has two points of inflection: $x = -0.331$ and $x = 2.128$.

3. A. We first use our calculators to determine the points of intersection, which are $x = -2.028$ and $x = 1.428$. Also, we can tell from the calculator that $y = -x^2 + 3$ is the top function. So, the area of R could be determined like this:

$$A = \int_{-2.028}^{1.428} \left(-x^2 + 3 - \tan^{-1}(x)\right)dx$$

$$= 7.243.$$

B. We are going to use the formula for arc length twice, once for each curve:

$$L = \int_{-2.028}^{1.428} \sqrt{1 + \left(-2x^2\right)^2}\, dx + \int_{-2.028}^{1.428} \sqrt{1 + \left(\frac{1}{\left(1 + x^2\right)}\right)^2}\, dx.$$

C. We need to integrate the area of a semicircle. Remember, the formula
for the area of a semicircle is $A = \dfrac{1}{2}\pi r^2$. First, we should determine r.
This should be $\dfrac{1}{2}$ the distance between the curves. So, $r = \dfrac{1}{2}(-x^2 + 3 - \tan^{-1} x)$. So, the volume of the solid is given by the following

$$V = \frac{1}{2}\pi \int_{-2.028}^{1.428} \frac{1}{4}\left(-x^2 + 3 - \tan^{-1} x\right)^2 dx$$

$$= \frac{\pi}{8}\int_{-2.028}^{1.428}\left(-x^2 + 3 - \tan^{-1} x\right)^2 dx.$$

PART B

4. A. A relative minimum exists wherever the value of the derivative changes
    from negative to positive. This happens twice: at $x = -2$ and at $x = 6$.

    B. A relative maximum exists wherever the derivative changes from
    positive to negative. This occurs at $x = 4$.

    C. There are four possible absolute minimums: $x = -3$, $x = -2$, $x = 6$,
    and $x = 8$. These are the relative minimums and the endpoints. We
    should examine the accumulated area under the derivative's graph for
    each one. Upon doing so, we see that the area between the
    derivative's graph and the $x -$ axis is least at $x = -2$. So, the absolute
    minimum occurs when $x = -2$.

    D. Since there is a vertical tangent line at $x = 3$, the derivative of the
    derivative does not exist there. Also, since there is a cusp at $x = 5$,
    $f''(5)$ does not exist either.

5. A.

    B. Point $(0,3)$: $\dfrac{dy}{dx} = x(y-2) = 0; \Delta y = (.2)(0) = 0$

    The new point will be $(0 + .2, 3 + 0) = (.2, 3)$

    Point $(.2, 3)$: $\dfrac{dy}{dx} = x(y-2) = (.2)(1) = .2; \Delta y = (.2)(.2) = .04.$

    The new point will be $(.2 + .2, 3 + .04) = (.4, 3.04)$

    Therefore, $f(0.4) \approx 3.04.$

C.
$$\frac{dy}{dx} = x(y-2)$$

$$\frac{dy}{y-2} = x\,dx$$

$$\int \frac{dy}{y-2} = \int x\,dx$$

$$\ln|y-2| = \frac{x^2}{2} + C$$

Now, we will substitute in our initial condition of $x = 0$ and $y = 3$:

$$\ln 1 = \frac{0}{2} + C$$
$$C = 0.$$

By substitution,

$$\ln|y-2| = \frac{x^2}{2}$$
$$y = e^{x^2/2} + 2.$$

6. A. This involves finding the antiderivatives of both components of the velocity vector:

$$x'(t) = 3 - \frac{3}{t^2} \text{ and } y'(t) = 4 + \frac{2}{t^2}$$

$$x(t) = 3t + \frac{3}{t} + C_1 \text{ and } y(t) = 4t - \frac{2}{t} + C_2$$
$$x(1) = 7 = 3 + 3 + C_1 \text{ and } y(1) = 0 = 4 - 2 + C_2$$
$$C_1 = 1 \text{ and } C_2 = 2$$
$$x(t) = 3t + \frac{3}{t} + 1 \text{ and } y(t) = 4t - \frac{2}{t} + 2$$
$$x(3) = 9 + 1 + 1 = 11 \text{ and } y(3) = 12 - \frac{2}{3} + 2 = \frac{40}{3}.$$

So, the position of the particle when t = 3 is $\left(11, \frac{40}{3}\right)$.

B. The slope of the tangent line is equal to

$$\frac{dy}{dx} = \frac{4 + \frac{2}{t^2}}{3 - \frac{3}{t^2}}.$$

In order for the slope to be zero, we would need the numerator of $\dfrac{dy}{dx}$ to be zero:

$$4 + \frac{2}{t^2} = 0.$$

However, there are no values for t that would make this equation true. Therefore, the line tangent to the path of the particle will never have a slope of zero.

Answer Sheet

SECOND MODEL AP CALCULUS BC EXAM

Section I Part A

1. Ⓐ Ⓑ Ⓒ Ⓓ Ⓔ   7. Ⓐ Ⓑ Ⓒ Ⓓ Ⓔ   13. Ⓐ Ⓑ Ⓒ Ⓓ Ⓔ   19. Ⓐ Ⓑ Ⓒ Ⓓ Ⓔ   25. Ⓐ Ⓑ Ⓒ Ⓓ Ⓔ

2. Ⓐ Ⓑ Ⓒ Ⓓ Ⓔ   8. Ⓐ Ⓑ Ⓒ Ⓓ Ⓔ   14. Ⓐ Ⓑ Ⓒ Ⓓ Ⓔ   20. Ⓐ Ⓑ Ⓒ Ⓓ Ⓔ   26. Ⓐ Ⓑ Ⓒ Ⓓ Ⓔ

3. Ⓐ Ⓑ Ⓒ Ⓓ Ⓔ   9. Ⓐ Ⓑ Ⓒ Ⓓ Ⓔ   15. Ⓐ Ⓑ Ⓒ Ⓓ Ⓔ   21. Ⓐ Ⓑ Ⓒ Ⓓ Ⓔ   27. Ⓐ Ⓑ Ⓒ Ⓓ Ⓔ

4. Ⓐ Ⓑ Ⓒ Ⓓ Ⓔ   10. Ⓐ Ⓑ Ⓒ Ⓓ Ⓔ   16. Ⓐ Ⓑ Ⓒ Ⓓ Ⓔ   22. Ⓐ Ⓑ Ⓒ Ⓓ Ⓔ   28. Ⓐ Ⓑ Ⓒ Ⓓ Ⓔ

5. Ⓐ Ⓑ Ⓒ Ⓓ Ⓔ   11. Ⓐ Ⓑ Ⓒ Ⓓ Ⓔ   17. Ⓐ Ⓑ Ⓒ Ⓓ Ⓔ   23. Ⓐ Ⓑ Ⓒ Ⓓ Ⓔ

6. Ⓐ Ⓑ Ⓒ Ⓓ Ⓔ   12. Ⓐ Ⓑ Ⓒ Ⓓ Ⓔ   18. Ⓐ Ⓑ Ⓒ Ⓓ Ⓔ   24. Ⓐ Ⓑ Ⓒ Ⓓ Ⓔ

Section I Part B

29. Ⓐ Ⓑ Ⓒ Ⓓ Ⓔ   32. Ⓐ Ⓑ Ⓒ Ⓓ Ⓔ   35. Ⓐ Ⓑ Ⓒ Ⓓ Ⓔ   38. Ⓐ Ⓑ Ⓒ Ⓓ Ⓔ   42. Ⓐ Ⓑ Ⓒ Ⓓ Ⓔ

30. Ⓐ Ⓑ Ⓒ Ⓓ Ⓔ   33. Ⓐ Ⓑ Ⓒ Ⓓ Ⓔ   36. Ⓐ Ⓑ Ⓒ Ⓓ Ⓔ   39. Ⓐ Ⓑ Ⓒ Ⓓ Ⓔ   43. Ⓐ Ⓑ Ⓒ Ⓓ Ⓔ

31. Ⓐ Ⓑ Ⓒ Ⓓ Ⓔ   34. Ⓐ Ⓑ Ⓒ Ⓓ Ⓔ   37. Ⓐ Ⓑ Ⓒ Ⓓ Ⓔ   40. Ⓐ Ⓑ Ⓒ Ⓓ Ⓔ   44. Ⓐ Ⓑ Ⓒ Ⓓ Ⓔ

41. Ⓐ Ⓑ Ⓒ Ⓓ Ⓔ   45. Ⓐ Ⓑ Ⓒ Ⓓ Ⓔ

Second Model AP Calculus BC Exam

SECTION I, PART A

Time – 55 Minutes

Number of Questions – 28

A CALCULATOR MAY NOT BE USED ON THIS PART OF THE EXAMINATION.

Directions: Solve each of the following problems, using the available space for scratchwork. After examining the form of the choices, decide which is the best of the choices given and fill in the corresponding oval on the answer sheet. No credit will be given for anything written in the test book. Do not spend too much time on any one problem.

In this test: Unless otherwise specified, the domain of a function f is assumed to be the set of all real numbers x for which $f(x)$ is a real number.

1. $\int_0^{\pi/4} \sin x \cos x \, dx =$

    (A) $-\dfrac{1}{4}$

    (B) $-\dfrac{1}{8}$

    (C) $\dfrac{1}{8}$

    (D) $\dfrac{1}{4}$

    (E) $\dfrac{3}{8}$

2. If $x = \ln t$ and $y = e^{2t}$ then $\dfrac{dy}{dx} =$

(A) $2e^{2t}$

(B) $\dfrac{2e^{2t}}{t}$

(C) te^{2t}

(D) $2te^{2t}$

(E) $\dfrac{te^{2t}}{2}$

3. The function $y = \dfrac{(x-2)^2}{x^2 - 8x + 7}$ has a local minimum at $x =$

(A) $-\dfrac{1}{2}$

(B) 1

(C) 2

(D) 7

(E) None of the above

4. $\dfrac{d}{dx}(e^x \ln(\cos e^x)) =$

(A) $- e^{2x}\tan e^x$

(B) $\dfrac{e^x}{\cos e^x} + e^x \ln\left(\cos e^x\right)$

(C) $e^{2x}\tan e^x$

(D) $- e^{2x}\tan e^x + e^x \ln(\cos e^x)$

(E) $e^x(e^x\tan e^x + \ln(\cos e^x))$

5. If $f(x) = \dfrac{\sin x}{x^2}$, then $f'(\pi) =$

(A) $\dfrac{1}{\pi^2}$

(B) π^2

(C) $-\dfrac{1}{\pi^2}$

(D) -1

(E) 0

6.

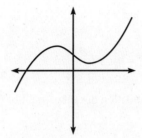

The graph of $y = h(x)$ is shown above. Which of the following could be the graph of $h''(x)$?

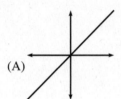

(A)

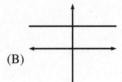

(B)

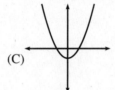

(C)

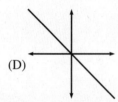

(D)

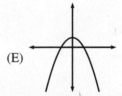

(E)

Questions 7 through 9 refer to the following graph and information.

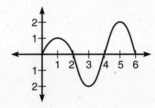

The function f is defined on the closed interval $[0,6]$. The graph of the derivative f' is shown above.

7. The point $(3,2)$ is on the graph of $y = f(x)$. An equation for the line tangent to the graph of f at $(3,2)$ is

    (A) $y = -2x + 4$

    (B) $y = 2x - 4$

    (C) $y + 2 = -2(x + 3)$

    (D) $y - 2 = -2(x - 3)$

    (E) $y = 2$

8. At what value of x does the absolute minimum value of f occur?

GO ON TO THE NEXT PAGE

(A) 0

(B) 2

(C) 3

(D) 4

(E) 6

9. How many points of inflection does the graph of f have?

(A) Two

(B) Three

(C) Four

(D) Five

(E) Six

10. If $6x^2 + 3y - 2xy^2 = 3$, then when $x = 0$, $\dfrac{dy}{dx} =$

(A) $\dfrac{1}{3}$

(B) $\dfrac{2}{3}$

(C) 1

(D) $\dfrac{4}{3}$

(E) $\dfrac{5}{3}$

11. $\displaystyle\int_3^\infty \dfrac{\ln x}{x^2}\, dx =$

(A) $\dfrac{1}{3}$

(B) $\dfrac{\ln 3 + 1}{3}$

(C) $\dfrac{\ln 3}{3}$

(D) $1 + \ln 3$

(E) It is divergent.

12. $\displaystyle\int x \sec^2 x\, dx =$

(A) $x\tan x - \dfrac{1}{2}\sec^2 x + C$

(B) $x\tan x + \ln\left|\sec x\right| + C$

(C) $x\tan x - \ln\left|\cos x\right| + C$

(D) $x\tan x + \ln\left|\cos x\right| + C$

(E) $x\tan x - \ln\left|\sec x + \tan x\right| + C$

13. $\displaystyle\lim_{x\to 1}\left(\dfrac{(\ln x)^2}{x^3 - 3x + 2}\right) =$

    (A) $\dfrac{1}{3}$

    (B) 0

    (C) 2

    (D) 6

    (E) It is nonexistent.

14. What is the approximation of the value of cos 2 obtained by using the sixth-degree Taylor polynomial about $x = 0$ for $\cos x$?

    (A) $1 - 2 + \dfrac{2}{3} - \dfrac{4}{45}$

    (B) $1 + 2 + \dfrac{16}{24} + \dfrac{64}{720}$

    (C) $1 - \dfrac{1}{2} + \dfrac{1}{24} - \dfrac{1}{720}$

    (D) $2 - \dfrac{4}{3} + \dfrac{4}{15} - \dfrac{8}{315}$

    (E) $2 + \dfrac{8}{6} + \dfrac{32}{120} + \dfrac{128}{5040}$

15. Which of the following sequence(s) converge?

I. $\left\{\dfrac{3n^2}{7n^3-1}\right\}$

II. $\left\{\dfrac{7}{n}\right\}$

III. $\left\{\dfrac{3n^4}{7n^2}\right\}$

(A) I only

(B) II only

(C) III only

(D) I and II only

(E) I, II, and III

16. A particle moves on a plane curve so that at any time $t > 0$ its position is defined by the parametric equations $x(t) = 3t^2 - 7$ and $y(t) = \dfrac{4t^2+1}{3t}$. The acceleration vector of the particle at $t = 2$ is

(A) $\left\langle 6, \dfrac{1}{12}\right\rangle$

(B) $\left\langle 17, \dfrac{17}{6}\right\rangle$

(C) $\left\langle 12, \dfrac{47}{12}\right\rangle$

(D) $\left\langle 12, \dfrac{33}{12}\right\rangle$

(E) $\left\langle 6, \dfrac{17}{6}\right\rangle$

17.

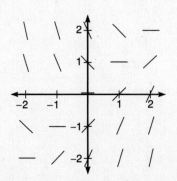

Shown above is the slopefield for which of the following differential equations?

(A) $\dfrac{dy}{dx} = 1 + x$

(B) $\dfrac{dy}{dx} = x - y$

(C) $\dfrac{dy}{dx} = \dfrac{x + y}{2}$

(D) $\dfrac{dy}{dx} = y - x$

(E) $\dfrac{dy}{dx} = y + 1$

18. $\displaystyle\lim_{x \to 2}\left(\dfrac{\displaystyle\int_{-2}^{x} t^3\,dt}{x^2 - 4}\right)$ is

   (A) 0

   (B) 2

   (C) 4

   (D) 8

   (E) nonexistent

19. $\displaystyle\int \dfrac{x^2 + 3}{x}\,dx =$

   (A) $\dfrac{1}{2}x^2 + 3x + C$

   (B) $\dfrac{1}{3}x^3 + 3x + C$

   (C) $\dfrac{3}{2}x^2 + C$

   (D) $\dfrac{x^2}{2} + 3\ln|x| + C$

   (E) $x + \dfrac{3}{x} + C$

GO ON TO THE NEXT PAGE

20. If $f(x) = \sec^2 x$, then $f'\left(\dfrac{\pi}{3}\right) =$

   (A) $\dfrac{\sqrt{3}}{2}$

   (B) $\dfrac{3\sqrt{3}}{2}$

   (C) $8\sqrt{3}$

   (D) $4\sqrt{3}$

   (E) $\dfrac{2\sqrt{3}}{3}$

21. What is the instantaneous rate of change of the derivative of the function $f(x) = \ln x^2$ when $x = 3$?

   (A) $-\dfrac{2}{3}$

   (B) $-\dfrac{2}{9}$

   (C) $\dfrac{2}{9}$

   (D) $\dfrac{2}{3}$

   (E) $\ln 9$

22. $\displaystyle\lim_{x \to \infty} \dfrac{x\left(x^2 + 7x - 9\right)}{(x-2)(2x+3)} =$

   (A) -7

   (B) 0

   (C) $\dfrac{1}{2}$

   (D) 2

   (E) It is nonexistent.

23. $\dfrac{d}{dx}(\sec x^2 \ln e^{\cos x^2})=$

    (A) $-2x \sec x^2 \sin x^2$

    (B) $2x \sec x^2 \tan x^2 \cos x^2$

    (C) -1

    (D) 0

    (E) 1

24. What is the approximation of the area under $y = x^2 - 2x + 1$ for $0 \le x \le 4$ using the Trapezoidal Rule with 4 subintervals?

    (A) $\dfrac{4}{3}$

    (B) 8

    (C) $\dfrac{28}{3}$

    (D) 10

    (E) 16

25. Let f be the function given by the first four nonzero terms of the Maclaurin polynomial used to approximate the value of e^x. Determine the area bounded by the graph and the x-axis for $0 \le x \le 2$.

    (A) 4

    (B) $\dfrac{64}{15}$

    (C) 5

    (D) 6

    (E) $\dfrac{20}{3}$

26.

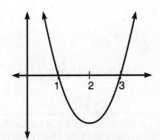

The graph of a twice-differentiable function f is shown in the above figure. Which of the following is true?

GO ON TO THE NEXT PAGE

(A) $f'(2) < f(2) < f''(2)$

(B) $f(2) < f'(2) < f''(2)$

(C) $f''(2) < f'(2) < f(2)$

(D) $f''(2) < f(2) < f'(2)$

(E) $f(2) < f''(2) < f'(2)$

27. $\int_1^9 \dfrac{e^{\sqrt{x}}}{\sqrt{x}}\, dx =$

(A) $\dfrac{e^3 - e}{2}$

(B) $e^3 - e$

(C) $2e(e^2 - 1)$

(D) $2e^3$

(E) $\dfrac{e^3}{3}$

28. The length of the path described by the parametric equations $x = \frac{4}{3}t^2$ and $y = \frac{1}{2}t^3$, where $0 \le t \le 2$, is

(A) $\displaystyle\int_0^2 \sqrt{\dfrac{64}{9}t^2 + 1}\; dt$

(B) $\displaystyle\int_0^2 \sqrt{\dfrac{9}{4}t^4 + 1}\; dt$

(C) $\displaystyle\int_0^2 \sqrt{\dfrac{64}{9}t^2 + \dfrac{9}{4}t^4}\; dt$

(D) $\dfrac{1}{2}\displaystyle\int_0^2 \sqrt{\dfrac{64}{9}t^2 - \dfrac{9}{4}t^4}\; dt$

(E) $\dfrac{1}{4}\displaystyle\int_0^2 \sqrt{\dfrac{16}{9}t^4 + \dfrac{1}{4}t^6}\; dt$

END OF PART A

STOP

END OF SECTION I PART A. IF YOU HAVE ANY TIME LEFT, GO OVER YOUR WORK IN THIS SECTION ONLY. DO NOT WORK IN ANY OTHER SECTION OF THE TEST.

PART B

Time – 50 Minutes

Number of Questions – 17

A GRAPHING CALCULATOR IS REQUIRED FOR SOME QUESTIONS ON THIS PART OF THE EXAMINATION.

Directions: Solve each of the following problems, using the available space for scratchwork. After examining the form of the choices, decide which is the best of the choices given and fill in the corresponding oval on the answer sheet. No credit will be given for anything written in the test book. Do not spend too much time on any one problem.

In this test: (1) The exact numerical value of the correct answer does not always appear among the choices given. When this happens, select from among the choices the number that best approximates the exact numerical value. (2) Unless otherwise specified, the domain of a function f is assumed to be the set of all real numbers x for which f(x) is a real number.

29. For what integer $k > 1$ will both $\sum_{n=1}^{\infty} \frac{(-1)^{kn}}{n^2}$ and $\sum_{n=1}^{\infty} \left(\frac{k}{3}\right)^n$ converge?

    (A) 2

    (B) 3

    (C) 4

    (D) 5

    (E) 6

30. The volume of the solid formed when the region bounded by

    $y = \sqrt{4 - x^2}$, $x = 0$, and $y = 0$ is rotated about the line $y = -2$ is given by which of these definite integrals?

    (A) $2\pi \int_0^2 x\sqrt{4 - x^2}\, dx$

    (B) $\pi \int_0^2 \left(4 - x^2\right) dx$

    (C) $\pi \int_0^2 \left(\sqrt{4 - x^2}\right)^2 dx$

    (D) $\pi \int_0^2 \left[\left(\sqrt{4 - x^2} + 2\right)^2 - 4\right] dx$

    (E) $2\pi \int_0^2 \left(x\sqrt{4 - x^2}\right)^2 dx$

31. If f is a vector-valued function defined by $f(t) = \left\langle e^{2t}, -\cos 2t \right\rangle$, then $f'(t) =$

    (A) $\left\langle 2e^{2t}, 2\sin 2t \right\rangle$

    (B) $\left\langle 4e^{2t}, 4\cos 2t \right\rangle$

    (C) $\left\langle 4e^{2t}, 2\sin 2t \right\rangle$

    (D) $\left\langle 4e^{2t}, -4\cos 2t \right\rangle$

    (E) $\left\langle e^{2t}, \cos 2t \right\rangle$

32. $\displaystyle\int e^x \sin x \, dx =$

    (A) $\dfrac{1}{2} e^x (\sin x - \cos x) + C$

    (B) $\dfrac{1}{2} e^x (\sin x + 2\cos x) + C$

    (C) $-e^x \cos x + C$

    (D) $e^x (\sin x - \cos x) + C$

    (E) $e^x \sin x + e^x \cos x + C$

33. The graph of the function represented by the Maclaurin series

    $$1 - 2x^2 + \frac{4}{3!} x^4 + \ldots = \frac{(-1)^n (2)^{2n}\left(x^{2n}\right)}{(2n)!} \quad \text{intersects the graph of } y = 3x^3 - 2x^2 + 7$$

    at $x =$

    (A) -1.248

    (B) -1.180

    (C) -1.109

    (D) -1.063

    (E) -1.056

34. The acceleration of a particle is described by the parametric equations $x''(t) = \frac{t^2}{4} + t$ and $y''(t) = \frac{1}{3t}$. If the velocity vector of the particle when $t = 2$ is $\langle 4, 1\text{n } 2 \rangle$, what is the velocity vector of the particle when $t = 1$?

    (A) $\left\langle \dfrac{5}{4}, \dfrac{1}{3} \right\rangle$

    (B) $\left\langle \dfrac{23}{12}, \dfrac{\ln 4}{3} \right\rangle$

    (C) $\left\langle \dfrac{23}{12}, \dfrac{\ln 2}{3} \right\rangle$

    (D) $\left\langle \dfrac{5}{4}, \dfrac{2}{3}\ln 2 \right\rangle$

    (E) $\left\langle \dfrac{23}{12}, \dfrac{1}{3}\ln 2 \right\rangle$

35. What is the average rate of change of $f(x) = \dfrac{x^2 - 3}{x - 1}$ over [2,5]?

    (A) $\dfrac{9}{8}$

    (B) $\dfrac{3}{2}$

    (C) 3

    (D) $\dfrac{9}{2}$

    (E) $\dfrac{11}{2}$

36. Let f be defined as the function $f(x) = x^2 + 4x - 8$. The tangent line to the graph of f at $x = 2$ is used to approximate values of f. Using this tangent line, which of the following best approximates a zero of f?

    (A) − 5.464

    (B) − 1.500

    (C) 0

    (D) 1.464

    (E) 1.500

37. $\displaystyle\int \dfrac{4x^2 - 3x + 3}{x^2 + 2x - 3}\, dx =$

    (A) $4x - 12 \ln |x + 3| + \ln |x - 1| + C$

    (B) $4x - 12 \ln |x + 3| - \ln |x - 1| + C$

    (C) $4x + 12 \ln |(x + 3)(x - 1)| + C$

    (D) $\ln |x^2 + 2x - 3| + C$

    (E) $\dfrac{8x^3 - 9x^2 + 18x}{2x^3 + 6x^2 - 18x} + C$

38. The revenue from the sale of the widgets is $108x + 1{,}000$ dollars, and the total production cost is $3x^2 + 16x - 500$ dollars, where x is the number of widgets produced. How many widgets should be made in order to maximize profits?

    (A) 0

    (B) 10

    (C) 15

    (D) 20

    (E) 24

GO ON TO THE NEXT PAGE

39. What are all the values of x for which the series $\displaystyle\sum_{n=1}^{\infty} \frac{(2x+3)^n}{\sqrt{n}}$ converges?

    (A) $-2 < x < -1$

    (B) $-2 \leq x \leq -1$

    (C) $-2 < x \leq -1$

    (D) $-2 \leq x < -1$

    (E) $-2 \leq x < 1$

40. If $f(x) = \begin{cases} e^x, & x < \ln 2 \\ 2, & x \geq \ln 2 \end{cases}$

  then $\displaystyle\lim_{x \to \ln 2} f(x) =$

    (A) $\dfrac{1}{2}$

    (B) $\ln 2$

    (C) 2

    (D) e^2

    (E) It is nonexistent.

41. $\displaystyle\lim_{x \to 1} \frac{\ln x^2}{x^2 - 1} =$

    (A) -1

    (B) 0

    (C) 1

    (D) e

    (E) It is nonexistent.

42. At which *point* is the graph of $f(x) = x^4 - 2x^3 - 2x^2 - 7$ decreasing and concave down?

    (A) $(1, -10)$

    (B) $(2, -15)$

    (C) $(3, 2)$

    (D) $(-1, -6)$

    (E) $(-2, 17)$

43. A population, $P(t)$ where t is in years, increases at a rate proportional to its size. If $P(0) = 40$ and $P(1) = 48.856$, how many years will it take the population to be double its original size?

   (A) 0.347 years

   (B) 3.466 years

   (C) 3.792 years

   (D) 34.657 years

   (E) 37.923 years

44. Let f be a continuous and differentiable function on the closed interval [1,5]. If $f(1) = f(5)$, then Rolle's Theorem guarantees which of the following?

   (A) $f(c) = 0$ for some c on (1,5).

   (B) $f'(c) = 0$ for some c on (1,5).

   (C) f is strictly monotonic.

   (D) If c is on [1,5], then $f(c) = f(1)$.

   (E) $f'(3) = 0$.

45. A particle starts from rest at the origin and moves along the x–axis with an increasing positive velocity. Which of the following could be the graph of the distance $s(t)$ that the particle travels as a function of time t?

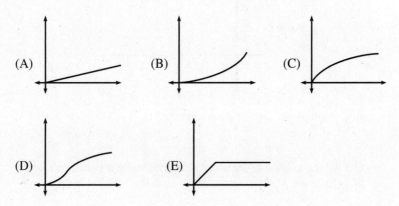

END OF SECTION I

STOP

END OF SECTION I. IF YOU HAVE ANY TIME LEFT, GO OVER YOUR WORK IN THIS SECTION ONLY. DO NOT WORK IN ANY OTHER SECTION OF THE TEST.

SECTION II, PART A

Time – 45 Minutes

Number of Questions – 3

A GRAPHING CALCULATOR IS REQUIRED FOR SOME
PROBLEMS OR PARTS OF PROBLEMS ON THIS PART OF
THE EXAMINATION.

SHOW ALL YOUR WORK. It is important to show your setups for
these problems because partial credit will be awarded. If you use
decimal approximations, they should be accurate to three decimal
places.

1. Let R be the region in the first quadrant enclosed by the graphs of $y = e^{-x} + 4$ and $y = \sqrt{3x}$.

    A. Sketch the region R on the axes provided.

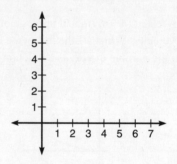

    B. Determine the area of the region R.

    C. Find the volume of the solid generated when R is rotated about the x–axis.

    D. The region R is the base of a solid. Each cross section perpendicular to the x-axis is an equilateral triangle. Find the volume of this solid.

2. The rate at which air is leaking out of a tire is proportional to the amount of air in the tire. The tire was originally filled to capacity with 1,500 cubic inches of air. After one hour, there were 1,400 cubic inches of air left in it.

    A. Express the amount of air in the tire in cubic inches as a function of time t in hours.

    B. A tire is said to be flat if it is holding $\frac{2}{3}$ of its capacity or less. After how many hours would this tire be flat?

3. Consider the curve defined by $9x^2 + 4y^2 - 54x + 16y + 61 = 0$.

    A. Verify that $\dfrac{dy}{dx} = \dfrac{27 - 9x}{4y + 8}$.

    B. Write the equation for each vertical tangent line of the curve.

    C. The points $(3,1)$ and $(1, -2)$ are on the curve. Write the equation for the secant line through these two points.

    D. Write the equation for a line tangent to the curve and parallel to the secant line from part C.

END OF PART A

STOP

> END OF SECTION II PART A. IF YOU HAVE ANY TIME LEFT, GO OVER YOUR WORK IN THIS SECTION ONLY. DO NOT WORK IN ANY OTHER SECTION OF THE TEST.

PART B

Time – 45 Minutes

Number of Questions – 3

A CALCULATOR IS NOT PERMITTED ON THIS PART OF THE
   EXAMINATION.

4.

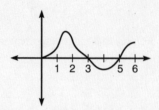

Above is the graph of the velocity of a bug crawling along the x–axis
   over a six-second interval.

   A. At what time(s) t, $0 < t < 6$, does the bug change directions? Explain
   your reasoning.

   B. At hat time t, $0 < t \le 6$, is the bug farthest from its starting point?
   Explain your reasoning.

   C. Over what interval(s) is the bug slowing down?

5. The path of a particle from $t = 0$ to $t = 10$ seconds is described by the
   parametric equations $x(t) = 4\cos\left(\frac{\pi}{2}t\right)$ and $y(t) = 3\sin\left(\frac{\pi}{2}t\right)$.

   A. Write a Cartesian equation for the curve defined by these parametric
      equations.

   B. Find $\dfrac{dy}{dx}$ for the equation in part A.

   C. Determine the velocity vector for the particle at any time t.

   D. Demonstrate that your answers for part A and part B are equivalent.

   E. Write, but do not evaluate, an integral expression that would give the
      distance the particle traveled from $t = 2$ to $t = 6$.

6. Let $P(x) = \ln 2 + (x-1) - \frac{(x-1)^2}{2} + \frac{(x-1)^3}{3} - \frac{(x-1)^4}{4}$ be the fourth-degree Taylor polynomial for the function f about $x = 1$. Assume that f has derivatives of all orders for all real numbers.

A. Find $f(1)$ and $f^{(4)}(1)$.

B. Write the third-degree Taylor polynomial for f' about $x = 1$, and use it to approximate $f'(1.2)$.

C. Write the fifth-degree Taylor polynomial for $g(x) = \int_1^x f(t)\, dt$ about $x=1$.

END OF SECTION II

STOP

END OF SECTION II. IF YOU HAVE ANY TIME LEFT, GO OVER YOUR WORK IN THIS SECTION ONLY. DO NOT WORK IN ANY OTHER SECTION OF THE TEST.

Answer Key

PART A

1. D	7. D	13. A	19. D	25. D
2. D	8. D	14. A	20. C	26. B
3. A	9. B	15. D	21. B	27. C
4. D	10. B	16. A	22. E	28. C
5. C	11. B	17. B	23. D	
6. A	12. D	18. B	24. D	

PART B

29. D	33. B	37. A	41. C	45. B
30. D	34. B	38. C	42. A	
31. B	35. B	39. D	43. B	
32. A	36. E	40. C	44. B	

SOLUTIONS FOR SECTION I

PART A

1. **The correct answer is (D).** This is a straight-forward u substitution integration problem. If we let $u = \sin x$, then $du = \cos x\, dx$ and

$$\int_0^{\pi/4} \sin x \cos x\, dx = \int_0^{\sqrt{2}/2} u\, du$$

$$= \frac{1}{4}$$

2. **The correct answer is (D).** Remember that $\frac{dy}{dx} = \frac{\frac{dy}{dt}}{\frac{dx}{dt}}$. First, we'll find $\frac{dx}{dt}$:

$$x = \ln t$$

$$\frac{dx}{dt} = \frac{1}{t}$$

Now, we'll find $\frac{dy}{dt}$:

$$y = e^{2t}$$

$$\frac{dy}{dt} = 2e^{2t}.$$

So,

$$\frac{dy}{dx} = \frac{2e^{2t}}{\frac{1}{t}} = 2te^{2t}.$$

3. **The correct answer is (A).** To find the local minimum, we need to determine when the derivative changes from negative to positive. First, we determine the derivative:

$$y = \frac{(x-2)^2}{x^2 - 8x + 7}$$

$$y' = \frac{\left(x^2 - 8x + 7\right)2(x-2) - (x-2)^2(2x-8)}{\left(x^2 - 8x + 7\right)^2}$$

$$= \frac{2(x-2)(-2x-1)}{\left(x^2 - 8x + 7\right)^2}$$

If we set $y' = 0$ and solve for x, we see that $x = -\frac{1}{2}$ and $x = 2$ are zeros of the derivative. By examining the wiggle graph below, we can see that the local minimum occurs at $x = -\frac{1}{2}$.

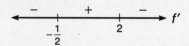

4. **The correct answer is (D).** This problem calls for the Product Rule. We must differentiate each term with respect to x.

$$\frac{d}{dx}\left(e^x \ln\left(\cos e^x\right)\right) = e^x \cdot \frac{-e^x \sin e^x}{\cos e^x} + e^x \ln \cos e^x$$

$$= -e^{2x} \tan e^x + e^x \ln \cos e^2$$

5. **The correct answer is (C).** Here, we use the Quotient Rule to determine the derivative; then, evaluate it at $x = \pi$.

$$f(x) = \frac{\sin x}{x^2}$$

$$f'(x) = \frac{x^2 \cos x - 2x \sin x}{x^4}$$

$$f'(\pi) = \frac{\pi^2 \cos \pi - 2\pi \sin \pi}{\pi^4}$$

$$= -\frac{1}{\pi^2}$$

6. **The correct answer is (A).** The graph of $h(x)$ is concave down for all $x < 0$ and concave up for all $x > 0$. This implies that the second derivative is negative for all $x < 0$ and positive for all $x > 0$. Choice (A) is the only graph that meets this requirement.

7. **The correct answer is (D).** To write the equation of a tangent line, we need a point and the slope. The point is given to us: (3,2). The slope is merely the y–coordinate that corresponds to $x = 3$ on the graph of f'. Since $f'(3) = -2$, then the slope of the tangent line is -2. In point–slope form, the equation of the tangent line is

$$y - 2 = -2(x - 3).$$

8. **The correct answer is (D).** This is an area accumulation problem. We can see that the accumulated area is least when $x = 4$.

9. **The correct answer is (B).** Points of inflection on the graph of a function correspond to horizontal tangents on the graph of the derivative. Since there are three, the function has three points of inflection.

10. **The correct answer is (B).** This is an implicit differentiation problem. Remember, we need to use the product rule to differentiate $2xy^2$.

$$6x^2 + 3y - 2xy^2 = 3$$

$$12x + 3\frac{dy}{dx} - 4xy\frac{dy}{dx} - 2y^2 = 0$$

$$\frac{dy}{dx} = \frac{2y^2 - 12x}{3 - 4xy}$$

Now, we determine the corresponding y value by substituting $x = 0$ into the original equation.

$$0 + 3y - 0 = 3$$

$$y = 1$$

Finally, we substitute $x = 0$ and $y = 1$ into $\frac{dy}{dx}$.

$$\frac{dy}{dx} = \frac{2}{3}.$$

11. **The correct answer is (B).** For an improper integral, we first change it to a limit of a definite integral.

$$\int_3^\infty \frac{\ln x}{x^2}\,dx = \lim_{p\to\infty}\int_3^p \frac{\ln x}{x^2}\,dx$$

Now, we have to address that tricky integrand. We do integration by parts and let $u = \ln x$ and $dv = x^{-2}\,dx$. So,

$$\lim_{p\to\infty}\int_3^p \frac{\ln x}{x^2}\,dx = \lim_{p\to\infty}\left[-\frac{\ln x}{x}\bigg|_3^p + \int_3^p x^{-2}\,dx\right]$$

$$= \lim_{p\to\infty}\left(-\frac{\ln x}{x} - \frac{1}{x}\right)\bigg|_3^p$$

$$= \lim_{p\to\infty}\left(-\frac{\ln p}{p} - \frac{1}{p} - \left(-\frac{\ln 3}{3} - \frac{1}{3}\right)\right)$$

$$= \frac{\ln 3 + 1}{3}.$$

Note: $\lim_{p\to\infty}\dfrac{\ln p}{p} = 0$ by L'Hôpital's Rule.

12. **The correct answer is (D).** This is an example of a straight-forward integration-by-parts problem. We let $u = x$ and $dv = \sec^2x \, dx$.

$$\int x\sec^2 x \, dx = x \tan x - \int \tan x \, dx$$

$$= x\tan x + \ln \left|\cos x\right| + C$$

13. **The correct answer is (A).** If we try to evaluate this limit using direct substitution, we will get an indeterminate form: $\frac{0}{0}$. So, we can use l'Hôpital's rule and take the derivative of the numerator and denominator; then, evaluate the limit.

$$\lim_{x \to 1} \frac{(\ln x)^2}{x^3 - 3x + 2} = \lim_{x \to 1} \frac{\frac{2\ln x}{x}}{3x^2 - 3}$$

If we evaluate the limit now, we still get $\frac{0}{0}$. So, we try l'Hôpital's rule again.

$$= \lim_{x \to 1} \frac{\frac{2 - 2\ln x}{x^2}}{6x}$$

$$= \frac{1}{3}$$

14. **The correct answer is (A).** $\cos x$ centered at $x = 0$ is one Taylor polynomial that we should be able to generate from memory. It goes like this:

$$\cos x = 1 - \frac{x^2}{2} + \frac{x^4}{4!} - \frac{x^6}{6!} + \dots$$

To find the value for $\cos 2$, we substitute 2 for x:

$$1 - 2 + \frac{2}{3} - \frac{4}{45}.$$

15. **The correct answer is (D).** Both I and II converge to 0, while III is divergent.

16. **The correct answer is (A).** Since acceleration is associated with the second derivative of position, we must determine the second derivative for each of these parametric equations and evaluate them at $x = 2$.

$$x(t) = 3t^2 - 7$$

$$x'(t) = 6t$$

$$x''(t) = 6$$

$$x''(2) = 6$$

Rewrite $y(t)$ as $y(t) = \frac{4}{3}t + \frac{1}{3}t^{-1}$.

$$y'(t) = \frac{4}{3} - \frac{1}{3}t^{-2}$$

$$y''(t) = \frac{2}{3}t^{-3}$$

$$y''(2) = \frac{1}{12}$$

The acceleration vector of the particle at $x = 2$ is $\left\langle 6, \frac{1}{12} \right\rangle$.

17. **The correct answer is (B).** Notice that all of the slopes on the line $y = x$ are zero.

18. **The correct answer is (B).** This is a well-disguised application of l'Hôpital's rule. We should take the derivative of the numerator and the derivative of the denominator and then evaluate the limit.

$$\lim_{x \to 2}\left(\frac{\int_{-2}^{x} t^3 dt}{x^2 - 4}\right) = \lim_{x \to 2}\frac{x^3}{2x} = 2$$

Notice we use the Fundamental Theorem Part Two to determine the derivative of the numerator.

19. **The correct answer is (D).** When integrating a rational expression with a numerator of greater degree than the denominator, we first divide and then integrate.

$$\int\left(\frac{x^2 + 3}{x}\right)dx = \int\left(x + \frac{3}{x}\right)dx$$

$$= \frac{x^2}{2} + 3\ln|x| + C$$

20. **The correct answer is (C).** We will determine the derivative of this function by using both the Power and Chain Rules. Then, we will evaluate it at $x = \frac{\pi}{3}$.

$$f'(x) = 2\sec x \cdot \sec x \tan x$$

$$f'\left(\frac{\pi}{3}\right) = 2\sec^2\frac{\pi}{3}\tan\frac{\pi}{3}$$

$$= 8\sqrt{3}$$

21. **The correct answer is (B).** This is asking for the derivative of the derivative when $x = 3$. So, we need the second derivative of the function.

$$f'(x) = \frac{2x}{x^2} = \frac{2}{x}$$

$$f''(x) = -\frac{2}{x^2}$$

$$f''(3) = -\frac{2}{9}$$

22. **The correct answer is (E).** Since the degree of the numerator is greater than the degree of the denominator, the limit as x approaches infinity does not exist because it is infinite.

23. **The correct answer is (D).** The trick to this problem is to recognize that

$$1ne^{\cos x^2} = \cos x^2.$$

So now all we need to find is the derivative of $\sec x^2\cos x^2$, which is equal to 1. The derivative of 1 is 0.

24. **The correct answer is (D).** Remember the Trapezoidal Rule:

$$A_T \approx \frac{b-a}{2n}\left(f(a)+2f(x_1)+\ldots 2f(x_{n-1})+f(b)\right).$$

Applying this to the function $y = x^2 - 2x + 1$ over [0,4] with $n = 4$ yields

$$A_T = \frac{4}{8}\left(1+2(0)+2(1)+2(4)+9\right)=10.$$

25. **The correct answer is (D).** This function is $f(x) = 1 + x + \frac{x^2}{2} + \frac{x^3}{6}$. We integrate this from $x = 0$ to $x = 2$.

$$\int_0^2\left(1+x+\frac{x^2}{2}+\frac{x^3}{6}\right)dx = \left(x+\frac{x^2}{2}+\frac{x^3}{6}+\frac{x^4}{24}\right)\Big|_0^2$$

$$=2+2+\frac{4}{3}+\frac{2}{3}$$

$$=6$$

26. **The correct answer is (B).** By reading the graph, we can tell that $f(2)<0$. Since there is a horizontal tangent line at $x = 2$, $f'(2) = 0$. $f''(2) > 0$ because the curve is concave up at $x = 2$. Therefore, $f(2) < f'(2) < f''(2)$.

27. **The correct answer is (C).** This is a rather complicated u-substitution integration problem. If we let $u = \sqrt{x}$, then $du = \frac{dx}{2\sqrt{x}}$.

$$\int_1^9\frac{e^{\sqrt{x}}}{\sqrt{x}}dx = 2\int_1^3 e^u\,du$$

$$= 2e^3 - 2e = 2e(e^2 - 1)$$

28. **The correct answer is (C).** We need to determine $\frac{dx}{dt}$ and $\frac{dy}{dt}$ first.

$$\frac{dx}{dt}=\frac{8}{3}t$$

$$\frac{dy}{dt}=\frac{3}{2}t^2$$

Now, we integrate from $x = 0$ to $x = 2$ the square root of the sum of the squares of $\frac{dx}{dt}$ and $\frac{dy}{dt}$.

$$l = \int_0^2 \sqrt{\frac{64}{9}t^2 + \frac{9}{4}t^4}\, dt$$

PART B

29. **The correct answer is (A).** If we let $k = 2$, the first series becomes $\sum_{n=1}^{\infty} \frac{1}{n^2}$ and converges since it is a p–series with $p > 1$. If $k = 2$, the second series becomes $\sum_{n=1}^{\infty} \left(\frac{2}{3}\right)^n$ and converges since it is a geometric series with $r < 1$.

30. **The correct answer is (D).** Begin by drawing a diagram.

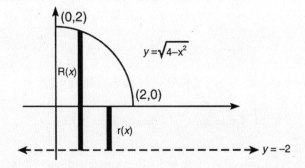

You could use the shell method, but we'll use the washer method. Use vertical rectangles, since they are perpendicular to the horizontal axis of rotation. $R(x)$ is the outer radius, and $r(x)$ is the inner radius.

$$R(x) = \sqrt{4 - x^2} - (-2) = \sqrt{4 - x^2} + 2$$
$$r(x) = 0 - (-2) = 2$$

Now, apply the washer method:

$$\pi \int_0^2 \left[\left(R(x)\right)^2 - \left(r(x)\right)^2 \right] dx$$

$$\pi \int_0^2 \left[\left(\sqrt{4 - x^2} + 2\right)^2 - 4 \right] dx$$

Use your graphing calculator to evaluate the integral. The volume will be 56.234.

31. **The correct answer is (B).** We must determine the second derivative for each component:

$$f(t) = \langle e^{2t}, -\cos 2t \rangle$$
$$f'(t) = \langle 2e^{2t}, 2\sin 2t \rangle$$
$$f''(t) = \langle 4e^{2t}, 4\cos 2t \rangle.$$

32. **The correct answer is (A).** This is an integration by parts with a twist toward the end. Let's let $u = \sin x$ and $dv = e^x\, dx$, so

$$\int e^x \sin x\, dx = e^x \sin x - \int e^x \cos x\, dx.$$

We need to integrate by parts again. We'll let $u = \cos x$ and $dv = e^x\, dx$, continuing:

$$\int e^x \sin x\, dx = e^x \sin x - e^x \cos x - \int e^x \sin x\, dx + C.$$

Here's the twist. We are going to add $\int(e^x\sin x)\, dx$ to both sides of the equation:

$$2\int e^x \sin x\, dx = e^x \sin x - e^x \cos x + C$$

To solve for $\int(e^x\sin x)\, dx$, we will divide both sides by 2:

$$\int e^x \sin x\, dx = \frac{1}{2}\left(e^x \sin x - e^x \cos x\right) + C.$$

33. **The correct answer is (B).** In order to succeed with this problem, we must readily recognize slight variations of series that we have previously memorized. Remember the Maclaurin series for $\cos x$:

$$\cos x = 1 - \frac{x^2}{2!} + \frac{x^3}{3!}.$$

The series in this problem is the Maclaurin series for $\cos 2x$. So, we are being asked to determine at what x value the graphs of $y = \cos 2x$ and $y = 3x^3 - 2x^2 + 7$ intersect. Our calculators will tell us that happens when $x = -1.180$.

34. **The correct answer is (B).** We are going to determine the antiderivative of each component of the acceleration vector, solve for the constants of integration, and plug and chug to determine the velocity vector when $t = 1$. First, we deal with the x component:

$$x''(t) = \frac{t^2}{4} + t$$

$$x'(t) = \frac{t^3}{12} + \frac{t^2}{2} + C_1$$

$$x'(2) = 4 = \frac{2}{3} + 2 + C_1$$

$$C_1 = \frac{4}{3}$$

$$x'(t) = \frac{t^3}{12} + \frac{t^2}{2} + \frac{4}{3}$$

$$x'(1) = \frac{1}{12} + \frac{1}{2} + \frac{4}{3} = \frac{23}{12}.$$

Now, we do it all again for y:

$$y''(t) = \frac{1}{3t}$$

$$y'(t) = \frac{1}{3} \ln |t| + C_2$$

$$y'(2) = \ln 2 = \frac{1}{3} \ln 2 + C_2$$

$$\frac{2}{3} \ln 2 = C_2$$

$$y'(t) = \frac{1}{3} \ln |t| + \frac{2}{3} \ln 2$$

$$y'(1) = \frac{1}{3} \ln |1| + \frac{2}{3} \ln 2$$

$$= \frac{2}{3} \ln 2$$

$$= \frac{\ln 4}{3}$$

Note that $\frac{2}{3} \ln 2 = \frac{1}{3}(2 \ln 2) = \frac{1}{3} \ln 4$ (by log properties).

Finally, the velocity vector of the particle when $t = 1$ is $\left\langle \frac{23}{12}, \frac{\ln 4}{3} \right\rangle$.

35. **The correct answer is (B).** To find the average rate of change of a function over an interval, we need the slope of the secant line over that interval.

$$m = \frac{f(b) - f(a)}{b - a}$$

$$= \frac{\frac{11}{2} - 1}{3}$$

$$= \frac{3}{2}$$

36. **The correct answer is (E).** We need to find the equation for the tangent line of the graph at $x = 2$ and use our calculator to determine where that line crosses the x–axis. Remember, to write an equation for a tangent line, we need a point on the line and the slope of the line. Since $f(2) = 4$, $(2,4)$ is on the line. The slope is

$$f'(x) = 2x + 4$$

$$f'(2) = 8.$$

Using point–slope form,

$$y - 4 = 8(x - 2)$$

$$y = 8x - 12.$$

Using the calculator (or maybe your head), $x = 1.5$ is a zero of $y = 8x - 12$.

37. **The correct answer is (A).** We have to use the method of partial fractions in order to get the integrand into a form that is integrable. To start, since the degrees of the numerator and denominator are equal, we use polynomial long division. So,

$$\int \left(\frac{4x^2 - 3x + 3}{x^2 + 2x - 3} \right) dx = 4x + \int \frac{-11x + 15}{(x + 3)(x - 1)} dx$$

To integrate $\int \frac{-11x + 15}{(x + 3)(x - 1)} dx$, use partial fractions:

$$\frac{-11x + 15}{(x + 3)(x - 1)} = \frac{A}{x + 3} + \frac{B}{x - 1}$$

Multiply through by $(x+3)(x-1)$ to get

$$-11x + 15 = A(x - 1) + B(x + 3)$$

$$= Ax - A + Bx + 3B$$

$$= x(A + B) + (-A + 3B)$$

This gives you the system of equations $A+B=-11$ and $-A+3B=15$. Solving simultaneously, we get:

$$A=-12 \text{ and } B=1$$

The integral can now take the easier from

$$\int \frac{-11x + 15}{(x + 3)(x - 1)} dx = \int \frac{-12x}{x + 3} dx + \int \frac{1}{x - 1} dx$$

Continuing with the integration from above:

$$4x + \int \frac{-12}{x + 3} + \frac{1}{x - 1} dx = 4x - 12 \ln|x + 3| + \ln|x - 1| + C.$$

38. **The correct answer is (C).** For this problem, we need to realize that profits = revenue – cost. So, to find profits,

$$P(x) = 108x + 1000 - 3x^2 - 16x + 500$$

$$= -3x^2 + 92x + 1500.$$

The derivative is $P'(x) = -6x + 92$. Set this equal to zero, and we find that $P(x)$ is maximized at $x = \frac{92}{6} = 15.333$.

39. **The correct answer is (D).** To determine the interval of convergence, we take the limit of the ratio test.

$$\lim_{n\to\infty}\left|\frac{(2x+3)^{n+1}}{\sqrt{n+1}}\cdot\frac{\sqrt{n}}{(2x+3)^n}\right|$$

$$= |2x+3|$$

$|2x+3|$ converges if it is less than one.

$$|2x+3| < 1$$

$$-1 < 2x+3 < 1$$

$$-2 < x < -1$$

By testing the endpoints, we find that the series converges when $x = -2$ and diverges when $x = -1$. So the interval of convergence is $-2 \le x < -1$.

40. **The correct answer is (C).** In order for the limit to exist, the left- and right-hand limits have to exist and be equal to each other. Since both of these are equal to 2, $\lim_{x\to\ln 2} f(x) = 2$.

41. **The correct answer is (C).** Because we get $\frac{0}{0}$ when we try to evaluate by direct substitution, we need to use l'Hôpital's rule on this limit.

$$\lim_{x\to 1}\frac{\ln x^2}{x^2-1} = \lim_{x\to 1}\frac{\frac{2x}{x^2}}{2x}$$

$$= \lim_{x\to 1}\frac{1}{x^2}$$

$$= 1$$

42. **The correct answer is (A).** The quickest and easiest way to attack this problem is by graphing it. Which x value makes both the first and second derivatives negative?

43. **The correct answer is (B).** Whenever the rate of a function increasing or decreasing is proportional to itself, it must be an exponential function of the form $P(t) = Ne^{kt}$. N is the initial value, so in this case, $N = 40$. We use $P(1) = 48.856$ to determine the value of k.

$$48.856 = 40e^k$$

$$k = \ln\frac{48.856}{40}$$

$$= 0.200$$

To determine how long it will take the population to double,

$$80 = 40e^{0.200t}$$

$$2 = e^{0.200t}$$

$$t = \frac{\ln 2}{0.200}$$

$$= 3.466 \text{ years.}$$

44. **The correct answer is (B).** Rolle's Theorem deals with the idea that if the function passes through the same y–coordinate twice, it must have a zero derivative somewhere between these two points.

45. **The correct answer is (B).** Since the velocity is positive, the position function must be increasing. Since the velocity is increasing, the position function must be concave up. The only choice to meet both of these requirements is choice (B).

SOLUTIONS FOR SECTION II

PART A

1. A.

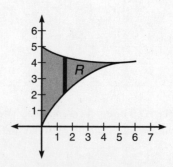

B. $\qquad A = \int_0^{5.346} \left(e^{-x} + 4 - \sqrt{3x}\right) dx$

$$= 8.106$$

C. We use the Washer Method to determine the volume:

$$= \pi \int_0^{5.346} \left[\left(e^{-x} + 4 \right)^2 - 3x \right] dx$$

$$= 160.624.$$

D. It would be good to know that the area of an equilateral triangle with side s is given by $A = \frac{\sqrt{3}}{4} s^2$. So, the volume of this solid would be given by

$$V = \frac{\sqrt{3}}{4} \int_0^{5.346} \left(e^{-x} + 4 - \sqrt{3x} \right)^2 dx$$

$$= 8.511.$$

2. A. Since the rate of decrease is proportional to the function itself, we have an exponential function of the following form:

$$A(t) = Ne^{kt}.$$

Since the tire initially had 1,500 cubic inches of air, $C = 1500$. We are given that $A(1) = 1400$:

$$1400 = 1500e^k.$$

Solving for k,

$$\frac{14}{15} = e^k$$

$$\ln \frac{14}{15} = k.$$

Subsituting this expression for k yields

$$A(t) = 1500^{t \ln (14/15)}.$$

B. Since $\frac{2}{3}$ of 1500 is 1000, we can substitute 1000 into the formula for $A(t)$ and solve for t:

$$1000 = 1500e^{t \ln (14/15)}$$

$$\frac{2}{3} = e^{t \ln (14/15)}$$

$$\ln \frac{2}{3} = t \ln \frac{14}{15}$$

$$t = \frac{\ln \frac{2}{3}}{\ln \frac{14}{15}}$$

$$= 5.877 \text{ hours}$$

3. We have to use implicit differentiation and differentiate with respect to x:

$$9x^2 + 4y^2 - 54x + 16y + 61 = 0$$

$$18x + 8y\frac{dy}{dx} - 54 + 16\frac{dy}{dx} = 0$$

$$\frac{dy}{dx} = \frac{54 - 18x}{8y + 16}$$

$$= \frac{27 - 9x}{4y + 8}.$$

B. Vertical tangent lines exist wherever the denominator of the derivative equals zero, and the numerator does not. So, we determine where the denominator is equal to zero.

$$4y + 8 = 0$$

$$y = -2$$

Since we are writing the equation for one or more vertical lines, we really need to know the corresponding x–coordinate(s). To this end, we will substitute $y = -2$ into the original equation and solve for x.

$$9x^2 + 4(-2)^2 - 54x + 16(-2) + 61 = 0$$

$$9x^2 - 54x + 45 = 0$$

$$x^2 - 6x + 5 = 0$$

$$x = 1 \text{ and } x = 5$$

So, the equations for the vertical tangent lines are $x = 1$ and $x = 5$.

C. We will first find the slope, write the equation in point–slope form, and then convert to slope–intercept form.

$$m = \frac{-2 - 1}{1 - 3}$$

$$= \frac{3}{2}$$

$$y - 1 = \frac{3}{2}(x - 3)$$

$$y = \frac{3}{2}x - \frac{7}{2}$$

D. Since the lines are parallel, they have equal slopes. So, the slope of the tangent line is $\frac{3}{2}$. Now, we need the point(s) on the curve where the derivative is equal to $\frac{3}{2}$. To determine this, we set the derivative equal to $\frac{3}{2}$, solve for y, substitute back into the original equation, and solve for x.

$$\frac{dy}{dx} = \frac{27-9x}{4y+8} = \frac{3}{2}$$

$$54 - 18x = 12y + 24$$

$$y = \frac{5}{2} - \frac{3}{2}x$$

Substituting this expression for y into the original equation and solving for x gives us

$$9x^2 + 4\left(\frac{5}{2}-\frac{3}{2}x\right)^2 - 54x + 16\left(\frac{5}{2}-\frac{3}{2}x\right) + 61 = 0.$$

With help from our calculators, $x = 1.586$ and $x = 4.414$. By substituting these x values into $y = \frac{5}{2} - \frac{3}{2}x$, we get the corresponding y values to be $y = 0.121$ and $y = -4.121$, respectively. So, there are two tangent lines parallel to the line from part C; they have the following equations:

$$y + 4.121 = \frac{3}{2}(x - 4.414)$$

$$y - 0.121 = \frac{3}{2}(x - 1.586).$$

PART B

4. A. The bug changes directions at t = 3 and t = 5. This is true because the velocity changes from positive to negative and negative to positive, respectively.

B. The bug is farthest from its starting point at time t = 3. The bug is moving in the positive direction (away from the starting point) from t = 0 to t = 3. Then, the bug turns around and moves toward the starting point for two seconds before changing directions again. By examining the area under the curve, we can see that the bug is closer to the starting point at t = 6 then it was at t = 3.

C. "Slowing down" means decreasing speed, not velocity. So, we need to include not only where the velocity is positive and decreasing, but also where the velocity is negative and increasing. The velocity is positive and decreasing over the interval (1.5,3), and it is negative and increasing over the interval (4,5). So, the bug is slowing down over these two intervals.

5. A. We want to try to isolate $\cos^2\left(\frac{\pi}{2}t\right)$ and $\sin^2\left(\frac{\pi}{2}t\right)$ in order to use the identity $\sin^2 x + \cos^2 x = 1$. Looking at the x component of the curve, we first square both sides:

$$x = 4\cos\left(\frac{\pi}{2}t\right)$$

$$x^2 = 16\cos^2\left(\frac{\pi}{2}t\right)$$

$$\frac{x^2}{16} = \cos^2\left(\frac{\pi}{2}t\right)$$

And now for the y component:

$$y = 3\sin\left(\frac{\pi}{2}t\right)$$

$$y^2 = 9\sin^2\left(\frac{\pi}{2}t\right)$$

$$\frac{y^2}{9} = \sin^2\left(\frac{\pi}{2}t\right)$$

By combining these equations, we get:

$$\frac{x^2}{16} + \frac{y^2}{9} = 1$$

B. Using implicit differentiation,

$$\frac{x}{8} + \frac{2y}{9}\left(\frac{dy}{dx}\right) = 0$$

$$\frac{dy}{dx} = -\frac{9x}{16y}.$$

C.
$$x'(t) = -2\pi\sin\left(\frac{\pi}{2}t\right)$$

$$y'(t) = \frac{3\pi}{2}\cos\left(\frac{\pi}{2}t\right)$$

$$V'(t) = \left\langle -2\pi\sin\left(\frac{\pi}{2}t\right), \frac{3\pi}{2}\cos\left(\frac{\pi}{2}\right)\right\rangle$$

D. From part C:

$$\frac{dy}{dx} = \frac{\frac{dy}{dt}}{\frac{dx}{dt}} = \frac{\frac{3\pi}{2}\cos\frac{\pi}{2}t}{-2\pi\sin\frac{\pi}{2}t}$$

$$= -\frac{3}{4}\tan\left(\frac{\pi}{2}t\right).$$

From part B:

$$-\frac{9x}{16y} = -\frac{9\left(4\cos\left(\frac{\pi}{2}t\right)\right)}{16\left(3\sin\left(\frac{\pi}{2}t\right)\right)}$$

$$= -\frac{3}{4}\tan\left(\frac{\pi}{2}t\right).$$

E. We will use the formula for arc length:

$$L = \int_a^b \sqrt{\left(\frac{dx}{dt}\right)^2 + \left(\frac{dy}{dt}\right)^2}\; dt$$

$$= \int_2^6 \sqrt{\left(-2\pi\sin\left(\frac{\pi}{2}t\right)\right)^2 + \left(\frac{3\pi}{2}\cos\frac{\pi}{2}t\right)^2}\; dt$$

6.   A. Recall the formula for a Taylor polynomial centered at $x = 1$:

$$f(x) = f(1) + f'(1)(x-1) + \frac{f''(1)(x-1)^2}{2} + \frac{f'''(1)(x-1)}{6} + \frac{f^{(4)}(1)(x-1)}{24}$$

This implies that $f(1) = \ln 2$ and $f^4(1) = -6$.

   B. $$f'(x) = 1 - (x-1) + (x-1)^2 - (x-1)^3$$

$$f'(1.2) = 1 - 0.2 + 0.04 - 0.008$$

$$= 0.832$$

   C.

$$g(x) = \int_1^x f(t)\; dt = (x-1)\ln 2 + \frac{(x-1)^2}{2} + \frac{(x-1)^3}{6} + \frac{(x-1)^4}{12} - \frac{(x-1)^5}{20}$$